RECENT TRENDS IN BIOTECHNOLOGY

APPLIED ALGAL BIOTECHNOLOGY

RECENT TRENDS IN BIOTECHNOLOGY

Additional books and e-books in this series can be found on Nova's website under the Series tab.

RECENT TRENDS IN BIOTECHNOLOGY

APPLIED ALGAL BIOTECHNOLOGY

MUTHU ARUMUGAM
SHANMUGAM KATHIRESAN
AND
NAGARAJ SUBRAMANI
EDITORS

nova
science publishers
New York

Library of Congress Cataloging-in-Publication Data

ISBN: 978-1-53617-524-0

Published by Nova Science Publishers, Inc. † New York

CONTENTS

PREFACE

In view of tremendous development in the area of Biotechnology, Algal Biotechnology is a fascinating field that has attracted many researchers in the past three decades. Considering its potential and future applications for human well being, studies related to basics and applied aspects of commercially important microalgae need to be focused. Keeping this in mind a comprehensive methodology starting from culture collection to metabolite production in microalgae need to be addressed and hence, our book *Applied Algal Biotechnology* will provide the valuable information and exciting results based explanation will easily guide the young researchers, PhD scholars and also for the PG and UG students.

In the present research scenario, every Plant Sciences laboratories in Universities/Institutes has a separate unit for microalgal biotechnology to understand the basic concepts that make microalgae as an alternative model system that competes with *Arabidopsis thaliana.* The strategies starting from isolation, identification, medium preparation, culturing condition, metabolite production, novel gene isolation, and its expression pattern under the influence of different biotic and abiotic stress condition, genetic transformation in homologous/heterologous host, etc., are very much essential for the fruitful execution of research. Here we have precisely well written the content for each of these sections that have been systematically arranged. There is a number of books that deal with different aspects of algal biotechnology, but, we expect it will be useful to have a book describing from basics i.e., from classifications, biology, and biochemistry, culturing to harvesting strategies, and potential of the algal system for many human applications. Each chapter has been framed based on the evidence of successful research discoveries cited from the peer-reviewed journals and that to familiarly followed/cited by many algal biotechnologists. We have also tried to provide suitable case studies with appropriate examples where the recent biotechnological approach has been implemented.

With this, we believe that the book will satisfy the teaching and research community who are dealing with algal biotechnology. We would appreciate your suggestions, criticism and research contributions which relate to different aspects relevant to this book. Please mention the errors that you find with page numbers, describe the mistakes and provide your assistance for improvements in the next edition.

We hope that this book would be of great interest to the students, researchers and teachers from academia, research institutes and commercial firms in the broad area of algae and cyanobacteria for energy, environment, therapeutic and nutraceuticals.

ACKNOWLEDGMENTS

We would like to express our sincere gratitude to all the authors from various institutes who have to contribute the chapters to this book, revised and submitted the final revisions. We also take this opportunity to extend our sincere appreciation to all the reviewers and research scholars who have devoted their valuable time and effort to ensure the scientific fact, English language and overall improvement of this book. As editors of the book, we welcome suggestions and critical inputs from the readers and the same will be revised in the next edition of the book for more clarity. We are extremely thankful to Ms. Nadya S. Columbus, Ms. Stella Rosa, and Ms. Carra Feagaiga and the entire "Nova Science Publishers team" for their timely support, cooperation and effort in producing this book.

M. Arumugam
S. Kathiresan
N. Subramani

INTRODUCTION

Algae are aquatic photosynthetic organisms ranging from tiny unicellular microalgae to large multicellular macroalgae (seaweed). They grow in a wide range of environments such as fresh, saltwater, brackish, marine and soil environments. Algae regained considerable attention in the recent past as it harbors a viable source of carbon compounds, which has tremendous scope in renewable energy, biopharmaceuticals, nutraceuticals, food, and feed supplement areas. They are often regarded as “Biological cell factories” due to its efficient fixation of atmospheric Carbon dioxide (CO_2) to many useful compounds as listed above. One Kg of algal biomass has the ability to fix about 1.83 Kg of CO_2 and it constitutes about 50% dry weight of algal biomass. Also, certain algal species are capable of using other minor flue gases such as Sulphur Oxides (SOx) and Nitrous Oxides (NOx) as its nutrient sources for their growth and development. Algae have the potential of converting 9-10% of solar energy into biomass with a theoretical yield of 77 g biomass/m^2/day. Likewise, 50 percent of atmospheric free oxygen is liberated from photosynthetic plants, algae, and cyanobacteria. Owing to the increase in demand but a decline in cultivable land, algae can be considered as a viable alternative as food and feed supplement.

Microalgae and cyanobacteria play a pivotal role in Agriculture in improving soil fertility as well as soil stabilizers. Microalgae give practically feasible solutions to treat the wastewater, selective heave metal removal from contaminated sites and the resulting biomass can be used as a biofuel feedstock. Taken together, algae can be regarded as wonder organisms as they provide valuable bioactive, liberate atmospheric Oxygen, reduce flue gases through CO_2 fixation, biofertilizers and also provide environmental protection by wastewater treatment. Having said that, microalgae are proven to be a natural gift and so many researchers are oriented towards understanding its biology as well as its therapeutic potentials, for commercialization. Keeping this in mind, the current book entitles "Applied Algal Biotechnology" was framed and it will serve as a useful reference for researchers involved in academics and industries.

For the convenience of the reader, the book is divided into three sections namely taxonomic classifications, physiology and metabolism, and applied aspects of microalgae. Further, it also provides guidelines to the undergraduates, postgraduates, and beginners for hands-on practical experience in algal sampling, culturing, purification and identification. It is noteworthy to mention that each algal species has it's nutrient and growth conditions and requirements and thus a fundamental principle involved in these processes is illustrated in the first section of the book.

The second section describes photosynthesis and metabolism of algae, as the basic understanding of these aspects is important to manipulate the culturing condition towards desirable metabolite production. The growth and metabolism of all living systems on the Earth directly or indirectly depend on photosynthesis for organic matter and energy, thus a brief account of Algal photosynthesis (similar to plant system) is described in detail. Also, the kind of secondary metabolites present in algae determines its economic importance as a nutraceutical, therapeutic, and other biological activities. A detailed account of the current status of various microalgal secondary metabolites and their commercial applications are described in this subsection. Some of the genetic transformation strategies attempted in different groups of algae and its impact on the biosynthetic pathways and their possibility of commercialization for the benefit of human welfare are also covered in this section.

The last section of the book describes applied aspects of algae on various commercial applications such as Agriculture, Bioenergy, Nutraceutical, and Environmental aspects. The specific topics covered in this section of the book include,

- Engineering strategies to scale-up the algal cultures
- Different strategies to enhance the lipid for energy and nutraceuticals application
- Harvesting and downstream processing of microalgae
- Nutraceuticals and Therapeutic Applications of Algae
- Applications of Algae and cyanobacteria in agriculture
- Biofuels and hydrocarbons
- CO_2 sequestration and biorefinery
- Environmental applications
- Phycoremidation of wastewater
- Green synthesis of metallic Nanoparticles

In: Applied Algal Biotechnology
Editors: M. Arumugam, S. Kathiresan et al.
ISBN: 978-1-53617-524-0

Chapter 1

CLASSIFICATION OF ALGAE: AN OVERVIEW

Nagaraj Subramani*
Centre for Advanced Studies in Botany, University of Madras,
Guindy Campus, Chennai, Tamil Nadu, India

ABSTRACT

The classification is a very significant aspect of the orderly study of organisms or materials in all areas of science. The classification of algae aids in the exact identification, grouping, and naming of the algae for further studies. It is carried out based on a standardized system depending on the morphological features primarily, ie., the similarities and differences, followed by the molecular studies such as genetic analyses recently. Through proper classification, the evolutionary history or phylogeny of the algae can also be elucidated thus understanding how the various classes of algae are related to each other as well as with the other organisms. The judicious employment of algae in diversified fields such as food, feed, pharmaceuticals, energy, agriculture, environmental biotechnology, etc. first requires them to be precisely classified. While speaking about classifying the algae, there were a number of classifications proposed, of which the algal classification by F. E. Fritsch is still widely accepted today. He classified algae into 11 classes, the features of which are discussed in the upcoming pages of this chapter. Although there are certain algal types which are yet to be assigned a class, Fritsch's classification is the basic foundation of algal classification to which additions, deletions, or merging can be carried out as and when required. Today, the classification of algae is based on a combination of molecular, ultrastructural, and biochemical characteristics rather than the old method solely based on the morphological features.

Keywords: classification, algae, Fritsch, molecular taxonomy, phylogeny

* Corresponding Author's Email: nagalilly@gmail.com.

1.1. INTRODUCTION

The classification of any group of organisms is subjective, varies with the classifier and his or her emphasis and acceptance or non-acceptance of criteria available at the time of classification. The authors hence view their own system of classification as tentative and subject to modification with time. When discussing the classification and phylogeny of organisms, it is appropriate to quote the famous, eminent Paleobotanist, Arnold, 1948,: "Once a system of classification becomes widely adapted, it takes on many of the attributes of a creed. Not only does it constitute the framework about which the botanist does his thinking but it rapidly becomes a substitute for it. To function properly, all systems must be kept in a fluid and flexible state. No system can be accepted as final so long as a single fact concerning any kind of plant remains unknown." Later in the year 1974, Leedale stated that: "The aim of modern systematics is that the classification shall be a 'natural' one, reflecting true relationships as far as we can ascertain them from biochemical, cytological, morphological, developmental, and behavioral features from the fossil records and from our understanding of heredity and evolution. Thus the more closely species are grouped in such a classification, the more closely related to one another they are thought to be. *The system is still artificial, of course* [italics ours]. Living organisms do not fit neatly into man-made categories; diversity is always continuous in some aspects." The early criteria considered for the classification of organisms is clearly evident from these quotes.

1.2. CLASSIFICATION OF ALGAE

When coming to the classification of algae, it was Papenfuss in 1955 who reviewed the history of classification of the major groups of algae and the history of the discovery of their sexual reproduction. Linnaeus in 1753 recognized 14 genera of "algae," but only four of them (*Conferva, Ulva, Fucus,* and *Chara*) were algae as we define them now. In 1836, Harvey had recognized four major groups of algae, the brown, red and green algae, and the diatoms; and color, as a manifestation of different pigmentation, continues to be significant in classifying the major groups of algae. Meanwhile, in the United States, the algae were grouped as a class coordinate with Fungi under the division Thallophyta of the plant kingdom until the publication of G. M. Smith's *Fresh-Water Algae of the United States* in 1933.

Investigations in the first quarter of the twentieth century subsequently revealed that the differences in pigmentation among the groups of algae were accompanied by the differences in cellular organization and storage products. It was, therefore, concluded that the former class Algae did not hold a closely related, cohesive, and natural association of organisms. Accordingly, Smith, 1933 and 1950, recognized 11 major groups of algae abandoning the former categories Thallophyta and Algae, and classified into seven major categories or divisions (in 1950), coordinate with the Bryophyta and other divisions of the plant kingdom. The 7 divisions he designated were Chlorophyta, Euglenophyta, Chrysophyta, Phaeophyta, Pyrrhophyta, Cyanophyta, and Rhodophyta in conformity with the International Code of Botanical Nomenclature.

Despite all these suggestions, Papenfuss, 1946 had pointed out that the use of the designation 'Chlorophyta,' meaning 'green plants,' for the green algae prevented its use for other members of the plant kingdom with identical pigmentation and storage products. Thus he suggested that the names of the algal divisions should include the term 'phyco,' and accordingly, the divisions were named as Chlorophycophyta, Euglenophycophyta, and so on. The inclusion of the term 'phyco' was to indicate that the members of some divisions were at the algal level of organization. The names as suggested by Papenfuss were although followed at first; it was been reconsidered and not followed later because it justified the term used with reference to green algae, but it is unnecessary to use the term in the divisional names of other groups of algae because there are no brown or red bryophytes, pteridophytes, and spermatophytes. As a result, the following divisions of algae are discussed: Cyanophyta, Prochlorophyta, Chlorophyta, Charophyta, Euglenophyta, Phacophyta, Chrysophyta, Pyrrhophyta, Cryptophyta, Rhodophyta, and in addition, several algae of uncertain affinities. These divisions of algae differ in their pigmentation, cellular organization, cell wall chemistry, flagellation (or its absence), and storage products, among other features.

However, the observations made on these aspects should be considered as tentative and may be subjected to change as in the case of many of the divisions, the data have been deduced from only a few representatives. In 1935, Tilden grouped the algae into five classes, whereas in the subsequent classification by other phycologists, the number of divisions varied from 4 to 13, and the number of classes varied from 5 to 24. Finally, Fritsch came up with his classification of algae in 1935 and 1945, in which there were 11 classes (Fritsch, 1975) which are described briefly below.

1. *Chlorophyceae* (*Isokontae*): As the name indicates, they are grass green and contain four pigments, two green and two yellow, with chromatophores and approximately in the same proportion as in higher plants. In general, most of the members are algal in nature and have one or a few chromatophores. Their cell wall mostly constitutes of cellulose and motile cells possess an equal number of flagella (2 or 4 usually) which arise from the anterior end in a similar orientation. The storage product is usually starch formed from photosynthesis, often accompanied by oil especially in resting stages. The pyrenoids commonly surrounded by a starch-sheath are frequently present in the chromatophores (chloroplast). They exhibit sexuality ranging from isogamy to advanced oogamy, usually with retention of the ovum. The highest type of structure observed is a heterotrichous filament, and bulky parenchymatous forms are not aware of. The majority of the Chlorophyceae are haploid with the zygote representing the only diploid phase, but some members exhibit a regular alternation of a similar haploid and diploid individual. On the contrary, the Siphonales exhibit types of thallus assembly similar to those found in red and brown algae and are probably in the main diploid. The class is more adapted to freshwater than in salt water, and there is a distinct terrestrial tendency.
2. *Xanthophyceae* (*Heterokontae*): They are characterized by a number of yellow-green, discoid chromatophores due to the presence of excess xanthophylls which is yellow in color, and hence the name. Their cell wall is rich in pectic compounds and is lapping at the edges, while the resting periods commonly possess a silicified membrane. Starch is absent and oil is the prime storage product. The pyrenoids are deficient or rarely evident. The motile cells possess two unequal flagella or sometimes only one flagellum arising from the anterior end. The majority of the members in this class are algal, but sexual reproduction is rare and always

isogamous. The most advanced forms have a simple filamentous habit and all are probably haploid. These algae are more abundantly seen in freshwater than in the sea.

3. *Chrysophyceae*: This class of algae possesses brown or orange-colored chromatophores containing one or more accessory pigments (phycochrysin). The cells typically contain one or two parietal chromatophores. They lack a distinct cell membrane yet the majority are flagellate. The food storage is in the form of round, whitish, opaque lumps composed of fat and a compound called leucosin. Starch is absent, but naked pyrenoid-like bodies are occasionally present. Another distinct feature is the presence of spherical silicified cysts that arise endogenously, usually provided with a very small aperture closed by a special plug. The motile cells have one or two or rarely three flagella attached at the front end and they are unequal. The nutrition is usually holozoic, and a branched filament forms the most advanced habit. Sexual reproduction is extremely rare and isogamy is only reported so far. Although this class is widely distributed in colder freshwaters, a few families are marine.
4. *Bacillariophyceae* (Diatoms): These algae have yellow or green-brown chromatophores containing an accessory brown pigment of disputed nature. They are unicellular or colonial with a cell wall partly composed of silica. It always consists of two halves; each composed of two or more pieces and is richly ornamented. The pyrenoids are often present and fat and volutin form the products of photosynthesis. One set of algae called Centrales is radial, while the other called Pennales is bilaterally symmetrical. The occurrence of flagella stages is likely to be high in Centrales but needs a complete elucidation. The Pennales are diploid and show a special type of sexual fusion between the protoplast. The diatoms are a highly differentiated group, widely distributed in a range of habitats including sea, all kinds of freshwaters, soil as well as in other terrestrial habitats.
5. *Cryptophyceae*: The special characteristics of these algae are the two larger parietal chromatophores showing a very diverse pigmentation, commonly a shade of brown. The cells have pyrenoids-like bodies, often independent from the chromatophores. The products of photosynthesis are carbohydrates, starch, or compounds similar to it. The motile cells have a specialized structure which is mostly dorsiventral, with two slightly unequal flagella. The majority are flagellate and have a complex vacuolar system. Isogamy has been reported in one form. The most advanced habit known is coccoid. The class is relatively small and appears to be sparse, distributed equally in the sea and in freshwaters.
6. *Dinophyceae* (*Peridinieae*): They possess numerous discoid chromatophores which are dark yellow, brown, etc., and contain a number of special pigments. The products of photosynthesis are starch and oil (fat). Most of the algae in this class are motile and unicellular, with an intricate cell wall made of cellulose arranged into richly sculptured plates. Many species are colorless saprophytes or exhibits holozoic nutrition, while one extensive series is parasitic. The motile cells have two furrows, a transverse furrow harboring the transverse flagellum which usually encircles the body, and a longitudinal flagellum directed backward. The algae produce resting cysts of characteristic form. The most advanced habit is that of a branched filament. Isogamous sexual reproduction is rare and not yet clearly established. This class

mainly has plankton organisms, more widely distributed in the sea than in freshwaters.

7. *Chloromonadineae*: This class possesses numerous discoid chromatophores having a bright green tint and an excess of xanthophylls. The algae lack pyrenoids and the assimilatory product is oil. Only a few members of this class are known and they are motile flagellates, with two almost equal flagella. The members appear similar to Xanthophyceae superficially, but the detailed structure of the cells is altogether different, with complex vacuolar apparatus, etc. The class is only recorded from freshwaters.
8. *Euglenineae*: The algae in this category possess several pure green chromatophores, some forms of pyrenoids-like bodies, a complex vacuolar system, and a large, prominent nucleus. The product of photosynthesis is a polysaccharide called paramylon, which occurs as solid grains of diverse and distinctive shapes. They are totally flagellate members and the majority are motile with one or two flagella which arise from the base of a canal-like invagination at the front end. Only a few isogamous cases are known and these are not completely authenticated. This class is highly specialized and simple forms are not known yet. The majority of the members probably inhabit freshwaters.
9. *Phaeophyceae*: As the name indicates, the members of this class have brown chromatophores containing the usual pigments along with the yellow fucoxanthin. The lower forms possess naked pyrenoid-like bodies. The assimilatory products include alcohol (mannitol), traces of sugars, as well as polysaccharides (laminarin) and fats, especially in the higher forms. One of the characteristic features of the cell-content of many forms is the so-called fucoidan-vesicles which probably represent waste products. The members range from simple filamentous forms to the bulky parenchymatous forms, attaining larger sizes with complex internal and external differentiation. The reproductive cells are motile with two laterally arising flagella, one directed forward and the other backward. These cells are always formed in a special unilocular or numerous small separately compartmentalized organs termed as plurilocular sporangia. Sexual reproduction is common ranging from isogamy to a primitive type oogamy, with the release of an ovum prior to fertilization and the zygote exhibits no resting periods. The life-cycle is very diverse, with different types of alternation of generation. All but a few are marine, for example, the brown seaweeds.
10. *Rhodophyceae*: This class exclusively contains additional red and blue pigments termed phycoerythrin and phycocyanin respectively, along with the usual pigments in their chromatophores. The freshwater forms especially possess phycocyanin. In this case, too, the lower groups have pyrenoids-like bodies and the assimilatory product is the Floridian starch, a solid polysaccharide similar to starch. But neither flagellate members nor any motile reproductive stages are known among them. They include both the simple filamentous forms as well as the highly complex forms with distinctive structures. Except in the case of the Bangiales, the majority of them have evident protoplasmic connections. All forms exhibit an advanced oogamous type of sexual reproduction, with a female organ having a long receptive neck and an antheridium producing a single non-motile male cell. The fertilization leads to the formation of special spores called carpospores from bunches of threads that arise

from the female organ or in the higher groups from other cells with which the female organ becomes connected after fertilization. The Rhodophyceae are either haploid or exhibit a regular alternation of similar haploid and diploid individuals, and the diploid ones bear characteristic tetrasporangia, and each producing four spores. Most of the Rhodophyceaen members are marine, for instance, the red seaweeds.

11. *Myxophyceae* (*Cyanophyceae*): These forms lack proper chromatophores and so the photosynthetic pigments lie diffused throughout the peripheral cytoplasm. The pigments present include chlorophyll, carotene, phycocyanin, and phycoerythrin, the latter two in varying proportions, and the cells being blue-green. They have a simple cell containing a very rudimentary nucleus or central body enclosed in a membrane. The photosynthetic products are sugars and glycogen. There are no motile stages and no sexual reproduction reported so far. The cell organization is simple and hence propagates by simple divisions or by vegetative means. The most advanced types are filamentous, many of them with typical "false" branching. They occur abundantly in freshwaters and terrestrial habitats and are also common in the sea.

Thus, the Fritsch's classification of algae into 11 divisions and respective classes has been considered as the pivotal one while discussing algal classification and taxonomy. Figure 1 shows examples of algae belonging to each class and F. E Fritsch, Father of Phycology. (Note: Except *Ulva lactuca*, *Laminaria*, and *Gracilaria*, the algal images are in microscopic view).

The evidence suggests that algae existed from the very earlier geological epochs and their evolution is still continuing ever since. Evolution is a process that works in two opposite ways; one part brings forth the similarities and the other, differences among various organisms. This sufficiently contributed to an algal diversity that ranges from the commonly known simpler unicellular, colonial and filamentous types to large, complex multicellular forms with distinctive body structures. Although efforts were made to include all possible forms of algae in this classification, still there are a considerable number of groups such as the colorless flagellates awaiting classification and elucidation of their exact relationship with the members of the other groups. Thus it is probable that new classes may be added in the future.

There are two colorless flagellates, namely *Cyanophora* and *Glaucocystis* the classification of which still remains debatable. These genera are believed to be originated through endosymbiosis as described by Gibbs in 1981. According to him, the colorless host organism which is eukaryotic and flagellate received the pigmented prokaryotic endosymbionts as chloroplasts or similar structures called cyanelles. The *Cyanophora* is a unicellular, biflagellate organism with multiple chloroplast-like structures called cyanelles containing one pyrenoid-like body. Its host cell is found to be cryptomonads. The flagella arise from the anterior and posterior end, and reproduction is through the division of motile cells. The *Glaucocystis* is also unicellular and possess a highly organized cellulosic microfibrillar cell wall, but are non-motile, with only two rudimentary flagella near the cell equator during cell division. The cyanelles are present as two distinct clusters, each cyanelle with a tapering end containing a pyrenoid-like body and separate thylakoids as in Rhodophyceae. The cells also contain floridean starch, the walls of cyanelles are rigid composing of peptidoglycan, and the reproduction is through the formation of autospores as in Cyanophyceae. Due to the similarity in the presence of cyanelles, Lee, 1980 suggested that these two genera must be classified in a separate division, naming it as Glaucophyta. But the varying affinities of the host cell types to the other classes like Rhodophyta, Cyanophyta, and Cryptomonads, oppose the former classification. So these algae are named as, 'algae with uncertain affinity' (Bold and Wynne, 1985).

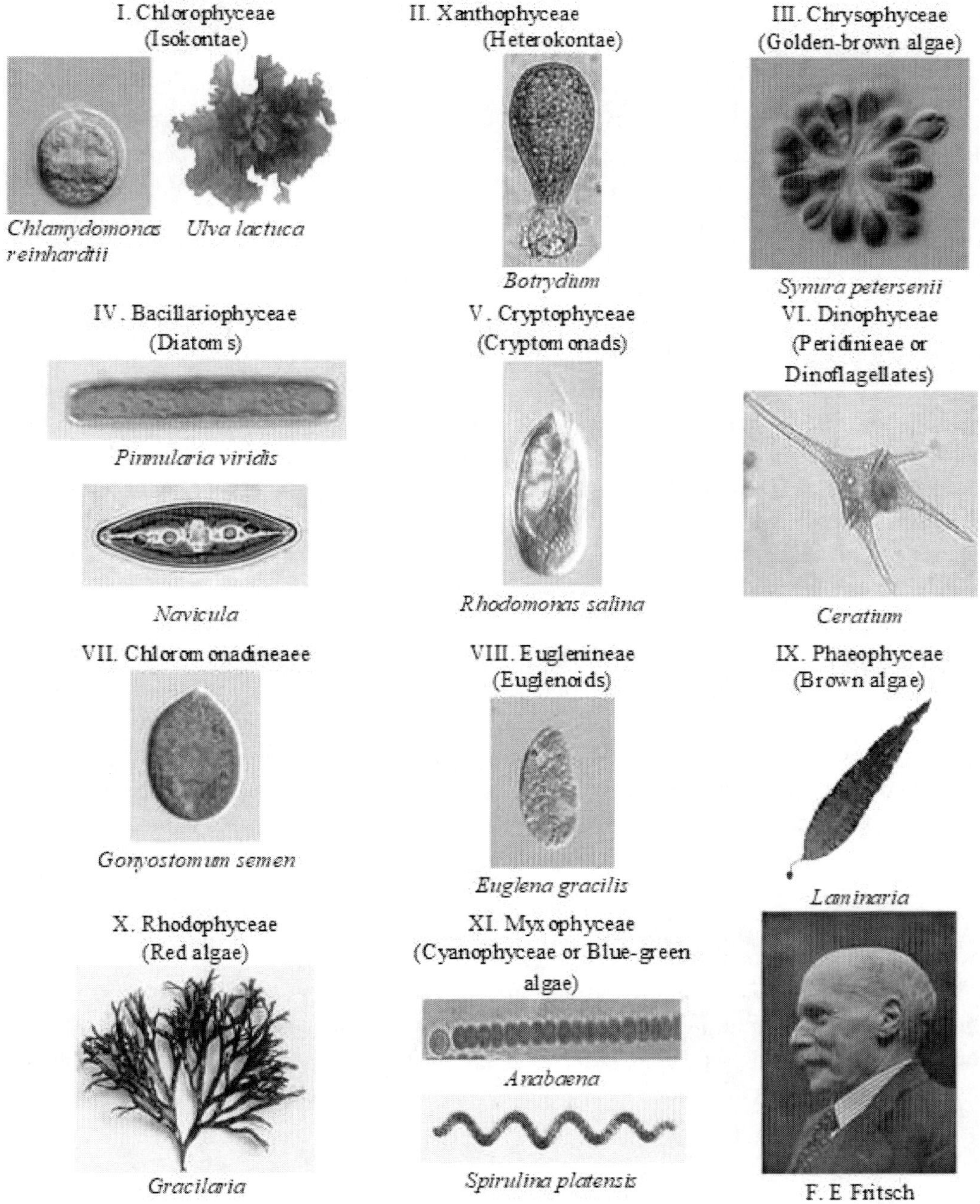

Figure 1.1. Some examples of algae belonging to each division in Fritsch's classification, and F. E Fritsch, the Father of Phycology.

On the contrary to this, some classes share a close relationship with the other classes. For instance, Pascher put forth evidence that the classes, Bacillariophyceae, Xanthophyceae, and Chrysophyceae have a common ancestry thus they may be classified under a single division Chrysophyta. Similarly, an affinity was observed between the classes Cryptophyceae and Dinophyceae and hence Pascher placed them under a single division Pyrrophyta. While with the other divisions, the Chlorophyta, Phaeophyta, Rhodophyta, Euglenophyta, and Myxophyta, their status as discrete divisions with a single class was retained. Yet another uncertainty that exists is in the apt classification of algal fossils and the larger calcified forms (cf. Dasycladaceae, Corallinaceae) discovered with time. With the lack of data related to their cell contents, the reproductive methods or any other class-specific feature, it becomes hard to assign a class for them. Thus taking all these points into account, it would be better to consider each of these 11 divisions as individual evolutionary series as of now, until the relationships among them if and when established.

Bold et al., 1978 proposed that the term "division" must be replaced by "phylum" but was not approved at the International Botanical Congress in Sydney, in 1981. The classes and families of living algae have also been summarized by Silva, 1980. During the 1980s, the approach to classification changed from the typological and phenetic angle to a classification based on comparative morphology and ultrastructure, and then to cladistic methods of classification based on homology and recognition of monophyletic groups. Thus the most recent aspect of classification of organisms is the advent of molecular taxonomy or molecular phylogenetic studies. There has been a notable increase in the study of molecular, biochemical and ultrastructural characteristics of algae during the last thirty years, which explored the minute level features of these organisms. The conclusions drawn from these studies, however, contradict certain points in the traditional system of classification (Barsanti and Gualtieri, 2014). Besides, the conformity of morphological and molecular classification leads to a belief that the molecular approach is unnecessary, while the non-conformity questioned the correctness of molecular data. Later it was believed that giving a due weightage to the molecular properties of organisms while classifying would help to bridge the gaps left by the traditional systems of classification founded on observing only the perceivable properties of organisms (Medlin et al., 2007). Though some of the features of algae differ with its adaptability over time, the overall general characteristics which are a property of a certain class or division still remain the same. Hence it is advisable to keep the fundamentals of Fritsch's classification as such and keep adding the finer details and changes that accumulate in the course of time.

The algae are ubiquitous and found in some unusual habitats in certain cases. Such algae particularly exhibit some unusual features. Some also have relationships or interactions with other organisms and this, in turn, reveals or suppresses some special traits in algae. These kinds of special features also require elaborate investigation to include them while classifying algae.

CONCLUSION

The classification of organisms, in general, allows the precise identification, grouping, and naming of organisms based on a standardized system developed on the basis of similarities and differences between various organisms. It must reflect the evolutionary history of organisms by descending through homology. A better classification system tries to include all features including the morphological, molecular, ultrastructural, adaptational as well as evolutionary traits of an organism, so that any new organism discovered may be accurately classified for the rest of the world to refer to. In this regard, the Fritsch's classification remains the most accepted classification system of algae even after eight decades of its proposal. Although the evolutionary process may sometimes demand the addition, deletion or merging of classes, the classification as proposed by Fritsch will stand as the foundation on which the necessary changes may be made from time to time. Besides, with the huge advancements achieved in the field of science, the unambiguous classification of organisms has become further feasible.

SUMMARY

This chapter gives an overview of the classification of algae – its various approaches and developments, from the classification proposed by Linnaeus in 1753 with 14 algal genera to the one proposed by F. E Fritsch in 1935 with 11 classes. The 11 classes as described by Fritsch include Chlorophyceae, Xantthophyceae, Chrysophyceae, Bacillariophyceae, Cryptophyceae, Dinophyceae, Chloromonadineae, Euglenineae, Phaeophyceae, Rhodophyceae, and Myxophyceae. The representative species of each of these classes has been given in Figure 1. The algae belonging to each class possess specific characteristics with respect to its motility, presence/absence and the number of flagella, level of cellular organization, type of reproductive mechanism, chromatophores and pyrenoids, and the presence of special characteristics if any. Although most of the organisms belonging to the group algae have been classified well, the classification of two colorless flagellates, namely *Cyanophora* and *Glaucocystis* still remains unclear. Another point is that some classes show affinities with others while some classes show variations with others, thus making it hard for the precise classification. The most recent approach towards algal classification involves not only the morphological features, but also the molecular, ultrastructural, adaptational, and evolutionary features of an alga. This has resulted in a far better understanding, grouping, and utilization of the algae.

FUTURE PERSPECTIVES

Despite the fact that the concept of molecular taxonomy and classification has become a hot topic nowadays, its application in the real research scenario is yet to be broadened. Especially, in algal research, its utility has to be fully tapped for solving the chaos existing regarding the unclassified algae and the algae with affinities/differences. Through proper classification, grouping, and naming, the full-fledged utilization of algae must be ensured for the welfare of humans as well as other organisms.

Acknowledgments

The author is grateful to the Director CAS in Botany, University of Madras, Chennai – 25 for providing laboratory facilities and also thank full to the MoEF&CC, AICOPTAX, for providing financial assistance through the project.

References

Arnold, C. A. (1948). Classification of the Gymnosperms from the Viewpoint of Paleobotany. *Bot. Gaz.* 110, 2-12.

Barsanti, L. and Gualtieri, P. (2014). *Algae – Anatomy, Biochemistry, and Biotechnology.* CRC Press, Taylor and Francis Group, Boca Raton, 2-46.

Bold, H. C. and Wynne, M. J. (1985). *Introduction to the Algae – Structure and Reproduction.* Prentice-Hall, Inc., Englewood Cliffs, N. J., 639-642.

Bold, H. C., Cronquist, A., Jeffrey, C., Johnson, L. A. S., Margulis, I., Merxmiller, H., Raven, P. H., and Takhtajan, A. I. (1978). Proposal (10) to substitute the term "Phylum" for "Division" for Groups Treated as Plants. *Taxon*, 27, 121-122.

Fritsch, F. E. (1975). *The Structure and Reproduction of the Algae.* Cambridge University Press, Cambridge., 1, 5-12.

Gibbs, S. P. (1981). The Chloroplasts of Some Alga Groups May Have Evolved from Endosymbiotic Eukaryotic Algae. *Ann. N. Y. Acad. Sci.* 361, 193-208.

Harvey, W. H. (1836). Algae. In *Flora Hibernica*, Pt. 3 (J. T. MacKay, Ed.). Wm. Curry, Dublin, 157-254.

Lee, R. E. (1980). *Phycology*. Cambridge University Press, Cambridge, 478.

Leedale, G. F. (1974). How Many are the Kingdoms of Organisms? *Taxon*, 23, 261-270.

Linnaeus, C. (1753). Species Plantarum. In *Carl Linnaeus' Species Plantarum* (W. T. Steearn, Ed.). Royal Society, London, 1957, 1 & 2, 1200.

Medlin, L. K., Metfies, K., John, U., and Olsen, J. L. (2007). *Algal molecular systematics: a review of the past and prospects for the future.* Unravelling the algae: the past, present, and future of algal systematics, CRC Press, Taylor and Francis Group, Boca Raton, 75, 341-353.

Papenfuss, G. F. (1955). Classification of the Algae. In a *Century of Progress in the Natural Sciences, 1853–1953*. California Academy of Sciences, San Fransico, 115-224.

Pappenfuss, G. F. (1946). *Proposed Names for the Phyla of Algae*. Bull. Torrey Bot. Club, 73, 217-218.

Silva, P. C. (1980). *Names of Classes and Families of Living Algae*. Regnum Veg., 103, 1-156.

Smith, G. M. (1950). *The Fresh-water Algae of the United States*. 2nd ed. McGraw-Hill, New York, 719.

Tilden, J. E. (1935). *The Algae and Their Life Relations; Fundamentals of Phycology.* University of Minnesota Press, Minneapolis, 550.

In: Applied Algal Biotechnology
Editors: M. Arumugam, S. Kathiresan et al.
ISBN: 978-1-53617-524-0

Chapter 2

INTERACTIONS BETWEEN ALGAE AND BACTERIA: ECOLOGY AND EVOLUTION

Karuppasamy Kattusamy[1,*] and Nagaraj Subramani[2]
[1]Institute of Soil, Water and Environmental Science,
Agriculture Research Organization,
Volcani Center, Rishon Lezion, Israel
[2]Center for Advanced Study in Botany, University of Madras,
Gundy Campus, Chennai, India

ABSTRACT

Algae and bacteria have coexisted from then on the early stages of evolution. This coevolution cannot be adequately understood if taken individually, but they influence ecosystems together and represent all conceivable modes of mutual interactions between different organisms, ranging from mutualism to parasitism. The algal and bacterial co-habitation could synergistically affect each other's physiology and metabolism, a classic case being algae-bacterial interaction. These interactions are ubiquitous and define primary productivity in most ecosystems. Moreover, in recent research, algae have received more attention for industrial exploitation and their interaction with bacteria is often considered as contamination during commercial application. Besides, a few recent reports highlight that bacterial occurrence is not only enhancing algal growth but also help in flocculation; both are essential processes in algal biotechnology. Hence, there is a need to understand algal-bacterial interactions in an evolutionary and ecological standpoint and to integrate this in further industrial applications. In this chapter, we reflect the diversity of bacteria-algae relationships and their associated mechanisms, as well as the habitats that they mutually influence. This chapter also outlines the role of these interactions in key evolutionary events such as endosymbiosis, besides their ecological role in biogeochemical cycles. Finally, we focus on extending such knowledge on algal–bacterial interactions to various environmental and biotechnological applications. This will help to create a better understanding of mechanisms underlying algae-bacteria interactions that will facilitate the development of more knowledge in biotechnology processes.

* Corresponding Author Email: kksamysnc@gmail.com.

Keywords: algae - bacteria association, biogeochemical cycle, ecology and environment

2.1. Introduction

Algae are the undisputed primary producers of the aquatic ecosystem and contribute to approximately half of the global net primary productivity. These photosynthetic organisms, along with cyanobacteria, live in the planktonic region of the aquatic habitats collectively called phytoplankton (Buchan et al., 2014). Ecological studies have identified the co-occurrence of particular species of algae and bacteria, suggesting the existence of their specific interactions. Those interactions are considered for nutrient exchange, signal transduction, and gene transfer. Phytoplankton and bacterioplankton are dominated in the ocean and freshwater ecosystem numerically, and which could actively influence the global carbon cycle. Therefore, the interactions between these two groups and the influence of their interaction with each other on global scale areas of recent research interests (Amin et al., 2015). In some studies reports shown that heterotrophic bacteria play a ubiquitous role in algal growth (Amin et al., 2015). Similarly, in earlier decades, it is shown that heterotrophic bacteria not only decompose plant and animal organic matter but also promote plant growth by complex communication and nutrient exchange (Philippot et al., 2013). Herewith, this chapter attempts to review on interactions between algae and bacteria through ecology and evolution, to the use of advanced knowledge to invigorate with their joined biotechnological application.

The evolution of life was transitionary, where self-replicating molecules and chemicals formed the basis of prokaryotes. Subsequently, aggregation of prokaryotes leads to create a complex form of eukaryotes. For example, cyanobacteria a prokaryote and their association with eukaryotes will evolve into algae. A group of single-celled algae and other ancestors led to multicellular organisms (Herron and Michod, 2008). Literature shows, efforts made to exploit algae for biotechnological applications, such as water treatment and bioenergy production, where bacteria influence algal activities in various ways (Hannon M et al, 2011). In the ecological and evolutionary hierarchy of life, the algal-bacterial association are considered highly significant for mass cultivation. Therefore, in order to complete the understanding of ecophysiology and symbiosis between algae and bacteria, thousands of years of time scale are needed to reach their evolution. This evolutionary study of algae and bacteria and their symbiosis took together shall be a fair opening deliberation in this chapter.

2.1.1. Why Algae Bacterial Interaction Is Important

Algae and bacteria have been living together in most of the aquatics ecosystems for millions of years, and both are primitive microorganisms. Among this, many of them are single-celled creatures that feed themselves through photosynthesis. In aquatic ecosystem, both algae and bacteria are essential parts of the aquatic food chain. Bacteria help to break down dead organic matter, so that it can become part of the soil. In the aquatic ecosystem, bacteria have evolved various ways to interact with algae through physical attachment, and secretion of small molecules. In the ocean, the metabolic exchange and spatial association between algae and bacteria are very hard to explore due to the more number changing factors. Although, algae and bacteria synergistically affect each other physiology and metabolism, although bacteria often considered as contamination of algal cultures in industrial applications. However, in the last few years research updates changed the fast scenario. It shown a positive effect of algae-bacteria interaction on algal growth and flocculation processes, which are the essential steps in algal biotechnology processes (Fuentes LJ et al, 2016). Consequently, the knowledge and control of the mechanisms involved in algae-bacteria interaction could help improve the algal biomass production processes.

2.1.2. Photosynthesis

Algae and some bacteria (cyanobacteria) are involved in photosynthesis to produce energy just like the plants. The key process of formation of life among primary producers is photosynthesis, using sunlight and CO_2 to form chemical energy (Figure 1). However, the mechanism of chemical energy formation is unique among each species in the food chain. Algae stores photosynthetic pigments (chlorophyll, xanthophyll, carotenoids, phycocyanine, and phycobilins, etc.,) inside cases like structure called chloroplasts. The formation of these pigments is distinctive among different strains of algae; depending on the molecules absorb light at specific wavelengths. However, bacteria do not have these chloroplasts, except for few of cyanobacterial strains. They can undergo photosynthesis anywhere on their bodies because their pigments are free-floating within the cellular membrane or 'skin' of the bacteria.

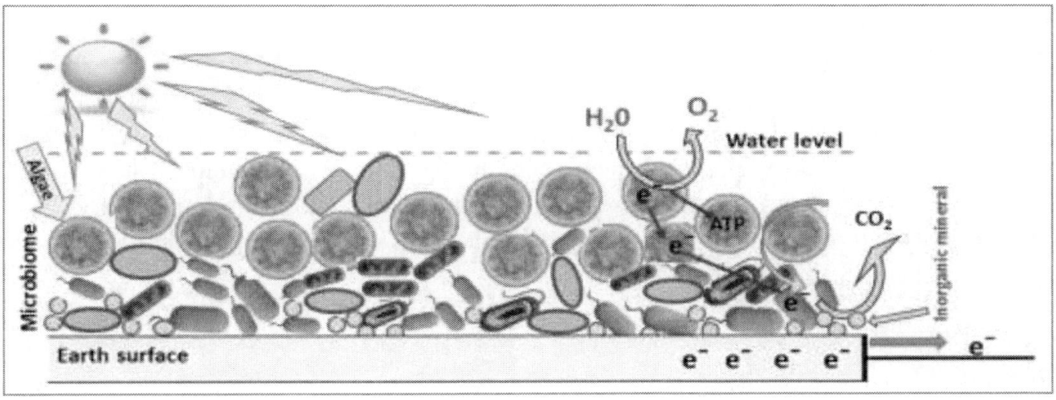

Figure 1. A hypothetical model of algae and bacteria photosynthesis to create energy just like plants.

2.1.3. Differences between Bacteria and Cyanobacteria

Cyanobacteria are known as a blue-green alga, and they differ from other bacteria by possessing chlorophyll-a photosynthetic pigment content. These pigments give them their characteristic blue-green colour. The detailed morphological and metabolic characteristics of bacteria and cyanobacteria are listed in the Table 1.

Table 1. Differences between Bacteria and Cyanobacteria list are given as follows

S.N.	Characteristics	Bacteria	Cyanobacteria
1	Size	Microscopic in size	Comparatively larger than bacteria
2	Distribution	Ubiquities nature	Grow only in the presence of sunlight and moisture-rich environment/water
3	Flagella	Some bacteria having flagella	No Flagella recorded
4	Cell wall	1-2 layered	4 layered
5	Cell wall composition	Glycolipids and peptidoglycan	Cellulose and pectin
6	Nutrition	Some are autotrophic or heterotrophic	Usually autotrophic
7	Photosynthetic pigments	The photosynthetic pigment is bacteriochlorophyll	Photosynthetic pigments are chlorophyll-a, xanthophyll, carotenoids, and phycocyanine
8	Accessory pigment	Not available	Pigment like phycocyanin and phycoerythrin are present
9	Reserve food	Glycogen is a storage food	Starch is a storage food
10	Spore formation	Is endogenous	Free spore formation
11	Hydrogen donor	The photosynthetic hydrogen donor is not water; as a result, oxygen is not evolved. Photosynthesis is anoxygenic	Hydrogen donor is water, and oxygen is usually evolved. This process is an anoxygenic condition
12	Locomotory organ	Flagella act as a locomotory organ	Flagella absent and other locomotory organs
13	Heterocyst	Heterocyst Absent	Heterocyst formation takes place
14	Sexual reproduction	Through conjugation, transformation, and transduction.	No sexual reproduction occur

2.1.4. Environment

Algal growth is found exclusively in the marine, ponds, pools, lakes, and aquariums and on any wet surfaces. Large species of algae found in the ocean resemble plants, serve as the basis of the ecosystem, and are served in oriental food dishes. Bacteria are found everywhere in nature. They can survive and thrive in water, on the skin, surfaces, carpet, earth, stone, and especially dead flesh, etc.

2.1.5. Size of Algae and Bacteria

The main difference between bacteria and cyanobacteria is that the bacteria are mainly heterotrophs and cyanobacteria are autotrophs. Furthermore, bacteria do not contain chlorophyll while cyanobacteria contain chlorophyll-a. Bacteria are single-celled microorganisms and cyanobacteria is also known as blue-green algae. They differ from other bacteria in that cyanobacteria possess chlorophyll-a, while most bacteria do not contain chlorophyll. Chlorophyll-a gives them their characteristic blue-green color.

2.1.6. Reproduction

Bacteria and algae reproduce asexually; sometimes, both of these microbes reproduce sexually. However, there is a difference in their methods of asexual reproduction. Bacteria reproduce through single-cell division. This means that a little copy of a bacterium grows within the cell and then divides into a separate cell. Algae can produce many copies at once through reproduction with spores formation and cell division.

2.1.7. Bacterial Evolution and Benefitted Algae

A simplistic view of algal–bacterial evolution and their role in endosymbiosis events is portrayed in Figure 2. Cyanobacterium retained as primary plastid over time in three distinct evolutionary lineages red algae, green algae, and glaucophytes. The study of plastid multi-gene phylogeny using molecular clock analyses placed the origin of alga before 1558 mya (Parfrey et al., 2011; Yoon et al., 2004). Later, a series of secondary endosymbiosis events led to the diversification of this ancestor (Curtis et al., 2012). Therefore, the role of bacteria in this ancestral algal genesis is not questioned. However, heterotrophic bacteria are always associated with algae in nature, the role of these bacteria during various secondary endosymbiosis events needs to be questioned. Thompson et al., 2012, have discussed the interplay between cyanobacteria, algae, bacteria, and protists in a series of endosymbiotic events in excellent reviews. Moreover, evidence of horizontal gene transfer from bacteria and archaea to algae to help adapt to extreme environments is also emerging (Schonknecht et al., 2013). Hence, the holistic role of ectosymbiotic heterotrophic bacteria that surround the present-day algae in these endosymbiosis events is not well documented, apart from a few studies reviewed below.

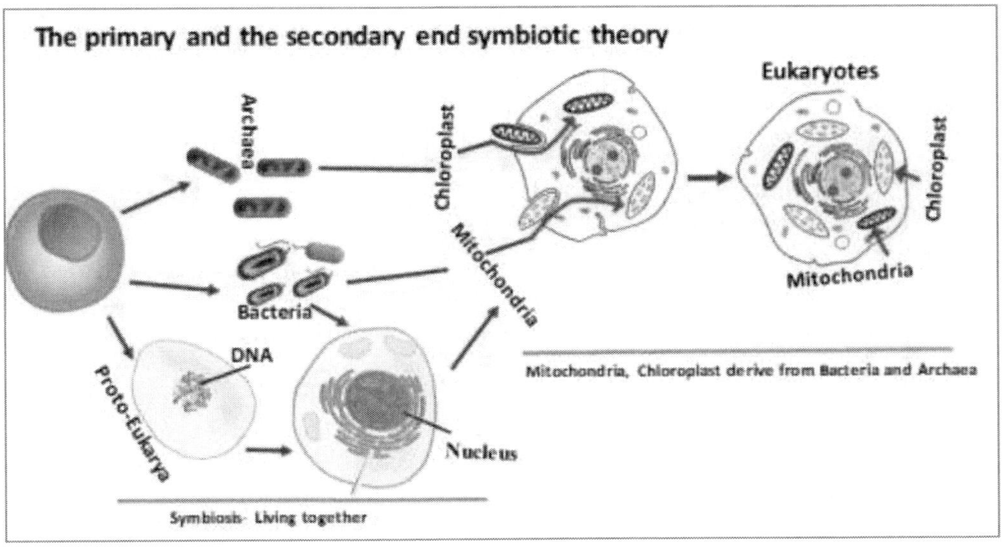

Figure 2. Diagram representing the evolution of photosynthetic eukaryotes. The primary and secondary endosymbiosis represented.

2.2. Ecology of Algal–Bacterial Interactions

Understanding the algal–bacterial interactions can be the direct way to cover the whole range of symbiotic relationships. Algae, heterotrophic bacteria, and archaea are the primary producers, as well as the decomposers, serving as structural pillars and the foremost functional entities of the ecosystem. However, these interactions between algae and bacteria in the planktonic zone are meagrely studied, because of the onerous task of separating the partners, which are naturally bound to each other. The algal-bacterial relationship is believed to be important for vertical flux of carbon and nitrogen cycle, through the mechanism of omnipresent mutualism (Thompson et al., 2012). The Figure 3 showing the role of energy production and the interaction between algae and bacteria in a complex environment. Although the primary function of heterotrophic bacteria is decomposition, it is now accepted that some bacteria can also play a part in algal growth promotion, establishing mutualistic interactions. Most of the emerging studies detailing the influencing factors of algal-bacterial interactions could raise a question on the conventional knowledge of the relational continuum. Hence, this section would dwell on those emerging studies on algal–bacterial interactions, which exhibit an enormous ecological significance for the future, as well as its relevance in growing algal-biotechnology industries.

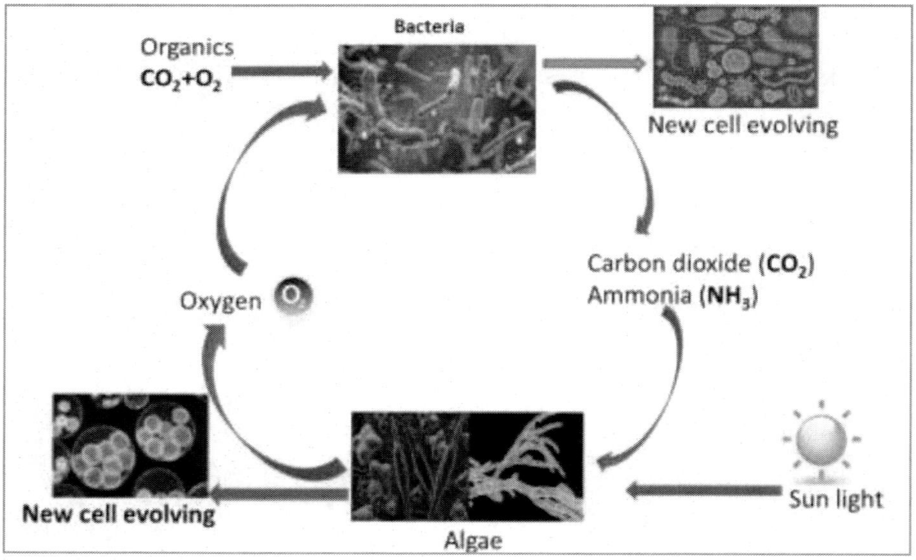

Figure 3. A hypothetical illustration of energy production and interaction of algae and bacteria in a complex environment.

2.2.1. Mutualism

Mutualism, as the word indicates, is not a one-way exchange; rather, the bacteria stand largely exploited by being associated with algae in an oligotrophic environment. There are many examples of mutualism between algae and bacteria. Thompson et al., 2012 is the first study to conclusively prove single-cell interactions and the symbiotic relation between algae and bacteria. Other studies have also revealed the role of mutualism, in some cases, obligate relationships, for each other's subsistence. In recently Croft et al., (2005) reported the evidence of symbiotic mutualism in Vitamin B12 auxotrophs, they proved that bacteria (*Halomonas* sp.) supplied Vitamin B12 to algae (*P. purpureum*) in exchange for fixed carbon. Another study by the same group validated the evolutionary importance of this symbiotic mutualism between algae and bacteria (Helliwell et al., 2011). A recent study conclusively proved the mutualism between a well known PGPB (plant growth-promoting bacteria), *Rhizobium* sp. and wastewater derived algae, *Chlorella vulgaris*, and highlighting its mutualism in freshwater (Kim et al., 2014a). This study demonstrated that algae (*C. vulgaris)* could supply fixed organic carbon to an artificial consortium of four PGPB mutualistic bacteria, and the bacteria in return, supply dissolved inorganic and organic carbon for algal consumption (Cho et al., 2015b). Therefore, it is clear that mutualism is not limited to unicellular microalgae but also prevalent in macroalgae; in some cases, they are endosymbiotic (Hollants et al., 2011).

2.2.2. Commensalism

Commensalism is a relationship in which only one species gains benefits, while the other species may neither benefit nor injured; unlike mutualism, both the species are benefitted. The most common utilized symbiotic relations like commensalism, mutualism, and parasitism are not only delineated but also determines these relationships with one another in an environment. These interactions are continual, but may not discrete its interface between them (Van Ommeren and Whitham, 2002). From this perspective, most algal–bacterial associations studied till the date are either mutualistic or parasitic.

2.2.3. Parasitism

Parasitism is a relation between two different classes of organisms, where one organism lives on another, i.e., the host or parasites by using them metabolically. Such parasites are accepted to be useful for its applications in algal and industrial biotechnology (Bhat, 2000; Dahiya et al., 2006). Many bacterial species can affect algae negatively (Lee et al., 2010; Wang et al., 2010), and several algal strains (red algae) are known to be parasitic (Sachs and Wilcox, 2006). In fact, red algae are considered a model parasite. About 10% of well-known red algal strains are parasitic, and the mechanism of parasitism exhibited by those are adequately established (Hancock et al., 2010). However, very few studies have been reported on algal parasitism on bacteria and vice-versa, along with their mechanism of elimination and its ecological reasoning. These studies demonstrate that the algal cell lysis is achieved through a mechanism similar to plant-pathogen interaction. Glucosidases, chitinases, cellulases, and other enzymes that are involved in plant cell wall degradation are also used for lysis of algal cell culture (Afi et al., 1996; Arora et al., 2012; Wang et al., 2010). This phenomenon of algal cell wall lysis by enzymes are not only limited to bacteria but also in fungal and mollusk (Nikolaeva et al., 1999). Apart from enzymatic lysis of cells and utilizing the intracellular compounds as nutrients by bacteria and fungi. There is another mechanism of parasitism that can occur in the ecosystem, where the competition for existing nutrients in algae leads to slow growth rates; eventually, after several generations, it ends up outcompeting for algal existence in the environment. Hence, it is imperative to specifically learn about the ecological and evolutionary significance of such algal–bacterial interaction, in general.

2.2.4. Lichens Symbiosis

The intimate symbiotic relation between filamentous fungi and photosynthetic blue-green algae or cyanobacterium in mitotic propagules is known as lichens. A classic example of algal symbiosis is considered to be lichens and the oldest well-known symbiotic relationships (Hodkinson et al., 2012). These lichens are stated to be one of the determinant compounds of ecosystems' health by providing specific signatures on the habitat they live in (Stengel et al., 2004). Aschenbrenner et al., (2014) reported that bacteria could colonize to symbiotic propagules of lichens. The lung lichen sampled from three different locations shares a similar core fraction of microbiome than the distinct population, which indicates that symbiotic functions are not isolated (Aschenbrenner et al., 2014). Bacterial symbionts in foliose lichen (*Peltigera membranacea*) are known to involve in phosphate solubilization and also possibly resulting in algal growth promotion. Also, the bacterial co-transmission in the symbiotic community of lichens is also strongly influenced by the nature of photobiont. Thus, the algal-bacterial symbiotic relationship could affect each other's physiology and existence, therefore finally affecting lichen survival in the environment.

2.2.5. Tolerance of Extreme Environments

Algae, cyanobacteria, and bacteria exhibit a strong resistance for a wide range of extreme habitats. A study showed that the survival ability of algae, bacteria, and cyanobacteria in the harsh environmental conditions and under UV radiation for a long time; 548 days (Cockell et al., 2011). This study demonstrated that phototrophs could survive in natural biofilms formed on the rocks when exposed to the lower-earth orbit without sufficient carbon and energy source, adding a new dimension on the survival and growth of algae and bacteria in the extreme environment. In the Antarctic sea-ice, it has been proved that algae and bacteria coexist to fight against high salinity, lower available free flowing-water, extreme low temperature, low sunlight, inorganic carbon source, and even at high UV-radiation. The adaptation of both the bacterial and algal species in such extreme conditions is by secreting high levels of EPS (extracellular polymeric substances), including organic carbon co Algae, cyanobacteria, and bacteria exhibit a strong resistance for a wide range of extreme habitats. A study showed that the survival ability of algae, bacteria, and cyanobacteria in the harsh environmental conditions and under UV radiation for a long time; 548 days (Cockell et al., 2011). This study demonstrated that phototrophs could survive in natural biofilms formed on the rocks when exposed to the lower-earth orbit without sufficient carbon and energy source, adding a new dimension on the survival and growth of algae and bacteria in the extreme environment. In the Antarctic sea-ice, it has been proved that algae and bacteria coexist to fight against high salinity, lower available free flowing-water, extreme low temperature, low sunlight, inorganic carbon source, and even at high UV-radiation. The adaptation of both the bacterial and algal species in such extreme conditions is by secreting high levels of EPS (extracellular polymeric substances), including organic carbon concentration. The secretion of special active substances like glycoproteins by algal and bacterial cells could also help for its survival in a very low-temperature zone. Similarly, algae and cyanobacteria adaptation to the extremely high-temperature zone at Atacama desert by switching to endolithic habitats within gypsum deposits present in organized succession. ncentration. The secretion of special active substances like glycoproteins by algal and bacterial cells could also help for its survival in a very low-temperature zone. Similarly, algae and cyanobacteria adaptation to the extremely high-temperature zone at Atacama desert by switching to endolithic habitats within gypsum deposits present in organized succession.

2.3. Mechanism of Interactions in the Ecosystem

The interaction between marine algae and single-celled phytoplankton are ubiquitously distributed from tropic to sub-Arctic waters. The algae are subjected to frequent upheaval and decline in population. When algal population decline, members of the population release cell wall degradation products, such as lignin and its degradation by-products, such as p-coumaric acid. However, this mechanism is not the only possible strategy prevalent in all interactions, since this mechanism of interactions are species specific as the microenvironment of each alga is different. Modes of interactions between algae and bacteria and their interrelation with the environment are depicted in Figure 4. In the mechanisms proven so far, carbon, macro, and micronutrients seem to play a central role. The micronutrients like vitamins and macronutrients like nitrogen and carbon, and other phytohormones were reported to exchange between algae and bacteria. A study conclusively proved that both algae and bacteria altered their metabolism to suit each other's needs, and this interaction is potentially very prevalent in the marine ecosystem (Amin et al., 2015). All this evidence demonstrates that the role bacteria in the enhancement of carbon storage in microalgae, which is particularly useful in algal production under heterotrophy. Table 2 provides examples of microalgae-bacteria interactions with positive effects on algal growth and the accumulation of valuable compounds. The microalgae-bacteria relationships are upwelling processes; this may even be more complex by changing from mutualism to parasitism according to the physiological circumstances of microalga.

2.3.1. Energy Production

Currently, there is not much available knowledge of H_2 production by algal-bacterial systems. It is well known that hydrogen production by microalgae depends on a hydrogenase enzyme activity that is highly sensitive to oxygen. Thus, stringent anaerobic conditions are required for the efficient production of hydrogen by microalgae. It has been reported that the bacteria, which consume the O_2 generated photosynthetic ally by the algae without damaging the photosynthetic apparatus, might confer such an anaerobic environment suitable for algal hydrogen production. With the help of bacteria that consume the O_2 evolved, the algae can capture light energy and produce H_2 at the same time without further manipulation in the system, such as sulfur (S) deprivation.

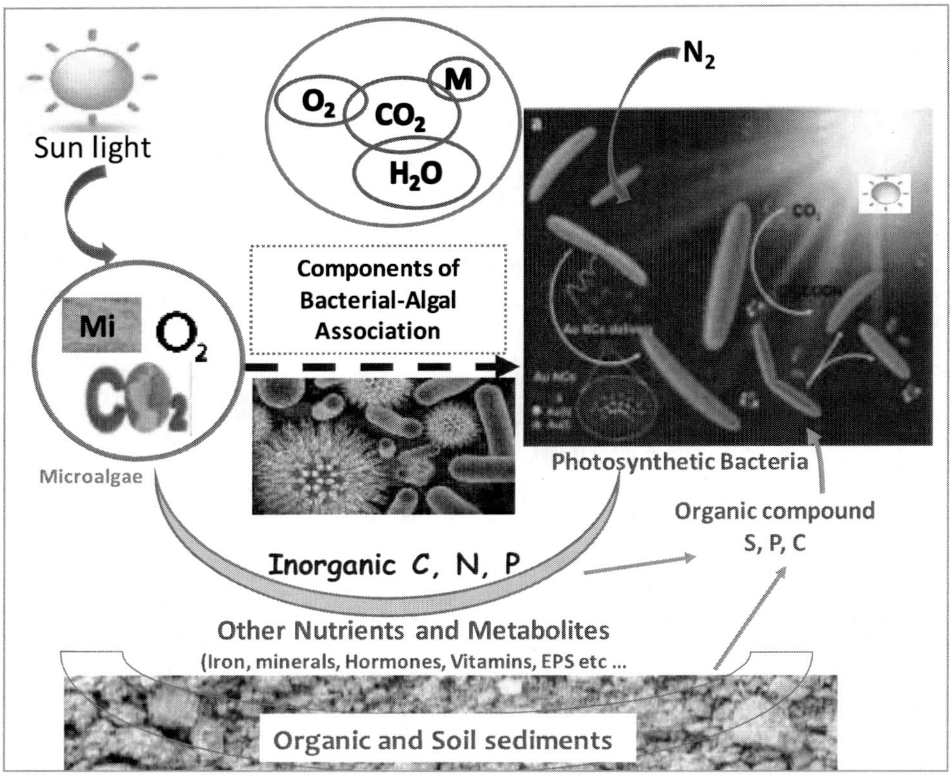

Figure 4. An illustration of components of algal–bacterial association in the aquatic ecosystem and some proven mechanisms. All the mechanisms of exchange have been described in the text.

Table. 2. Examples of microalgae-bacteria interactions having positive effects on algal growth and accumulation of valuable compounds

	Microalga	Bacterium	Mediators from Microalgae	Mediators from Bacteria	Reference
Algal growth improvement/production cost decrease					
1	*E. huxleyi*	*P. gallaeciensis*	Dimethyl sulphonic-propionate	Promoters and antibiotics	Seyedsayamdost et al., (2011)
2	*B. braunii*	*Rhizobium* sp.		AHL	Rivas et al., (2010)
3	*L. rostrate*	*M. loti*		Vitamin B_{12}	Kazamia et al., (2012)
4	*T. seudonana* CCMP1335	*R. pomeroyi* DSS-3	2, 3-dihydroxy-propane-1-sulfonate	Vitamin B_{12}	Durham et al., (2015)
5	*S. trochoidea*	*Marinobacter*	Organic molecules	Vibrioferrin	Amin et al., (2009)
6	*S. trochoidea*	*Roseobacter*	Organic molecules	Vibrioferrin	Amin et al., (2009)
7	*N. oleoabundans*	*A. vinelandii*		Siderophore	Santos et al., (2014)
8	*Scenedesmus sp.*	*A. vinelandii*		Siderophore	Santos et al., (2014)

	Microalga	Bacterium	Mediators from Microalgae	Mediators from Bacteria	Reference
Accumulation of fatty acids and lipids					
9	*C. vulgaris*	*A. brasilense*		Siderophore mediated nitrogen fixation	Leyva et al., (2014)
Heterotrophic accumulation of starch and carbohydrates					
10	*C. vulgaris*	*A. brasilense*		Siderophore mediated nitrogen fixation	Choix et al., (2012)
11	*C. sorokiniana*	*A. brasilense*		Siderophore mediated nitrogen fixation	Choix et al., (2012)
Photoautotrophic accumulation of starch and carbohydrates					
12	*C. vulgaris*	*A. brasilense*		Siderophore mediated nitrogen fixation	Choix et al., (2012)
13	*C. sorokiniana*	*A. brasilense*		Siderophore mediated nitrogen fixation	Choix et al., (2012)

2.4. Omics Approach to Illuminate Interactions

The omics approach to study microbial ecology has transformed our understanding of microbial communities into the next level (Jansson et al., 2012). Although the metagenomics and meta-transcriptomic studies (Moran et al., 2013), were performed in complex microbial communities predominated by algae and bacteria specific studies on microbiome of algal phycosphere in natural systems and artificial systems like photobioreactors using high-throughput sequencing were only recently established (Krohn-Molt et al., 2013; Ramanan et al., 2015). A recent study demonstrated that the power of using multi-omics data to illuminate the interaction and mechanism of natural assemblages (Amin et al., 2015) of algae and bacteria. Taken together, the omics approach and NGS platforms would help our scientific community to answer some basic questions on the algal–bacterial association. As microbiome is being explored as an operational concept, such analyses using advanced tools inherited from highly studied communities like gut microbiota are being performed, and this knowledge would further unravel algae and dependent microbial communities, their active drivers, functionalities and implications, and its resulting applications.

2.5. Environmental Applications

2.5.1. Nutrient Removal and Wastewater Treatment

Algae depend on nitrogen and phosphorus from the environment for growth, as they are non-diazotrophic. Macronutrients like N (nitrogen), P (phosphorus), and S (sulfur), deprivation in algae for a prolonged period could pose them in severe stress, which could lead to stagnation and eventually death (Ramanan et al., 2013; Schmollinger et al., 2014). On the other hand, nutrient-rich wastewaters, when discharged in natural surface waters, might result in blooms of toxic algae and cyanobacteria (Srivastava et al., 2014). Heterotrophic bacteria require carbon and other nutrients for growth and are widely used for the treatment of wastewater. Naturally, algal–bacterial systems have been extensively used in the treatment of nutrient-rich wastewaters since the 1950s. Therefore, the treatment efficiency achieved with these systems is a fraction of what could be achieved with ponds or systems developed later (Benemann et al., 1977; Hoffmann, 1998). As early as 1955, it was proposed that in oxidation ponds, algal–bacterial symbiosis results in sewage treatment with the exchange of O_2 and CO_2, and NH^+ ions. Thus, it was proved technically that most nutrient-rich, low oxygen-demand environments should be conducive to both algal and bacterial growth (Oswald and Gotaas, 1957). However, this conjunction between algae and bacteria holds enormous environmental potential even today. The schematic diagram of algal-bacterial symbiotic interactions in wastewater treatment was given in Figure 5.

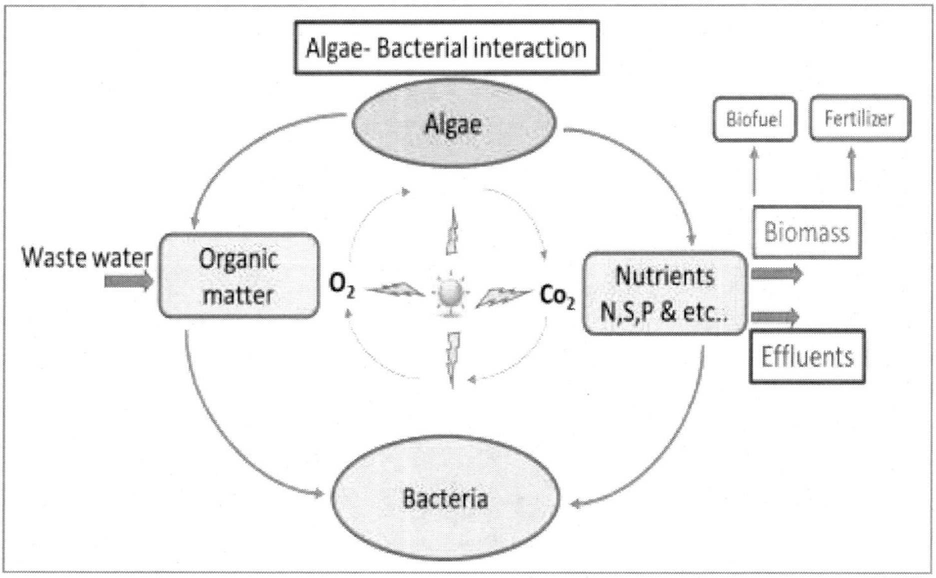

Figure 5. Schematic diagram of algal-bacterial symbiotic interactions in wastewater treatment.

2.6. BIOREMEDIATION

Many studies have dealt with algae–bacteria consortium for metal bioremediation and degradation of organic pollutants. The effective use of algal–bacterial interactions in the degradation of organophosphate insecticides such as monocrotophos, quinalphos, and methyl parathion was also demonstrated (Subashchandrabose et al., 2011). Several studies have shown that the involvement of bacteria-cyanobacteria and algae in the treatment of organic pollutants, including black oil, acetonitrile, phenol, naphthalene, benzopyrene, dibenzofuran, azo compounds, among others. A recent study by Ryu et al., (2015), stated the role of algal–bacterial interactions in the degradation of thiocyanate wastewater, which provides a glimpse of the use of this interaction for degradation of the toxic substrate. In the other hand, algae require several metals in trivial concentrations for its normal growth and metabolism, however, higher levels of same metals are found to be toxic. In this sense, an algal-bacterial community in mutualistic interactions can detoxify and assimilate metals from metal-rich environments.

2.7. Biotechnological Application Potential

Both the algae and bacteria can synergistically affect each other's physiology and metabolism, a classic case being algae-*roseobacter* interactions. These interactions are abundant and define primary productivity in most ecosystems. Algae are known to produce a variety of compounds from fuels to cosmetics. The future bio-refineries would not only look to capitalize on this enormous potential but also augment this approach to produce more compounds and enhance their respective amounts by using an ecological engineering approach (Figure 6). Ecological engineering or synthetic ecology is a broad term used for artificial biomimetic systems that use a multi-organism approach for present-day solutions (Cho et al., 2015b). Any bio-refinery system would benefit from these beneficial bacteria for algal growth promotion and harvesting, as discussed in the above sections. This approach has also been demonstrated in a single stage by co-cultivation. Such an algal–bacterial PHB (polyhydroxybutyrate) production process is also studied in laboratory photobioreactors using inoculum from natural environments (Fradinho et al., 2013).

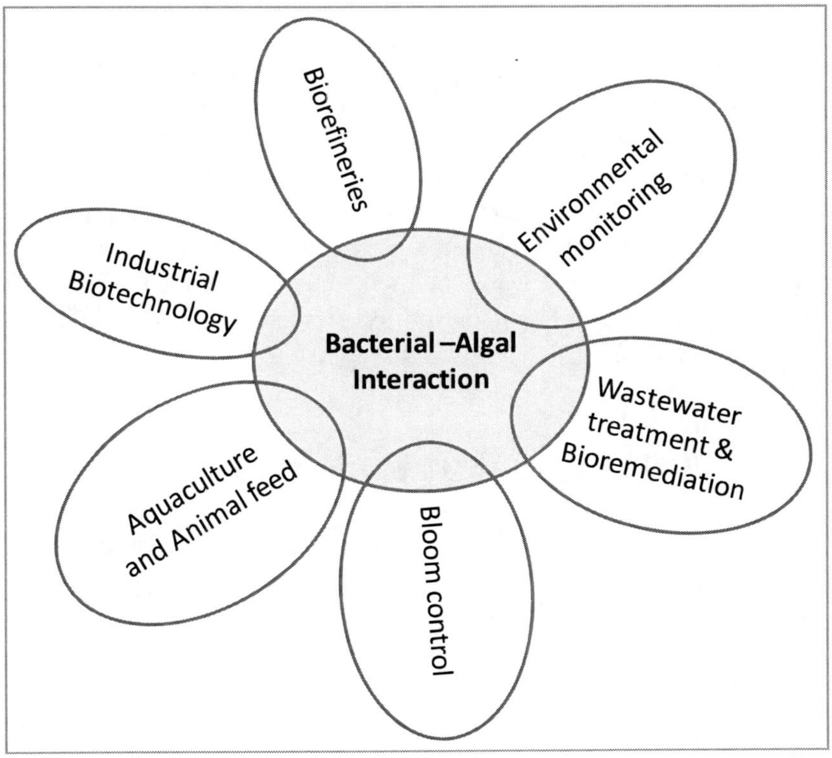

Figure 6. Schematic illustration of various applications of algal–bacterial interactions for biotechnology and environmental sectors.

2.8. Bio-Ethanol Production

Bio-ethanol production from algal–bacterial co-culture is an enriching advance, which is yet to be fully explored. Some algae can produce up to 38% of starch granules (dry cell weight basis), and various marine bacteria can utilize this starch to produce ethanol, these process can be performed by two-step or one-step process, based on the harmony between the species used (Matsumoto et al., 2003). Another promising approach that unites the ecological and genetic engineering approach is the use of photosynthetically fixed carbon from algae to produce the desired product, such as succinate by a genetically engineered bacteria, and *Corynebacterium glutamicum* (Lee et al., 2014b). This approach promises unlimited possibilities for the production of high-value compounds from light energy. In summary, the use of ecological engineering approaches opens a new era for exciting possibilities for algae-based bio-refineries for the sustainable production of fuels and chemicals.

2.9. Sustainable Aquaculture Uses

So far, very little attention has been paid to the use of bacteria in aquaculture, and their presence usually is associated with the control of bacterial diseases. The interaction between bacteria and microalgae involves different mechanisms, including growth stimulatory or inhibitory compound production, cross signaling, and the natural capacity of microalgae to adhere to associated specific microorganisms. In this context, it is essential to get a more in-depth insight into the specific bacterial species naturally associated with algae. This involves aquaculture, the diversity of the bacteria-microalgae interaction mechanisms, and the understanding of the chemistry involved. Several algal species were also used for controlling pathogenic bacteria in aquaculture systems. Many studies have dealt with the algal-bacterial treatment of aquaculture wastewater, but the resounding demonstration that these flocs harvested after wastewater treatment would be used as feed again for Pacific white shrimps, *Litopenaeus vannamei,* throws light on the utility of such an integrated, sustainable and recyclable aquaculture system.

2.10. Present and Future Prospects

At present day, we have more questions than answers in algal–bacterial interactions from its evolution to applications. The understanding of ecology and evolution of algal–bacterial interactions have not been exploited in algae-based technologies (Shurin et al., 2013). Table 3 summarizes the need for such understanding and integration in algal biotechnology. Moreover, further studies in ecological engineering about the use of algae and bacteria as a production platform for industrially important chemicals and fuels in future bio-refineries. As prices of traditionally cultivated food products rise with a burgeoning population and reduced land area, the use of alternative food products such as an algae-based diet could gain prominence (Kim et al., 2014a). Thus, there is a need to look at algal–bacterial interactions for both high-value products such as nutraceuticals and cosmetics, low-value food products for aquaculture and animal feed, as well as medium to high-value chemicals such as fuels and others. Also, algal–bacterial interactions are of potential use in environmental technologies. Algae offers photosynthetically produced oxygen, which could be used for algae–bacteria-based wastewater treatment (Praveen and Loh, 2015). It has been demonstrated adequately that algae and bacteria combine to remediate toxic chemicals and metals (Subashchandrabose et al., 2011). This makes for an exciting new era with a paradigm shift from a single species-based approach to community-based integration, and this integration closely mimics the natural ecosystem, therefore driving us towards sustainable production and development of the beneficial application.

Table 3. Summary of the biotechnological potential of algal–bacterial interactions and current understanding of evolutionary and ecological roles

S. No	Biotechnological process	Evolutionary and Ecological role	Application in Biotechnology	References
1	Strain selection	Algae harbouring PGPB are known to evade pathogens like harmful bacteria and fungi.	Help maintain desirable microbial communities and avoid frequent cultivation crashes.	Cho et al. (2015); Egan et al. (2000)
2	Cultivation	Bacteria aid algal growth by supplementing various major and minor nutrients in oligotrophic environments. See Figure 6 for details.	Enhanced growth rate and algal productivity. Reduced dependence on supplied nutrients.	Cho et al. (2015); Gonzalez and Bashan (2000)
3	Harvesting	Bacteria initiate algal flocculation, possibly for two reasons. Firstly, large algal–bacterial flocs help algae evade predators like zooplankton, as large flocs are difficult to consume. Secondly, bacteria willingly settle algae resulting in algal death and subsequent bacterial degradation.	Harvesting accounts for 30% of overall costs in the algal bio-product industry. Large algal–bacterial flocs settle readily, resulting in reduced use of flocculants and costs, and better yield.	Lee et al. (2013); Montemezzani et al. (2015), Powell and Hill (2013)
4	Extraction	Pathogenic bacteria weaken the algal cell wall resulting in disruption and cell death, playing a leading role in decomposition.	A study showed enhanced lipid recovery from bacteria-infested algae, thereby reducing the cost of organic solvent extraction.	Halim et al. (2012); Lenneman et al. (2014)

CONCLUSION

Microalgae-bacteria interactions are very complex, and needs further research to understand its beneficial applications. At present, fragmentary knowledge has already been gathered on the chemical nature of an ecosystem containing the number of mediator molecules, including nutrients, which regulate the relationship between microalgae and bacteria. Essential amino acids and vitamins have been identified among the main mediator molecules that regulate the relationship between microalgae and bacteria. However, limited information only available at the molecular level. The chemical complexity of the microalgae-bacteria interactions includes a wide variety of molecular signals, exchanged metabolites, transporters, and the molecules whose functions still have to be investigated. A greater insight at the molecular level in the regulation of microalgae-bacteria interactions with sequenced organisms will enable the driving of specific algal-bacterial systems to produce the desired effects. The current development in the understanding of these interactions is leading to specific biotechnological applications, for instance, in the fields of wastewater treatment, bioremediation, and sustainable aquaculture. The new emerging technologies derived from microalgae-bacteria interactions are being developed in the field of energy generation. Among them, phototrophic microbial fuel cells are worth mentioning, although they still in its nascent stage of development to be commercialized. Currently, it needs an exponentially increasing demand for microalgae biomass for novel applications, one of the key challenges would be the controlled integration of specific bacteria in the massive production processes of a specific microalga. These integrations should be aimed at the cost reduction of nutrients, algal biomass harvesting, and intracellular algal product recovery. The overall process should hopefully become more sustainable using reducing the use of synthetic chemicals and energy demand.

Acknowledgments

The author Dr. KKS would like to great acknowledge to the Agriculture Research Organization, Israel for providing Postdoctoral Fellowship and Dr. Eddie Cytryn, Researcher, Institute of Soil Water and Environmental Science, Volcani center for his engorgement and support.

References

Afi, L., Metzger, P., Largeau, C., Connan, J., Berkaloff, C., and Rousseau, B. (1996). Bacterial degradation of green microalgae: incubation of Chlorella emersonii and Chlorella vulgaris with Pseudomonas oleovorans and *Flavobacterium aquatile*. *Org. Geochem*. 25, 117–130.

Amin, S. A., Green, D. H., Hart, M. C., Küpper, F. C., Sunda, W. G., and Carrano, C. J. (2009). Photolysis of iron-siderophore chelates promotes bacterial-algal mutualism. *Proc. Natl. Acad. Sci. USA*. 106, 17071–17076.

Amin, S. A., Hmelo, L. R., van Tol, H. M., Durham, B. P., Carlson, L. T., and Heal, K. R. (2015). Interaction and signaling between a cosmopolitan phytoplankton and associated bacteria. *Nature*. 522, 98–101.

Arora, M., Anil, A. C., Delany, J., Rajarajan, N., Emami, K., and Mesbahi, E. (2012). Carbohydrate degrading bacteria closely associated with Tetraselmis indica: Influence on algal growth. *Aquat. Biol.* 15 (1), 61–71.

Aschenbrenner, I. A., Cardinale, M., Berg, G., and Grube, M. (2014). Microbial cargo: do bacteria on symbiotic propagules reinforce the microbiome of lichens? *Environ. Microbiol.* 16, 3743–3752.

Benemann, J. R., Weissman, J. C., Koopman, B. L., and Oswald, W. J. (1977). Energy production by microbial photosynthesis. *Nature* 268, 19–23.

Bhat, M. K. (2000). Cellulases and related enzymes in biotechnology. *Biotechnol. Adv*. 18, 355–383.

Buchan, A., LeClair, G. R., Gulvik, C. A., and Gonzalez, J. M. (2014). Master recyclers: features and functions of bacteria associated with phytoplankton blooms. *Nat. Rev. Microbiol.* 12, 686–698.

Cho, D. H., Ramanan, R., Heo, J., Kang, Z., Kim, B. H., and Ahn, C. Y. (2015a). Organic carbon, influent microbial diversity, and temperature strongly influence algal diversity and biomass in raceway ponds treating raw municipal wastewater. *Bioresour. Technol.* 191, 481–487.

Cho, D. H., Ramanan, R., Heo, J., Lee, J., Kim, B. H., Oh, H. M. (2015b). Enhancing microalgal biomass productivity by engineering a microalgal–bacterial community. *Bioresour. Technol.* 175, 578–585.

Choix, F. J., de-Bashan, L. E., and Bashan, Y. (2012). Enhanced accumulation of starch and total carbohydrates in alginate-immobilized *Chlorella* spp. induced by *Azospirillum brasilense*: II. Heterotrophic conditions. *Enzym. Microb. Technol.* 51, 300–309.

Choix, F. J., de-Bashan, L. E., and Bashan, Y. (2012). Enhanced accumulation of starch and total carbohydrates in alginate-immobilized *Chlorella* spp. induced by *Azospirillum brasilense*: I. Autotrophic conditions. *Enzym. Microb. Technol.* 51, 294–299.

Cockell, C. S., Rettberg, P., Rabbow, E., and Olsson-Francis, K. (2011). Exposure of phototrophs to 548 days in low Earth orbit: microbial selection pressures in outer space and on early *Earth. ISME J.* 5, 1671–1682.

Cooper, M. B., and Smith, A. G. (2015). Exploring mutualistic interactions between microalgae and bacteria in the omics age. *Curr. Opin. Plant Biol.* 26, 147–153.

Croft, M. T., Lawrence, A. D., Raux-Deery, E., Warren, M. J., and Smith, A. G. (2005). Algae acquire vitamin B12 through a symbiotic relationship with bacteria. *Nature*, 438, 90–93.

Curtis, B. A., Tanifuji, G., Burki, F., Gruber, A., Irimia, M., and Maruyama, S. (2012). Algal genomes reveal evolutionary mosaicism and the fate of nucleomorphs. *Nature*, 492, 59–65.

Dahiya, N., Tewari, R., and Hoondal, G. (2006). Biotechnological aspects of chitinolytic enzymes: a review. *Appl. Microbiol. Biotechnol.* 71, 773–78t2.

Durham, B. P., Sharma, S., Luo, H., Smith, C. B., Amin, S. A., Bender, S. J., Dearth, S. P., Van Mooy, B. A., Campagna, S. R., and Kujawinski, E. B. (2015). Cryptic carbon and sulfur cycling between ocean surface plankton. *Proc. Natl. Acad. Sci. USA*. 112, 453–457.

Egan, S., Thomas, T., Holmström, C., and Kjelleberg, S. (2000). Phylogenetic relationship and antifouling activity of bacterial epiphytes from the marine alga *Ulva lactuca*. *Environ. Microbiol.* 2, 343–347.

Fradinho, J., Domingos, J., Carvalho, G., Oehmen, A., and Reis, M. (2013). Polyhydroxyalkanoates production by a mixed photosynthetic consortium of bacteria and algae. *Bioresour. Technol.* 132, 146–153.

Fuentes JL, Garbayo I, Cuaresma M, Montero Z, Manuel GV and Vilchez C (2016). Impact of Microalgae-Bacteria Interactions on the Production of Algal Biomass and Associated Compounds. *Mar. Drugs*, 14, 100.

Gonzalez, L. E., and Bashan, Y. (2000). Increased growth of the microalga Chlorella vulgaris when immobilized and cocultured in alginate beads with the plant-growth-promoting bacterium Azospirillum brasilense. *Appl. Environ. Microbiol.* 66, 1527–1531.

Halim, R., Danquah, M. K., and Webley, P. A. (2012). Extraction of oil from microalgae for biodiesel production: a review. *Biotechnol. Adv*. 30, 709–732.

Hancock, L., Goff, L., and Lane, C. (2010). Red algae lose key mitochondrial genes in response to becoming parasitic. *Genome Biol. Evol*. 2, 897–910.

Hannon M, Gimpel J, Miller T, Rasala B, and Mayfield S (2010). Biofuels from algae: challenges and potential. *Biofuels* 1(5), 763–784.

Helliwell, K. E., Wheeler, G. L., Leptos, K. C., Goldstein, R. E., and Smith, A. G. (2011). Insights into the evolution of vitamin B12 auxotrophy from sequenced algal genomes. *Mol. Biol. Evol*. 28, 2921–2933.

Herron, M. D., and Michod, R. E. (2008). Evolution of complexity in the volvocine algae: transitions in individuality through Darwin's eye. *Evolution*. 62, 436–451.

Hodkinson, B. P., Gottel, N. R., Schadt, C. W., Lutzoni, F. (2012). Photoautotrophic symbiont and geography are major factors affecting highly structured and diverse bacterial communities in the lichen microbiome. *Environ. Microbiol*. 14, 147–161.

Hoffmann, J. P. (1998). Wastewater treatment with suspended and nonsuspended algae. *J. Phycol*. 34, 757–763.

Hollants, J., Leroux, O., Leliaert, F., Decleyre, H., De Clerck, O., and Willems, A. (2011). Who is in there? Exploration of endophytic bacteria within the siphonous green seaweed *Bryopsis* (Bryopsidales, Chlorophyta). *PLoS One* 6, 426 - 458.

Jansson, J. K., Neufeld, J. D., Moran, M. A., and Gilbert, J. A. (2012). Omics for understanding microbial functional dynamics. *Environ. Microbiol*. 14, 1–3.

Kazamia, E., Czesnick, H., Nguyen, T. T., Croft, M. T., Sherwood, E., Sasso, S., Hodson, S. J., Warren, M. J., and Smith, A. G. (2012). Mutualistic interactions between vitamin B12-dependent algae and heterotrophic bacteria exhibit regulation. *Environ. Microbiol*. 14, 1466–1476.

Kim, B. H., Ramanan, R., Cho, D. H., Oh, H. M., Kim, H. S. (2014a). Role of *Rhizobium*, a plant growth-promoting bacterium, in enhancing algal biomass through mutualistic interaction. *Biomass Bioenergy*. 69, 95–105.

Klepac-Ceraj, V., Hayes, C. A., Gilhooly, W. P., Lyons, T. W., Kolter, R., and Pearson, A. (2012). Microbial diversity under extreme euxinia: Mahoney Lake, Canada. *Geobiology*. 10, 223–235.

Krohn-Molt, I., Wemheuer, B., Alawi, M., Poehlein, A., Güllert, S., Schmeisser, C. (2013). Metagenome survey of a multispecies and alga-associated biofilm revealed key elements of bacterial–algal interactions in photobioreactors. *Appl. Environ. Microbiol*. 79, 6196–6206.

Landa, M., Blain, S., Christaki, U., Monchy, S., Obernosterer, I. (2015). Shifts in bacterial community composition associated with increased carbon cycling in a mosaic of phytoplankton blooms. *ISME J*. 10(1), 39-50.

Lee, J., Cho, D.-H., Ramanan, R., Kim, B. H., Oh, H. M., and Kim, H. S. (2013). Microalgae associated bacteria play a key role in the flocculation of *Chlorella vulgaris*. *Bioresour. Technol*. 131, 195–201.

Lee, S. H., Oh, H. M., Jo, B. H., Lee, S. A., Shin, S. Y., and Kim, H. S. (2014b). Higher biomass productivity of microalgae in an attached growth system using wastewater. *J. Microbiol. Biotechnol*. 24, 1566–1573.

Lee, Y. K., Ahn, C. Y., Kim, H. S., and Oh, H. M. (2010). Cyanobactericidal effect of *Rhodococcus* sp. isolated from eutrophic lake on *Microcystis* sp. *Biotechnol. Lett.* 32, 1673–1678.

Lenneman, E. M., Wang, P., and Barney, B. M. (2014). A potential application of algicidal bacteria for improved lipid recovery with specific algae. *FEMS Microbiol. Lett.* 354 (2), 102–110.

Leyva, L. A., Bashan, Y., Mendoza, A., de-Bashan, L. E. (2014). Accumulation fatty acids of in *Chlorella vulgaris* under heterotrophic conditions in relation to the activity of acetyl-CoA carboxylase, temperature, and co-immobilization with *Azospirillum brasilense*. *Naturwissen schaften.* 101, 819–830.

Matsumoto, M., Yokouchi, H., Suzuki, N., Ohata, H., and Matsunaga, T. (2003). Saccharification of marine microalgae using marine bacteria for ethanol production. *Biotechnology for Fuels and Chemicals*. Humana Press, pp. 247–254.

Montemezzani, V., Duggan, I. C., Hogg, I. D., and Craggs, R. J. (2015). A review of potential methods for zooplankton control in wastewater treatment high rate algal ponds and algal production raceways. *Algal Res.* 11, 211–226.

Moran, M. A., Satinsky, B., Gifford, S. M., Luo, H., Rivers, A., and Chan, L. K. (2013). Sizing up metatranscriptomics. *ISME J.* 7, 237–243.

Nikolaeva, E. V., Usov, A. I., Sinitsyn, A. P., and Tambiev, A. H. (1999). Degradation of agarophytic red algal cell wall components by new crude enzyme preparations. *J. Appl. Phycol.* 11, 385–389.

Oswald, W. J., and Gotaas, H. B. (1957). Photosynthesis in sewage treatment. *Trans. Am. Soc. Civ. Eng*. 122, 73–105.

Parfrey, L. W., Lahr, D. J. G., Knoll, A. H., and Katz, L. A. (2011). Estimating the timing of early eukaryotic diversification with multigene molecular clocks. *Proc. Natl. Acad. Sci.* 108, 13624–13629.

Powell, R. J., and Hill, R. T. (20130. Rapid aggregation of biofuel producing algae by the bacterium *Bacillus* sp. Strain RP1137. *Applied Environ. Microbial.* 79 (19), 6093-6101.

Praveen, P., and Loh, K. C. (2015). Photosynthetic aeration in biological wastewater treatment using immobilized microalgae–bacteria symbiosis. *Appl. Microbiol. Biotechnol.* 99(23), 10345-54.

Ramanan, R., Kang, Z., Kim, B. H., Cho, D. H., Jin, L., Oh, H. M. (2015). Phycosphere bacterial diversity in green algae reveals an apparent similarity across habitats. *Algal Res.* 8, 140–144.

Ramanan, R., Kim, B.-H., Cho, D. H., Ko, S. R., Oh, H. M., Kim, H. S. (2013). Lipid droplet synthesis is limited by acetate availability in a starchless mutant of *Chlamydomonas reinhardtii*. *FEBS Lett.* 587, 370–377.

Rivas, M. O., Vargas, P., and Riquelme, C. E. (2010). Interactions of *Botryococcus braunii* cultures with bacterial biofilms. *Microb. Ecol.* 60, 628–635.

Ryu, B. G., Kim, W., Nam, K., Kim, S., Lee, B., and Park, M. S. (2015). A comprehensive study on algal–bacterial community's shift during thiocyanate degradation in a microalga - mediated process. *Bioresour. Technol.* 191, 496–504.

Sachs, J. L., and Wilcox, T. P. (2006). A shift to parasitism in the jellyfish symbiont *Symbiodinium microadriaticum*. *Proceedings biological sciences/The Rroyal. Society.* 273, 425–429.

Santos, C. A., and Reis, A. (2014). Microalgal symbiosis in biotechnology. *Appl. Microbiol. Biotechnol.* 98, 5839–5846.

Schmollinger, S., Mühlhaus, T., Boyle, N. R., Blaby, I. K., Casero, D., and Mettler, T. (2014). Nitrogen-sparing mechanisms in *Chlamydomonas* affect the transcriptome, the proteome, and photosynthetic metabolism. *Plant Cell Online*. 26, 1410–1435.

Schönknecht, G., Chen, W.-H., Ternes, C. M., Barbier, G. G., Shrestha, R. P., and Stanke, M. (2013). Gene transfer from bacteria and archaea facilitated the evolution of an extremophilic eukaryote. *Science*. 339, 1207–1210.

Seyedsayamdost, M. R., Carr, G., Kolter, R., and Clardy, J. (2011). Roseobacticides: Small molecule modulators of an algal-bacterial symbiosis. *J. Am. Chem. Soc.* 133, 18343–18349.

Shurin, J. B., Abbott, R. L., Deal, M. S., Kwan, G. T., Litchman, E., and McBride, R. C. (2013). Industrial strength ecology: trade-offs and opportunities in algal biofuel production. *Ecol. Lett.* 16, 1393–1404.

Srivastava, A., Ahn, C. Y., Asthana, R. K., Lee, H. G., and Oh, H. M. (2015). Status, alert system, and prediction of cyanobacterial bloom in South Korea. *Bio. Med. Res. Int.* 2015, 1-8.

Stengel, D. B., Macken, A., Morrison, L., Morley, N. (2004). Zinc concentrations in marine macroalgae and lichen from western Ireland in relation to phylogenetic grouping, habitat, and morphology. *Mar. Pollut. Bull.* 48, 902–909.

Subash Chandrabose, S. R., Ramakrishnan, B., Megharaj, M., Venkateswarlu, and K., Naidu, R. (2011). Consortia of cyanobacteria/microalgae and bacteria: Biotechnological potential. *Biotechnol. Adv*. 29, 896–907.

Thompson, A. W., Foster, R. A., Krupke, A., Carter, B. J., Musat, N., and Vaulot, D. (2012). Uni- cellular cyanobacterium symbiotic with a single-celled eukaryotic alga. *Science*. 337, 1546–1550.

Van Ommeren, R., and Whitham, T. (2002). Changes in interactions between juniper and mistletoe mediated by shared avian frugivores: Parasitism to potential mutualism. *Oecologia*. 130, 281–288.

Wang, H., Hill, R. T., Zheng, T., Hu, X., and Wang, B. (2016). Effects of bacterial communities on biofuel producing microalgae: Stimulation, inhibition, and harvesting. *Crit. Rev. Biotechnol*. 36(2), 341-52.

Wang, X., Li, Z., Su, J., Tian, Y., Ning, X., and Hong, H. (2010). Lysis of a red-tide causing alga, *Alexandrium tamarense*, caused by bacteria from its phycosphere. *Biol. Control*. 52, 123–130.

Yoon, H. S., Hackett, J. D., Ciniglia, C., Pinto, G., and Bhattacharya, D. (2004). A molecular timeline for the origin of photosynthetic eukaryotes. *Mol. Biol. Evol*. 21, 809–818.

In: Applied Algal Biotechnology
Editors: M. Arumugam, S. Kathiresan et al.
ISBN: 978-1-53617-524-0

Chapter 3

COLLECTION, ISOLATION, AND PURIFICATION OF MICROALGAE AND CYANOBACTERIA

Sreekala Kannikulathel Gopidas and Nagaraj Subramani*
Centre for Advanced Studies in Botany, University of Madras,
Guindy Campus, Chennai, Tamil Nadu, India

ABSTRACT

The microalgae are microscopic, photosynthetic organisms with highly varied forms and are found in almost all types of habitats around the world. Cyanophyceae or blue-green algae are one of the most impotant classes of microalgae, the members of which have been reported with multiple utility. They are believed to be the most primitive forms that were originated on earth. The microalgae needs to be collected from their sources, isolated, identified, and cultured in the laboratory for the better understanding of the algae itself and of its bioactive potential. The physico-chemical parameters of the collection site also has to be noted on the spot, as they have influence on the algal growth and characteristics. The samples containing algae are carefully collected using various available techniques, brought to the lab and isolated under a microscope. The microalga in the samples are preserved in formalin or Lugol's solution, or as permanent slides for later use. The identification and taxonomy of tthe microalgae are carried out by referring to the various well-known monographs and literatures. The isolated algae can be cultured in the laboratory under optimized conditions in selected media. The pure cultures or unialgal cultures can be developed by streaking, serial dilution, or single cell isolation techniques. Such cultures may be further mass cultivated indoor or outdoor as desired, for recovering algal biomass or its bioactive components for application in different fields.

Keywords: microalgae cyanobacteria, unialgal culture, axenic culture, defined medium

---

* Corresponding Author's Email: nagalilly@gmail.com.

3.1. Introduction

The term 'microalgae' refers to a class of microscopic, chlorophyll-containing, photosynthetic organisms which may be pigmented or non-pigmented. It is a heterogeneous group of prokaryotic and eukaryotic organisms that exhibit varied habitats, shapes, morphology, cell features, types of pigmentation, structural composition, reserve food products, and life cycles. This clearly indicates the varied evolutionary origins of the organisms classified under this term. Microalgae lack stems, roots, leaves, or vascular systems with phloem and xylem, and are unicellular or multicellular, colonial, siphonaceous, or filamentous forms which may be branched or unbranched. Their reproduction ranges from simple, asexual cell division to complex reproductive methods like oogamy involving specialized, motile or non-motile cells, and exhibits monogenetic or digenetic life cycle with alternation of generations (Van der Hoek, 1995). In the most widely accepted Fritsch's classification of algae, out of the eleven classes, all except three are microalgae; some members of Chlorophyceae, majority of Rhodophyceae, and Phaeophyceae are macroalgae. Among the microalgae, the Cyanophyceae or the blue-green algae are the most significant forms due to the fact that they are considered one of the most primitive forms that appeared on earth about three billion years ago, during the Precambrian era as indicated by the fossil records. Thus it is believed that they are the progenitors of other higher plants, showing similarity in pigmentation and starch as the storage product.

The microalgae in the form of marine phytoplankton form the base of the marine food chain and produce about fifty percent of the earth's total oxygen. Microalgae occur in a wide range of habitats, which can be widely classified into three as, aquatic, terrestrial, and unusual habitats. Aquatic algae are present either in freshwater (≤10% salinity) or marine (30-40% salinity). The freshwater algae can be planktonic (surface dwellers), benthic (bottom dwellers), or neustonic (air-water interface dwellers). The marine microalgae are of three types as, supralittoral (sub-aerial or in spray zone), eulittoral (intertidal), and sublittoral (submerged). The terrestrial algae includes cryptophytes (under soil surface), lithophytes (moist surface of rocks or stones; hypolithic –on lower surface of stones, chasmolithic – in rock fissures, and endolithic – penetrating rocks) (Whitton, 2012), epipelic (attached to mud or sand), endedaphic (on soil), epidaphic (on soil surface), and epiphytic (on other plants including algae) (Cantonati and Lowe, 2014). Besides this, the microalgae are found to grow in some unusual habitats like salt pans or salt lakes (halophytes), ice or snow (cryophytes), and the hot springs (thermophytes). They also exhibit interaction or association with other organisms, such as the symbiotic association with fungi in the roots of angiosperms, grows on animals like fish or snail (epizoic), or in the tissues of animals (endozoic). Another interesting group is the algae that grow on the tree barks mostly in association with lichen, termed as corticolous algae, and are mainly found to be blue-green or green algae. In addition to this, algae have also been found growing in man-made habitats like walls, dams, reservoirs, pools, plant pots, fountains, etc. and are called as aerophytes in general (Sahoo and Seckbach, 2015). The examples of microalgae belonging to various habitats are given in Table 1.

The microalgae also exhibit diversity in its nutrition, although they are basically considered as photoautotrophs. It has been observed that they use a combination of photoautotrophy and heterotrophy referred to as 'mixotrophy'. On requirement, they shift between these strategies especially to survive the adverse or extreme environment or nutrient deficiency. On the basis of nutrition, microalgae are of four types as, obligate phototrophs which are primarily phototrophs but turn into phagotrophy or osmotrophy when light is limiting, obligate heterotrophs which are primarily heterotrophs but turn to phototrophy when prey concentration is limited, obligate mixotrophs which are also primarily phototrophs but phagotrophy and/or osmotrophy also aids in nutrient uptake, and finally, the facultative mixotrophs which are equally photoautotrophs as well as heterotrophs.

Altogether, microalgae are unique organisms with specific characteristics and form a very significant part of the environment. Their ability to grow in diverse habitats and the flexibility in modes of nutrition makes them easily available for collection, investigation, observation, culturing, and utility. This chapter is an overview of the basic steps involved in the collection, isolation, identification, cultivation, and maintenance of the microalgae and cyanobacteria.

Table 3.1. Examples of microalgae belonging to various habitats

Sl. No.	Habitats	Types of Habitats	Sub-types	Examples
1.	Aquatic	Freshwater	Planktonic	*Volvox*
			Benthic	*Tolypothrix penicillata*
			Neustonic	*Botrydiopsis*
		Marine	Supralittoral	*Placoma vesiculosa*
			Eulittoral	*Lithophyllum lichenoides*
			Sublittoral	*Ceramium rubum*
2.	Terrestrial	Cryptophytes	-	Cyanobacteria
		Lithophytes	Hypolithic	*Nostoc, Gloeocapsa*
			Chasmolithic	Filamentous Cyanophyta
			Endolithic	*Chroococcidiopsis*
			Epilithic	*Ulothrix tenuissima*
		Endedaphic	-	*Fritschiella*
		Epidaphic	-	*Vaucheria*
		Epiphytic	-	*Trentepohlia*
		Epipsammic	-	*Eunotia vanheurckii*
		Epipelic	-	*Oedogonium sp.*
3.	Unusual habitats	Halophytic	-	*Dunaliella*
		Thermophytic	-	*Osciillatoria brevis*
		Cryophytic	-	*Chlamydomonas yellowstonensis*
		Symbiotic	-	*Nostoc*
		Endozoic	-	*Zoochlorella*
		Epizoic	-	*Stigeocladium*
		Corticolous or Aerophytes	-	*Trebouxia, Chroococcus*

3.2. Collection of Samples

As described earlier, the microalgae inhabit a wide variety of habitats and hence a phycologist has to explore as many types of habitats as possible while attempting to investigate the microalgal diversity of his/her choice of location or sampling site. The water, soil, or material sample has to be collected carefully in glass or plastic vials, containers, bottles or bags as required, and the containers should not be fully filled or closed for long durations. Always wide-mouthed, shallow containers are preferable than narrow, deeper ones. The samples have to be brought to the laboratory as early as possible and further processed. Each sample has to be labeled with proper information regarding the sampling site, date of collection, type of habitat, and the physicochemical parameters of the site such as pH, temperature, salinity, and light intensity.

The water samples are collected by immersing the container slowly in the aquatic source. The soil sediments at the bottom are also collected, as algae (diatoms) also grow in it. The floating microalgae can be collected using mesh or net with a pore size of 25-30 μm. The phytoplankton is collected using a phytoplankton net or mesh of the size 1 μm commonly; various mesh sizes are available for size-selective collection. The direct mass collection is done if the algae are larger. The aquatic plants can be squeezed to get the epiphytic algae growing on them. In the case of soil samples, as such, it is unable to isolate algae from it, and requires further processing and culturing of the sample to obtain a sufficient amount of algae. The soil samples are collected from a quadrate of one square foot or more either by pressing a pre-sterilized test tube or petri dish into the soil or using a spatula to transfer the soil into a petri dish or plastic bags. The desmids and diatoms are collected using a single bulb pipette attached to a 25 cm long plastic tube by sucking up the detritus from aquatic habitats (Williamson, 1999). The sample collection from other terrestrial habitats such as the tree barks, rocks, walls, etc. has to be carried out using hands, knives, spatula, blade, forceps, needle, and cellophane paper, scrapping down the algal growth.

A portion of each sample has to be fixed in formalin for the microscopic observation and taxonomic study, while the other portion will serve as the live material for isolation and culturing of the algae. Some preliminary observations can also be made on the collection site with regard to the macroscopic mat, scum, or net forming, branched multicellular algae. The parameters such as the pH and temperature of the water samples can be determined on-site using strips of pH papers of the range 6.5 – 9.0, and any mercurial centigrade thermometer respectively. In the laboratory, the pH has to be re-checked in a digital pH meter for accuracy.

3.3. Preservation of Samples

The algae in the samples are preserved in liquids or as permanent slides, for long term storage. Once brought to the laboratory, the samples have to be immediately preserved to prevent the loss of the original features of the algae present. The samples are washed in freshwater, shaken, and kept undisturbed for the debris to settle down. Then the floating algae are collected and washed further. The filamentous algae are isolated by using a mesh or a fine cloth to drain the sample through, leaving the algae on the mesh or cloth. The mesh or cloth is further rubbed carefully to remove other particles and may be washed under a running tap. If the sample is a hard substance, such as a soil crust, soft brush is used to clean the debris. The soil sample, on the other hand, is treated differently. It is wetted with distilled water or the corresponding medium and is kept in optimal culture conditions for the microalgae to grow. The growth obtained at the end of 10-15 days is further studied and cultured (Pal and Ray, 2013). The most commonly used preservatives are a 4-5% solution of formalin and 0.05 – 1% by volume of Lugol's solution or Lugol's Iodine (Table – 2). Lugol's solution is often preferred as it imparts color to the algae, makes them denser for easy sedimentation, and also helps in maintaining the delicate structures like flagella intact.

If the live sample is not observed within 48 hours of collection, it has to be fixed in Lugol's solution and stored in amber bottles at room temperature. The formalin solution is ideal for the preservation of cyanobacteria, dinoflagellates, and chlorophytes as it maintains the color of these algae when stored in dark, while the Lugol's solution is ideal for preserving chrysophytes. The formalin solution can preserve a specimen for a long term (one year or more) but Lugol's solution being sensitive to light can preserve the specimen for a period of six months or one year at the most. Lugol's solution also stains the starch in algal cells blue-back, thus enabling the separation of Chlorophyta (contains true starch) from Cyanophyta, Chrysophyta, and Cryptophyta (do not contain true starch, but morphologically similar to the former group).

Note: The formalin is carcinogenic and hence proper protective measures have to be taken while handling of this chemical.

3.4. Cleaning of Diatom Frustules

The diatom frustules are cleaned by the method of Sridharan (1989). Five milliliters of the sample containing diatoms is to be centrifuged at 2000 rpm for 5 minutes and wash the pellet twice with distilled water. Re-suspend the pellet in 10 mL acetone and centrifuge at 1500 rpm for 5 minutes to remove pigments. Discard the supernatant, add equal quantities of sulfuric acid and nitric acid, and gently warm the mixture, cool and incubate at room temperature for 2 hours. Further, centrifuge the mixture at 2000 rpm for 5 minutes, and wash the pellet several times with distilled water to remove dust particles. The pellet can be stored in 70% ethanol. To remove the remaining dust, density gradient centrifugation may be employed, in which the diatoms stored in alcohol can be poured over a layer of 100% glycerol taken in a centrifuge tube and centrifuged up to 500 rpm for 3-5 minutes. The dust particles form a ring at the interface between the glycerol and alcohol layers, while the diatoms being heavier settle to the bottom. The supernatant can be totally discarded saving the pellet which should be washed several times with distilled water to remove the residual glycerol before being stored in 70% ethanol, until further use.

3.5. Preparation of Permanent Slides

Two drops of the cleaned sample containing algal material may be added on to a coverslip (zero number) placed over a clean glass slide, add a few drops of distilled water using a Pasteur pipette and thoroughly spread for an even distribution of microalgae. Dry the slide over the heat to evaporate off any moisture content present. After proper drying, the cover glass has to be carefully inverted over the glass slide and sealed using Dystyrene-Phthalate-Xylene (DPX) mounting medium and allow drying completely. Such slide preparation of all the algal species has to be prepared, labeled and stored for microscopic observation.

3.6. Microscopic Observation and Microphotography

The collected samples are observed under a microscope to enumerate the morphological characteristics of the algae such as the color, size, shape of cell or colony, type of filament, branching patterns, vegetative cells, spores, and heterocysts. A wet mount of sample is used. A drop of water is placed on a clean glass slide and covered by a coverslip before examination under the microscope. Another method is called the hanging drop method wherein a drop of the sample with the algae is placed on the coverslip and is inverted onto a glass slide with a paraffin wax ring or slide ring. Bacillariophyceae members are studied after a cleaning protocol called 'Mixgen' (Prasad and Singh, 1996). The Cyanophyceae are stained by methylene blue and Chlorophyceae using iodine. Glycerin is used while mounting the material. A compound microscope with 10X, 40X, and 100X oil immersion objectives is a very essential tool for a phycologist in the observation and initial identification of microalgae. For identifying and studying the smaller or bleached algae, a phase contrast or interference microscope may be useful. The dissecting microscope is also commonly used. Apart from this, the scanning and transmission electron microscopes can be used for identifying still smaller algae and for understanding their ultrastructure. The most advanced and recent methods of identification include flow cytometry, absorbance spectroscopy, high-performance liquid chromatography, fluorescence spectroscopy, and gene probe method. Following the observation, microphotographs of the algae are taken, usually with a camera lucida attached to the microscope, and the size of the algae is determined using a calibrated eyepiece or a micrometer attached eyepiece. In certain cases, drawings are also created for a better understanding

3.7. Identification, Taxonomy and Documentation of Microalgae

Through microscopic observations, the morphological features and the ultrastructure of the microalgae are determined. The identification and taxonomy of isolated microalgae are achieved by referring to the descriptions and illustrations in various well-known monographs and literature, *viz.* Volvocales - Iyengar and Desikachary, 1981; Cyanophyta – Desikachary, 1959; Bacillariophyta – Hustedt, 1930; Zygonemaceae – Randhawa, 1936; Ulotrichales – Ramanathan, 1964; Chlorococcales – Philipose, 1967, and Oedogoniales – Gonzalves, 1981. Some others which may be helpful include Smith, 1920, 1933, 1950; Fritsch, 1935, 1945; Prescott, 1951, 1964; Chapman and Chapman, 1973, Prasad and Srivastava, 1992, Anand, 1998, Gupta, 2005; and Mahendraperumal and Anand, 2009.

Table 3.2. Composition of commonly used sample preservatives

(a). Composition of Formalin solution		**(b). Composition of Lugol's solution**	
Commercial formalin (40%)	: 100 mL	Pure Iodine	: 10 g
Glacial acetic acid	: 50 mL	Potassium iodide (KI)	: 20 g
Ethanol (50%)	: 500 mL	Distilled water	: 180 mL
Distilled water	: 350 mL	Glacial acetic acid	: 20 mL

3.8. Isolation and Purification of Microalgae

The isolation and purification of microalgae involve the laboratory culturing of the algae under optimized conditions. The algae need to be provided with optimal pH, temperature, light intensity and nutrients which resemble their existing environment, at the laboratory level. This also demands a thorough understanding of the physiology and nutrition of the algae. It is already known that they are photoautotrophic, synthesizing their own materials using light energy and inorganic substances from the environment. In the laboratory, these requirements are artificially met and the desired algae are grown under controlled conditions, with or without the interactions from other organisms. The growing of microalgae in the laboratory has immensely contributed to the precise comprehension of the taxonomy, ecology, ultrastructure, genetics, physiology, and biochemistry of the algae (Bold and Wynne, 1985).

3.9. Types of Cultures

There are several types of cultures based on the presence or absence of organisms, and the requirement of the phycologist. They are:

- **Maintenance cultures**, which are nothing but the fresh algal cultures kept under a light source showing continuous growth of many species of algae with time.
- **Enrichment cultures**, as the name indicates are cultures added with special nutrients like phosphorus or nitrogen to bring about the competitive establishment of certain desired species alone.
- **Unialgal cultures** grow only a single species, but the contamination from other bacteria, fungi, or protozoa may be present.
- **Axenic** or **Pure culture** is the growth of only a single algal species but without any contamination from other organisms.
- **Clonal cultures** raise genetically homologous populations of algae using single cells or fragments as the clones.

3.10. Preparation of Unialgal Cultures

In order to develop a unialgal culture, the alga of interest should be isolated from all other living forms. This can be achieved by mainly three techniques:

- **Streak plate method**, in which the algal sample is streaked on agar plates allowing single colonies of algae to grow and subsequently cultures are developed from these colonies. This method is employed for unicellular, colonial, or filamentous algae which are able to grow on agar medium. Single filaments can be held at the tip of a curved pipette tip and dragged on the surface of soft agar to get rid of the other algae or debris.

- **Serial dilution** is the stepwise dilution of the algal sample to obtain a fewer number of individuals per sample. Unialgal cultures of flagellates are developed by this method.
- **Single-cell isolation** using capillary pipettes is applied for some algae.

The best way to obtain unialgal cultures is to choose young filaments or branches which are less colonized by epiphytes or to choose the newly released zoospores which will be free of contaminations. But both these methods require expertise.

3.11. Preparation of Axenic Cultures

It is often difficult to obtain a totally axenic or pure culture since the removal of many bacteria attached to the surface of microalgae without harming the algae seems hard, though the removal of other organisms may be easily achieved. Hence, the axenic culture is often considered as a culture "without demonstrable unwanted prokaryotes or eukaryotes" (Barsanti and Gualtieri, 2014). The various ways of obtaining axenic culture are as follows:

- The direct method is by getting pure cultures from specialized algal culture collections.
- The best method is to isolate a single cell from the culture with the aid of a micropipette and place it fresh, sterilized medium.
- The most common method is to treat the algal culture with a cocktail of selected antibiotics in different combinations and concentrations for a period of 12 hours. The best combination that provides maximum removal of contaminants can be selected through multiple trials. The antibiotics which are usually used include streptomycin, penicillin, gentamycin, tetracycline, chloramphenicol, bacitracin, etc. Some algae are susceptible to antibiotics and therefore a range of concentration from 50 to 500% w/v of antibiotics is often tested. The antibiotic solutions are prepared in sterilized distilled water and require to be filtered through a 0.2μm filter and stored frozen prior to use.
- A range of concentrations of antibiotics may also be added to a series of algal subcultures and allowed to grow for a period. The cultures are to be checked under a phase-contrast microscope at the end of this period, to find out the culture with minimum contamination and maximum live algal cells, which can be further raised.
- Zoospores and akinetes should be washed with a bleach solution to remove epiphytes before transferring to agar plates. The bleach concentration and the period of exposure can be fixed through trials.
- If the contamination is diatoms, the algal culture has to be treated with low concentrations of germanium dioxide (GeO_2, 5 mg L^{-1}) as it can disrupt the silica structure of the diatoms.
- Another method is the streaking of the algal culture on agar plates to raise colonies of single algal cells, which can be further inoculated into the fresh, sterilized medium.

- In order to remove closely adhering epiphytes, the above-said methods are insufficient and hence a different method is adopted. A small quantity of algal culture transferred to a test tube is subjected to a low intensity (90 kHz s^{-1}) ultrasonic sound in a water bath for about a few seconds to ten minutes. This effectively separates the epiphytes from the algal cell surface without killing it.

In spite of the availability of all these techniques, proper care during the handling of the algal cultures is mandatory to develop axenic cultures. Maintaining two stock cultures at any time, one for maintenance of species and the other for starter cultures is a wise practice. Another concern is that it has been observed that certain species of algae fails to grow in absence of its epiphytes, probably due to the symbiotic association between them, and this makes it even harder to develop pure cultures of such species.

3.12. CULTURE PARAMETERS

The different parameters that regulate the growth of microalgae are temperature, light, pH, salinity, nutrients, and mixing. The optimal parameters and the range of tolerance of each parameter depend on the species of alga cultured. It also depends on the nutrition mode of the alga. These parameters should be closer to that available in the environment from which the desired alga is collected. The ideal and often employed range of temperature is between 18-20°C. A temperature below 16°C and above 35°C is usually a negative factor for the growth of many algae. The intensity, quality and period of light influence algal growth, especially through photosynthesis. The commonly used intensity is 100 -200 μE s^{-1} m^{-2} which is about 5-10% of the full daylight. It can be provided from a natural source or fluorescent tubes with red and blue light filters. Full-Time lighting leads to photo-inhibition and overheating thus adversely affecting the algal growth, hence a light/dark photoperiod (14:10 or 12:12 usually) is used. The optimum pH range is 8.2-8.7 although a range of 7-9 is acceptable. Variations in pH lead to the collapse of cellular structures and functions thus affecting algal growth. Aeration or CO_2 bubbling can help in maintaining the pH of densely grown algal culture. Salinity as a parameter of growth influences the marine algal species. The optimal salinity range is 20-24 gL^{-1}, which is attained by diluting the seawater used in the culture medium for marine microalgae. Apart from these physical parameters, the microalgae require some nutrients in the form of chemicals to grow. These include macronutrients such as sodium, potassium, magnesium, calcium, chlorine, sulfur, carbon, hydrogen, nitrogen, oxygen, phosphorus, and iron, and micronutrients such as manganese, copper, zinc, cobalt, molybdenum, chromium, nickel, vanadium, selenium, silica, bromine, fluorine, and iodine, along with certain vitamins (biotin, thiamin HCl, cobalamine, etc.). The gentle mixing of the cultures has to be ensured once a day to prevent sedimentation of the algae. The sedimentation and lack of mixing lead to cell death from nutrient limitation and poor gaseous exchange. Mixing allows equal exposure of all cells to light, nutrients as well as CO_2 required for photosynthesis.

3.13. Selection and Preparation of Media

First of all, ideal vessels are needed for media preparation and culturing of microalgae. These vessels should be non-toxic, chemically inert, transparent to light, easily sterilizable, and provide a large surface-to-volume ratio. The materials recommended for vessels include borosilicate glass (conical flasks and test tubes), polycarbonate, Teflon, and polystyrene. The first three are autoclavable while the last one is not. Non-adsorbent cotton wool plugs, foam plugs, o silicone bubble stoppers, glass covers, polypropylene covers, or metal covers and caps may be used to close the vessels. The materials that need to be avoided during microalgal culture include rubber and polyvinyl chloride (PVC) (Anderson, 2005).

The proper choice and careful preparation of medium are inevitable for the development of the cultures of the desired microalga. The medium is expected to provide the alga with all the nutrients and parameters similar to its natural habitat, in the laboratory under controlled conditions. Although many culture media recipes are available, it is always preferred to refine it according to the requirements of the alga under study. The media can be broadly classified as a defined medium and undefined medium. The defined medium is the one in which all the constituents are known and will be in a known or fixed proportion, whereas in the undefined medium, the composition is unknown and may also contain some natural ingredients. While preparing a medium, initially stock solutions of the nutrients are made in less volume of double distilled water and stored in a refrigerator at 4°C. The vitamins stock solution has to be stored at -20°C and thawed prior to use. Later measured amounts of stock solutions are mixed and the volume is made up to one liter. This is done to prevent the precipitation of some ingredients. The pH is adjusted at the end of mixing. Solid and ready-made media mixtures are also available with the chemical dealers recently. The medium has to be sterilized before use to prevent contamination. Autoclaving at 126°C (20 min, 1 atm pressure) is the often used method of sterilization, but for thermolabile components such as vitamins, membrane filtration (0.2 μm pore, cellulose nitrate or acetate filters) is preferable.

The media can also be of two types depending on the type of water used in the preparation. They are freshwater media and marine media, the former for growing freshwater microalgae and the latter for marine microalgae. The freshwater media uses double distilled or membrane filtered water for the dissolution of media components, whereas in marine media, the water base is the filtered seawater. The freshwater media may be defined or undefined. The seawater medium has to be enriched with nitrogen, phosphorus, and iron and salinity should be maintained between 30-35%. The other general media additions include soil extracts, buffers, and natural ingredients such as agar or liver extract. The media containing such additional nutrients may be referred to as the enrichment media (Anderson, 2005). The compositions of some commonly used media are given in the following Tables 3 to 10.

The microalgae including cyanobacteria can be cultured in the laboratory following the principles and methods discussed above. Many of these are applicable to the large scale production of microalgal biomass in the outdoor ponds as well. The various types of culture methods used worldwide are batch cultures, continuous cultures, semi-continuous cultures, pond cultures, and photobioreactors. Batch cultures are a closed culture system where the medium is finite and there is no input or out of resources. The algae show sigmoid growth, with a decline in growth and subsequent death due to nutrient depletion. Further propagation is by subculture. While in continuous cultures, there is an equal input of new nutrient medium as well as the removal of spent medium, thereby the system is open type, the nutrients are infinite and the algae are maintained in a continuous growth stage. In semi-continuous culture, the nutrient medium is fed after a particular time period, and an equal amount is removed. It is a combination of the batch and continuous culture methods (Richmond, 1995). The commercial cultivation of algae employs extensive outdoor open raceway ponds or closed bioreactors. Each of these methods has its own advantages and disadvantages and directly influences the quantity and quality of the microalgal biomass or the products derived from them. Hence a judicious choice of the method has to be undertaken.

Table 3.3. List of some commonly used media and the respective groups of microalgae cultured

Name of the Medium	Cultured Algal Groups
BG11 Medium	Freshwater & soil Cyanophyceae
Diatom Medium	Freshwater Bacillariophyceae
Bold Basal Medium	Broad-spectrum medium for freshwater Chlorophyceae, Xanthophyceae, Chrysophyceae, Cyanophyceae
Zarrouk's Medium	Mass cultivation of *Arthrospira* sp.
Walne's Medium	Broad-spectrum medium for marine algae mass culture
CHU-11 Medium	Marine Cyanobacteria
f/2 Medium	Broad spectrum medium for coastal algae

Table 3.4. Composition of BG11 medium

Reagents	Per Liter
$NaNO_3$*	1.5 g
$K_2HPO_4.3H_2O$	0.004 g
$MgSO_4.7H_2O$	0.075 g
$CaCl_2.2H_2O$	0.027 g
Citric acid ($C_6H_8O_7$)	0.006 g
Ammonium ferric citrate ($C_6H_8O_7.xFe.yNH_3$)	0.006 g
Na_2Mg-EDTA	0.001 g
Na_2CO_3	0.02 g
Microelement stock solution	1 mL
Microelement stock solution	**Per Liter**
H_3BO_3	2.860 g
$MnCl_2.4H_2O$	1.810 g
$ZnSO_4.7H_2O$	0.222 g
$Na_2MoO_4.2H_2O$	0.390 g
$CuSO_4.5H_2O$	0.079 g
Co $(NO_3)_2$. $6H_2O$	0.0494 g
pH 7.4	

*To be avoided for N_2-fixing Cyanobacteria.

Table 3.5. Composition of Diatom medium

Reagents	Per Liter
$Ca(NO_3)_2.4H_2O$	20 mg
KH_2PO_4	12.4 mg
$MgSO_4.7H_2O$	25 mg
$NaHCO_3$	15.9 mg
FeNa-EDTA	2.25 mg
Na_2-EDTA	2.25 mg
H_3BO_3	2.48 mg
$MnCl_2.4H_2O$	1.39 mg
$(NH_4)_6Mo_7O_{24}.4H_2O$	1.0 mg
Biotin (Vitamin B_7)	0.04 mg
Thiamine HCl (Vitamin B_1)	0.04 mg
Cobalamine (Vitamin B_{12})	0.04 mg
$Na_2SiO_3.9H_2O$	57 mg
pH 6.9	

Table 3.6. Composition of Zarrouk's medium

Reagents	Per Liter
$NaHCO_3$	16.8 g
K_2HPO_4	0.5 g
$NaNO_3$	2.5 g
K_2SO_4	1.0 g
NaCl	1.0 g
$MgSO_4.7H_2O$	0.2 g
$CaCl_2.2H_2O$	0.04 g
$FeSO_4.7H_2O$	0.01 g
EDTA	0.08 g
Microelement stock solution	1 mL
Microelement stock solution	**Per Liter**
H_3BO_3	2.86 g
$MnCl_2.4H_2O$	1.81 g
$ZnSO_4.4H_2O$	0.222 g
Na_2MoO_4	0.0177 g
$CuSO_4.5H_2O$	0.079 g
pH 9	

Table 3.7. Composition of Bold Basal medium

Reagents	Per Liter
KH_2PO_4	175 mg
$CaCl_2.2H_2O$	25 mg
$MgSO_4.7H_2O$	75 mg
$NaNO_3$	250 mg
K_2HPO_4	75 mg
NaCl	25 mg
H_3BO_3	11.42 mg
Microelement stock solution	1 mL

Reagents	**Per Liter**
Solution 1	1 mL
Solution 2	1 mL
Microelement stock solution	**Per Liter**
$ZnSO_4.7H_2O$	8.82 g
$MnCl_2.4H_2O$	1.44 g
Microelement stock solution	**Per Liter**
MoO_3	0.71 g
$CuSO_4.5H_2O$	1.57 g
$Co (NO_3)_2. 6H_2O$	0.49 g
Solution 1	**Per Liter**
Na_2-EDTA	50 g
KOH	3.1 g
Solution 2	**Per Liter**
$FeSO_4$	4.98 g
H_2SO_4 (conc.)	1 mL
pH 6.8	

Table 3.8. Composition of Walne's medium

Reagents	**Per Liter Seawater**
Solution A	1.0 mL
Solution B	0.1 mL
Solution C	2.0 mL
Solution A	**Per Liter**
$FeCl_3.6H_2O$	1.3 g
$MnCl_2.4H_2O$	0.4 g
H_3BO_3	33.6 g
Na_2-EDTA	45.0 g
$NaH_2PO_4.2H_2O$	20.0 g
$NaNO_3$	100.0 g
Solution B	1 mL
Solution B	**Per 100 mL**
$ZnCl_2$	2.1 g
$CoCl_2.6H_2O$	2.0 g
$(NH_4)_6Mo_7O_{24}.4H_2O$	0.9 g
$CuSO_4.5H_2O$	2.0 g
Concentrated HCl	10 mL
Solution C	**Per 200 mL**
Thiamine HCl (Vitamin B_1)	0.2 g
Cobalamine (Vitamin B_{12})	10 mg
Solution D	**Per Liter**
$Na_2SiO_3.5H_2O$	40.0 g

Table 3.9. Composition of CHU-11 medium

Reagents	Per Liter Seawater
$MgSO_4.7H_2O$	0.075 g
$CaCl_2.2H_2O$	0.036 g
$NaNO_3$	1.5 g
$K_2HPO_4.3H_2O$	0.04 g
Na_2CO_3	0.02 g
$Na_2SiO_3.9H_2O$	0.58 g
Citric acid ($C_6H_8O_7$)	0.006 g
Ammonium ferric citrate ($C_6H_8O_7.xFe.yNH_3$)	0.006 g
Na_2Mg-EDTA	0.001 g
Microelement stock solution	1 mL
Microelement stock solution	**Per Liter**
H_3BO_3	0.5 g
Zn $(NO_3)_2.6H_2O$	2 g
$Na_2MoO_4.2H_2O$	0.5 g
$CuCl_2.2H_2O$	0.025 g
Co $(NO_3)_2$. $6H_2O$	0.025 g
$VOSO_4.6H_2O$	0.025 g
HCl 1M	3 mL
pH 7.5	

Table 3.10. Composition of f/2 medium

Reagents	Per Liter Seawater
$NaNO_3$	0.075 g
$NaH_2PO_4.H_2O$	0.005 g
Microelement stock solution	1 mL
Vitamin solution	1 mL
Microelement stock solution	**Per Liter**
$FeCl_3.6H_2O$	3.150 g
Na_2-EDTA	4.160 g
$MnCl_2.4H_2O$	0.180 g
$CoCl_2.6H_2O$	0.010 g
$CuSO_4.5H_2O$	0.010 g
$ZnSO_4.H_2O$	0.022 g
$Na_2MoO_4.2H_2O$	0.006 g
Vitamin solution	**Per Liter**
Biotin (Vitamin B_7)	0.5 mg
Thiamine HCl (Vitamin B_1)	100 mg
Cobalamine (Vitamin B_{12})	0.5 mg
pH 8.0 adjusted with 1M NaOH or HCl	

CONCLUSION

To conclude, the collection, isolation, and purification of microalgae and cyanobacteria involve many steps as aforementioned. It requires careful handling and practice for the successful purification of algae. The practice of isolating microalgae and its detailed investigations have improved our knowledge of various aspects of phycology. It was through such investigations over the years, the scientific community realized the ecological and economic significance of microalgae. As a result of which, numerous new species were reported, and also various bioactives that could be applied in different fields were identified and utilized.

SUMMARY

The microalgae and Cyanophyceae are ubiquitous organisms with no stem, leaves or roots differentiation, but are diverse in forms and are significant due to their adaptability and employability in various applications. For the study and preservation of microalgae, they have to be collected from their habitats and isolated, identified and grown in laboratory under optimized conditions. This involves a series of aseptic techniques some of which are similar to the microbiological techniques. The various parameters of the collection site such as the temperature, pH, and salinity of the sample collection site has to be determined. The samples with algae are collected using various techniques such as using bottles or vials for water samples, net/mesh for phytoplanktons, and scrapping off with knife, spatula, forceps, blade, or needle in case of material surfaces. The samples are brought to the lab and isolated and identified under a microscope – compound, phase contrast, or interference. The samples can be preserved in formalin or Lugol's solution, or as permanent slides. The identification and taxonomy of tthe microalgae is done referring to the well-known monographs and literatures. The isolated algae then has to be cultured in a selected medium, defined or undefined, under optimized conditions. The pure/axenic cultures and unialgal cultures of the algae is raised by streak plate method, serial dilution, or single cell isolation using micropipettes, using antibiotics, bleaching solutions, germanium oxide, or ultrasonication. Various culture parameters such as temperature, light intensity and duration, pH, salinity, nutrients, aeration, and mixing have to be keenly maintained for the successful production of algal cultures. The algal cultures can be maintained aseptically in a culture collection centre for their preservation and further studies.

FUTURE PERSPECTIVES

The practice of collection, isolation, identification, and investigation of microalgae has to be extensively carried out for identifying potential natural candidates for procuring useful products that find application in various arenas. Although the aseptic condition is followed throughout the algal culturing process, the contamination of the cultures is a common hindrance in raising axenic cultures. So it is necessary to develop some improved techniques to solve this issue. The microalgal biodiversity is an indicator of the climatic and the environmental status of a location. Hence it is possible to develop a bioindicator technique based on this relationship to assess the impact of climatic variations and anthropogenic activities on the ecosystem and environment.

Acknowledgment

The authors immensely thank to the Director, Centre for Advanced Studies in Botany, University of Madras, Guindy Campus, Chennai, Tamil Nadu, India, for providing the necessary laboratory facilities for carrying out the research.

References

Andersen, R. A. (2005). *Algal Culturing Techniques.* Burlington, USA: Elsevier Academic Press, 14-33.

Barsanti, L. & Gualtieri, P. (2014). *Algae – Anatomy, Biochemistry, and Biotechnology.* Boca Raton: CRC Press, Taylor and Francis Group, 221-259.

Bold, H. C. & Wynne, M. J. (1985). *Introduction to the Algae – Structure, and Reproduction.* Englewood Cliffs, N.J: Prentice-Hall, Inc., 15-17.

Cantonati, M. & Lowe, R. L. (2014). Lake Benthic Algae: Toward an understanding of their ecology. *Freshwater Science*, *33*(2), 475-486.

Pal, R. & Ray, S. (2013). *Manual of Cryptogamic Botany.* College Street, Kolkata: Probir Chatterjee Research Foundation, University of Calcutta, 1-2, 24-25, 103-105.

Prasad, B. N. & Singh, Y. (1996). *Algal Indicator of Water Pollution.* Bishen Singh and Mahendra Pal Singh, Dehradun, India, pp 263.

Richmond, A. (Ed.). (2004). *Handbook of Microalgae Mass Culture and Biotechnology and Applied Phycology.* Malden, USA: Blackwell Publishing, 47-54.

Sahoo, D. & Sechbach, J. (2015). *The Algae World* (*Cellular Origin, Life in Extreme Habitats and Astrobiology*). New York: Springer Dordrecht Hiedelberg, 4-9.

Sridharan, V. T. (1989). *Phytoplankton and Algae Studies. Techniques of Plankton Methodology.* Prepared for Training workshop on Integrated Environmental Research program on Kaveri River, 1-69.

Van der Hoek, C., Mann, D. G. & Jahns, H. M. (1995). *Algae – An Introduction to Phycology.* Cambridge, UK: Cambridge University Press, pp 623.

Whitton, B. A. (2012). *Ecology of Cyanobacteria II: Their Diversity in Space and Time.* New York, London: Springer Dordrecht Hiedelberg, 291-370.

Williamson, D. B. (1999). A proposed new desmid genus *Cruciangulum* and descriptions of three new desmid species from rock pools in the Western Cape Town, South Africa. *Algological Studies*, *93*, 55-62.

In: Applied Algal Biotechnology
Editors: M. Arumugam, S. Kathiresan et al.
ISBN: 978-1-53617-524-0

Chapter 4

PHOTOSYNTHESIS IN ALGAE

Gour Gopal Satpati[1] and Ruma Pal[2]
[1]Department of Botany, Bangabasi Evening College,
University of Calcutta, West Bengal, India
[2]Department of Botany, University of Calcutta, West Bengal, India

Keywords: photosynthesis, algae, cyanobacteria, endosymbiosis, pigment, light energy, electron flow, rubisco, carbon fixation

1. INTRODUCTION

Photosynthesis is a phenomenon of carbon fixation from inorganic to organic form and producing carbohydrates by photoautotrophs with the help of sunlight, water and chlorophyll. The growth and metabolism of all living systems on Earth either directly or indirectly depends on photosynthesis for organic matter and energy (Masojídek, Torzillo, and Koblížek 2013). In anoxygenic photosynthesis, light is trapped and converted to energy in the form of Adenosine tri phosphate (ATP) without producing oxygen (Figure 1). About 3 billion years ago, some photoautotrophic microorganisms and anoxygenic photosynthetic bacteria used light energy to produce protons and electrons from external donors like hydrogen sulfide (H_2S), Fe^{2+}, $S_2O_3^{2-}$ as cytochrome c_2, and FeS as cytochrome bc_1; reducing carbon dioxide (CO_2) into organic compounds. The process of converting or reducing CO_2 into organic molecules with the help of external electron donors is known as cyclic photophosphorylation, which commonly takes place in the anoxygenic process. On the other hand, oxygenic photosynthesis takes place in oxygen producing photosynthetic microorganisms like cyanoprokaryotes (cyanobacteria) or blue green algae (BGA) and eukaryotes including algae and higher plants (Figure 1). The oxygenic environment on Earth was formed by such oxygenic microorganisms about 2 billions year ago (Masojídek, Torzillo, and Koblížek 2013). Oxygenic photosynthesis in eukaryotic microorganisms can be divided into two stages: light dependent reaction or light reaction and light independent reaction or dark reaction. Both the reactions take place in special photosynthetic apparatus known as chloroplast, which contains stacks of lipoprotein membranes or thylakoids and aqueous matrix or stroma. These thylakoids are the light-harvesting sites for the production of biochemical reductant $NADPH_2$ by NADP reductase enzyme. The ATP and NADPH are the cofactors used to perform the light independent reactions to prepare sugar as storage product. The universal plant enzyme Ribulose-bis-phosphate carboxylase/oxygenase (Rubisco) helps to fix atmospheric CO_2 into storage sugar. Eukaryotic algae evolved through endosymbiotic events became photosynthetically more active than the prokaryotes (cyanobacteria). Different algal groups evolved from time to time through endosymbiotic process.

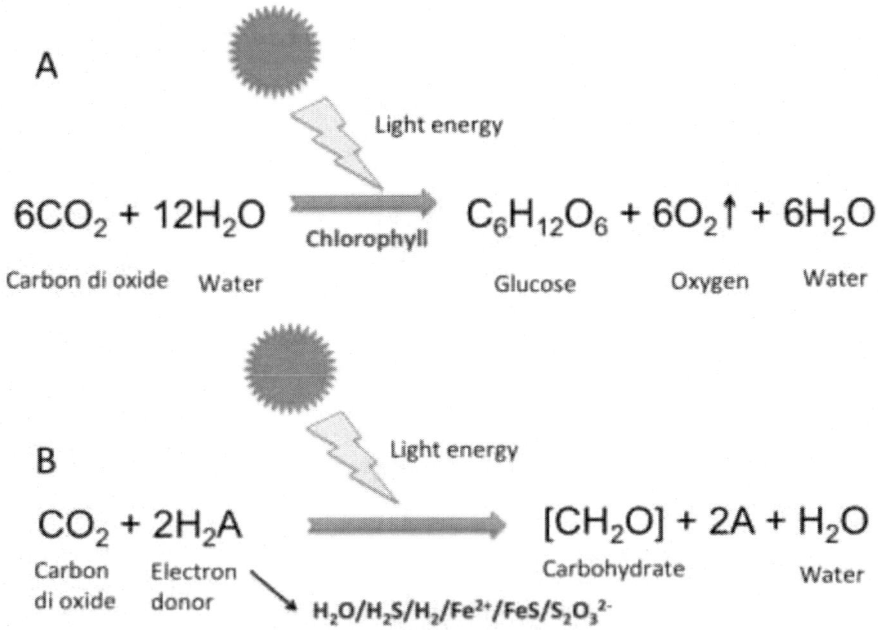

Figure 1. A. Oxygenic photosynthesis and B. Anoxygenic photosynthesis in photoautotrophic microorganisms.

Photosynthetic pigments are the primary light absorbing units for photosynthesis. They absorb light energy and convert them into chemical energy through electron transport system generated by the oxygen-evolving complex. Electron transport system consists of several photosynthetic apparatus, which helps to carry electrons and protons to produce ATP and $NADPH_2$. Algae can capture CO_2 naturally and via cultivation systems like tanks and photobioreactor culture mode. In environment there are many naturally growing species, which can consume high amount of CO_2 to convert them into storage products. In tropical countries, different algal and cyanobacterial blooms and other filamentous green algal growth like, *Cladophora*, *Pithophora*, *Ulva*, *Enteromorpha*, etc. are very common. In this book the authors have demonstrated the photosynthetic process of algae including pigment composition and light trapping mechanism, photosynthetic apparatus, activity of Rubisco, carbon fixation mechanisms, etc.

2. PLASTID EVOLUTION AND ENDOSYMBIOTIC THEORY

The evolution of eukaryotes from prokaryotic microorganisms is believed to have happened by the process of 'endosymbiosis'. Konstantin Mereschkowski first proposed the idea of 'endosymbiosis' in the year 1905 and subsequently coined the term "symbiogenesis" when he found the symbiotic relationship between algae and fungi (Mereschkowski 1905). Endosymbiosis is defined as the phenomenon of an organism living within another organism. The photosynthetic bacteria or cyanobacteria are engulfed by the modern heterotrophic eukaryotes resulting in the formation of chloroplast in algae, which further evolved into higher plant chloroplast. In the year 1960, Lynn Margulis worked on endosymbiosis and attempted to explain the origin of eukaryotic cell organelles such as mitochondria and chloroplast (Figure 2). She established her results based on biochemical, cytological and paleontological evidences (Margulis 1970).

The symbiotic relationship between the two species is a crucial element in the evolution of higher order of eukaryotes. The primary and secondary endosymbiosis resulted in the formation of simple and complex plastids. In primary endosymbiosis, a heterotrophic eukaryote engulfs a photosynthetic cyanobacterium to form simple plastid bearing eukaryotes such as chlorophytes, glaucophytes and rhodophytes (Figure 2). In secondary and tertiary endosymbiosis, the simple plastid or plastid bearing eukaryotes are engulfed by one or two heterotrophic eukaryotes, which leads to the formation of complex plastid bearing eukaryotic organisms; belonging to the group of Euglenophyta, Dinophyta Apicomplexa, Cryptophyta, Chlorarachniophyta, Heterokontophyta and Haptophyta. The chloroplast endoplasmic reticulum is formed by the continuation of the outer plastid membrane and outer membrane of the nuclear envelope. Among the evolved algal groups; Chlorophyta and Rhodophyta do not contain any chloroplast endoplasmic reticulum. In association to this, the chloroplast of the Rhodophycean algae contains one thylakoid per band whereas two to six thylakoids per band are present in Chlorophytes. During evolution, two membranes of the chloroplast E.R. and two thylakoids per band were formed in Cryptophytic algae. Three thylakoids per band and one membrane of the chloroplast E. R. was noticed in Dinophytes and Euglenoids. The two membranes of chloroplast E. R. and three thylakoids per band evolved in the group of Prymnesiophyta and Heterokontophyta. The evolution had been carried out with primary and secondary endosymbionts, and the evolution of chromalveolates was noticed (Figure 2) (Nosenko 2006). The recent discovery of green and red algae genes in diatom has suggested the presence of additional secondary endosymbiosis or tertiary endosymbiosis during the early evolution of chromalveolates (Moustafa et al. 2009; Chan and Bhattacharya 2010). Chromalveolates are the organisms that descended from a single secondary endosymbiosis involving a red alga and a bikont i.e., biflagellate eukaryotic microorganism. They were separated from the Kingdom Chromista and first coined by Thomas Cavalier- Smith in 1981.

Rapid advancement in the fields of molecular biology and bioinformatics, has helped in finding the evidence of cyanobacterial gene transfer in the algal plastid and to the plastid of higher plants (Reyes-Prieto, Weber and Bhattacharya 2008). The biochemistry of plastids and protein transport mechanism in algae provided several independent lines of evidence of endosymbiosis (Weber, Linka and Bhattacharya 2006). About 1.5 billion years ago, the evolution of primary plastids was established in eukaryotic algae but molecular data remains controversial due to limited fossil records (Blair, Shah and Hedges 2005). The first cyanobacterial donor recorded was *Prochlorococcus-Synechococcus* with unusual pigmentation of chlorophyll a_2 and chlorophyll b_2 (Yoon et al. 2006). The plastid genomes are highly reduced in the red tide forming taxa *Alexandrium* due to substantial gene transfer from the endosymbionts to the host nucleus (Hackett et al. 2004). The transition to obligate mode of parasitism subsequently led to the loss of photosynthetic capability of Apicoplasts as they have shared common origin with secondary plastids of the closely related Dinoflagellates (Waller and McFadden 2005). The photosynthetic excavates like *Euglena* and chlorarachniophyte amoebae *Rhizaria* have acquired their plastids in secondary endosymbiosis during which their common ancestors independently captured green alga (Rogers et al. 2007). Independent acquisition hypothesis and Chromalveolate hypothesis are the alternative explanations for the origin of secondary plastids (Baurain et al. 2010). The origin of secondary plastids is still controversial whereas the evolution of primary plastids by endosymbiosis involving a cyanobacterium is well established. In numerous empirical studies, the chromalveolate hypothesis or the secondary endosymbiosis involving a red alga is the best-known hypothesis till date. Plastid origin in other eukaryotic lineages by other free-living eukaryotes involves tertiary endosymbiosis. Advanced studies related to biochemical, cytological and phylogenetic relationship in terms of molecular validation is needed for better understanding of the origin of plastids in eukaryotes.

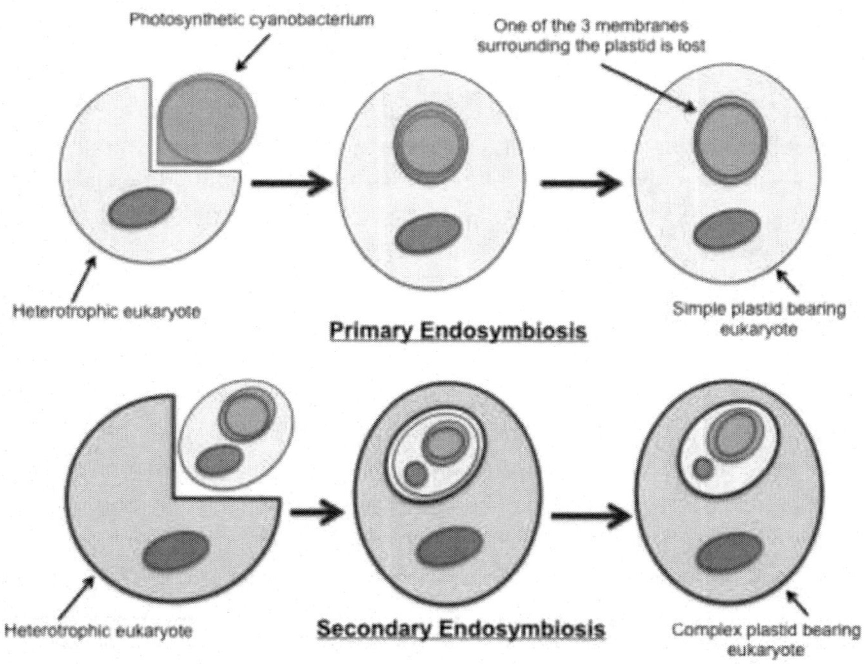

Figure 2. Primary and Secondary endosymbiosis.

3. Pigment Compositions in Algae

A pigment is a chemical or compound or material that changes the color of reflected or transmitted light as a result of wavelength-selective absorption. Photosynthetic pigments are very useful for autotrophic microorganisms including algae and plants that make their own food through photosynthesis. In photosynthetic algae and cyanobacteria, chlorophyll is the main pigment distributed in the chloroplasts. Chlorophyll is mainly composed of a porphyrin head and a phytol chain, which is very similar to animal chromoprotein and haemoglobin but have magnesium atom instead of an iron atom in their chemical structure. The major chlorophyll present in algae is of 4 types: a, b, c (c_1 and c_2) and d (Figures 3-7). An additional chlorophyll i.e., chlorophyll f is also found in some cyanobacteria (Figure 8). Chlorophylls absorb light mostly in the blue and red portion of the electromagnetic spectrum. The light energy is captured by chlorophyll and transferred to other parts of the photosystem during photosynthesis. This is followed by charge separation i.e., energy absorbed by photon is transferred to the next electron. The oxidation reaction triggers the removal of electrons from chlorophylls. A series of molecular intermediate carries electron through the electron transport system. Both the photosystems (PS II/P_{680}^+ and PS I/P_{700}^+) become charged and pumps H^+ ions across the thylakoid membranes to form ATP and NADPH. The NADPH is then directed to fix or reduce CO_2 into sugars and other biosynthetic products.

The accessory photosynthetic pigments are the associated molecules that help to serve photosynthesis in conjunction with universal pigment chlorophyll a (Table 1). They are of two categories: chlorophyll accessory pigments such as chlorophyll b, c, d, f, etc. and non-chlorophyll accessory pigments like carotenoids and phycobiliproteins (Table 2). Carotenoids and phycobiliproteins also serve photosynthetic functions like light energy transfer to photosystem chlorophyll. Carotenoids are tetraterpenoids, which are present in algae and cyanobacteria as yellow, orange and red organic molecules (Figure 9-13). They give color to carrot, pumpkins, tomatoes, corn, and also to the different groups of algae. Structurally, carotenoids comprise 40 branched carbon units bonded together by alternating single and double bonds. Carotenoids also absorb and dissipate excess light energy and work as antioxidants. There are different types of carotenoids present in different algal classes serving as non-chlorophyll accessory pigments for photosynthesis. Another group of pigments like Peridinin and Phycobilins are unique to dinoflagellates and cryptophytes (Table 2). Peridinin is an apocarotenoid type pigment associated with chlorophyll and functions as Peridinin-chlorophyll-protein (PCP) light harvesting complex in dinoflagellates (Figure 14). These protein complexes absorb blue-green light in the range of 470-550 nm to perform photosynthesis. In cyanobacteria, glaucophytes and red algae, some aggregates of light harvesting protein complexes are abundantly found in the stroma of the thylakoid membrane known as phycobilisomes (Figure 17). Phycobilisomes are composed of trimeric protein complexes namely: phycocyanin (PC), phycoerythrin (PE) and allophycocyanin (APC) and are covalently bound to the apoprotein (Table 2) (Figure 17). The core of the phycobilisomes is structured by APCs from which outwardly directed stacks of PCs and PEs are formed. Phycobilins are group of light harvesting bilin proteins abundant in the phycobilisomes complexes (Figure 15). They function by bonding with water-soluble proteins, phycobiliproteins that pass the light energy to chlorophylls for photosynthesis. They are composed of chromophore, which absorb red, orange, green and yellow wavelengths of light that usually is not absorbed by chlorophyll a. The major phycobilins are grouped into phycocyanobilin, phycoerythrobilin, phycourobilin and phycobiliviolin (Table 2) (Figure 14-17).The antenna like arrangements of phycobilisomes results in 95% efficiency in energy transfer to the chlorophyll molecules for photosynthesis (Glazer 1985).

Table 1. Major photosynthetic pigments in algae and their properties

Major Chlorophylls in Algae	Algal groups	Solubility	Molecular formula	Absorbance	Function
Chlorophyll a	All photosynthetic algae	Organic solvents- alcohol, diethyl ether, benzene, acetone except petroleum ether	$C_{55}H_{72}O_5N_4Mg$	Red light absorption band-663 nm Other- 430 nm	Light receptor in photosystem I of the light reaction
Chlorophyll b	Chlorophyta Euglenophyta Charophyta	Maximum in acetone and Methanol	$C_{55}H_{70}O_6N_4Mg$	645 nm and 435 nm	Light harvesting pigment transferring absorbed light energy to chlorophyll a
Chlorophyll c_1	Heterokontophyta	Ether, acetone, methanol, ethyl acetate	$C_{35}H_{30}O_5N_4Mg$	634 nm, 583 nm and 440 nm	Accessory pigment to photosystem II
Chlorophyll c_2	Dinophyta Cryptophyta Phaeophyta Heterokontophyta	Ether, acetone, methanol, ethyl acetate	$C_{35}H_{28}O_5N_4Mg$	635 nm, 586 nm and 452 nm	Accessory pigment to photosystem II
Chlorophyll d	Cyanobacteria Rhodophyta	Diethyl ether, benzene, acetone, Methanol	$C_{54}H_{70}O_6N_4Mg$	696 nm, 456 nm and 400 nm and 710 nm (Infrared light)	Energy capturing from sunlight
Chlorophyll f	Cyanobacteria Xanthophyta	Diethyl ether, benzene, acetone, Methanol	$C_{55}H_{70}O_6N_4Mg$	>700 nm (Infrared light)	Energy transfer and charge separation

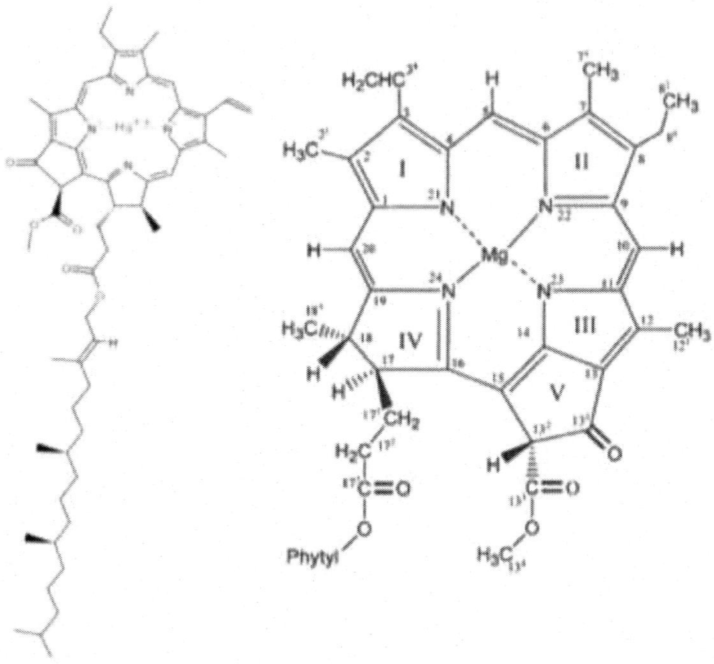

Figure 3.Chlorophyll a.

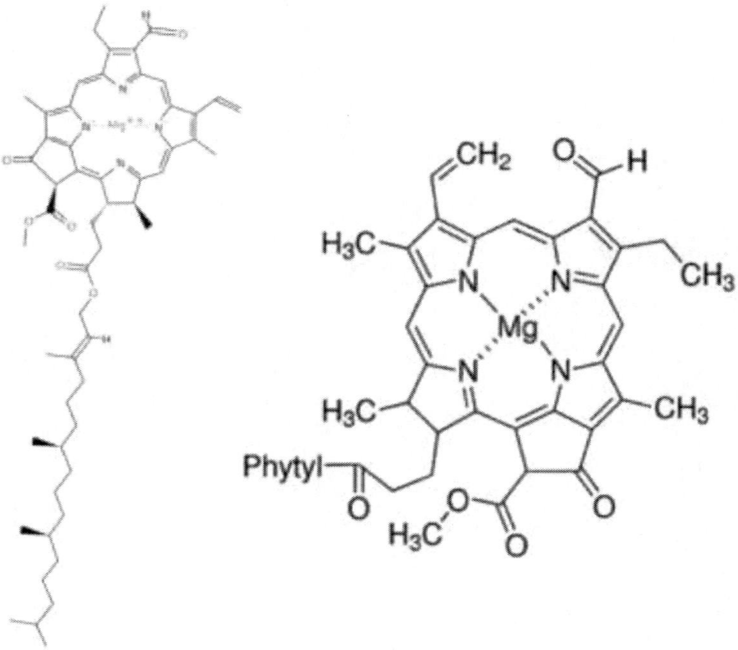

Figure 4.Chlorophyll b.

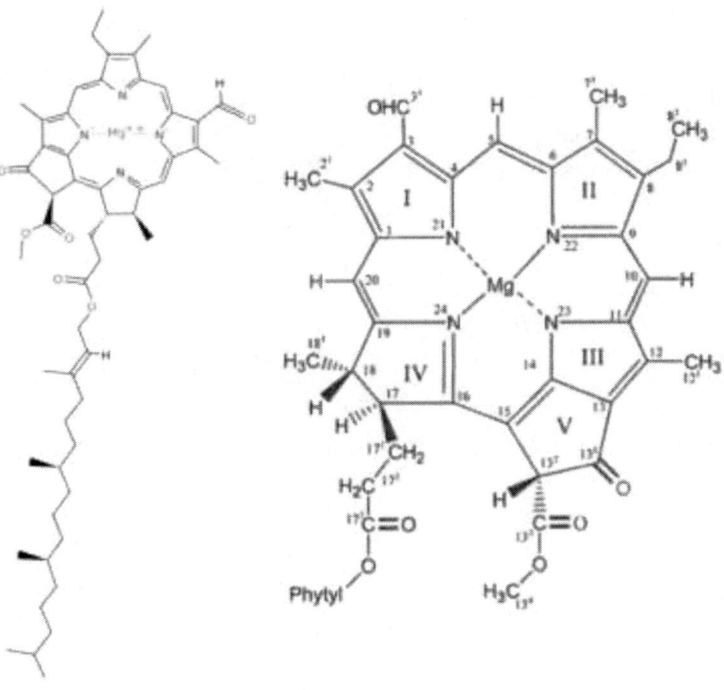

Figure 5. Chlorophyll d.

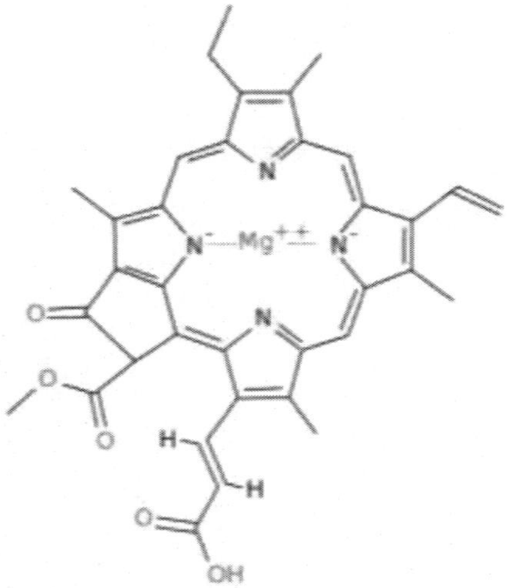

Figure 6.Chlorophyll c_1.

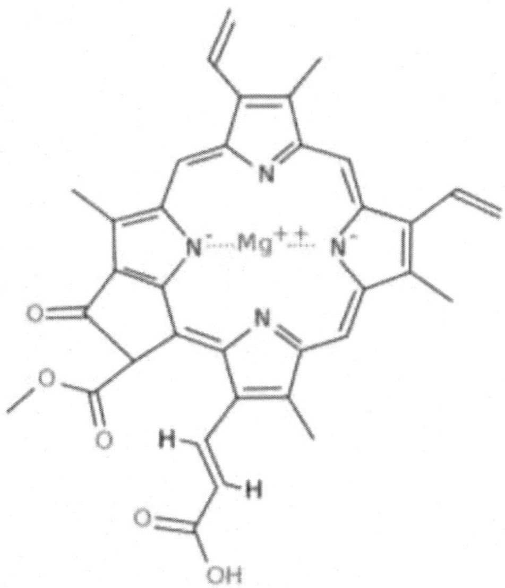

Figure 7. Chlorophyll c_2.

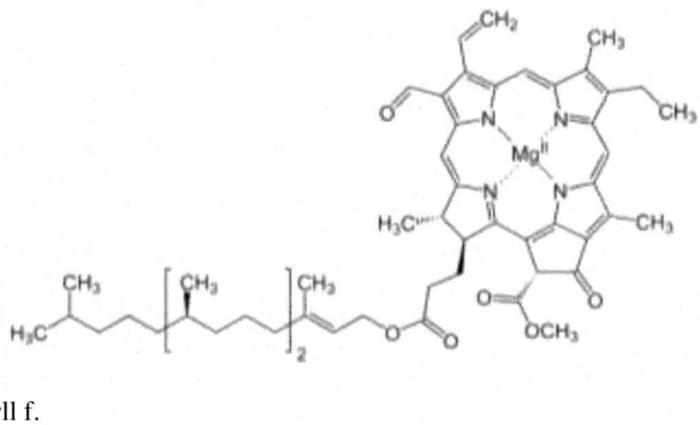

Figure 8. Chlorophyll f.

Table 2. Accessory non-chlorophyll pigments in different algal groups

Algal groups	Common name	Non-Chlorophyll accessory pigments
Chlorophyta	Green algae	β- carotene and xanthophyll
Charophyta	Charophytes	γ- carotene, lycopene, xanthophyll
Euglenophyta	Euglenoids or Euglenophytes	β- carotene, xanthophyll (Antheraxanthin)
Chrysophyta	Yellow brown or Golden brown algae	β- carotene, xanthophyll (Fucoxanthin)
Pyrrhophyta/Dinophyta	Dinoflagellates	Peridinin
Cryptophyta	Cryptomonads	Phycobilins
Xanthophyta	Yellow green algae	β- carotene, xanthophyll (Diadinoxanthin)
Bacillariophyta	Diatoms	β- carotene, ε- carotene, xanthophyll (Fucoxanthin)
Phaeophyta	Brown algae	β- carotene, xanthophyll (Fucoxanthin)
Rhodophyta	Red algae	r-phycocyanin, r-phycoerythrin, α-carotene, β- carotene, xanthophyll (Lutein, tetraxanthin)
Cyanophyta/ Cyanoprokaryota/ Cyanobacteria	Blue green algae	Phycocyanin, phycoerythrin, phycocyanobilin, β- carotene, myxoxanthin, myxoxanthophyll

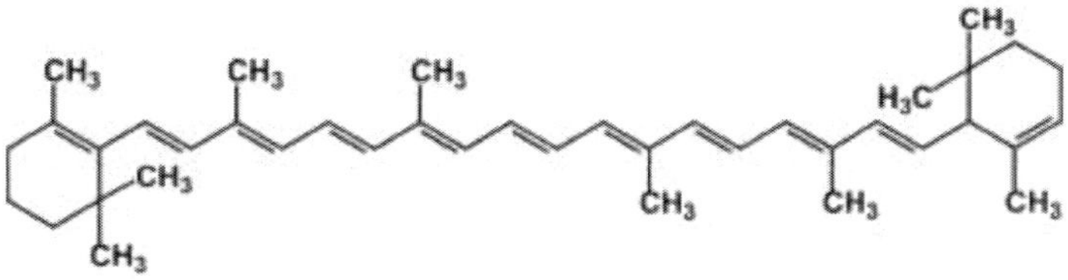

Figure 9. α-carotene.

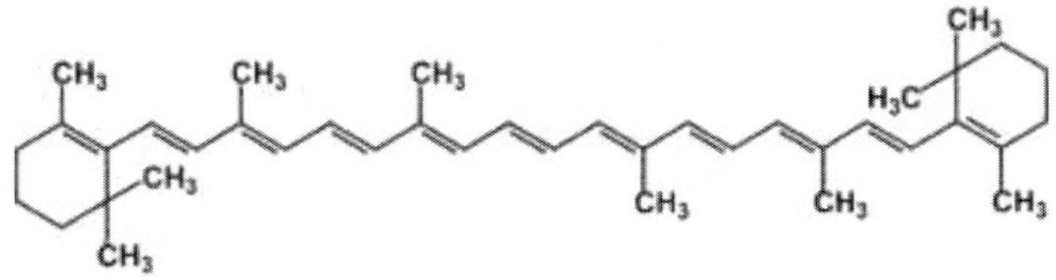

Figure10. β-carotene.

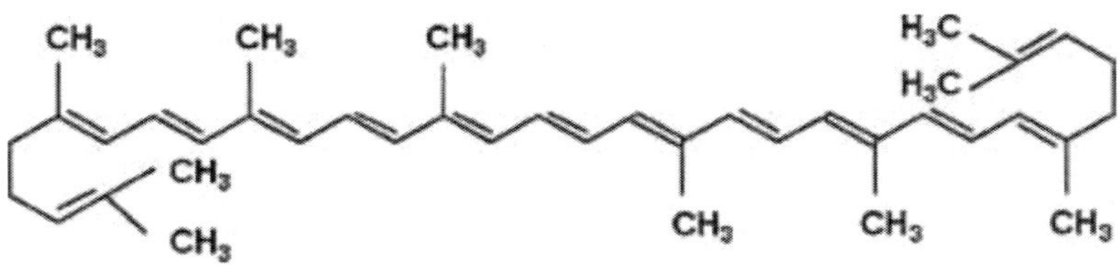

Figure 11. Lycopene.

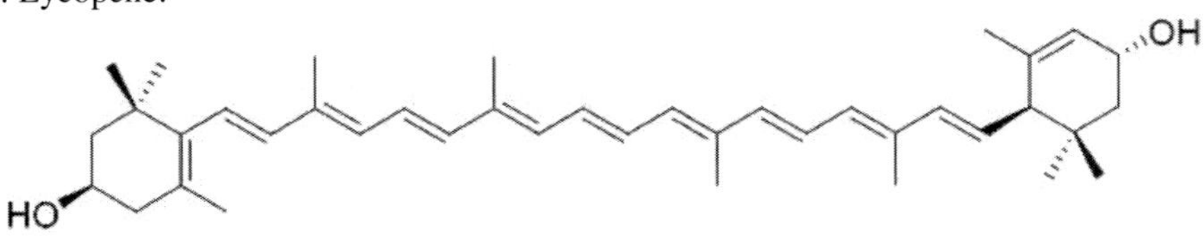

Figure 12. Lutein.

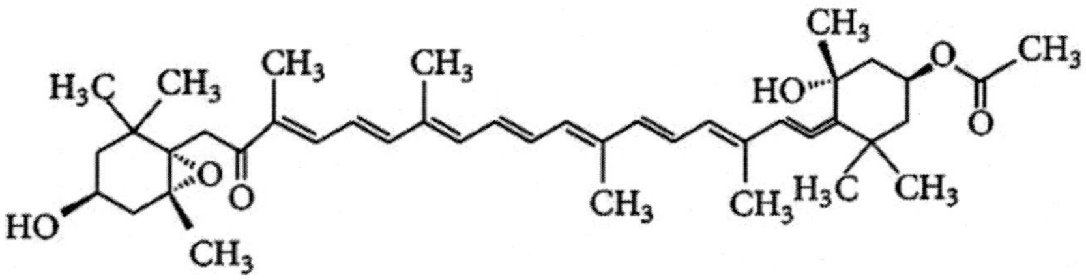

Figure 13. Fucoxanthin.

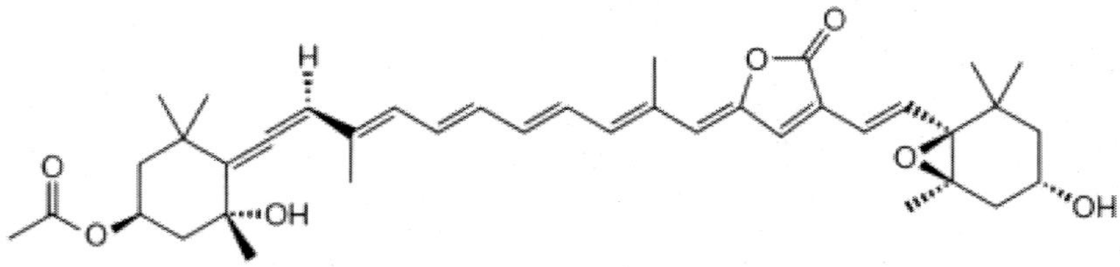

Figure 14. Peridinin.

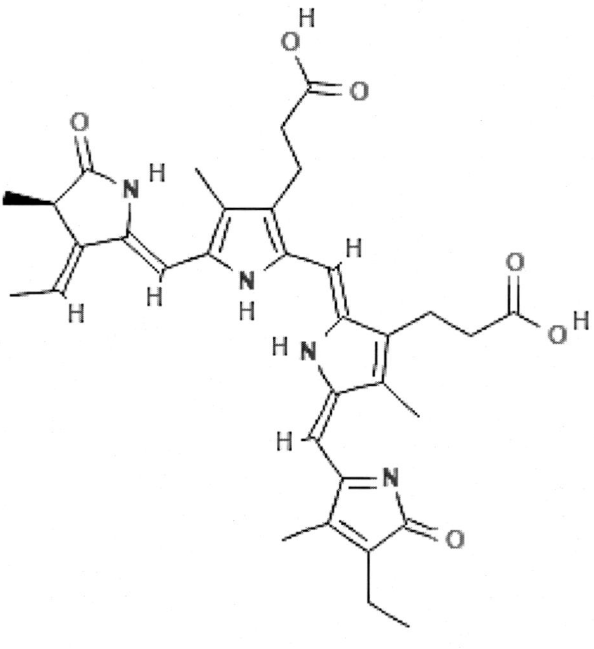

Figure 15. Phycobilin or phycobiliprotein.

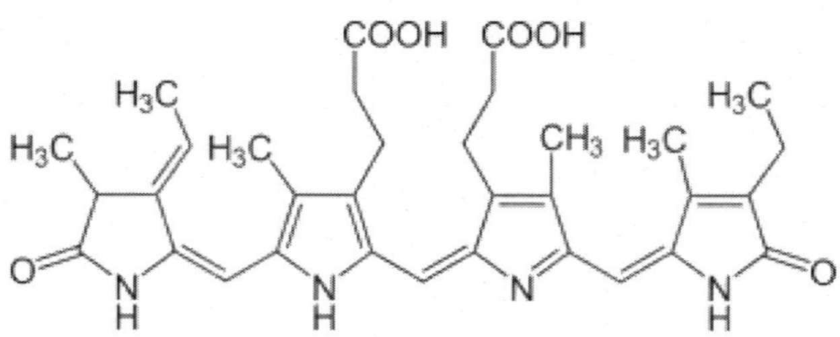

Figure 16. Phycocyanobilin.

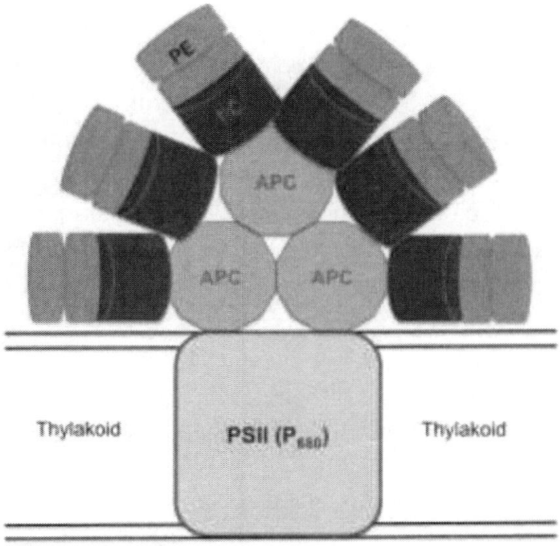

Figure 17. Phycobilisome protein complexes (PSII- Photosystem II; APC- Allophycocyanin; PC- Phycocyanin; PE-Phycoerythrin).

4. Photosynthetic Apparatus

Photosynthesis in algae and cyanobacteria mainly occurs through energy transfer mechanism in many compartments. Cyanobacteria differ from bacteria in terms of performing oxygenic photosynthesis. In cyanobacteria, the photosynthetic compartments are embedded in the internal membrane system whereas in eukaryotic algae, primary reactions take place in the thylakoid membrane protein complexes of chloroplast (Rochaix 2011).

The cyanobacterial photosynthetic apparatus is composed of five supramolecular, multiprotein assemblies like photosystem II/P_{680}/oxygen evolution complex (water-plastoquinone-photooxidoreductase); Cytochrome b_6/f complex (plastoquinol-plastocyanine oxidoreductase); photosystem I/P_{700}/plastocyanin-ferredoxin-photooxidoreductase; ATP synthase complex/coupling factor and lastly light harvesting antennae complexes (Bryant et al. 1987). The thylakoid-extrinsic protein complexes or phycobilisomes triggers the light harvesting mechanism in cyanobacteria, some red algae and enigmatic dinoflagellates like *Cyanophora paradoxa*). Phycobilisomes deliver excitations to the photosystem II present in the intrinsic membrane of the thylakoids. Light intensity and light wavelength can alter the rate of photosynthesis by changing the composition of phycobiliproteins in cyanobacteria (Bryant et al. 1987). The cpcA, cpcB and cpcC gene encoded proteins of plastocyanin are the major sites for capturing light.Different colourless linker polypeptide can covalently link with the phycobiliproteins like APC, PC, PE and maintain their structural rigidity to optimize their absorbance and energy characteristics. The PCs and PEs together form hexameric proteins to collect the radiating light energy to the reaction centre protein. Abiotic factors like temperature, light and nutrient availability can alter the composition of cyanobacterial photosynthetic apparatus and photosynthetic activity (Miskiewicz, Ivanov and Huner, 2002).The energy transfer mechanism from PC to chlorophyll follows three mechanisms: (1) Collisions of one molecule to the other, (2) emission of radiation from one molecule and its re-absorption by other, and (3) Resonance transfer of energy from one oscillator to another through internal conversion (Arnold and Oppenheimer 1949).

The modern eukaryotic algae can be selected by nature for their rapid growth rate, lack of mobility and simple developmental cycle (Zallen 1993). The photosynthetic apparatus of eukaryotic algae largely consists of two photosystems viz., PSI and PSII with several membrane protein complexes such as cytochrome b_6/f, plastocyanin, plastoquinone and ATP synthase complex as coupling factor similar to cyanobacteria. Most important characteristic feature of photosynthetic complexes is that they all consist of numerous chloroplast and nucleus encoded subunits. Some redox cofactors such as iron-sulfur protein (Fe-S canter), quinone molecules (A_0, A_1), hemes, xanthophylls and chlorophylls are associated with PSI, PSII and cytochrome b_6/f as major photosynthetic components. The photosynthetic systems are not evenly distributed through the thylakoid membrane system with exception of cytochrome b_6/f. The PSII is confined to the grana region whereas PSI and ATPase are restricted to stroma lamellae. The diffusible electron carrier plastoquinone and plastocyanin mediate the transfer of electrons through these complexes (Rochaix 2011). A concerted interplay between the chloroplast and nucleocytosolic genetic systems as well as pigment and protein synthesis in thylakoid triggers the biogenesis of the photosynthetic apparatus. *Chlamydomonas reinhardtii* is well-known model taxa providing many new insights into the photosynthesis apparatus (Eberhard et al. 2008).The coordination of photosynthetic apparatus totally depends on the RNA processing, translation mechanism and protein assembly that are nucleus encoded. These nuclear encoded proteins act on specific chloroplast encoded genes or RNAs (Eberhard et al. 2008). The PSI core subunits are encoded by the three chloroplast genes *psaA*, *psaB* and *psaC*. The nuclear encoded transcription factors like Raa1, Raa2, Raa3, Rat1 and Rat2 triggers the formation of 70S chloroplast ribosomes and finally PsaA protein in *Chlamydomonas* (Eberhard et al. 2008). The PSII assembly depends on the D1 and D2 reaction centre proteins that bind the redox cofactors CP43, CP47, α- and β-subunits of cytochrome b_{559} and PsbI. These proteins constitute inner antenna and coordinate chlorophyll-a molecule for the light reaction. Oxygen-evolving complex consists of PsbO, PsbP, and PsBQ genes on the lumen side of the chloroplast. These protein complexes trigger the electron transport mechanism for the formation of ATP and $NADPH_2$, which has been discussed in the next section.

5. Light Trapping Mechanism and Electron Transfer

Photosynthesis is a light dependent process where light is an electromagnetic radiation which travels at the speed $c \sim 3 \times 10^8 m\ s^{-1}$. Based on its wavelength, the electromagnetic spectrum can be divided into several components. The visible portion ranges from 400 nm to 700 nm, violet to infrared region of the spectrum. This region corresponds to the photosynthetically active radiation (PAR) that is useful in performing photosynthetic reactions. The smallest unit of light energy is known as photon or quanta, which is absorbed by chloroplast and nuclear encoded proteins in the form of wavelengths during photosynthesis. A photon of blue light (400-450 nm) is more energetic than the photon of red light (680-700 nm). The energy of photon is absorbed by the pigments and transferred to the reaction center for photochemical reactions. A single photon is sufficient to excite a single electron from one pigment molecule to another for charge separation. The radiant flux energy or irradiance might be expressed as the number of photons reaching unit surface area in unit time. In photosynthetic algae, chlorophyll and non-chlorophyll, accessory pigments absorb light energy to transfer them to the reaction center protein.

The light energy is trapped in two photochemical reactions carried out by two complexes PS II and PS I (Figure 18). The light harvesting antenna proteins channels the light energy to the reaction centers placed energetically downhill. A very less amount of heat is released during energy transfer. Photosynthetic pigments and redox cofactors are associated for specific photosynthetic functions such as light harvesting and energy transfer (Miskiewicz, Ivanov and Huner, 2002). There are two light harvesting pigment protein complexes, which are commonly found in prokaryotic and eukaryotic algae. One of them is hydrophilic phycobiliproteins that is common in cyanobacteria, red algae and some glaucophytes. The other group involves hydrophobic pigment-protein complexes such as LHC II and LHC I containing chlorophyll a, b and carotenoids for other eukaryotic algae.

The energy transfer mechanism both in cyanobacteria and eukaryotic algae follows the same direction by a chain of electron carriers, commonly known as non-cyclic photophosphorylation usually visualized in a so called "Z" scheme (Hill and Bendall 1960) (Figures 18-19). The electron transport reaction proceeds from more negative (lower) to more positive (higher) redox potential. Upon illumination of light, the LHC II of P680 (PS II) is charged and excited to breakdown the water molecules into electron and proton. The photochemical reaction results in evolution of oxygen (O_2) through oxygen-evolving complex proteins composed of manganese cluster (Mn^{2+}) and tyrosine TyrZ (Figure 19). The electrons are involved in the formation of $NADPH_2$ through electron transport system and protons are transported from stroma to thylakoid lumen forming a pH gradient, which drives ATP synthesis (Figure 18). The $NADPH_2$ and ATP are formed by the enzymes NADP reductase and ATP synthase respectively (Figure 18). Upon excitation the multimeric protein complex PS II transfers electron to the phaeophytin, the first electron acceptor in "Z" scheme (Figure 19). The D1 and D2 proteins present in PS II carry all essential prosthetic groups necessary for the electron transfer and charge separation. The charged phaeophytin transfers electron to the primary and secondary quinone molecules, Q_A and Q_B (Figure 19). The CP43 and CP47 proteins present on either side of the D1-D2 proteins are associated with electron transfer mechanism. After receiving electrons from quinone molecules, the lipophilic benzoquinone/plastoquinone becomes charged and transfers electron to cytochrome b_6f. Simultaneously, plastoquinone molecules translocate two protons from the stroma into the lumen. The X-ray crystallographic structure of cytochrome b_6f was studied from *Mastigocladus* and *Chlamydomonas* (Kurisu et al. 2003; Stroebel et al. 2003; Masojídek, Torzillo, and Koblížek 2013). From the cytochrome b_6f electron is channeled to the plastocyanin and it carries one electron to the PS I reaction center protein. Upon illumination by light and acceptance of electron, PSI becomes excited. Charged PS I releases electrons and are transported to some electron carriers like A_0 (Chl a), A_1 (phylloquinone) and Fx (4Fe-4S) molecules. The generated electrons are further transported to the terminal mobile electron acceptor ferredoxin (Fd) through 4Fe-4S electron acceptors F_A and F_B molecules resulting finally in the reduction of NADP to $NADPH_2$. ATPase or ATP synthase is a membrane bound enzymatic complex of hydrophobic CF_0 and hydrophilic CF_1. The subunit CF_0 spans the thylakoid membrane and CF_1 is attached to CF_0 on the stromal side of the membrane. CF0 functions as proton channel and CF1 acts as catalytic site for ATP synthesis. For synthesis of one ATP molecule, 4 protons are needed that have been derived from proton motive force and water splitting mechanism (Masojídek, Torzillo, and Koblížek 2013). In natural environment, photosynthetic algae have adopted acclimation mechanism that aims to balance the light and dark reactions of photosynthesis.

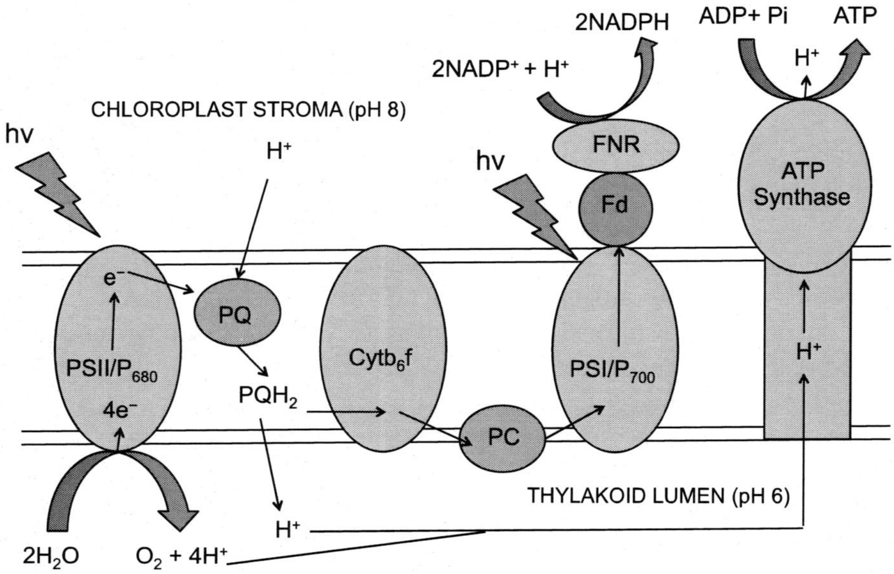

Figure 18. Electron transport mechanism in photosynthetic algae.

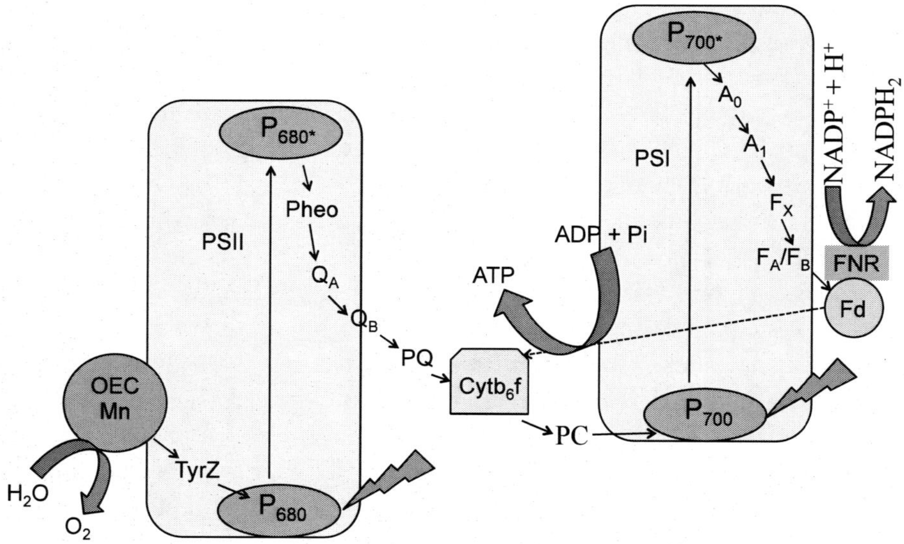

Figure 19. "Z" scheme of photosynthetic electron transport system in algae (OEC- Oxygen evolving complex; Mn- Manganese; TyrZ- Tyrosine Z; Pheo- Pheophytin; PQ- Plastoquinone; Cytb$_6$f- Cytochrome b$_6$f; PC- Plastocyanin; Fd- Ferredoxin; FNR- Ferredoxin NADP reductase; PSI and PSII- Photosystem I and Photosystem II).

6. Structure and Activity of Rubisco

Rubisco is a universal enzyme present in photosynthetic microorganisms including algae and cyanobacteria that catalyzes the carbon fixation reactions. In cyanobacteria, Rubisco is concentrated to carboxysome and pyrenoids of algae. Immunolocalization studies have suggested that the primary location of Rubisco is carboxysome in *Synechococcus* (McKay, Gibbs & Espie 1993). Many studies have been performed for purification of pyrenoids so far, for isolation of Rubisco. *Eremosphera* and *Chlamydomonas reinhardtii* are two microalgae studied so far containing Rubisco in their pyrenoids (Moroney & Somanchi 1999).The Rubisco of a red alga (Red-type-Rubisco), *Cyanidioschyzon merolae* possess superior kinetic property to the green-type-Rubisco in algae and higher plants (Loganathan, Candace Tsai & Muller- Cajar 2016). Structurally the enzyme consists of two protein subunits, known as large chain (L, 55KD) and small chain (S, 13KD). The large chain is encoded by the chloroplast DNA and designated as *rbcL*. There are many small chain genes present in the nucleus of eukaryotic algae and they are imported to the stromal compartment of chloroplast from the cytosol by traversing the outer chloroplast membrane. Amino acids from each large chain dimer serve as enzymatically active substrate binding site for ribulose 1, 5- bisphosphate (RuBP). A total of eight large chains and eight small chains assemble into a large functional enzyme Rubisco. In algae, Rubisco exists in two forms designated as Form I and Form II. Form I is composed of eight large and eight small subunits. Functionally, form I has a high affinity for CO_2 and a low catalytic efficiency. In green algae and Euglenoids, the large subunit is encoded by chloroplast DNA and small subunit by nuclear DNA. In primitive endosymbiotic cyanobacterium (cyanelle) like *Cyanophora paradoxa* and in some non-green algae, both large and small subunits are coded by chloroplast DNA (Lee 2008). The other form, form II has low affinity for CO_2 and high catalytic efficiency just opposite to form I. Form II is composed of two subunits and very common in dinoflagellates. Unfortunately, some other forms like Form III and Form IV are dominant in the group of Methanogens, Thermophilic bacteria, Proteobacteria, Chlorobia and only in one eukaryote *Ostreococcus tauri* (Tabita et al. 2008).

The active site of enzyme is mainly composed of magnesium (Mg^{2+}) and involves addition of CO_2 molecule to the amino acid lysine and histidine. Thus, CO_2 and Mg^{2+} activate Rubisco. The activation of Rubisco releases protons (H^+) in the stroma and thus increases the concentration of stromal pH and Mg^{2+}. The pH and Mg^{2+} ion concentration in stroma can regulate the enzymatic activity of Rubisco. Sugar phosphates also bind to Rubisco and affect the activation reaction with CO_2. The enzyme Rubisco activase triggers the activation reaction of Rubisco by carbamylation. Hydrolysis of ATP into ADP and inorganic phosphate (Pi) facilitates the binding of activase and removal of sugar phosphate RuBP from the active site of Rubisco (Buchanan, Gruissem & Jones 2004). The free Rubisco then can be activated by carbamylation through binding of CO_2 and Mg^{2+}.

The dual activity of Rubisco proceeds through carboxylation and oxygenation reaction (Figure 20). Carboxylation reaction involves photosynthetic carbon fixation into C_3 and C_4 compounds by Calvin cycle and Hatch-Slack pathway. The carboxylation reaction of RuBP in C_3 plants results in the formation of two molecules of 3- phosphoglyceric acid (3-PGA). Most of the eukaryotic algae possess this C_3 mechanism to produce 3-PGA, as the first stable product in the multistep conversion of CO_2 into carbohydrate. In addition to the carboxylase activity of Rubisco, it also has oxygenase activity that uses molecular O_2 as a substrate instead of CO_2. The enzyme catalyzes the oxygenation reaction with RuBP, producing two-carbon molecule, 2-phosphoglycolate (2-PG) in association with 3-PGA. The oxygenic activity is intrinsic to Rubisco and evolves in an anaerobic atmosphere. The relative amount of CO_2 and O_2 dictates the activity of the enzyme as they compete for binding at the same active site. In natural environment, the carboxylation reaction proceeds three times faster than oxygenation due the presence of large number of C_3 and C_4 plants. The overall efficiency of CO_2 fixation depends on the oxygenase activity of Rubisco. Rubisco can alter the photosynthetic activity by introducing itself into other organisms. The Rubisco isolated from a red alga, *Galdieria partita* may improve the photosynthetic efficiency in crop plants (Whitney & Andrews 2001). Some transplastomic tobacco can be formed with functional Rubisco from the cyanobacterium, *Synechococcus elongatus* PCC7942 (Lin et al. 2014). Since, photosynthesis is the main event of regulating CO_2 in the atmosphere, Rubisco reactions are used as dynamic pathway for climate change through CO_2 mitigation.

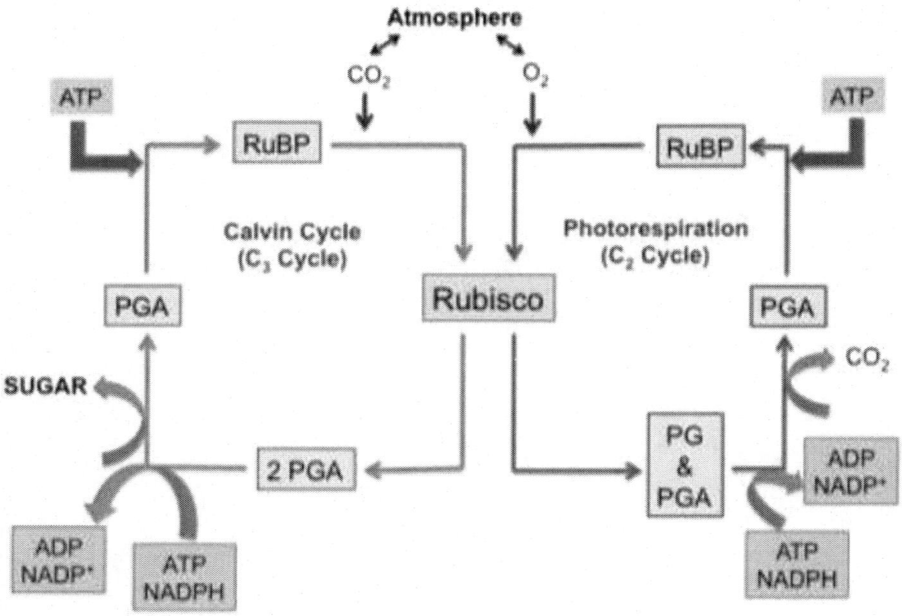

Figure 20. Dual activity of Rubisco- CO_2 fixation through Calvin cycle and Photorespiration (RuBP- Ribulose-1,5-bisphosphate; PGA- Phosphoglyceric acid; PG- Phosphoglycolate; ADP- Adenosine disphosphate; ATP- Adenosine triphosphate; NADPH- Nicotinamide adenine dinucleotide phosphate; Rubisco- Ribulose-1,5-bisphosphate carboxylase oxygenase).

7. Carbon Fixation through CO_2 Concentrating Mechanisms

Carbon fixation in algae and cyanobacteria follows oxygenic photosynthesis. They contain chlorophyll and C_3 mechanisms to fix carbon autotrophically. They are grouped into photoautotrophs, which synthesize organic compounds using light energy. The heterotrophs use these organic compounds to produce energy and building blocks of living system. Carbon assimilation happens in the light independent phase of photosynthesis popularly known as dark reaction. Like higher plants, algae and cyanobacteria use $NADPH_2$ and ATP to produce organic molecules in the form of carbohydrates (Figure 20). $NADPH_2$ and ATP are produced in the light reaction of photosynthesis as discussed previously. In order to fix one molecule of CO_2, 3 molecules of ATP and 2 molecules of $NADPH_2$ are required (Figure 21).

Carbon fixation mechanism in algae and cyanobacteria involves four major phases, commonly known as Calvin cycle or C_3 cycle. The four phases are carboxylation, reduction, regeneration and product synthesis phase. In carboxylation, CO2 molecule binds to the active site of the Rubisco and triggers the carboxylase activity of Rubisco to convert RuBP to 3 PGA. Reduction reaction involves two complementary steps: phosphorylation of phosphoglycerate to diphosphoglycerate and ADP; and reduction of diphosphoglycerate to phosphoglyceraldehyde (PGAld) and $NADPH_2$.The RuBP is regenerated in the regeneration phase for further CO2 fixation. Regeneration involves a series of complex chemical reactions combining 4C, 5C, 6C and 7C sugar phosphates. Transketolase and Aldolase are the two major enzymes that help in the regeneration of RuBP (5C sugar phosphate) from 6C and 3C sugars. Since the end product of photosynthesis is concerned with carbohydrates, other compounds like amino acids, organic acids and fats can also be synthesized in photosynthetic carbon assimilation (Masojídek, Torzillo, and Koblížek 2013). Many environmental variables like nutrition, light intensity, O_2 and CO_2 concentrations effect the formation of various end products in carbon assimilation reactions (Satpati & Pal 2015; Satpati et al. 2015; Satpati et al. 2016).

$$CO_2 + 4^{H+} + 4^{e-} \xrightarrow[\text{Enzymes}]{2\ NADPH_2,\ 3\ ATP} \underset{\text{Carbohydrate}}{CH_2O + H_2O}$$

Figure 21. Chemical equation of carbon fixation in light independent phase.

The CO_2 concentrating mechanism in eukaryotic algae and cyanobacteria has been discussed below. The most important point to be considered is that CO_2 can easily diffuse through the biological membranes. The bundle sheath cells of C_4 plants contain large amount of Rubisco, which can accumulate HCO_3^- that must be converted to CO_2 molecules before carbon fixation. In comparison, microalgae possess greater capacity to fix CO_2 than C_4 plants. Algae can fix carbon in gaseous (as CO_2) and non-gaseous forms (as bicarbonate or carbonate). In microalgae, HCO_3^- diffuses through the biological membrane slowly. Since HCO_3^- is diffused, it can again be converted to CO_2 as Rubisco can only bind CO_2 not HCO_3^-. Algae possess active bicarbonate pump to concentrate the level of bicarbonates in the cell. In eukaryotic algae, atmospheric CO_2 and HCO_3^- both can diffuse through the periplasmic space, cytosol and chloroplast. In pyrenoid of the chloroplast the major carbon assimilation takes place by the conversion of CO_2 to 3-carbon compound, PGA. It has been found that the rate of carbon assimilation in *Chlamydomonas reinhardtii* is enhanced with the large number of pyrenoids and Rubisco in the chloroplast (Borkhsenious, Mason & Moroney 1998).

The report has suggested that the plants on land can absorb 52% of the total atmospheric CO_2, while ocean-based algae can absorb 45% to 50% because of their small size and short life cycles (Haoyang 2018). Algae can even reduce the amount of CO_2 in the atmosphere more if they are provided a suitable environment, including light, nutrients and temperature (Satpati, Mallick & Pal 2015; Satpati, Gorain & Pal 2016). Thus, mass cultivation of algae is a promising factor for carbon assimilation and mitigation (Satpati & Pal 2018). The major microalgal species like *Chlorella vulgaris*, *C. ellipsoidea*, *Scenedesmus*, *Chroococcus*, *Chlamydomonas*, *Chlorococcum* can rapidly fix CO_2 into their cellular system as storage organic compounds (Satpati et al. 2016).

REFERENCES

Arnold, W. & Oppenheimer, J. R. (1949).Internal conversion in the photosynthetic mechanism of blue-green algae.*The Journal of General Physiology*,pp. 423-435.

Baurain, D., Brinkmann, H., Petersen, J., Rodríguez-Ezpeleta, N., Stechmann, A., Demoulin, V., Roger, A.J., Burger, G., Lang, B.F. & Philippe, H. (2010). Phylogenomic Evidence for Separate Acquisition of Plastids in Cryptophytes, Haptophytes, and Stramenopiles, *Molecular Biology and Evolution*, *27*(7), 1698–1709.

Blair, J. E., Shah, P. & Hedges, S. B. (2005).Evolutionary sequence analysis of complete eukaryote genomes. *BMC Bioinformatics*, *6*, 53.

Borkhsenious, O.N., Mason, C.B. & Moroney, J.V. (1998). The intracellular localization of ribulose-1,5-bisphosphate carboxylase/oxygenase in Chlamydomonas reinhardtii. *Plant Physiology*, *116*, 1585-1591.

Bryant, D.A., De Lorimier, R., Guglielmi, G., Stirewalt, V.L., Cantrell, A. & Stevens, S.E. (1987). The Cyanobacterial Photosynthetic Apparatus: A Structural and Functional Analysis Employing Molecular Genetics. In: Biggins J. (eds) *Progress inPhotosynthesis Research*. Springer, Dordrecht.

Buchanan, B. B., Gruissem, W. & Jones, R. L. (2004). *Biochemistry & Molecular Biology of Plants*. I. K. International Pvt. Ltd. New Delhi, India., pp. 616-619.

Chan, C.X. & Bhattacharya, D. (2010). The Origin of Plastids.*Nature Education*, *3*(9), 84.

Eberhard, S., Finazzi, G. & Wollman, F. A. (2008). The dynamics of photosynthesis.*Annual Review of Genetics*, *42*, 463–515.

Glazer, A. N. (1985).Light harvesting by phycobilisomes.*Annual Review of Biophysics and Biophysical Chemistry*, *14*, 47-77.

Hackett, J.D., Anderson, D.M., Erdner, D.L. & Bhattacharya, D. (2004). Dinoflagellates: a remarkable evolutionary experiment. *American Journal of Botany*, *91*(10), 1523-1534.

Haoyang, C. (2018). Algae-Based Carbon Sequestration.*IOP Conference Series: Earth and Environmental Science*, *120*(012011), 1-10.

Hill, R. & Bendall, R. (1960). Function of the two cytochrome components in chloroplasts: a working hypothesis. *Nature*, *186*, 136–137.

Kurisu, G., Zhang, H., Smith, J.L. & Cramer, W.A. (2003). Complex of oxygenic photosynthesis: tuning the cavity. *Science*, *302*, 1009–1014.

Lee, R. E. (2008). *Phycology*. Fourth Edition. Cambridge University Press India Pvt. Ltd., Cambridge House, New Delhi, India., pp. 12-13.

Lin, M.T., Occhialini, A., Andralojc, P.J., Parry, M.A. & Hanson, M.R. (2014). "A faster Rubisco with potential to increase photosynthesis in crops".*Nature*, *513*(7519), 547–50.

Loganathan, N., Candace Tsai, Y.C.& Muller-Cajar, O. (2016).Characterization of the heterooligomeric red-type rubisco activase from red algae.*Proceedings of the National Academy of Sciences of the United States of America.*, *113*(49), 14019-14024.

Margulis, L. (1970). Origin of Eukaryotic Cells. New Haven: Yale University Press.

Masojídek, J., Torzillo, G. & Koblížek, M. (2013). Photosynthesis in Microalgae.Handbook of Microalgal Culture: Applied Phycology and Biotechnology, Second Edition. Edited by Richmond, A. & Hu, Q. John Wiley & Sons, Ltd. Published 2013 by Blackwell Publishing Ltd. pp. 21-36.

McKay, R.M.L., Gibbs, S.P. & Espie, G.S. (1993). Effect of dissolved inorganic carbon on the expression of carboxysomes, localization of Rubisco and the mode of inorganic carbon transport in cells of the cyanobacterium *Synechococcus* UTEX 625. *Archives of Microbiology*, *159*, 21–29.

Mereschkowski, C. (1905). ÜberNatur und Ursprung der ChromatophorenimPflanzenreiche [About Nature and Origin of Chromatophores in the Plant Kingdom]. *Biologisches Centralblatt*, *25*, 593–604.

Miskiewicz, E., Ivanov, A.G. & Huner, N.P. (2002). Stoichiometry of the photosynthetic apparatus and phycobilisome structure of the cyanobacterium Plectonema boryanum UTEX 485 are regulated by both light and temperature. *Plant physiology*, *130*(3), 1414–1425.

Moroney, J.V., & Somanchi, A. (1999). How Do Algae Concentrate CO2 to Increase the Efficiency of Photosynthetic Carbon Fixation? *Plant Physiology*, *119*, 9-16.

Moustafa, A., Beszteri, B., Maier, U.G., Bowler, C., Valentin, K. &Bhattacharya, D. (2009). Genomic Footprints of a Cryptic Plastid Endosymbiosis in Diatoms.*Science*, *324*(5935), 1724-1726.

Musumeci, M.A., Ceccarelli, E.A. & Catalano-Dupuy, D.L. (2012).The Plant- Type Ferredoxin-NADP$^+$Reductases.In *Advances in Photosynthesis- Fundamental Aspects.* Edited by Najafpour, M. M., InTech, Croatia, pp. 539-562.

Nosenko, T., Lidie, K. L., Van Dolah, F. M., Lindquist, E. &Cheng, J. F. (2006). US Department of Energy–Joint Genome Institute, Debashish Bhattacharya, Chimeric Plastid Proteome in the Florida "Red Tide" Dinoflagellate *Kareniabrevis*, *Molecular Biology and Evolution*, *23*(11), 2026–2038.

Reyes-Prieto, A., Moustafa, A. & Bhattacharya, D. (2008). Multiple genes of apparent algal origin suggest ciliates may once have been photosynthetic. *Current Biology*, *18*, 956–962.

Rochaix, J.D. (2011). Assembly of the photosynthetic apparatus.*Plant Physiology*, *155*, 1493-1500.

Rochaix, J.D. (2011). Regulation of photosynthetic electron transport.*BiochimicaBiophysica Acta*, *1807*, 375-383.

Rogers, M.B., Gilson, P.R., Su, V., McFadden, G.I. & Keeling, P.J. (2007). The Complete Chloroplast Genome of the Chlorarachniophyte *Bigelowiellanatans*: Evidence for Independent Origins of Chlorarachniophyte and Euglenid Secondary Endosymbionts, *Molecular Biology and Evolution*, *24*(1), 54–62.

Satpati, G.G. & Pal, R. (2015). Rapid detection of neutral lipid in green microalgae by flow cytometry in combination with Nile red staining- and improved technique.*Annals of Microbiology*, *65*, 937-949.

Satpati, G.G., Mallick, S.K. & Pal, R. (2015). An Alternative High-throughput Staining Method for Detection of Neutral Lipids in Green Microalgae for Biodiesel Applications.*Biotechnology and Bioprocess Engineering*, *20*,1044-1055.

Satpati, G.G., Kanjilal, S., Prasad, R.B.N. & Pal, R. (2015). Rapid Accumulation of Total Lipid in *Rhizoclonium africanum* Kutzing as Biodiesel Feedstock Under Nutrient Limitations and the Associated Changes at Cellular Level. *International Journal of Microbiology*, *Hindawi*, *275035*, 1-13.

Satpati, G.G., Gorain, P.C. & Pal, R. (2016). Efficacy of EDTA and Phosphorous on Biomass Yield and Total Lipid Accumulation in Two Green Microalgae with Special Emphasis on Neutral Lipid Detection by Flow Cytometry.*Advances in Biology*, *Hindawi*, *8712470*, 1-12.

Satpati, G.G., Gorain, P.C., Paul, I. & Pal, R. (2016). An integrated salinity-driven workflow for lipid enhancement in green microalgae for biodiesel application.*RSC Advances*, 6, 112340-112355.

Satpati, G.G. & Pal, R. (2018). Microalgae-Biomass to Biodiesel: A Review. *Journal of algal Biomass Utilization*, *9*(4), 11-37.

Stroebel, D., Choquet, Y., Popot, J.L. & Picot, D. (2003). An atypical haem in the cytochrome b6f complex.*Nature*, *426*, 413–418.

Tabita, F. R., Satagopan, S., Hanson, T. E., Kreel, N. E. & Scott, S. S. (2008). Distinct form I, II, III, and IV Rubisco proteins from the three kingdoms of life provide clues about Rubisco evolution and structure/function relationships. *Journal of Experimental Botany*, *59*(7), 1515-1524.

Waller, R. F. & McFadden, G. I. (2005). The apicoplast: A review of the derived plastid of Apicomplexan parasites. *Current Issues in Molecular Biology*, *7*, 57–80.

Weber, A. P., Linka, M. & Bhattacharya, D. (2006). Single, ancient origin of a plastid metabolite translocator family in Plantae from an endomembrane-derived ancestor. *Eukaryotic Cell*, *5*, 609–612.

Whitney, S.M. & Andrews, T.J. (2001). "Plastome-encoded bacterial ribulose-1, 5-bisphosphate carboxylase/oxygenase (RubisCO) supports photosynthesis and growth in tobacco". *Proceedings of the National Academy of Sciences of the United States of America*, *98*(25), 14738–14743.

Yoon, H.S., Reyes-Prieto, A., Melkonian, M. & Bhattacharya, D. (2006).Minimal plastid genome evolution in the *Paulinella*endosymbiont.*Current Biology*, *16*(17), R670–R672.

Zallen, D. T. (1993). The "Light" Organism for the Job: Green Algae and Photosynthesis Research. *Journal of the History of Biology*, *26*(2), 269-279.

In: Applied Algal Biotechnology
Editors: M. Arumugam, S. Kathiresan et al.
ISBN: 978-1-53617-524-0

Chapter 5

SECONDARY METABOLITES OF MICROALGAE

C. K. Madhubalaji*[1,2]*, P. Ajana*[1,2]*,
***V. S. Chauhan*[1,2] *and R. Sarada*[1,2,*]**

[1]Plant Cell Biotechnology Department,
CSIR – Central Food Technological Research Institute,
Mysuru, Karnataka, India
[2]Academy of Scientific and Innovative Research (AcSIR),
Ghaziabad, India

ABSTRACT

Microalgae are varied group of photosynthetic organisms, exist in different habitats such as freshwater, marine, brackish, polar regions, acidic/alkali waters, hot water springs, and wastewaters. Microalgae produce secondary metabolites as an adaptive mechanism to thrive in extreme environments. In recent years the bioprospecting of microalgae from different habitats opened up a wide scope for identification of new metabolites and understanding the mechanism of their synthesis, bioactivities, and applications. In addition, the advancements in culturing methods and techniques are contributing to the exploration of new species having metabolites of commercial importance. The secondary metabolites are of varied chemical nature *viz.*, isoprenoids (carotenoids, terpenoids, sterols), polyketides, amino-acid-derived natural products, phenolic compounds, polyunsaturated fatty acids, etc. The secondary metabolites found importance in nutraceutical, therapeutic, and biological activities. The present chapter discusses the current status of various microalgal secondary metabolites and their commercial applications.

Keywords: microalgae, metabolites, carotenoids, terpenoids, sterols, phenolics

[*] Corresponding Author's Email: sarada@cftri.res.in.

ABBREVIATIONS

ALA:	Alpha-linolenic acid
AMD:	Age-related Macular Degeneration
ARA:	Arachidonic acid
CAGR:	Compound Annual Growth Rate
DHA:	Docosahexaenoic acid
EPA:	Eicosapentaenoic acid
GLA:	Gamma-Linolenic acid
LDL:	Low-Density Lipoprotein
MAA:	Mycosporin like amino acids
MIB:	Methyl-isoborneol
PUFA:	Poly Unsaturated Fatty Acid
USFDA:	United States Food and Drug Administration

5.1. INTRODUCTION

Microalgae are unicellular organisms and are primary producers in the ecological chain of the aquatic environment. These primary producers of aquatic environment utilize the natural light energy (solar) and CO_2 for photosynthetic production of various compounds, which includes commercially useful metabolites. It has been estimated that the existence of microalgal species would be about 200,000-800,000 out of which only 50000 species of microalgae are studied (Kirby and Keasling, 2009; Stark and O'Gara, 2012). This tremendous biodiversity of microalgae makes it a potential group to be exploited.

5.2. MICROALGAE SECONDARY METABOLITES

Microalgae produce the compounds, which are not essential for their growth and metabolic process. However, these compounds influence the ecological interaction between microalgae and its environment (biotic, abiotic factors). Secondary metabolites of microalgae are not necessary for their survival but have essential functions like protection, competition, and species interaction.

Microalgae are a potential natural source for the production of nutraceutical compounds and secondary metabolites. Owing to their multiple applications in various fields, microalgae global market size is expected to exceed the US $3700 million by 2024, at 3.5% CAGR. Whereas, for algal products global market has announced to reach $6.09 billion by 2026 at a CAGR of 6.7%.

Microalgae produce a vast and diverse assortment of compounds, and around 15,000 novel compounds/metabolites have been chemically determined (Cardozo et al., 2007). However, most of the compounds and their functions are remaining unknown. Microalgae are omnipresent can sustain and grow even in deserts as well as in Antarctica. For habituation of such harsh and extreme environments, microalgae have to adapt strategies of defence. One of the strategies is the production of high-value secondary metabolites (having various biological activities) through different metabolic pathways. Microalgal secondary metabolites include isoprenoids (e.g., terpenes, terpenoids, carotenoids, steroids), shikimates (e.g., phenolics, flavonoids), polyketides (e.g., macrolides, antibiotics, phlorotannins), amino-acid-derived compounds (e.g., alkaloids, MAAs, scytonemin), fatty acids (ALA, GLA, EPA, DHA).

Secondary metabolites have prominence as nutraceuticals and bio-actives performing in various biological activities such as anticancer, antimicrobial, antitumor, antiviral, and anti-inflammatory activities. Earlier times, due to lack of knowledge on the metabolites of microalgae, very less research has been done. Recent advancements in analysis, using high-end instruments opened the gate for discovering of commercially useful, and highly sustainable microalgal metabolites. Similarly, improved characterization and toxicology studies of secondary metabolites are widening their applications in various sectors such as animal feed, nutritional supplements, pharmaceutical, agricultural, biotechnology, bioremediation, water treatment, renewable energy, cosmetic, medicinal and other industrial sectors as shown in Figure 5.1.

In microalgae, along with eukaryotic microalgal species, prokaryotic microalgae such as Cyanobacteria (*Spirulina, Anabaena, Nostoc,* and *Oscillatoria*) have been reported to produce a variety of secondary metabolites. Cryptophycin, which was firstly isolated from a blue-green alga, *Nostoc* sp., has been confirmed as broad-spectrum anti-tumour activity in both *in vivo* and *invitro* studies. Microalgae secondary metabolites have extensive applications as photoprotective compounds and are being used in antiaging, anti-irritating, and skincare products (Shilpa et al., 2010). Recent studies showed that microalgae are promising organisms for the production of novel bioactive compounds, including nutraceuticals/secondary metabolites, which have great importance commercially. The present chapter mainly deals with the secondary metabolites of the microalgae, their commercial importance, and applications.

The secondary metabolite production in microalgae increases under the following conditions.

1) When microalgae are exposed to adverse biotic conditions and lack of defence mechanism. It will follow the chemical strategy by producing secondary metabolites to fight against other microbes
2) Specific abiotic conditions will enhance the specific secondary metabolite production
3) Gene modification mechanisms
4) When specific mutations are induced in the microalgae

It is challenging to classify microalgal secondary metabolites based on the biosynthetic pathway as it is puzzling and confusing. For example, the same secondary metabolites in different microalgae are produced through different pathways (In cyanobacteria, carotenoid production is through the non-mevalonate pathway, whereas, in *Chlorella,* it follows the mevalonate pathway).

There are several secondary metabolites, which are produced by the microalgae. Though the functions and biological activities of these compounds were determined, their commercial importance was not exploited (Table 5.1). In the present chapter, we have focused only on, commercially exploited few microalgal secondary metabolites (Table 5.2).Microalgae secondary metabolites and their possible biosynthesis production pathways are represented in

Figure 5.2. Among all secondary metabolites reported from microalgae, isoprenoids provide major contribution followed by polyketides, amino acid-derived natural compounds.

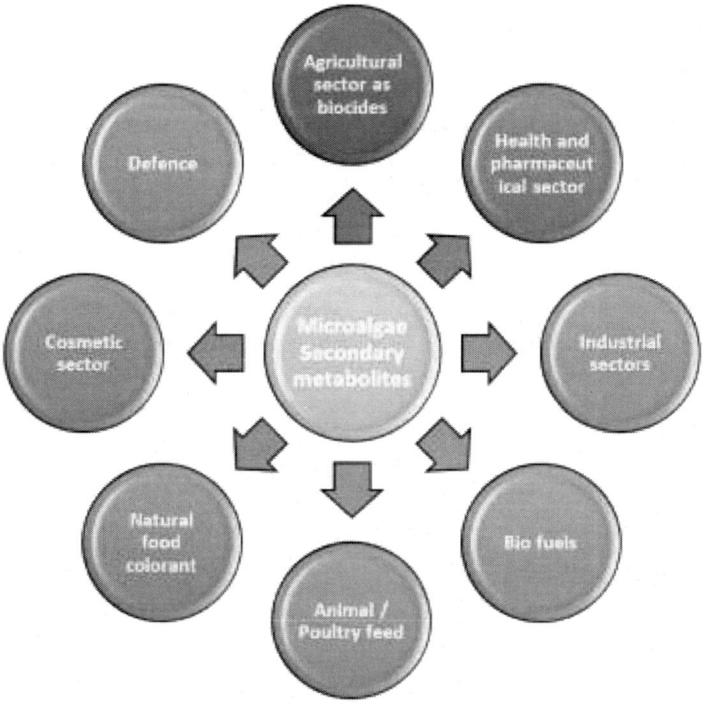

Figure 5.1. Applications of microalgal secondary metabolites in various sectors.

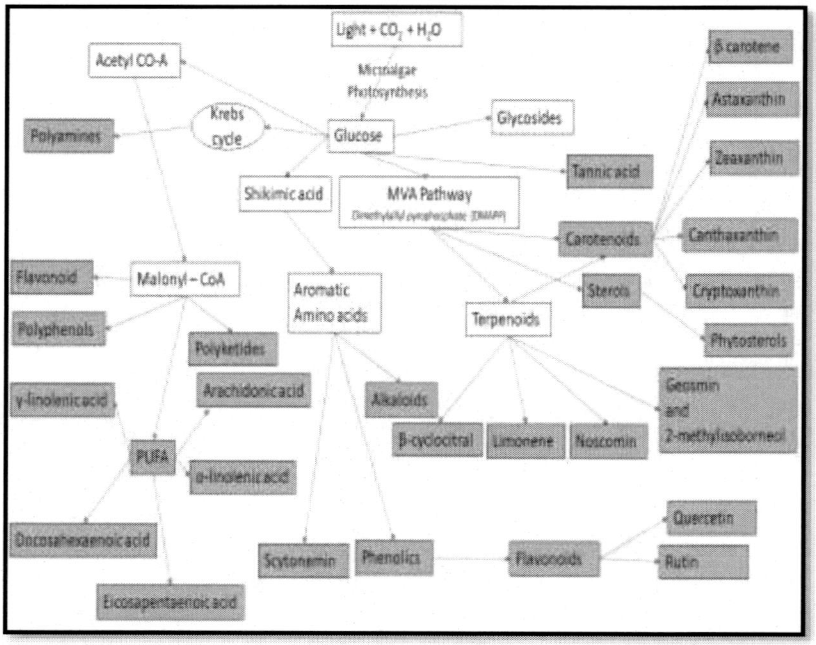

Figure 5.2. Possible mechanisms for the production for secondary metabolites (Green boxes indicates secondary metabolites).

Table 5.1. Biological activities of secondary metabolites produced by various microalgae

Bioactivity	Secondary metabolite	Microalgae	Reference
Anticancer	Acutiphycin	*Oscillatoria acutissima*	(Barchi, Moore, and Patterson 1984)
	Ankaraholide A	*Lyngbya majuscula*	(Mynderse et al., 1977)`
	Apratoxin D	*Lyngbya majuscule*	(Gutiérrez et al., 2008)
		Lyngbya sordida	
	Aurilide B	*Lyngbya majuscula*	(Han et al., 2006)
	Aurilide C		
	Bisebromoamide	*Lyngbya sp*	(Teruya et al., 2009)
	Biselyngbyaside	*Lyngbya sp*	
	Borophycin	*Nostoc linckia*	(Hemscheidt et al., 1994)
		Nostoc spongiaeforme	
	Calothrixins A	*Calothrix sp*	(Khan, Lu, and Hecht 2009)
	Calothrixins B		
	Coibamide A	*Leptolyngbya sp*	(Medina et al., 2008)
	Cryptophycins	*Nostoc sp*	(Moore et al., 1996)
	Curacin A	*Lyngbya sp*	(Simmons et al., 2005)
	Dragonamide C	*Lyngbya polychroa*	(Gunasekera et al., 2008)
	Dragonamide D		
	Hoiamide A	*Lyngbya majuscule*	(Choi et al., 2010)
		Phormidium gracile	
	Hoiamide B	*Cyanobacteria*	
	Homodolastatin 16	*Lyngbya majuscula*	(Davies-Coleman et al., 2003)
	Lagunamide C	*Lyngbya majuscula*	(Tripathi et al., 2011)
	Largazole	*Symploca sp*	(Zeng et al., 2010)
	Lyngbyatoxins	*Lyngbya majuscula*	(Shimizu 2003)
	Majusculamide C	*Lyngbya majuscula*	(Pettit et al., 2008)
	Malevamide D	*Symploca hydnoides*	(Horgen et al., 2002)
	Malyngamide 2	*Lyngbya sordida*	(Malloy et al., 2011)
	Palmyramide A	*Lyngbya majuscula*	(Taniguchi et al., 2010)
	Pitiprolamide		(Montaser et al., 2011)
	Somocystinamide A	*Lyngbya majuscula*	(Wrasidlo et al., 2008)
	Symplostatin 3	*Symploca sp*	(Luesch et al., 2002)
	Tjipanazoles	*Tolypothrix tjipanasensis*	(Bonjouklian et al., 1991)
	Wewakazole	*Lyngbya sordida*	(Malloy et al., 2011)
	Wewakpeptins	*Lyngbya semiplena*	(Han et al., 2005)
Antibacterial	Ambiguine H isonitriles	*Fischerella sp*	(Raveh and Carmeli 2007)
	Ambiguine I isonitriles		
	Ambigol A	*Fischerella ambigua*	(Falch et al., 1993)
	Ambigol B		
	Butanoic acid	*Haematococcus pluvialis*	(Santoyo et al., 2009)
	Methyl lactate		
	Diterpenoid	*Nostoc commune*	(Asthana et al., 2009)
	Eicosapentaenoic acid	*Phaeodactylum tricornutum*	(Smith, Desbois, and Dyrynda 2010)
	Coriolic acid	*Oscillatoria redekei*	(Mundt, Kreitlow, and Jansen 2003)
	α-dimorphecolic acid		
	Diterpenoid	*Nostoc commune*	(Asthana et al., 2009)
	Coriolic acid	*Oscillatoria redekei*	(Mundt, Kreitlow, and Jansen 2003)
	α-dimorphecolic acid		
	γ-linolenic acid	*Fischerella sp*	(Asthana et al., 2006)
	Hormothamnins	*Hormothamnion enteromorphoides*	(Gerwick et al., 1989)
	Noscomin	*Nostoc commune*	(Jaki, Orjala, and Sticher 1999)
	Belamide A	*Anabena variabilis*	(Ma and Led 2000)
	Muscoride	*Nostoc muscorum*	(Nagatsu, Kajitani, and Sakakibara 1995)

Table 5.1. (Continued)

Bioactivity	Secondary metabolite	Microalage	Reference
Antiviral	Acetylated sulfoglyco-lipids	*Oscillatoria raoi*	(Reshef et al., 1997)
	Calcium spirulan	*Spirulina platensis*	(Hayashi, Hayashi, and Kojima 1996)
	β-carbolines	*Dichothrix baueriana*	(Larsen, Moore, and Patterson 1994)
	Cyanovirin N	*Nostoc ellipsosporum*	(Gustafson et al., 1997)
Antifungal	Ambigol A	*Fischerella ambigua*	(Falch et al., 1993)
	Ambigol B		
	Anhydrohapaloxindole	*Hapalosiphon fontinalis*	(Moore, Yang, and Patterson 1987)
	Fischerindole L	*Fischerella muscicola*	(Park, Moore, and Patterson 1992)
	Fisherellin	*Fischerella muscicola*	(Dahms, Ying, and Pfeiffer 2006)
	Hapalindole	*Hapalosiphon fontinalis*	(Burja et al., 2001)
	Hormothamnins	*Hormothamnion enteromorphoides*	(Gerwick et al., 1989)
	Laxaphycins	*Anabaena laxa*	(Frankkmolle et al., 1992)
	Lobocyclamide A	*Lyngbya confervoides*	(MacMillan and Molinski 2002)
	Nostodione	*Nostoc commune*	(Bhadury and Wright 2004)
	Quercetin	*Haematococcus pluvialis*	(Goiris et al., 2014)
	Scytophycins	*Scytonema pseudohofmanni*	(Burja et al., 2001)
	Scytophycins	*Scytonema sp,*	(Ishibashi et al., 1986; Carmeli, Moore, and Patterson 1990)
		Tolypothrix sp	
	Phytoalexin	*Scytonema ocellatum*	(Patterson and Bolis 1997)
	Tolytoxin	*Scytonema ocellatum*	(Patterson and Carmeli 1992)
	Tubercidin	*Plectonema radiosum*	(Stewart et al., 1988)
	Toyocamycin	*Tolypothrix tenuis*	
Enzyme inhibitors	Aeruginosins	*Microcystis*	(Shin et al., 1997)
		Nodularia,	
		Oscillatoria	
	Nostocarboline	*Nostoc sp*	(Barbaras et al., 2008)
	Rutin	*Haematococcus pluvialis*	(Goiris et al., 2014)
	Circinamide	*Anabaena circinalis*	(Shin et al., 1997)
	Dehydroradiosumin	*Anabaena cylindrica*	(Kodani, Ishida, and Murakami 1998)
	Anabaenopeptin B	*Oscillatoria agardhii*	(Murakami, Shin, et al., 1997)
	Oscillapeptin G	*Oscillatoria agardhii*	(Sano and Kaya 1996)
	Nostopeptin A	*Nostoc minutum*	(Okino et al., 1997)
	Nostopeptin B		
	Microviridin	*Nostoc minutum*	(Murakami, Sun, et al., 1997)
	Micropeptin 103	*Microcystis viridis*	(Murakami, Kodani, et al., 1997)
	Aeruginosin 102 A	*Microcystis viridis*	(Matsuda et al., 1996)
	Aeruginosin 102 B		
	Microginin 299-A	*Microcystis aeruginosa*	(Ishida et al., 1997b)
	Microginin 299-B		
	Micropeptin 478-A	*Microcystis aeruginosa*	(Ishida et al., 1997a)
	Micropeptin 478-B		
	Lyngbyastatin	*Lyngbya confervoides*	(Matthew et al., 2007)
Antimalarial	Calothrixins A	*Calothrix sp*	(Bernardo et al., 2007)
	Calothrixins B		
	Dolastatins	*Lyngbya sp*	(Fennell 2003)
Bioactivity	Secondary metabolite	Microalage	Reference
		Symploca sp	
	Dragonamide A	*Lyngbya majuscula*	(McPhail et al., 2007)
	Nostocarboline	*Nostoc sp*	(Barbaras et al., 2008)
Inflammatory	Anatoxin-a	*Anabaena circinalis*	(Jha and Zi-rong 2004)

Bioactivity	Secondary metabolite	Microalage	Reference
Algicide/larvicide/ herbicide	Anatoxin-a	*Anabaena sp*	(Berry et al., 2008)
	Cylindrospermopsin	*Cylindrospermopsis sp*	
	Microcystin	*Microcystis aeruginosa*	
Sodium channel blocker	Kalkitoxin	*Lyngbya majuscula*	(Shimizu 2003)
Microtubulin assembly inhibitors	Curacin A	*Lyngbya majuscula*	(Shimizu 2003)
Predecessor of arachidonic acid	γ- linolenic acid	*Spirulina platensis*	(Z. Cohen 1999)
Anti -proliferative	Grassypeptolide	*Lyngbya confervoides*	(Kwan et al., 2008)
Cytotoxic	Tubercidin	*Plectonema radiosum*	(Stewart et al., 1988)
	Toyocamycin	*Tolypothrix tenuis*	
	Scytocyclamides	*Scytonema hofmanni*	(Grewe 2005)
	Lyngbyacyclamides A	*Lyngbya sp.*	(Maru, Ohno, and Uemura 2010)
	Lyngbyacyclamides B		
	Laxaphycins A	*Lyngbya majuscula*	(Bonnard et al., 1997)
	Laxaphycins B		
Cytostatic	symplostatin 1	*Symploca hydnoides*	(Harrigan et al., 1998)
PKC activator	Lyngbyatoxins	*Lyngbya majuscula*	(Shimizu 2003)
Atioxidants	β-carotene	*Spirulina*	(Deng and Chow 2010)
	Astaxanthin	*Haematococcus pluvialis*	(Capelli, Bagchi, and Cysewski 2013)
Reversing multidrug resistance	Dendroamids	*Stigonema dendroideum*	(Ogino et al., 1996)

Table 5.2. Commercially important secondary metabolites of microalgae and their applications in various fields

S.No	Secondary metabolites	Microalgae	Application in various sectors	References
1	β-carotene	*Dunaliella* sp., *Chlorella* spp., *Coelastrella striolata var. multistriata Chlorella ellipsodea, Coccomyxa acidophila*	Health food, food supplement, feed Pharmaceutical, food colorant, nutrition, diagnosis of cancer,	(Sathasivam et al., 2012; Sathasivam and Juntawong 2013)
2	Astaxanthin	*Haematococcus pluvialis, C. zofingiensis, Chlorococcum spp., Neospongiococcum* spp., *Monoraphidium sp, H. lacustris, Chlamydocapsa* spp., *S. obliquus, C. nivalis, C. fusa, C. vulgaris, Chlorococcum* spp., *C. striolata, Haematococcus* spp.,	Health sector, food, pharmaceuticals, feed additives, coloring (pigmentation)	(MA 1988; Tso and Lam 1996; Liu and Lee 2000; Lorenz and Cysewski 2000; J.-P. Yuan et al., 2002; Jin, Lee, and Polle 2006; Abe, Hattori, and Hirano 2007; Kamath et al., 2008; Ambati et al., 2014; Kamath et al., 2008)
3	Canthaxanthin	*C. nivalis, C. fusa, Coelastrella striolata var. multistriata Chlamydocapsa* spp., *C. zofingiensis, Chlorococcum* spp., *C. emersonii, C. vulgaris, Neospongiococcum* sp *C. striolata, H. lacustris,*	Pharmaceuticals, nutrition, cosmetics, poultry and aquatic feed,	(MA 1988; L. Gouveia et al., 1996; J.-P. Yuan et al., 2002; Mendes et al., 2003; Pelah, Sintov, and Cohen 2004; Bhosale and Bernstein 2005; Abe, Hattori, and Hirano 2007; Esfahani-Mashhour et al., 2009; Brizio et al., 2013);
4	Zeaxanthin	*C. ellipsiodea, D. salina, C. nivalis, D. salina*	Therapeutic, applications	(Bhosale and Bernstein 2005; Cha, Koo, and Lee 2008; Leya et al., 2009)

Table 5.2. (Continued)

S.No	Secondary metabolites	Microalgae	Application in various sectors	References
5	Limonene	*Anabaena* sp. PCC 7120	Fragrance and flavoring agent in chewing gum, perfumes, soaps, foods, and beverages.	(Jahandideh et al., 2017)
6	Phytosterol	*Dunaliella* sp. *Isochrysis galbana, Nannochloropis*	Pharmaceutical and functional food industry	(Sioen et al., 2011)
7	Flavonoids	*ChlorellaSpirulina*	Pharmaceutical industry	(Chassany et al., 2007)
8	Arachidonic acid	*Porphyridium* spp., *Parietochloris incise N. atomus, Porphyridium cruentum, P. boryanum*	Nutritional supplement	(Bigogno et al., 2002; Clandinin et al., 2005))
9	α-Linolenic acid	*C. vulgaris, Botryococcus spp., Ankitrodesmus spp., C. moewusii, Chlamydomonas spp., D. salina, D. Tertiolecta D. bardawil*	Nutritional supplement	(Piorreck, Baasch, and Pohl 1984; Ben-Amotz, Tornabene, and Thomas 1985; Ahlgren, Gustafsson, and Boberg 1992; Chiang, Huang, and Wu 2004; Poerschmann, Spijkerman, and Langer 2004)
10	γ-Linolenic acid,	*Spirulina platensis Nostoc, Spirulina spp., C. homosphaera, Chlorococcum spp., D. primolecta*	Nutritional supplement	(Tanticharoen et al., 1994; Ohta et al., 1995; Chiang, Huang, and Wu 2004)
11	Eicosapentaenoic acid	*Nannochloropsis gaditana, Phaedactylum tricornutum, Monochrysis lutheri, Pseudopedinella* sp., *Coccolithus huxleyi, Nannochloropsis salina, Hetermastrix rotundra, Cryptomonas maculate, Asterionella japonica*	Pharmaceuticals, nutrition, nutritional supplement	(Yongmanithai 1989)
10	Docosahexaenoic acid	*Schizochytrium, Thraustochytrids, Isochrysis galbana, Crypthecodinium conhi Schizochytrium* sp., *Ulkenia* spp.	Pharmaceuticals, nutrition	(Jiang, Chen, and Liang 1999; Ward and Singh 2005; Ratledge 2004)
12	MAAs	*Chlamydomonas nivalis C. nivalis, Scenedesmus* spp., *C. minutissima, C. sphaerica, C. luteo-viridis, C. sorokiniana, Stichococcus* spp	Cosmetic field	(Xiong, Kopecky, and Nedbal 1999; Duval, Shetty, and Thomas 1999; U. Karsten, Lembcke, and Schumann 2007; De la Coba, Aguilera, De Galvez, et al., 2009)

5.3. Isoprenoids

Isoprenoids are the natural organic compounds having two to many isoprene (five-carbon unit of hydrocarbons) units, which includes both primary and secondary metabolites (Rodríguez et al., 2002). Isoprenoids are widely used in the manufacturing of perfumes, incense, flavors, spices and also in medicines. Backbone of isoprenoid (isoprene) can vary based on the attachment of functional groups (hydroxyl, carbonyl), which allowing their greater diversity. Isoprenoids cover a variety of compounds such as carotenoids, sterols, volatile compounds, precursor of sex hormones, and also vitamins (E, A, K, Gibberellins).

5.4. Carotenoids

Microalgae produce several carotenoid pigments as secondary metabolites having yellow/orange/red colors. Carotenoids have higher antioxidant potential and protect the microalgae against oxidative stress (Sarada et al., 2006). Endogenous production of carotenoids in humans/animals is absent; hence, carotenoids need to be obtained through their diet. Different types of carotenoids of more than 600 are identified in nature, and among the very few, like staxanthin, lutein, canthaxanthin, and β-carotene are highly commercialized, followed by zeaxanthin and lycopene (Chen et al., 1993; Raja et al., 2004; Sarada et al., 2006; Spolaore et al., 2006; Ambati et al., 2014).

5.4.1. β-Carotene

Dunaliella salina, a halophilic and biflagellate green microalga is widely reported for the high concentration of β-carotene (10-14% of the dry weight) and is cultivated in commercial scales for β-carotene production (Sathasivam et al., 2012; Simon et al., 2016). In general, β-carotene production significantly favors at higher salinity, higher temperatures, higher light intensities, and nutrient limitation conditions (Raja et al., 2004; Henley et al., 2002). As *Dunaliella* contains both isomeric forms of β-carotene (cis and trans forms), the cis form of β –carotene is good antihepatotoxic compared to synthetic all-trans β-carotene (Chidambara Murthy, 2005). β-Carotene is one of the most active pro-vitamin A; as it can be converted to retinol. Natural β-carotene can be easily absorbed by the animal body, and it has higher biological activity than synthetic form (Chen et al., 1993; Sijil et al., 2019). The current global market for β-carotene is US$423 million, at a CAGR of 2.3%. β-Carotene has a strong antioxidant capacity to scavenge harmful free radicals, which have a role in various degenerative diseases (macular degeneration), and cancer prevention (Leon et al., 2003; Siems et al., 2005), hinderance of oxidative DNA damage, modifications in insulin-like growth factors, cell cycle regulatory proteins (Cooper, 2004). β-Carotene is used as an additive in health food products. It has several applications in pharmaceutical, food, feed and cosmetic industries; thereby, its global demand has increased. It has been reported that β-carotene from microalgae have shown protective effects against atherosclerosis in both mice and humans. *Dunaliella* biomass enriched with β-Carotene supplementation influences the plasma cholesterol, triglycerides and high-density lipoprotein. β-Carotene inhibits the low-density lipoprotein oxidation and have an important role in delaying the atherosclerosis development (Harari et al., 2008). Intake of *Dunaliella* can also have potential effects like protection from UV, UV-induced erythema, oxidative damage and premature ageing in humans (Heinrich et al., 2003; Ben-Amotz, 2019).

5.4.2. Astaxanthin

Haematococcus pluvialis is a green freshwater microalga and a commercial source for the production of astaxanthin. *H. pluvialis* is a single-celled, motile chlorophyceae microalga and it has the highest accumulating capability of astaxanthin (up to 4% of its dry biomass) (Kamath et al., 2008; Ambati et al., 2014). Astaxanthin ($C_{40}H_{52}O_4$) is a ketocarotenoid, having the highest antioxidant activity and reported to have ten times more antioxidant activity than β-carotene and 500 times more activity than α-tocopherol. It is extensively used as a pigmentation source in aquaculture and poultry feed. It also has application in cosmetics, food colorants, nutritional supplement, and pharmaceutical industries (Ambati et al., 2014). Among all carotenoids, astaxanthin has special unique attributes in its structure having two ketone and hydroxy moieties present on the ionone rings, which gives the most definite antioxidant potential. Astaxanthin shows potent radical scavenging activity, and it protects against oxidative damage of LDL cholesterol and lipid peroxidation. It has been reported that astaxanthin has activities against the various diseases like diabetics, diabetic neuropathy, heart disease, inflammatory diseases, *Helicobacterpylori*-induced gastric ulcer, age-related cognitive function, obesity, cancers, metabolic syndrome, eye-related and neurodegenerative diseases (Tso and Lam, 1996; Ikeuchi et al., 2007; Yuan et al., 2008; Satoh et al., 2009; Yuan et al., 2011; Ambati et al., 2014). It has been reported that the astaxanthin production from *Haematococcus* varies with culture age, sodium chloride and sodium acetate concentrations (Sarada et al., 2002). The global market potential of the astaxanthin is projected at US$ 1.1 billion by 2020. Ranga Rao et al., (2015) showed that astaxanthin and astaxanthin esters isolated from *H. pluvialis* have higher potency of anti-hepatoprotective activity and anti-oxidant activity in rats (Ambati et al., 2014). The USFDA has approved astaxanthin to use as a food colorant in animal and fish feed (Ambati et al., 2014). The most challenging aspect of *H. pluvailis* is outdoor cultivation and protection against predators. Other microalgae identified for production of astaxanthin are *Chlorococcum* sp., *Chlorella zofingiensis*, and *Chlamydomonas nivalis*. There are reported studies on genetic improvement for the better production of the commercially important astaxanthin from *Haematococcus pluvialis* (Kathiresan and Sarada, 2009; Kamath et al., 2008; Vidhyavathi et al., 2008; 2007).

5.4.3. Canthaxanthin

Canthaxanthin ($C_{40}H_{52}O$) is a β-carotene derived xanthophyll pigment reported to be present in microalgae such as *Chlorella zofingiensis* and *Dactylococcus dissociates*. Canthaxanthin has been used in fish feed as pigmentation source for coloration of the salmonid fish (Brizio et al., 2013) and in poultry feed to impart a dark yellow color to egg yolk (Esfahani-Mashhour et al., 2009). Canthaxanthin shows antioxidant and anticancer activities (Palozza et al., 2000; Brizio et al., 2013). The market for canthaxanthin is fetching high, but there is a dearth of studies on appropriate microbial sources. Studies with a new microalga strain of *Dactylococcus dissociates* shows suitability for industrial production of canthaxanthin (Grama et al., 2014). *Chlorella zofingiensis* also found to be a good source of canthaxanthin (Li et al., 2006). Commercial-scale production of canthaxanthin from microalgal source is yet to be established.

5.4.4. Cryptoxanthin

Few microalgae are reported to contain cryptoxanthin as one of the carotenoids. ...rella *pyrenoidosa* contains a variety of carotenoids, and one among them is β-...nthin (Inbaraj et al., 2006). Cryptoxanthin ($C_{40}H_{56}O$) is a xanthophyll derived from ... It is considered as provitamin A, and it is converted to vitamin A in the animal ... been reported that cryptoxanthin possesses antioxidant activity and bone ...t al., 2006; Yamaguchi, 2012). This pigment has an effect on bone ... bone loss in woman menopausal stage (Wright et al., 1993). It has ... microalgae such as *Chlorococcum*, *Haematococcus pluvialis*, ... *Spirulina platensis* may contain in lower quantities (Granado-...l., 2005; Yuan et al., 2002; Jaime et al., 2010). The ... from microalgae is yet to be established.

...*irulina*, and *Synechocystis* sp. are ...ls (Jin et al., 2003; Bhosale et ...ment from β-carotene/ β-...roalgae from the photo ...nthin plays a role ...lated macular ...hino et al., ...athy in

...e

... *Mycrocystis*, the ...ooms impart flavor

and odor to water bodies, and in *Mycrocystis*, oxidative cleavage of β-carotene produces β-cyclocitral (Jüttner, 2014). *Dunaliella salina* is also found to be producing this terpene compound. β- cyclocitral is very similar to the β-ionone part of β-carotene, and there are some arguments that β- cyclocitral is formed by a degradation process of carotenes.

5.5.2. Limonene

The chemical name of D-limonene is 1-methyl-4-(1-methylethenyl) cyclohexane [illegible] limonene has a lemon-like odor and one of the most common terpenes (Sun, 2007). It [illegible] monocyclic monoterpene having a pleasant fragrance, and hence, it is widely used as a [illegible] and fragrancing agent in beverages, perfumes, foods, soaps, and chewing gum For lar[illegible] production of limonene genetically engineered filamentous *cyanobacteria*, *Anabaena* [illegible] 7120 could be used (Jahandideh et al., 2017).

5.5.3. Geosmin and 2-Methylisoborneol (2-MIB)

Cyanobacteria namely, *Aphanizomenon* flos-aquae (L.) Ralfs and *Os*[illegible] produce geosmin; and *O. agardhii* produces 2-MIB as well. Geosmin and [illegible] (2-MIB) are the terpenoid compounds with cyclic tertiary alcohol. [illegible] dimethyl-trans-9-decalol) is an irregular sesquiterpene and 2-MIB ((1,[illegible] bicycle [2.2.l] heptan-2-01) is a monoterpene). Geosmin and 2-MIB [illegible] are very resistant to chemical and biological degradation, and the[illegible] together or separately. Though Butakova (2013) studied this com[illegible] and ecological roles are unclear. Their causes of synthesis [illegible] discovered. Both metabolites deteriorate water quality, and [illegible] intense 'earthy' or 'musty' odor (Bentley and Meganathan, 19[illegible]

5.5.4. Noscomin

A novel diterpenoid compound with an unpreceden[illegible] noscomin(8-[(5-carboxy-2-hydroxy)benzyl]-2-hydrox[illegible] 8,8a,9,10,10a-dodecahydrophenanthrene). This c[illegible] property against *Bacillus cereus, Staphylococ*[illegible] Terrestrial cyanobacterium *Nostoc commune* is f[illegible] metabolite (Jaki et al., 1999).

5.5.5. Tolypodiol

Tolypodiol is a potential anti-inf[illegible] cyanobacterium *Tolypothrix nodosa* ([illegible] are yet to be explored.

5.5.6. Bacteriohopanetetrol Glucuronisides

Bacteriohopanetetrol glucuronisides (C35 triterpenoids) reported in the thermophilic cyanobacterium *Synechococcus*. It is a first of a kind that a rare bacteriohopanetetrol glycosides containing glucuronic moieties from a cyanobacterium are reported (Llopiz et al., 1996).

5.6. Sterols

Sterols or steroid alcohols are secondary metabolites and are a part of steroids. Sterols are by products produced in isoprenoids biosynthesis. Most of the sterols have a common structure containing planar structure consist of methyl or ethyl substituted hydrocarbon chain of three β-hydroxy tetracycle. Sterol structure contains a fused ring, which gives rigidity to structure, and this helps in maintaining integrity, the stability of the cell membrane to hold membranes together. Sterols have been found to be present in almost all major groups of living organisms. They are essential components of cell membranes responsible for functioning in fluidity and permeability. Steroids play a significant role in various biochemical processes (steroid hormones production) and also in the regulation of membrane fluidity (Volkman, 2003; Piironen et al., 2000; Silvestro et al., 2013). *Chaetoceros, Thalassiosira, Pavlova,* and *Skeletonema* are reported to contain cholesterol majorly. Apart from Cholesterol, other sterols such as rassicasterol, campesterol, stigmasterol, and sitosterol have been reported from *Pavlova lutheri*. For the first time, the sistosterol, cholesterol in *Anacystis nidulans* and *Fremyella diplosiphon*as reported in 1968 (Reitz and Hamilton 1968). Kohlhase and Pohl (1988) reported saturated and unsaturated sterols from cyanobacteria (Kohlhase and Pohl, 1988). In cyanobacterial blooms, sterols concentration of 2.25 mg/L was reported (Procházková et al., 2017). Sterols are also used in many other fields such as pharmaceuticals and cosmetic industry. In microalgae, some sterols are widespread, whereas in other sterols are limited to a few genera (Görs et al., 2010). Ergosterol detected as a principal sterol in *Chlorella vulgaris*. *Chlorella* species exhibit wide biochemical variety in their sterol composition. These sterols act as good dietary sources in aquaculture (Barros et al., 2007).

5.6.1. Phytosterol

Phytosterols are naturally occurring steroid alcohol with economic value. Tetracyclic cyclopenta (α) phenanthrene is the characteristic structure of phytosterols and is amphiphilic. Microalgae can produce an extensive range of phytosterols such as brassica-sterol, sitosterol, and stigma-sterol (Volkman, 2003; Patterson and Carmeli, 1992; Marshall, 2002). Phyto sterols are reported to use in functional foods and pharmaceutical applications (Sioen et al., 2011). The worldwide market for phytosterols is increasing annually by approx. 7-9% (Borowitzka, 2013). Besides, microalgae have also been used for growth promotion of juveniles, specifical oysters for their content of sterols. *D. tertiolecta* and *D. salina* produce abundant phytosterols such as (22E, 24R)-methylcholesta5,7,22-trien-3b-ol (ergosterol) and

(22E, 24R)-ethylcholesta-5,7,22-trien-3b-ol (7- dehydroporiferasterol) at various NaCl concentrations (Francavilla et al., 2012). Other microalgae like *Isochrysis galbana, Nannochloropis gaditana, Nannochloropsis* sp. and *Phaeodactylum tricornutum* are also found to be a good source of phytosterol with content in the range of 7 to 34 g per kg (Ryckebosch et al., 2014). Phytosterol producing *Pavlova lutheri, Tetraselimis* sp. M8 and *Nannochloropsis* sp. (0.4%–2.6% dry weight) from Australia have been reported (Ahmed, Zhou and Schenk, 2015). Phytosterol also shows protection against certain types of cancer, such as colon, breast, and prostate.

5.7. Phenolic Compounds

Polyphenols are a major group of phytochemical molecules. Phenolic compounds are small phytochemical molecules, which are having hydroxyl group attached to phenolic ring. Phenolic compounds will form through either the shikimic acid pathway or malonate-acetate pathways. Phenolic compounds are stress compounds produced in microalgae as secondary metabolites, and they have the defense mechanisms like resistance against the abiotic stress produced by UV-rays (De la Coba et al., 2009), metal toxicity, biotic stress by grazing, bacterial settlement (Lau and Qian, 2000). Phenolics which includes phenolic acids, stilbenes, tannins, coumarins, and flavonoids (flavones, chalcones, flavanone, flavonols) (Thomas and Kim, 2011). These compounds mainly take part in the defense system of an organism. Based upon their structure have been divided into two main categories, i.e., flavonoids and non-flavonoids (Crozier et al., 2008). Most of the flavonoids have common 3 -ring structure, in which A and B are aromatic rings, and C is heterocyclic ring. Depends on variation in the heterocyclic C ring the flavonoids are divided into various subclasses such as flavonols, flavanones, flavones, isoflavones, flavanals, anthocyanidins and chalcones (Scalbert and Williamson, 2000). Phenolic acids are non-flavonoid polyphenolic compounds. It can be divided into benzoic acid derivative, cinnamic acid derivatives.

The two edible microalgae *Chlorella* and *Spirulina* are reported as rich sources of phenolic compounds, which are attributed to have antioxidant activities, nutritional, and health benefits (Mallikarjun et al., 2015). Their phenolic composition includes phloroglucinol, p-coumaric acid, ferulic acid, apigenin (Goiris et al., 2014). *Chlorella* is found to have more content of phloroglucinol which is used by the pharmaceutical industry for the treatment of gastrointestinal disorders (Chassany et al., 2007). Apigenin is found to be inducing autophagy in leukemia cells (Ruela-de-Sousa et al., 2010). On the other hand, p-coumaric, acid, and ferulic acid found to possess antioxidant properties (Ferguson et al., 2005). Phenolic compounds of microalgae help to protect against adverse conditions (Connan and Stengel, 2011).

Microcystis aeruginosa produced polyphenols ellagic and gallic acids and catechin, which are reported to contain growth inhibitory effects (Nakai et al., 2005). Red algae produced phenolic compounds (catechol, caffeic acid, catechin, morin, epigallocatechin gallate, rutin, and hesperidin) that have been reported to contain anti-inflammatory activity (Ibañez and Cifuentes, 2013).

5.7.1. Flavonoids

Polyphenolic compounds containing 15-carbon atoms with two aromatic rings connected through a 3-carbon bridge are coming under this category. Natural flavonoids are well known for their significant role in scavenging properties on oxygen radicals *in vivo* and *in vitro*. The flavonoid synthesis enzymes and their gene homologs are reported in both micro and macroalgae. (May et al., 2008).

5.7.2. Quercetin

Quercetin is one of the main representatives of the flavonol subclass, shown to be potential functions such as cell cycle regulating agent, tyrosine kinase inhibition, anti-cancer agent, and including interaction with type II estrogen binding sites (Lamson and Brignall, 2000). These flavonoids are ubiquitous in nature and are large group over 4000 naturally available plant phenolics. A wide range of pharmacological activity has shown by these compounds such as antiallergic, anti-inflammatory, and vasoactive. *H. pluvialis* is found to be a rich source of different types of quercetin (Goiris et al., 2014).

5.7.3. Rutin (Quercetin-3-Rhamnosyl Glucoside)

Rutin is a flavonoid. Rutin was first discovered in buckwheat in the 19th century and known as vitamin P (Yang et al., 2008). Rutin has potential activities in cell cycle regulation, interaction with type II estrogen binding sites, tyrosine kinase inhibition, and also as potential anti-cancer agent. The presence of rutin is detected in the microalga *Hematococcus* (Goiris et al., 2014).

5.8. Halogenated Compounds

Halogenated compounds have been characterized from microalgae (including cyanobacteria). There are various types of halogenated compounds such as fatty acids, indoles, terpenes acetogenins, and volatiles (chloroform, bromoform) (Butler and Carter-Franklin, 2004). These halogenated compounds are widely used in pharmacological sectors due to their varied biological activities such as anti-inflammatory, antibacterial, cytotoxic, anti-proliferative, antifungal, antiviral, insecticidal, antifeedant, antifouling, and antitumoral activities (Vairappan et al., 2001). *Egregia menziesii, Macrocystis pyrifeara, Custoseria osmundacea, Eisenia arborea, Laminaria farlowii,* and *Prochlorococcus marinus* are reported to produce a diverse range of halogenated compounds (Manley et al., 1992; Dembitsky and Tolstikov, 2003; Hughes et al., 2011). It has been reported that *Laurencia,* a marine microalga produces diverse halogenated compounds such as diterpene, triterpene, sesquiterpene, brominated indoles and fatty acids (John Faulkner, 2001; Wright et al., 2003). The red microalgae (*Plocamium, Chondrococcus,* and *Ochtodes*) reported containing polyhalogenated monoterpene having pharmacological activities (Fuller et al., 1992; 1994;

Darias et al., 2001; Blunt et al., 2003; Knott et al., 2005). It was reported that halogenated phenolic compounds show strong biological activities (Cabrita et al., 2010). Halogenated phenolic compounds have higher biological activities like wise antioxidant, antiviral, anticarcinogenic, antifungal, and antitumor properties (Munawer and Mazharuddin, 2011; Chauhan and Kasture, 2014; Kumar et al., 2016).

5.9. Fatty Acid Derivatives

5.9.1. Polyunsaturated Fatty Acids (PUFA)

Microalgae have the potential for production of essential fatty acids such as α-linolenic acid, γ-linolenic acid (GLA), arachidonic acid (AA), eicosapentaenoic acid (EPA) and docosahexaenoic acid (DHA) (Borowitzka, 2013; Vidyashankar et al., 2015). PUFAs play a vital role in cellular/ tissue metabolism, thermal adaption, membrane fluidity, electron/oxygen transport, and also as eicosanoid precursors (Funk, 2001). Recent reports showed that essential fatty acids have health benefits against various disorders of coronary heart disease, brain-related diseases and neurological functions like neurogenesis, neurotransmission, and protection against oxidative stress (Barreira et al., 2015; Morgese et al., 2016; Sahu et al., 2013). As animals and humans cannot produce the enzymes responsible for the production of PUFA having more than 18 carbons (Parsaeimehr et al., 2015) so they are to be provided through diet, and these are termed as essential fatty acids.

5.9.2. Arachidonic Acid (AA)

The unicellular rhodophyte *Porphyridium cruentum* and the chlorophyte *Parietochloris incisa* are the rich sources of long-chain polyunsaturated fatty acid (LC-PUFA) arachidonic acid (Cohen, 1990; Bigogno et al., 2002; Kavitha et al., 2016). AA and DHA are to contribute the major PUFAs of brain cell phospholipids (Hansen et al., 1997). Addition of AA to formula milk are recommended since it had shown to improve infant development (Bigogno et al., 2002; Clandinin et al., 2005).

5.9.3. γ-Linolenic Acid (GLA)

γ-linolenic acid (GLA) is most widely found in the *Spirulina* and *Nostoc* strains of the cyanobacteria (Tanticharoen et al., 1994). GLA is an omega-6 polyunsaturated fatty acid. It is a vital precursor for the prostaglandins synthesis. It has been reported that GLA plays an essential role in the treatment of various diseases like obesity, heart disease, arthritis, schizophrenia, alcoholism, schizophrenia, zinc deficiency, parkinson's disorder of elderly persons (Kerby et al., 1987). Because of these aspects, it is considered an active molecule for human health (Kerby et al., 1987).

5.9.4. α-Linolenic Acid (ALA)

The α-linolenic acid (ALA) is reported to contain in various microalgae such as *Ankitrodesmus* sp., *Botryococcus* sp., *Chlamydomonas* sp., *C. moewusii, N.atomus, C. vulgaris, S.acutus, D. bardawil, D. salina, D. tertiolecta, M. pusilla, Muriellopsis* sp., *P. subcapitata., S. obliquus, S. quadricauda, Tetraselmis* sp., *T. suecica* (Ben-Amotz et al., 1985; Chiang et al., 2004; Ahlgren et al., 1992; Piorreck et al., 1984). ALA is one of the essential polyunsaturated omega-3 fatty acids and has nutritional importance. ALA acts as a precursor for the production of eicosapentaenoic acid (EPA) and docosahexaenoic acid (DHA). ALA has been reported to contain preventative effect against cardiovascular diseases by decreasing the serum triglyceride level, blood pressure, and blood clot formation. ALA reported to have a crucial role in the inflammation-related mechanisms and regulation of their genes. ALA used as nutritional supplementation for treating dietary imbalance(Poerschmann et al., 2004).

5.9.5. Eicosapentaenoic Acid (EPA)

Eicosapentaenoic acid (EPA) has been reported in various microalgal species *Monochrysis lutheri* (19%), *Pseudopedinella* sp. (27%), *Coccolithus huxleyi* (17%), *Nannochloropsis salina* (15%), *Hetermastrix rotundra* (28%), *Cryptomonas maculata* (17%), *Asterionella japonica* (20%) (Yongmanithai, 1989). EPA is omega-3 higher carbon fatty acid. EPA production with an autotrophic nutritional mode of microalgae cultivation face challenges such as low growth rates and fewer cell densities (Wen and Chen, 2003). Many recent reports have shown that higher growth rates of microalgae along with EPA production obtained using organic carbon source as heterotrophic nutritional mode (Wen and Chen, 2003). EPA plays an essential role in the treatment of many diseases such as cancer, atherosclerosis, rheumatoid arthritis, Alzheimer's, and psoriasis (Asgharpour et al., 2015). EPA is also identified to present in *Chlorella minutissima, Isochrysis galbana, Diacronema vlkianum* (Seto et al., 1984), *Nanochloropsis* sp, and *Nitsschia* (Spolaore et al., 2006).

5.9.6. Docosahexaenoic Acid (DHA)

Crypthecodinium cohnii a microalga reported to contain 40-50% of Docosahexaenoic acid (DHA) (Jiang et al., 1999; Ward and Singh, 2005; Ratledge, 2004). DHA is an omega-3 fatty acid having high nutritional prominence. In general, available sources of docosahexaenoic acids are very limited and available in breast milk, fatty fish, organic meat, and microalgae. DHA reported to use in the development of brain and eye in infants, and heart-related diseases in adults (Kroes et al., 2003; Ward and Singh, 2005). It is also reported to play an important role in the treatment of diseases such as cancer, atherosclerosis, Alzheimer's, rheumatoid arthritis, and psoriasis (Asgharpour et al., 2015). DHA produced by *Schizochytrium* strains are currently using as a dietary supplement in various foods and beverages, animal feeds and maricultural products (cheese, spreads, and yogurts). These supplemented foods are used for nursing and pregnant woman (Ward and Singh, 2005). Microalgal species *viz Schizochytrium, Thraustochytrids, Isochrysis galbana, Diacronema vlkianum Crypthecodinium,* and *Ulkenia* are reported to produce DHA (Ratledge, 2004; Luísa Gouveia et al., 2008; Barclay et al., 2010; Wynn et al., 2010).

5.10. Polyketides

Polyketides are the polymers of acetate or propionate. Polyketides are widely reported secondary metabolites in microalgae (Blunt W et al., 2007). Polyketides have high commercial value in the pharmaceutical industry (Dos Santos et al., 2005). Polyketides have antimicrobial, immunosuppressive properties (Fjaervik and Zotchev, 2005). In microalgae polyketides present as the open-chain form and polycyclic ether macrolides form (Kobayashi et al., 1989). Due to a wide range of polyketides, depending on polyketide synthases (PKS) these were divided into two types. Type I Polyketide synthases are responsible for the production of algal toxins (i.e., brevetoxins) as well as antibiotics (i.e., erythromycin) (Staunton and Weissman, 2001). Type II PKS responsible for aromatic rings (Hertweck et al., 2007) which includes flavonoids, algal phlorotannins, and antibiotic tetracycline. Polyketides are widely being used in veterinary medicine (as anti-coccidiosis agents and antibiotics), feed supplements in poultry and cattle (Lopes et al., 2001; 2002). Previous reports also showed that these polyketides could not be used in human therapy because of algal toxicogenicity, which cause more attention to controversy (Dos Santos et al., 2005). But few patents are showing that macrolides from microalgae were used in human therapy (Kobayashi et al., 1988).

5.11. Amino Acid-Derived Microalgae Secondary Metabolites

Secondary metabolites derived from the amino acids are divergent and broad, which includes alkaloids, polypeptides, and their derivatives. These alkaloids and polypeptides are very complex in their structure.

5.11.1. Alkaloids

Alkaloids are nitrogen-containing natural compound and synthesized from amino acids, which predominantly possess toxic properties. Structural diversity of the alkaloids are due to oxygenation, cyclisation, methylation, and chlorination of terpene precursors. Alkaloids are condensed amino acids, bitter in taste, and aromatic in structures. All related compounds are not derived from the unifying pathway. Indole alkaloids are a class of alkaloids containing an indole moiety such as the hapalindoles hapalindolinones, ambiguines, fischambiguines, fisherindoles, welwitindolinones, etc. (Kultschar and Llewellyn, 2018). Fisher indoles produced by *Fischerella,* which possess antifungal property; ambiguin H and ambiguine I are potential antibacterial agents from the same microalgae. Hapalindole reported to have antibacterial activity against Gram-negative and Gram-positive bacteria (*Escherichia coli* and

Staphylococcus aureus respectively), and it is isolated from *Fischerella* sp. (Dixit and Suseela, 2013).

5.11.2. Mycosporin-Like Amino Acids (MAAs)

Mycosporine like amino acids (MAAs) are low molecular weight (<400Kd), water-soluble compounds. MAAs are molecules consist of an amino cyclohexenimine ring, or amino cyclohexanone, carrying nitrogen or amino alcohol substituents (Nakamura et al., 1982) that absorb light of low wavelength. MAAs are isolated and characterized from various fresh and marine microalgae. MAAs have the capability to protect the microalgae from UV rays. When microalgae exposed to UV radiation, which enabled higher amounts of MAAs production in green microalgae (e.g., *Chlamydomonas nivalis*) (Ulf Karsten et al., 2005; Xiong et al., 1999). It has been reported that MAAs can also protect the human/animals skin from UV radiation damage (De la Coba, Aguilera, De Galvez, et al., 2009). Exploration of the microalgal MAAs widened applications in the cosmetic sector for commercial production and its usage in skin care products against UV radiation. MAAs are known to protect skin from UV radiations and for prevention of wrinkles, rough texture, and skin flaccidity.

5.11.3. Scytonemin

Cyanobacterial species *Nostoc, Scytonema, Calothrix, Lyngbya, Rivularia, Chlorogloeopsis,* and *Hyella* are reported to produce the scytonemin. Scytonemin is highly stable, hydrophobic, alkaloid compound synthesized by many cyanobacterial species. It acts against UV radiation and temperature stresses. Scytonemin comes under the photoprotective compounds, and it is a dimer of phenolic and indolic subunits. Initially, scytonemin is identified in cyanobacterial sp, which was located at exopolysaccharide sheath as a yellowish-brown lipid-soluble pigment (Garcia Pichel and Castenholz, 1991; Rothrock Jr and Garcia-Pichel, 2005; Wada et al., 2013; Rastogi et al., 2015). Scytonemin exists in three forms - oxidized form (yellow color, i.e., fuscochlorin), oxidized form (green color) and reduced form (red in color) (Garcia Pichel and Castenholz, 1991; Wada et al., 2013). Among the three it is mostly found as in oxidized green form. Various forms of scytonemin such as dimethoxyscytonemin, teteramethoxyscytonemin, and scytonin have been isolated from organic extracts (Bultel-Poncé et al., 2004; Grant and Louda, 2013; Rastogi et al., 2014). Scytonemin has various absorption wavelengths such as 252, 278, 300, and 370nm (in vivo), 386nm (pure form). This may be a typical characteristic of the compound that allows cyanobacteria to survive under lethal UV-radiations. Recent studies showed that scytonemin has the ability to lessen the most lethal UV-C radiation risk damage (Dillon and Castenholz, 1999; Rastogi et al., 2013). It also acts effectively against UV-A radiation as it reduces the inhibition of photosynthesis and also reduces chlorophyll-a photobleaching (Cockell and Knowland, 1999; Gao and Garcia-Pichel, 2011). Scytonemin is widely used in the cosmetics sector due to its higher screening potential (Rastogi et al., 2014; 2015). Scytonemin from *Chlorogloeopsis* sp. is shown to have UV-sunscreen activity (Garcia Pichel and Castenholz, 1991).

5.12. Polyamines

Polyamines are reported to produce in various microalgae such as *Chlamydomonas* sp., *Bracteacoccus giganteus, Chlorococcum echinozygotum, Chlorogonium elongatum, Coccomyxa dispar, Errerella bornhemiensis, Eudorina illinoisensis, Gonium pectoral, Haematococcus lacustris, Lobomonas piriformis, Oedogonium obesum, Dunaliella* sp., *Volvox aureus, Scenedesmus* sp., *Mesostigma viride, Monomastix minuta, Nephroselmis olivacea, Nannochloropsis oculata, Vischeria* sp., *Chattonella antiqua* (Hada), *Heterosigma akashiwo, Cyanidium caldarium* (Lin and Lin, 2019). Polyamines are the derivatives of multiple amino groups, which are connected either linear or branched to aliphatic compounds. Microalgae can produce a variety of polyamines, and it is necessary to maintain polyamine content to meet physiological needs of microalgae. Most common polyamines and typically found in most of the microalgae are putrescine, spermidine, and Spermine. Whereas some microalgae have uncommon polyamines thermos spermine, di-aminopropane, nor-spermidine, cadaverine, long-chain polyamines, and branched-chain polyamines (Lin and Lin, 2019). Polyamines have been reported to play important roles in physiological functions such as regulation of gene and protein expressions, promotion of growth, cell division, photosynthesis aid, modulation of signal transduction, cell membrane stabilization and nucleic acid protection against the stress conditions. There are some polyamines produced by microalgae have medical nanotechnological applications.

5.13. Toxins

Well-known cyanobacteria such as *Anabaena, Nostoc, Oscillatoria,* and *Microcystis* reported to produce these compounds. Cyanobacteria and microalgae produce some toxic compounds also to protect themselves from foreign microbes. Cyanobacteria reported to produce a huge number of allopathic compounds, cyano-toxins. Even though the secondary metabolites of microalgae have toxic effects, and these are exploited in various ways such as herbicides, fungicide, algicide, weedicide. These compounds are widely used for therapeutic applications such as anti-tumor, antibiotic, immunosuppressant, and neurotoxic properties. Most of the algal toxins compounds are neurotoxic and proved in the investigations of neurodegenerative diseases. For example, okadaic acid, which is used as a typical, antipsychotic drugs in the treatment of schizophrenia and cognitive impairment (He et al., 2005).

Conclusion

Microalgae are prospective source for the natural production of various secondary metabolites with specific biological activity. Secondary metabolites from microalgae have been reported to have applications in food, feed, cosmetic, pharmaceutical and health sectors. The demand for microalgal secondary metabolites is increasing globally. Low concentration of the secondary metabolites in the native microalgae has limited their isolation, identification, and functional characterization. Only a few secondary metabolites (such as β-carotene, astaxanthin, DHA, EPA, and MAAs, etc.) from microalgae are successfully commercialized. In order to use microalgae as a reliable source for the production of metabolites having ameliorative effects for human health and diseases, cultivation methods of microalgae need to be improved. Further exploration of the biodiversity of microalgae, especially from extreme environments, would provide ample scope for identification of newer metabolites with novel functions.

REFERENCES

Abe, K., Hattori, H., & Hirano, M. (2007). Accumulation and antioxidant activity of secondary carotenoids in the aerial microalga Coelastrella striolata var. multistriata. *Food Chem.* 100(2), 656-661.

Ahlgren, G., Gustafsson, I. B., & Boberg, M. (1992). Fatty acid content and chemical composition of freshwater microalgae 1. *J. Phycol.* 28(1), 37-50.

Ahmed, F., Zhou, W., & Schenk, P. M. (2015). *Pavlova lutheri* is a high-level producer of phytosterols. *Algal Res.* 10, 210-217.

Ambati, R. R., Phang, S. M., Ravi, S., & Aswathanarayana, R. G. (2014). Astaxanthin: sources, extraction, stability, biological activities and its commercial applications-a review. *Mar. Drugs.* 12(1), 128-152.

Asgharpour, M., Rodgers, B., & Hestekin, J. (2015). Eicosapentaenoic acid from *Porphyridium cruentum*: Increasing growth and productivity of microalgae for pharmaceutical products. *Energies.* 8(9), 10487-10503.

Asthana, R. K., Tripathi, M. K., Srivastava, A., Singh, A. P., Singh, S. P., Nath, G., & Srivastava, B. S. (2009). Isolation and identification of a new antibacterial entity from the Antarctic cyanobacterium *Nostoc* CCC 537. *J. Appl. Phycol.* 21(1), 81.

Asthana, R. K., Srivastava, A., Kayastha, A. M., Nath, G., & Singh, S. P. (2006). Antibacterial potential of γ-linolenic acid from *Fischerella* sp. colonizing Neem tree bark. *W. J. Microbiol. Biotechnol.* 22(5), 443-448.

Barbaras, D., Kaiser, M., Brun, R., & Gademann, K. (2008). Potent and selective antiplasmodial activity of the cyanobacterial alkaloid nostocarboline and its dimers. Bioorg. *Med. Chem. Lett.* 18(15), 4413-4415.

Barchi Jr, J. J., Moore, R. E., & Patterson, G. M. (1984). Acutiphycin and 20, 21-didehydroacutiphycin, new antineoplastic agents from the cyanophyte *Oscillatoria acutissima*. *J. Am. Chem. Soc.* 106(26), 8193-8197.

Barclay, W., Weaver, C., Metz, J., & Hansen, J. (2010). Development of a docosahexaenoic acid production technology using *Schizochytrium*: historical perspective and update. In *Single cell oils* (pp. 75-96). AOCS Press.

Barreira, L., Pereira, H., Gangadhar, K. N., Custódio, L., & Varela, J. (2015). Medicinal Effects of Microalgae-Derived Fatty Acids. In *Handbook of Marine Microalgae* (pp. 209-231). Academic Press.

Cardozo, K. H., Guaratini, T., Barros, M. P., Falcão, V. R., Tonon, A. P., Lopes, N. P., & Pinto, E. (2007). Metabolites from algae with economical impact. *Comparative Biochemistry and Physiology Part C: Toxicol. Pharmacol.* 146(1-2), 60-78.

Ben-Amotz, A. (Ed.). (2019). *The alga Dunaliella*. CRC Press.

Ben-Amotz, A., Tornabene, T. G., & Thomas, W. H. (1985). Chemical profile of selected species of microalgae with emphasis on lipids 1. *J. Phycol.* 21(1), 72-81.

Bentley, R., & Meganathan, R. (1981). Geosmin and methylisoborneol biosynthesis in streptomycetes: Evidence for an isoprenoid pathway and its absence in non-differentiating isolates. *FEBS. lett*, 125(2), 220-222.

Bernardo, P. H., Chai, C. L., Le Guen, M., Smith, G. D., & Waring, P. (2007). Structure activity delineation of quinones related to the biologically active Calothrixin B. *Bioorg. Med. Chem. Lett.* 17(1), 82-85.

Berry, J. P., Gantar, M., Perez, M. H., Berry, G., & Noriega, F. G. (2008). Cyanobacterial toxins as allelochemicals with potential applications as algaecides, herbicides and insecticides. *Mar. Drugs.* 6(2), 117-146.

Bhadury, P., & Wright, P. C. (2004). Exploitation of marine algae: biogenic compounds for potential antifouling applications. *Planta.* 219(4), 561-578.

Bhosale, P., & Bernstein, P. S. (2005). Microbial xanthophylls. *Appl. Microbiol. Biotechnol.* 68(4), 445-455.

Bigogno, C., Khozin-Goldberg, I., Boussiba, S., Vonshak, A., & Cohen, Z. (2002). Lipid and fatty acid composition of the green oleaginous alga *Parietochloris incisa*, the richest plant source of arachidonic acid. *Phytochem.* 60(5), 497-503.

Blunt, J. W., Copp, B. R., Munro, M. H., Northcote, P. T., & Prinsep, M. R. (2011). Marine natural products. *Nat. Prod. Rep.* 28(2), 196-268.

Blunt, J. W., Copp, B. R., Wan, P. Hu., Murray, H. G. M., Peter T. N., and Michèle R. P. (2007). Marine Natural Products. *Nat. Prod. Rep.* 24 (1), 31–86.

Bonjouklian, R., Smitka, T. A., Doolin, L. E., Molloy, R. M., Debono, M., Shaffer, S. A., & Patterson, G. M. (1991). Tjipanazoles, new antifungal agents from the blue-green alga *Tolypothrix tjipanasensis*. *Tetrahedron.* 47(37), 7739-7750.

Bonnard, I., Rolland, M., Francisco, C., & Banaigs, B. (1997). Total structure and biological properties of laxaphycins A and B, cyclic lipopeptides from the marine cyanobacterium *Lyngbya majuscula*. *Lett. Pep. Sci.* 4(4-6), 289-292.

Borowitzka, M. A. (2013). High-value products from microalgae their development and commercialisation. *J. Appl. Phycol.* 25(3), 743-756.

Borowitzka, M. A. (1988). Vitamins and fine chemicals from micro-algae. Cambridge University Press.

Brizio, P., Benedetto, A., Righetti, M., Prearo, M., Gasco, L., Squadrone, S., & Abete, M. C. (2013). Astaxanthin and canthaxanthin (xanthophyll) as supplements in rainbow trout diet: in vivo assessment of residual levels and contributions to human health. *J. Agri. Food Chem.* 61(46), 10954-10959.

Bultel-Poncé, V., Félix-Théodose, F., Sarthou, C., Ponge, J. F., & Bodo, B. (2004). New pigments from the terrestrial cyanobacterium *Scytonema* sp. collected on the Mitaraka Inselberg, French Guyana. *J. Nat. Prod.* 67(4), 678-681.

Burja, A. M., Banaigs, B., Abou-Mansour, E., Burgess, J. G., & Wright, P. C. (2001). Marine cyanobacteria a prolific source of natural products. *Tetrahedron.* 57(46), 9347-9377.

Butakova, E. A. (2013). Specific features of odor-causing compounds (geosmin and 2-methylisoborneol) as secondary metabolites of cyanobacteria. *Russ. J. Plant Physiol.* 60(4), 507-510.

Butler, A., & Carter-Franklin, J. N. (2004). The role of vanadium bromoperoxidase in the biosynthesis of halogenated marine natural products. *Nat. Prod. Rep.* 21(1), 180-188.

Cabrita, M. T., Vale, C., & Rauter, A. P. (2010). Halogenated compounds from marine algae. *Mar. Drugs.* 8(8), 2301-2317.

Capelli, B., Bagchi, D., & Cysewski, G. R. (2013). Synthetic astaxanthin is significantly inferior to algal-based astaxanthin as an antioxidant and may not be suitable as a human nutraceutical supplement. *Nut. Foods.* 12(4), 145-152.

Carmeli, S., Moore, R. E., & Patterson, G. M. (1990). Tolytoxin and new scytophycins from three species of *Scytonema*. *J. Nat. Prod.* 53(6), 1533-1542.

Cha, K. H., Koo, S. Y., & Lee, D. U. (2008). Antiproliferative effects of carotenoids extracted from *Chlorella ellipsoidea* and *Chlorella vulgaris* on human colon cancer cells. *J. Agri. Food chem.* 56(22), 10521-10526.

Chassany, O., Bonaz, B., Bruley Des Varannes, S., Bueno, L., Cargill, G., Coffin, B., & Grangé, V. (2007). Acute exacerbation of pain in irritable bowel syndrome: efficacy of phloroglucinol/trimethylphloroglucinol–a randomized, double-blind, placebo-controlled study. *Aliment. Pharmacol. Ther.* 25(9), 1115-1123.

Chauhan, J., & Kasture, A. (2014). Antimicrobial compounds of marine algae from Indian coast. *Int. J. Curr. Microbiol. Appl. Sci*, 7, 526-532.

Chen, B. H., Chuang, J. R., Lin, J. H., & Chiu, C. P. (1993). Quantification of provitamin A compounds in Chinese vegetables by high-performance liquid chromatography. *J. Food Protect.* 56(1), 51-54.

Chiang, I. Z., Huang, W. Y., & Wu, J. T. (2004). Allelochemicals of *Botryococcus braunii* (chlorophyceae) 1. *J. Phycol.* 40(3), 474-480.

Chidambara Murthy, K. N. (2005). Production of Beta-Carotene from Cultured *Dunaliella* Sp. and Evaluation Of Biological Activities. (Doctoral dissertation, University of Mysore).

Choi, H., Pereira, A. R., Cao, Z., Shuman, C. F., Engene, N., Byrum, T., & Gerwick, W. H. (2010). The hoiamides, structurally intriguing neurotoxic lipopeptides from Papua New Guinea marine cyanobacteria. *J. Nat. Prod.* 73(8), 1411-1421.

Clandinin, M. T., Van Aerde, J. E., Merkel, K. L., Harris, C. L., Springer, M. A., Hansen, J. W., & Diersen-Schade, D. A. (2005). Growth and development of preterm infants fed infant formulas containing docosahexaenoic acid and arachidonic acid. *The J. Pediat.*, 146(4), 461-468.

Cockell, C. S., & Knowland, J. (1999). Ultraviolet radiation screening compounds. *Biol. Rev.* 74(3), 311-345.

Cohen, Z. (1999). *Porphyridium cruentum. Chemicals from microalgae*, 1-24.

Cohen, Z. (1990). The production potential of eicosapentaenoic and arachidonic acids by the red alga *Porphyridium cruentum*. *J. Am. Oil Chem. Soc.* 67(12), 916-920.

Connan, S., & Stengel, D. B. (2011). Impacts of ambient salinity and copper on brown algae: 2. Interactive effects on phenolic pool and assessment of metal binding capacity of phlorotannin. *Aqua. Toxicol.* 104(1-2), 1-13.

Cooper, D. A. (2004). Carotenoids in health and disease: recent scientific evaluations, research recommendations and the consumer. *The J. Nutri.* 134(1), 221S-224S.

Crozier, A., Clifford, M. N., & Ashihara, H. (Eds.). (2008). *Plant secondary metabolites: occurrence, structure and role in the human diet.* John Wiley & Sons.

Dahms, H. U., Ying, X., & Pfeiffer, C. (2006). Antifouling potential of cyanobacteria: a mini-review. *Biofoul.* 22(5), 317-327.

Darias, J., Rovirosa, J., San Martin, A., Díaz, A. R., Dorta, E., & Cueto, M. (2001). Furoplocamioids A− C, Novel Polyhalogenated Furanoid Monoterpenes from *Plocamium cartilagineum*. *J. Nat. Prod.* 64(11), 1383-1387.

Davies-Coleman, M. T., Dzeha, T. M., Gray, C. A., Hess, S., Pannell, L. K., Hendricks, D. T., & Arendse, C. E. (2003). Isolation of Homodolastatin 16, a New Cyclic Depsipeptide from a Kenyan Collection of *Lyngbya majuscula*. *J. Nat. Prod.* 66(5), 712-715.

De la Coba, F., Aguilera, J., De Galvez, M. V., Alvarez, M., Gallego, E., Figueroa, F. L., & Herrera, E. (2009). Prevention of the ultraviolet effects on clinical and histopathological changes, as well as the heat shock protein-70 expression in mouse skin by topical application of algal UV-absorbing compounds. *J. Derm. Sci*, 55(3), 161-169.

De la Coba, F., Aguilera, J., Figueroa, F. L., De Gálvez, M. V., & Herrera, E. (2009). Antioxidant activity of mycosporine-like amino acids isolated from three red macroalgae and one marine lichen. *J. Appl. Phycol.* 21(2), 161-169.

Dembitsky, V. M., & Tolstikov, G. A. (2003). Natural halogenated organic compounds. *Geo-Science, Novosibirsk*, 366.

Deng, R., & Chow, T. J. (2010). Hypolipidemic, antioxidant, and antiinflammatory activities of microalgae *Spirulina*. *Cardio. Ther.* 28(4), e33-e45.

Dillon, J. G., & Castenholz, R. W. (1999). Scytonemin, a cyanobacterial sheath pigment, protects against UVC radiation: implications for early photosynthetic life. *J. Phycol.* 35(4), 673-681.

Dixit, R. B., & Suseela, M. R. (2013). Cyanobacteria: Potential candidates for drug discovery. *Antonie Van Leeuwenhoek*. 103(5), 947-961.

Dos Santos, M. D., Guaratini, T., Lopes, J. L. C., Colepicolo, P., & Lopes, N. P. (2005). Plant cell and microalgae culture. *Mod. Biotechnol. Med. Chem. Ind.*

Duval, B., Shetty, K., & Thomas, W. H. (1999). Phenolic compounds and antioxidant properties in the snow alga *Chlamydomonas nivalis* after exposure to UV light. *J. Appl. Phycol.* 11(6), 559.

Esfahani-Mashhour, M., Moravej, H., Mehrabani-Yeganeh, H., & Razavi, S. H. (2009). Evaluation of coloring potential of *Dietzia natronolimnaea* biomass as source of canthaxanthin for egg yolk pigmentation. *Asian. Austral. J. Anim.* 22(2), 254-259.

Falch, B. S., Koenig, G. M., Wright, A. D., Sticher, O., Ruegger, H., & Bernardinelli, G. (1993). Ambigol A and B: new biologically active polychlorinated aromatic compounds from the terrestrial blue-green alga *Fischerella ambigua*. *The J. Org. Chem.* 58(24), 6570-6575.

Fennell, B. J., Carolan, S., Pettit, G. R., & Bell, A. (2003). Effects of the antimitotic natural product dolastatin 10, and related peptides, on the human malarial parasite *Plasmodium falciparum*. *J. Antimicrobial. Chemother.* 51(4), 833-841.

Ferguson, L. R., Zhu, S. T., & Harris, P. J. (2005). Antioxidant and antigenotoxic effects of plant cell wall hydroxycinnamic acids in cultured HT-29 cells. *Mol. Nutr. Food. Res.* 49(6), 585-593.

Fjærvik, E., & Zotchev, S. B. (2005). Biosynthesis of the polyene macrolide antibiotic nystatin in *Streptomyces noursei*. *Appl. Microbiol. Biotechnol.* 67(4), 436-443.

Francavilla, M., Colaianna, M., Zotti, M., G Morgese, M., Trotta, P., Tucci, P., & Trabace, L. (2012). Extraction, characterization and in vivo neuromodulatory activity of phytosterols from microalga *Dunaliella tertiolecta*. *Curr. Med. Chem.* 19(18), 3058-3067.

Frankmölle, W. P., Knübel, G., Moore, R. E., & Patterson, G. M. (1992). Antifungal cyclic peptides from the terrestrial blue-green alga *Anabaena laxa*. II. Structures of laxaphycins A, B, D and E. *The. J. Antibio*, 45(9), 1458-1466.

Fuller, R. W., Cardellina, J. H., Jurek, J., Scheuer, P. J., Alvarado-Lindner, B., McGuire, M., & Clardy, J. (1994). Isolation and structure/activity features of halomon-related antitumor monoterpenes from the red alga *Portieria hornemannii*. *J. Med. Chem.* 37(25), 4407-4411.

Fuller, R. W., Cardellina, J. H., Kato, Y., Brinen, L. S., Clardy, J., Snader, K. M., & Boyd, M. R. (1992). A pentahalogenated monoterpene from the red alga *Portieria hornemannii* produces a novel cytotoxicity profile against a diverse panel of human tumor cell lines. *J. Med. Chem.* 35(16), 3007-3011.

Funk, C. D. (2001). Prostaglandins and leukotrienes: advances in eicosanoid biology. *Science*. 294(5548), 1871-1875.

Gao, Q., & Garcia-Pichel, F. (2011). Microbial ultraviolet sunscreens. *Nat. Rev. Microbiol.* 9(11), 791.

Garcia-Pichel, F., & Castenholz, R. W. (1991). Characterization and biological implications of scytonemin, a cyanobacterial sheath pigment 1. *J. Phycol.* 27(3), 395-409.

Gerwick, W. H., Mrozek, C., Moghaddam, M. F., & Agarwal, S. K. (1989). Novel cytotoxic peptides from the tropical marine cyanobacterium *Hormothamnion enteromorphoides* 1. Discovery, isolation and initial chemical and biological characterization of the hormothamnins from wild and cultured material. *Experientia*. 45(2), 115-121.

Goiris, K., Muylaert, K., Voorspoels, S., Noten, B., De Paepe, D., E Baart, G. J., & De Cooman, L. (2014). Detection of flavonoids in microalgae from different evolutionary lineages. *J. Phycol.* 50(3), 483-492.

Görs, M., Schumann, R., Gustavs, L., & Karsten, U. (2010). The potential of ergosterol as chemotaxonomic marker to differentiate between "chlorella" species (chlorophyta) 1. *J. Phycol.* 46(6), 1296-1300.

Gouveia, L., Veloso, V., Reis, A., Fernandes, H., Novais, J., & Empis, J. (1996). Evolution of pigment composition in *Chlorella vulgaris*. *Biores. Technol.* 57(2), 157-163.

Gouveia, L., Coutinho, C., Mendonça, E., Batista, A. P., Sousa, I., Bandarra, N. M., & Raymundo, A. (2008). Functional biscuits with PUFA-ω3 from *Isochrysis galbana*. *J. The. Sci. Food. Agri.* 88(5), 891-896.

Grama, B. S., Chader, S., Khelifi, D., Agathos, S. N., & Jeffryes, C. (2014). Induction of canthaxanthin production in a *Dactylococcus* microalga isolated from the Algerian Sahara. *Biores. Technol.* 151, 297-305.

Granado-Lorencio, F., Herrero-Barbudo, C., Acién-Fernández, G., Molina-Grima, E., Fernández-Sevilla, J. M., Pérez-Sacristán, B., & Blanco-Navarro, I. (2009). In vitro bioaccesibility of lutein and zeaxanthin from the microalgae *Scenedesmus almeriensis*. *Food. Chem.* 114(2), 747-752.

Grant, C. S., & Louda, J. W. (2013). Scytoneminimine, a mahogany-colored UV/Vis sunscreen of cyanobacteria exposed to intense solar radiation. *Org. Geochem.* 65, 29-36.

Grewe, J. C. (2005). Cyanopeptoline Und Scytocyclamide: Zyklische Peptide Aus *Scytonema hofmanni* PCC7110; Struktur Und Biologische Aktivität."

Gunasekera, S. P., Ross, C., Paul, V. J., Matthew, S., & Luesch, H. (2008). Dragonamides C and D, linear lipopeptides from the marine cyanobacterium brown *Lyngbya polychroa*. *J. Nat. Prod.* 71(5), 887-890.

Gustafson, K. R., Sowder II, R. C., Henderson, L. E., Cardellina II, J. H., McMahon, J. B., Rajamani, U., & Boyd, M. R. (1997). Isolation, primary sequence determination, and disulfide bond structure of cyanovirin, an anti-hiv (human immunodeficiency virus) protein from the cyanobacterium *Nostoc ellipsosporum*. *Biochem. Biophy. Res. Comm.* 238(1), 223-228.

Gutiérrez, M., Suyama, T. L., Engene, N., Wingerd, J. S., Matainaho, T., & Gerwick, W. H. (2008). Apratoxin D, a potent cytotoxic cyclodepsipeptide from Papua New Guinea collections of the marine cyanobacteria *Lyngbya majuscula* and *Lyngbya sordida*. *J. Nat. Prod.* 71(6), 1099-1103.

Han, B., Goeger, D., Maier, C. S., & Gerwick, W. H. (2005). The Wewakpeptins, Cyclic Depsipeptides from a Papua New Guinea Collection of the Marine Cyanobacterium *Lyngbya semiplena*. *The J. Org. Chem.* 70(8), 3133-3139.

Han, B., Gross, H., Goeger, D. E., Mooberry, S. L., & Gerwick, W. H. (2006). Aurilides B and C, Cancer Cell Toxins from a Papua New Guinea Collection of the Marine Cyanobacterium *Lyngbya majuscula*. *J. Nat. Prod.* 69(4), 572-575.

Hansen, J., Schade, D., Harris, C., Merkel, K., Adamkin, D., Hall, R., & Twist, P. (1997). Docosahexaenoic acid plus arachidonic acid enhance preterm infant growth. *Prostaglandins Leukot Essent Fatty Acids*. 57(196), 303-315.

Harari, A., Harats, D., Marko, D., Cohen, H., Barshack, I., Kamari, Y., & Shaish, A. (2008). A 9-cis β-carotene–enriched diet inhibits atherogenesis and fatty liver formation in LDL receptor knockout mice. *The J. Nutr.* 138(10), 1923-1930.

Harrigan, G. G., Luesch, H., Yoshida, W. Y., Moore, R. E., Nagle, D. G., Paul, V. J., & Valeriote, F. A. (1998). Symplostatin 1: A Dolastatin 10 Analogue from the Marine Cyanobacterium *Symploca hydnoides*. *J. Nat. Prod.* 61(9), 1075-1077.

Hayashi, K., Hayashi, T., & Kojima, I. (1996). A natural sulfated polysaccharide, calcium spirulan, isolated from *Spirulina platensis*: in vitro and ex vivo evaluation of anti-herpes simplex virus and anti-human immunodeficiency virus activities. *AIDS. Res. Hum. Retrovir.* 12(15), 1463-1471.

He, J., Yang, Y., Xu, H., Zhang, X., & Li, X. M. (2005). Olanzapine attenuates the okadaic acid-induced spatial memory impairment and hippocampal cell death in rats. *Neuropsychopharmacol.* 30(8), 1511.

Heinrich, U., Gärtner, C., Wiebusch, M., Eichler, O., Sies, H., Tronnier, H., & Stahl, W. (2003). Supplementation with β-carotene or a similar amount of mixed carotenoids protects humans from UV-induced erythema. *The J. Nutr.* 133(1), 98-101.

Hemscheidt, T., Puglisi, M. P., Larsen, L. K., Patterson, G. M., Moore, R. E., Rios, J. L., & Clardy, J. (1994). Structure and biosynthesis of borophycin, a new boeseken complex of boric acid from a marine strain of the blue-green alga *Nostoc linckia*. *The J. Org. Chem.* 59(12), 3467-3471.

Henley, W. J., Major, K. M., & Hironaka, J. L. (2002). Response to salinity and heat stress in two halotolerant chlorophyte algae1. *J. Phycol.* 38(4), 757-766.

Hertweck, C., Luzhetskyy, A., Rebets, Y., & Bechthold, A. (2007). Type II polyketide synthases: gaining a deeper insight into enzymatic teamwork. *Nat. Product. Rep.* 24(1), 162-190.

Horgen, F. D., Kazmierski, E. B., Westenburg, H. E., Yoshida, W. Y., & Scheuer, P. J. (2002). Malevamide D: isolation and structure determination of an isodolastatin H

Analogue from the Marine Cyanobacterium *Symploca hydnoides*. *J. Nat. Prod.* 65(4), 487-491.

Hughes, C., Franklin, D. J., & Malin, G. (2011). Iodomethane production by two important marine cyanobacteria: *Prochlorococcus marinus* (CCMP 2389) and *Synechococcus* sp.(CCMP 2370). *Mar. Chem.* 125(1-4), 19-25.

Ibañez, E., & Cifuentes, A. (2013). Benefits of using algae as natural sources of functional ingredients. *J. The. Sci. Food. Agri.* 93(4), 703-709.

Ikeuchi, M., Koyama, T., Takahashi, J., & Yazawa, K. A. Z. U. N. A. G. A. (2007). Effects of astaxanthin in obese mice fed a high-fat diet. *Biosci. Biotechnol. Biochem.* 0703070340-0703070340.

Inbaraj, B. S., Chien, J. T., & Chen, B. H. (2006). Improved high performance liquid chromatographic method for determination of carotenoids in the microalga *Chlorella pyrenoidosa*. *J. Chromatograph. A*, 1102(1-2), 193-199.

Ishibashi, M., Moore, R. E., Patterson, G. M., Xu, C., & Clardy, J. (1986). Scytophycins, cytotoxic and antimycotic agents from the cyanophyte *Scytonema pseudohofmanni*. *The J. Org. Chem.* 51(26), 5300-5306.

Ishida, K., Matsuda, H., Murakami, M., & Yamaguchi, K. (1997). Micropeptins 478-A and-B, plasmin inhibitors from the cyanobacterium *Microcystis aeruginosa*. *J. Nat. Prod.* 60(2), 184-187.

Ishida, K., Matsuda, H., Murakami, M., & Yamaguchi, K. (1997). Microginins 299-A and-B, leucine aminopeptidase inhibitors from the cyanobacterium *Microcystis aeruginosa* (NIES-299). *Tetrahedron.* 53(30), 10281-10288.

Jahandideh, A., Johnson, T. J., Esmaeili, N., Johnson, M. D., Richardson, J. W., Muthukumarappan, K., & Gibbons, W. R. (2017). Life cycle analysis of a large-scale limonene production facility utilizing filamentous N2-fixing cyanobacteria. *Algal Res.* 23, 1-11.

Jaime, L., Mendiola, J. A., Herrero, M., Soler-Rivas, C., Santoyo, S., Señorans, F. J., & Ibáñez, E. (2005). Separation and characterization of antioxidants from *Spirulina platensis* microalga combining pressurized liquid extraction, TLC, and HPLC-DAD. *J. Sep. Sci.* 28(16), 2111-2119.

Jaime, L., Rodríguez-Meizoso, I., Cifuentes, A., Santoyo, S., Suarez, S., Ibáñez, E., & Señorans, F. J. (2010). Pressurized liquids as an alternative process to antioxidant carotenoids extraction from *Haematococcus pluvialis* microalgae. *LWT-Food. Sci. Technol.* 43(1), 105-112.

Jaki, B., Orjala, J., & Sticher, O. (1999). A novel extracellular diterpenoid with antibacterial activity from the cyanobacterium *Nostoc commune*. *J. Nat. Prod.* 62(3), 502-503.

Jha, R. K., & Zi-Rong, X. (2004). Biomedical compounds from marine organisms. *Mar. Drugs*. 2(3), 123-146.

Jiang, Y., Chen, F., & Liang, S. Z. (1999). Production potential of docosahexaenoic acid by the heterotrophic marine dinoflagellate *Crypthecodinium cohnii*. *Process. Biochem.* 34(6-7), 633-637.

Jin, E. S., Lee, C. G., & Polle, J. E. (2006). Secondary carotenoid accumulation in *Haematococcus* (Chlorophyceae): biosynthesis, regulation, and biotechnology. *J. Microbiol. Biotechnol.* 16(6), 821-831.

Faulkner, D. J. (2001). Marine natural products. *Nat. Prod. Rep.* 18(1), 1R-49R.

Jüttner, F. (1976). β-Cyclocitral and alkanes in Microcystis (Cyanophyceae). *Zeitschrift für Naturforschung C*, 31(9-10), 491-495.

Kadian, S. S., & Garg, M. (2012). Pharmacological effects of carotenoids: a review. *Int. J. Pharma. Sci. Research*. 3(1), 42.

Kamath, B. S., Vidhyavathi, R., Sarada, R., & Ravishankar, G. A. (2008). Enhancement of carotenoids by mutation and stress induced carotenogenic genes in *Haematococcus pluvialis* mutants. *Biores. Technol.* 99(18), 8667-8673.

Kamath, B. S., Srikanta, B. M., Dharmesh, S. M., Sarada, R., & Ravishankar, G. A. (2008). Ulcer preventive and antioxidative properties of astaxanthin from *Haematococcus pluvialis*. *Euro. J. Pharmacol.* 590(1-3), 387-395.

Karsten, U., Lembcke, S., & Schumann, R. (2007). The effects of ultraviolet radiation on photosynthetic performance, growth and sunscreen compounds in aeroterrestrial biofilm algae isolated from building facades. *Planta*. *225*(4), 991-1000.

Karsten, U., Friedl, T., Schumann, R., Hoyer, K., & Lembcke, S. (2005). Mycosporine-like amino acids and phylogenies in green algae: prasiola and its relatives from the trebouxiophyceae (chlorophyta) 1. *J. Phycol.* 41(3), 557-566.

Kathiresan, S., & Sarada, R. (2009). Towards genetic improvement of commercially important microalga *Haematococcus pluvialis* for biotech applications. *J. Appl. Phycol.* 21(5), 553.

Kavitha, M. D., Kathiresan, S., Bhattacharya, S., & Sarada, R. (2016). Culture media optimization of *Porphyridium purpureum*: production potential of biomass, total lipids, arachidonic and eicosapentaenoic acid. *J. Food. Sci. Technol.* 53(5), 2270-2278.

Kerby, G. R., Mayer, L. S., & Pingleton, S. K. (1987). Nocturnal positive pressure ventilation via nasal mask. *Am. Rev. Res. Dis.* 135(3), 738-740.

Khan, Q. A., Lu, J., & Hecht, S. M. (2009). Calothrixins, a new class of human DNA topoisomerase I poisons. *J. Nat. Prod.* 72(3), 438-442.

Kirby, J., & Keasling, J. D. (2009). Biosynthesis of plant isoprenoids: perspectives for microbial engineering. *Ann. Rev. Plant. Biol.* 60, 335-355.

Knott, M. G., Mkwananzi, H., Arendse, C. E., Hendricks, D. T., Bolton, J. J., & Beukes, D. R. (2005). Plocoralides A–C, polyhalogenated monoterpenes from the marine alga *Plocamium corallorhiza*. *Phytochem.* 66(10), 1108-1112.

Kobayashi, J., Cheng, J., Ohta, T., Nakamura, H., Nozoe, S., Hirata, Y., & Sasaki, T. (1988). Iejimalides A and B, novel 24-membered macrolides with potent antileukemic activity from the Okinawan tunicate *Eudistoma cf. rigida*. *The J. Org. Chem.* 53(26), 6147-6150.

Kobayashi, J. I., Ishibashi, M., Nakamura, H., Ohizumi, Y., Yamasu, T., Hirata, Y., & Nozoe, S. (1989). Cytotoxic macrolides from a cultured marine dinoflagellate of the genus *Amphidinium*. *J. Nat. Prod.* 52(5), 1036-1041.

Kodani, S., Ishida, K., & Murakami, M. (1998). Dehydroradiosumin, a trypsin inhibitor from the cyanobacterium *Anabaena cylindrica*. *J. Nat. Prod.* 61(6), 854-856.

Kohlhase, M., & Pohl, P. (1988). Saturated and unsaturated sterols of nitrogen-fixing blue-green algae (cyanobacteria). *Phytochem.* 27(6), 1735-1740.

Kowluru, R. A., Menon, B., & Gierhart, D. L. (2008). Beneficial effect of zeaxanthin on retinal metabolic abnormalities in diabetic rats. *Invest. Ophthalmol. Vis. Sci.* 49(4), 1645-1651.

Kroes, R., Schaefer, E. J., Squire, R. A., & Williams, G. M. (2003). A review of the safety of DHA45-oil. *Food. Chem. Toxicol.* 41(11), 1433-1446.

Kultschar, B., & Llewellyn, C. (2018). Secondary Metabolites in Cyanobacteria. *Secondary Metabolites: Sources and Applications*, 23.

Kumar, R., Yang, S., Kumar, V., & Cholli, A. L. (2010). *U.S. Patent No. 7,799,948*. Washington, DC: U.S. Patent and Trademark Office.

Kwan, J. C., Rocca, J. R., Abboud, K. A., Paul, V. J., & Luesch, H. (2008). Total structure determination of grassypeptolide, a new marine cyanobacterial cytotoxin. *Org. Lett.* 10(5), 789-792.

Lamson, D. W., & Brignall, M. S. (2000). Antioxidants and cancer, part 3: quercetin. *Alt. Med. Rev. J. Clin. Ther.* 5(3), 196-208.

Larsen, L. K., Richard, E. M, and Gregory M. L. P. (1994). β-Carbolines from the Blue-Green Alga *Dichothrix baueriana. J. Nat. Prod.* 57 (3), 419–421.

Lau, S. C. K, and Pei, Y. Q. (2000). Inhibitory Effect of Phenolic Compounds and Marine Bacteria on Larval Settlement of the Barnacle *Balanus amphitrite amphitrite* Darwin. *Biofoul.* 16 (1), 47–58.

Leon, R., Martín, M., Vigara, J., Vilchez, C., & Vega, J. M. (2003). Microalgae mediated photoproduction of β-carotene in aqueous–organic two phase systems. *Biomol. Engg.* 20(4-6), 177-182.

Leya, T., Rahn, A., Lütz, C., & Remias, D. (2009). Response of arctic snow and permafrost algae to high light and nitrogen stress by changes in pigment composition and applied aspects for biotechnology. *FEMS. Microbiol. Ecol.* 67(3), 432-443.

Li, H. B., Fan, K. W., & Chen, F. (2006). Isolation and purification of canthaxanthin from the microalga *Chlorella zofingiensis* by high-speed counter-current chromatography. *J. Sep. Sci.* 29(5), 699-703.

Lian, F., Hu, K. Q., Russell, R. M., & Wang, X. D. (2006). β-Cryptoxanthin suppresses the growth of immortalized human bronchial epithelial cells and non-small-cell lung cancer cells and up-regulates retinoic acid receptor β expression. *Int. J. Cancer.* 119(9), 2084-2089.

Lin, H. Y., & Lin, H. J. (2019). Polyamines in microalgae: something borrowed, something new. *Mar. Drugs.* 17(1), 1.

Liu, B. H., & Lee, Y. K. (2000). Secondary carotenoids formation by the green alga *Chlorococcum* sp. *J. Appl. Phycol.* 12(3-5), 301-307.

Llopiz, P., Jürgens, U. J., & Rohmer, M. (1996). Prokaryotic triterpenoids: Bacteriohopanetetrol glycuronosides from the thermophilic cyanobacterium *Synechococcus* PCC 6907. *FEMS. Microbiol. Lett.* 140(2-3), 199-202.

Lopes, N. P., Stark, C. B., Hong, H., Gates, P. J., & Staunton, J. (2001). A study of the effect of pH, solvent system, cone potential and the addition of crown ethers on the formation of the monensin protonated parent ion in electrospray mass spectrometry. *Analyst.* 126(10), 1630-1632.

Lopes, N. P., Stark, C. B., Hong, H., Gates, P. J., & Staunton, J. (2002). Fragmentation studies on monensin A and B by accurate-mass electrospray tandem mass spectrometry. *Rap. Comm. Mass. Spec.* 16(5), 414-420.

Lorenz, R. T., & Cysewski, G. R. (2000). Commercial potential for *Haematococcus* microalgae as a natural source of astaxanthin. *T. Biotechnol.* 18(4), 160-167.

Luesch, H., Yoshida, W. Y., Moore, R. E., Paul, V. J., Mooberry, S. L., & Corbett, T. H. (2002). Symplostatin 3, a new dolastatin 10 analogue from the marine cyanobacterium *Symploca* sp. VP452. *J. Nat. Prod.* 65(1), 16-20.

Ma, L., & Led, J. J. (2000). Determination by high field NMR spectroscopy of the longitudinal electron relaxation rate in Cu (II) plastocyanin from *Anabaena variabilis*. *J. Am. Chem. Soc.* 122(32), 7823-7824.

MacMillan, J. B., & Molinski, T. F. (2002). Lobocyclamide B from lyngbya confervoides. Configuration and asymmetric synthesis of β-hydroxy-α-amino acids by (−)-sparteine-mediated aldol addition. *Org. Lett.* 4(11), 1883-1886.

Mallikarjun Gouda, K. G., Kavitha, M. D., & Sarada, R. (2015). Antihyperglycemic, Antioxidant and Antimicrobial Activities of the Butanol Extract from *Spirulina Platensis*. *J. Food. Biochem.* 39(5), 594-602.

Malloy, K. L., Villa, F. A., Engene, N., Matainaho, T., Gerwick, L., & Gerwick, W. H. (2010). Malyngamide 2, an oxidized lipopeptide with nitric oxide inhibiting activity from a Papua New Guinea marine cyanobacterium. *J. Nat. Prod.* 74(1), 95-98.

Manley, S. L., Goodwin, K., & North, W. J. (1992). Laboratory production of bromoform, methylene bromide, and methyl iodide by macroalgae and distribution in nearshore southern California waters. *Limnol. Ocean.* 37(8), 1652-1659.

Marshall, J. G. (2002). Consideration of steroids for endodontic pain. *Endodontic topics*, 3(1), 41-51.

Maru, N., Ohno, O., & Uemura, D. (2010). Lyngbyacyclamides A and B, novel cytotoxic peptides from marine cyanobacteria *Lyngbya* sp. *Tetrahed. Lett. 51*(49), 6384-6387.

Matsuda, H., Okino, T., Murakami, M., & Yamaguchi, K. (1996). Aeruginosins 102-A and B, new thrombin inhibitors from the cyanobacterium *Microcystis viridis* (NIES-102). *Tetrahedron.* 52(46), 14501-14506.

Matthew, S., Ross, C., Rocca, J. R., Paul, V. J., & Luesch, H. (2007). Lyngbyastatin 4, a dolastatin 13 analogue with elastase and chymotrypsin inhibitory activity from the marine cyanobacterium *Lyngbya confervoides*. *J. Nat. Prod.* 70(1), 124-127.

May, P., Wienkoop, S., Kempa, S., Usadel, B., Christian, N., Rupprecht, J., & Walther, D. (2008). Metabolomics-and proteomics-assisted genome annotation and analysis of the draft metabolic network of *Chlamydomonas reinhardtii*. *Genetics*. 179(1), 157-166.

McPhail, K. L., Correa, J., Linington, R. G., González, J., Ortega-Barría, E., Capson, T. L., & Gerwick, W. H. (2007). Antimalarial linear lipopeptides from a Panamanian strain of the marine cyanobacterium *Lyngbya majuscula*. *J. Nat. Prod.* 70(6), 984-988.

Medina, R. A., Goeger, D. E., Hills, P., Mooberry, S. L., Huang, N., Romero, L. I., & McPhail, K. L. (2008). Coibamide A, a potent antiproliferative cyclic depsipeptide from the Panamanian marine cyanobacterium *Leptolyngbya* sp. *J. Am. Chem. Soc.* 130(20), 6324-6325.

Mendes, R. L., Nobre, B. P., Cardoso, M. T., Pereira, A. P., & Palavra, A. F. (2003). Supercritical carbon dioxide extraction of compounds with pharmaceutical importance from microalgae. *Inorg. Chim. Acta.* 356, 328-334.

Montaser, R., Abboud, K. A., Paul, V. J., & Luesch, H. (2010). Pitiprolamide, a proline-rich dolastatin 16 analogue from the marine cyanobacterium *Lyngbya majuscula* from Guam. *J. Nat. Prod.* 74(1), 109-112.

Moore, R. E., Corbett, T. H., Patterson, G. M., & Valeriote, F. A. (1996). The search for new antitumor drugs from blue-green algae. *Curr. Pharm. Des.* 2(3), 317-330.

Moore, R. E., Yang, X. Q. G., & Patterson, G. M. (1987). Fontonamide and anhydrohapaloxindole A, two new alkaloids from the blue-green alga *Hapalosiphon fontinalis*. *The J. Org. Chem.* 52(17), 3773-3777.

Morgese, M., Mhillaj, E., Francavilla, M., Bove, M., Morgano, L., Tucci, P., & Schiavone, S. (2016). *Chlorella sorokiniana* extract improves short-term memory in rats. *Molecules*. 21(10), 1311.

Mohammed, M. K., & Mohd, M. K. (2011). Enhancement in Production of Phenolic Compounds (AntiOxidants) in *Spirulina Plantensis* under Different IAA Regimes.

Mundt, S., Kreitlow, S., & Jansen, R. (2003). Fatty acids with antibacterial activity from the cyanobacterium *Oscillatoria redekei* HUB 051. *J. Appl. Phycol.* 15(2-3), 263-267.

Murakami, M., Kodani, S., Ishida, K., Matsuda, H., & Yamaguchi, K. (1997). Micropeptin 103, a chymotrypsin inhibitor from the cyanobacterium *Microcystis viridis* (NIES-103). *Tetrahedron lett.* 38(17), 3035-3038.

Murakami, M., Shin, H. J., Matsuda, H., Ishida, K., & Yamaguchi, K. (1997). A cyclic peptide, anabaenopeptin B, from the cyanobacterium *Oscillatoria agardhii*. *Phytochem.* 44(3), 449-452.

Murakami, M., Sun, Q., Ishida, K., Matsuda, H., Okino, T., & Yamaguchi, K. (1997). Microviridins, elastase inhibitors from the cyanobacterium *Nostoc minutum* (NIES-26). *Phytochem.* 45(6), 1197-1202.

Mynderse, J. S., Moore, R. E., Kashiwagi, M., & Norton, T. R. (1977). Antileukemia activity in the Osillatoriaceae: isolation of Debromoaplysiatoxin from *Lyngbya*. *Science*. 196(4289), 538-540.

Nagatsu, A., Kajitani, H., & Sakakibara, J. (1995). Muscoride A: A new oxazole peptide alkaloid from freshwater cyanobacterium *Nostoc muscorum*. *Tetrahedron lett.* 36(23), 4097-4100.

Nakai, S., Yamada, S., & Hosomi, M. (2005). Anti-cyanobacterial fatty acids released from *Myriophyllum spicatum*. *Hydrobiologia*. 543(1), 71-78.

Nakamura, H., Kobayashi, J. I., & Hirata, Y. (1982). Separation of mycosporine-like amino acids in marine organisms using reversed-phase high-performance liquid chromatography. *J. Chromatography. A*, 250, 113-118.

Nishino, H., Murakoshi, M., Tokuda, H., & Satomi, Y. (2009). Cancer prevention by carotenoids. *Arch. Biochem. Biophy.* 483(2), 165-168.

Ogino, J., Moore, R. E., Patterson, G. M., & Smith, C. D. (1996). Dendroamides, new cyclic hexapeptides from a blue-green alga. Multidrug-resistance reversing activity of dendroamide A. *J. Nat. Prod.* 59(6), 581-586.

Ohta, S., Shiomi, Y., Kawashima, A., Aozasa, O., Nakao, T., Nagate, T., & Miyata, H. (1995). Antibiotic effect of linolenic acid from *Chlorococcum* strain HS-101 and *Dunaliella primolecta* on methicillin-resistant *Staphylococcus aureus*. *J. Appl. Phycol.* 7(2), 121-127.

Okino, T., Qi, S., Matsuda, H., Murakami, M., & Yamaguchi, K. (1997). Nostopeptins A and B, elastase inhibitors from the cyanobacterium *Nostoc minutum*. *J. Nat. Prod.* 60(2), 158-161.

Palozza, P., Calviello, G., Emilia De Leo, M., Serini, S., & Bartoli, G. M. (2000). Canthaxanthin supplementation alters antioxidant enzymes and iron concentration in liver of Balb/c mice. *The J. Nutr.* 130(5), 1303-1308.

Park, A., Moore, R. E., & Patterson, G. M. (1992). Fischerindole L, a new isonitrile from the terrestrial blue-green alga *Fischerella muscicola*. *Tetrahedron lett.* 33(23), 3257-3260.

Parsaeimehr, A., Sun, Z., Dou, X., & Chen, Y. F. (2015). Simultaneous improvement in production of microalgal biodiesel and high-value alpha-linolenic acid by a single regulator acetylcholine. *Biotechnol. Biofuels.* 8(1), 11.

Patterson, G. M., & Bolis, C. M. (1997). Fungal cell wall polysaccharides elicit an antifungal secondary metabolite (phytoalexin) in the cyanobacterium *Scytonema ocelutum* 2. *J. Phycol.* 33(1), 54-60.

Patterson, G. M., & Carmeli, S. (1992). Biological effects of tolytoxin (6-hydroxy-7-O-methyl-scytophycin b), a potent bioactive metabolite from cyanobacteria. *Arch. Microbiol.* 157(5), 406-410.

Pelah, D., Sintov, A., & Cohen, E. (2004). The effect of salt stress on the production of canthaxanthin and astaxanthin by *Chlorella zofingiensis* grown under limited light intensity. *World. J. Microbiol. Biotechnol.* 20(5), 483-486.

Pettit, G. R., Hogan, F., Xu, J. P., Tan, R., Nogawa, T., Cichacz, Z., & Herald, C. L. (2008). Antineoplastic agents. 536. New sources of naturally occurring cancer cell growth inhibitors from marine organisms, terrestrial plants, and microorganisms. *J. Nat. Prod.* 71(3), 438-444.

Piironen, V., Lindsay, D. G., Miettinen, T. A., Toivo, J., & Lampi, A. M. (2000). Plant sterols: biosynthesis, biological function and their importance to human nutrition. *J. Sci. Food. Agri.* 80(7), 939-966.

Piorreck, M., Baasch, K. H., & Pohl, P. (1984). Biomass production, total protein, chlorophylls, lipids and fatty acids of freshwater green and blue-green algae under different nitrogen regimes. *Phytochem.* 23(2), 207-216.

Poerschmann, J., Spijkerman, E., & Langer, U. (2004). Fatty acid patterns in Chlamydomonas sp. as a marker for nutritional regimes and temperature under extremely acidic conditions. *Microbial. Ecol.* 48(1), 78-89.

Prinsep, M. R., Thomson, R. A., West, M. L., & Wylie, B. L. (1996). Tolypodiol, an antiinflammatory diterpenoid from the cyanobacterium *Tolypothrix nodosa. J. Nat. Prod. 59*(8), 786-788.

Procházková, T., Sychrová, E., Javůrková, B., Večerková, J., Kohoutek, J., Lepšová-Skácelová, O., & Hilscherová, K. (2017). Phytoestrogens and sterols in waters with cyanobacterial blooms-Analytical methods and estrogenic potencies. *Chemosphere.* 170, 104-112.

Raja, R., Anbazhagan, C., Lakshmi, D., & Rengasamy, R. (2004). Nutritional studies on *Dunaliella salina* (Volvocales, Chlorophyta) under laboratory conditions. *Seaweed Res Utili.* 26, 127-146.

Rastogi, R. P., Sinha, R. P., & Incharoensakdi, A. (2013). Partial characterization, UV-induction and photoprotective function of sunscreen pigment, scytonemin from *Rivularia* sp. HKAR-4. *Chemosphere.* 93(9), 1874-1878.

Rastogi, R. P., Sonani, R. R., & Madamwar, D. (2014). The high-energy radiation protectant extracellular sheath pigment scytonemin and its reduced counterpart in the cyanobacterium *Scytonema* sp. R77DM. *Biores. Technol.* 171, 396-400.

Rastogi, R. P., Sonani, R. R., & Madamwar, D. (2015). Cyanobacterial sunscreen scytonemin: role in photoprotection and biomedical research. *Appl. Biochem. Biotechnol.* 176(6), 1551-1563.

Ratledge, C. (2004). Fatty acid biosynthesis in microorganisms being used for single cell oil production. *Biochimie.* 86(11), 807-815.

Raveh, A., & Carmeli, S. (2007). Antimicrobial ambiguines from the cyanobacterium *Fischerella* sp. collected in Israel. *J. Nat. Prod.* 70(2), 196-201.

Reitz, R. C., & Hamilton, J. G. (1968). The isolation and identification of two sterols from two species of blue-green algae. *Comp. Biochem. Physiol.* 25(2), 401-416.

Reshef, V., Mizrachi, E., Maretzki, T., Silberstein, C., Loya, S., Hizi, A., & Carmeli, S. (1997). New acylated sulfoglycolipids and digalactolipids and related known glycolipids from cyanobacteria with a potential to inhibit the reverse transcriptase of HIV-1. *J. Nat. Prod.* 60(12), 1251-1260.

Rodrıguez-Concepción, M., & Boronat, A. (2002). Elucidation of the methylerythritol phosphate pathway for isoprenoid biosynthesis in bacteria and plastids. A metabolic milestone achieved through genomics. *Plant. physiol.* 130(3), 1079-1089.

Rothrock Jr, M. J., & Garcia-Pichel, F. (2005). Microbial diversity of benthic mats along a tidal desiccation gradient. *Environ. Microbiol.* 7(4), 593-601.

Ruela-de-Sousa, R. R., Fuhler, G. M., Blom, N., Ferreira, C. V., Aoyama, H., & Peppelenbosch, M. P. (2010). Cytotoxicity of apigenin on leukemia cell lines: implications for prevention and therapy. *Cell death & disease*. 1(1), e19.

Ryckebosch, E., Bruneel, C., Termote-Verhalle, R., Muylaert, K., & Foubert, I. (2014). Influence of extraction solvent system on extractability of lipid components from different microalgae species. *Algal. Res.* 3, 36-43.

Sahu, A., Pancha, I., Jain, D., Paliwal, C., Ghosh, T., Patidar, S., & Mishra, S. (2013). Fatty acids as biomarkers of microalgae. *Phytochem. 89*, 53-58.

Sajilata, M. G., Singhal, R. S., & Kamat, M. Y. (2008). The carotenoid pigment zeaxanthin a review. *Comprehensive reviews in food science and food safety*, 7(1), 29-49.

Sano, T., & Kaya, K. (1996). Oscillapeptin G, a tyrosinase inhibitor from toxic *Oscillatoria agardhii. J. Nat. Prod.* 59(1), 90-92.

Santoyo, S., Rodríguez-Meizoso, I., Cifuentes, A., Jaime, L., Reina, G. G. B., Señorans, F. J., & Ibáñez, E. (2009). Green processes based on the extraction with pressurized fluids to obtain potent antimicrobials from *Haematococcus pluvialis* microalgae. *LWT-Food. Sci. Technol.* 42(7), 1213-1218.

Sarada, R., Bhattacharya, S., Bhattacharya, S., & Ravishankar, G. A. (2002). A response surface approach for the production of natural pigment astaxanthin from green alga, *Haematococcus pluvialis*: effect of sodium acetate, culture age, and sodium chloride. *Food. Biotechnol.* 16(2), 107-120.

Sarada, R., Vidhyavathi, R., Usha, D., & Ravishankar, G. A. (2006). An efficient method for extraction of astaxanthin from green alga *Haematococcus pluvialis. J. Agri. Food chem.* 54(20), 7585-7588.

Sathasivam, R., & Juntawong, N. (2013). Modified medium for enhanced growth of *Dunaliella* strains. *Int J Curr Sci.* 5, 67-73.

Sathasivam, R., Kermanee, P., Roytrakul, S., & Juntawong, N. (2012). Isolation and molecular identification of β-carotene producing strains of *Dunaliella salina* and *Dunaliella bardawil* from salt soil samples by using species-specific primers and internal transcribed spacer (ITS) primers. *Afr. J. Biotechnol.* 11(102), 16677-16687.

Satoh, A., Tsuji, S., Okada, Y., Murakami, N., Urami, M., Nakagawa, K., & Shirasawa, T. (2009). Preliminary clinical evaluation of toxicity and efficacy of a new astaxanthin-rich *Haematococcus pluvialis* extract. *J. Clin. Biochem. Nutr.* 44(3), 280-284.

Scalbert, A., & Williamson, G. (2000). Dietary intake and bioavailability of polyphenols. *The J. Nutr.* 130(8), 2073S-2085S.

Seto, A., Wang, H. L., & Hesseltine, C. W. (1984). Culture conditions affect eicosapentaenoic acid content of *Chlorella minutissima*. *J. Am. Oil. Chem. Soc.* 61(5), 892-894.

Shilpa, K., Varun, K., & Lakshmi, B. S. (2010). An alternate method of natural drug production: Eliciting secondary metabolite production using plant cell culture. *J. Plant. Sci.* 5(3), 222-247.

Shimizu, Y. (2003). Microalgal metabolites. *Curr. Op. Microbiol.* *6*(3), 236-243.

Shin, H. J., Matsuda, H., Murakami, M., & Yamaguchi, K. (1997). Aeruginosins 205A and-B, serine protease inhibitory glycopeptides from the cyanobacterium *Oscillatoria agardhii* (NIES-205). *The J. Org. Chem.* 62(6), 1810-1813.

Siems, W., Wiswedel, I., Salerno, C., Crifò, C., Augustin, W., Schild, L., & Sommerburg, O. (2005). β-Carotene breakdown products may impair mitochondrial functions potential side effects of high-dose β-carotene supplementation. *The J. Nutr. Biochem.* 16(7), 385-397.

Sijil, P. V., Sarada, R., & Chauhan, V. S. (2019). Enhanced accumulation of alpha-linolenic acid rich lipids in indigenous freshwater microalga *Desmodesmus* sp.: The effect of low-temperature on nutrient replete, UV treated and nutrient stressed cultures. *Biores. Technol.* 273, 404-415.

Silvestro, D., Andersen, T. G., Schaller, H., & Jensen, P. E. (2013). Plant sterol metabolism. Δ7-sterol-C5-desaturase (STE1/DWARF7), Δ5, 7-sterol-Δ7-reductase (DWARF5) and Δ24-sterol-Δ24-reductase (DIMINUTO/DWARF1) show multiple subcellular localizations in *Arabidopsis thaliana* (Heynh) L. *PloS one*. 8(2), e56429.

Simmons, T. L., Andrianasolo, E., McPhail, K., Flatt, P., & Gerwick, W. H. (2005). Marine natural products as anticancer drugs. *Mol. Can. Ther.* 4(2), 333-342.

Simon, D. P., Anila, N., Gayathri, K., & Sarada, R. (2016). Heterologous expression of β-carotene hydroxylase in *Dunaliella* salina by *Agrobacterium*-mediated genetic transformation. *Algal. Res.* 18, 257-265.

Sioen, I., Matthys, C., Huybrechts, I., Van Camp, J., & De Henauw, S. (2011). Consumption of plant sterols in Belgium: estimated intakes and sources of naturally occurring plant sterols and β-carotene. *British. J. Nutr.* 105(6), 960-966.

Smith, V. J., Desbois, A. P., & Dyrynda, E. A. (2010). Conventional and unconventional antimicrobials from fish, marine invertebrates and micro-algae. *Mar. Drugs.* 8(4), 1213-1262.

Spolaore, P., Joannis-Cassan, C., Duran, E., & Isambert, A. (2006). Commercial applications of microalgae. *J. Biosci. Bioengg.* 101(2), 87-96.

Stark, M., & O'Gara, I. (2012). An Introduction to Photosynthetic Microalgae. *Dis. Sci. Technol.* 1(2), 65-67.

Staunton, J., & Weissman, K. J. (2001). Polyketide biosynthesis: a millennium review. *Nat. Prod. Res.* 18(4), 380-416.

Stewart, J. B., Bornemann, V., JIAN, L., Moore, R. E., Caplan, F. R., Karuso, H., & Patterson, G. M. (1988). Cytotoxic, fungicidal nucleosides from blue green algae belonging to the Scytonemataceae. *The J. Antibiotics*. 41(8), 1048-1056.

Sun, J. (2007). D-Limonene: safety and clinical applications. *Alt. Med. Rev.* 12(3), 259.

Taniguchi, M., Nunnery, J. K., Engene, N., Esquenazi, E., Byrum, T., Dorrestein, P. C., & Gerwick, W. H. (2009). Palmyramide A, a cyclic depsipeptide from a Palmyra Atoll collection of the marine cyanobacterium *Lyngbya majuscula*. *J. Nat. Prod.* 73(3), 393-398.

Tanticharoen, M., Reungjitchachawali, M., Boonag, B., Vonktaveesuk, P., Vonshak, A., & Cohen, Z. (1994). Optimization of γ-linolenic acid (GLA) production in *Spirulina platensis*. *J. Appl. Phycol.* 6(3), 295-300.

Teruya, T., Sasaki, H., Fukazawa, H., & Suenaga, K. (2009). Bisebromoamide, a potent cytotoxic peptide from the marine cyanobacterium *Lyngbya* sp.: isolation, stereostructure, and biological activity. *Org. lett.* 11(21), 5062-5065.

Thomas, N. V., & Kim, S. K. (2011). Potential pharmacological applications of polyphenolic derivatives from marine brown algae. *Environ. Toxicol. Pharmacol.* 32(3), 325-335.

Tripathi, A., Puddick, J., Prinsep, M. R., Rottmann, M., Chan, K. P., Chen, D. Y. K., & Tan, L. T. (2011). Lagunamide C, a cytotoxic cyclodepsipeptide from the marine cyanobacterium *Lyngbya majuscula*. *Phytochem.* 72(18), 2369-2375.

Tso, M. O., & Lam, T. T. (1996). *U.S. Patent No. 5,527,533*. Washington, DC: U.S. Patent and Trademark Office.

Vairappan, C. S., Suzuki, M., Abe, T., & Masuda, M. (2001). Halogenated metabolites with antibacterial activity from the Okinawan Laurencia species. *Phytochem.* 58(3), 517-523.

Vidhyavathi, R., Venkatachalam, L., Kamath, B. S., Sarada, R., & Ravishankar, G. A. (2007). Differential expression of carotenogenic genes and associated changes in pigment profile during regeneration of *Haematococcus pluvialis* cysts. *Appl. Microbiol. Biotechnol.* 75(4), 879-887.

Vidhyavathi, R., Venkatachalam, L., Sarada, R., & Ravishankar, G. A. (2008). Regulation of carotenoid biosynthetic genes expression and carotenoid accumulation in the green alga *Haematococcus pluvialis* under nutrient stress conditions. *J. Exp. Bot.* 59(6), 1409-1418.

Vidyashankar, S., Sireesha, E., Chauhan, V. S., & Sarada, R. (2015). Evaluation of microalgae as vegetarian source of dietary polyunsaturated fatty acids under autotrophic growth conditions. *J. Food. Sci. Technol.* 52(11), 7070-7080.

Volkman, J. (2003). Sterols in microorganisms. *Appl. Microbiol. Biotechnol.* 60(5), 495-506.

Wada, N., Sakamoto, T., & Matsugo, S. (2013). Multiple roles of photosynthetic and sunscreen pigments in cyanobacteria focusing on the oxidative stress. *Metabolites.* 3(2), 463-483.

Ward, O. P., & Singh, A. (2005). Omega-3/6 fatty acids: alternative sources of production. *Pro. Biochem.* 40(12), 3627-3652.

Wen, Z. Y., & Chen, F. (2003). Heterotrophic production of eicosapentaenoic acid by microalgae. *Biotechnol. Adv.* 21(4), 273-294.

Wrasidlo, W., Mielgo, A., Torres, V. A., Barbero, S., Stoletov, K., Suyama, T. L., & Stupack, D. G. (2008). The marine lipopeptide somocystinamide A triggers apoptosis via caspase 8. *Proc. Natl. Acad. Sci.* 105(7), 2313-2318.

Wright, A. D., Goclik, E., & König, G. M. (2003). Three new sesquiterpenes from the red alga *Laurencia perforata*. *J. Nat. Prod.* 66(3), 435-437.

Wright, A. D., König, G. M., de Nys, R., & Sticher, O. (1993). Seven new metabolites from the marine red alga *Laurencia majuscula*. *J. Nat. Prod.* 56(3), 394-401.

Wynn, J., Behrens, P., Sundararajan, A., Hansen, J., & Apt, K. (2010). Production of single cell oils by dinoflagellates. In *Single cell oils* (pp. 115-129). AOCS Press.

Xiong, F., Kopecky, J., & Nedbal, L. (1999). The occurrence of UV-B absorbing mycosporine-like amino acids in freshwater and terrestrial microalgae (Chlorophyta). *Aqua. Bot.* 63(1), 37-49.

Yamaguchi, M. (2012). Role of carotenoid β-cryptoxanthin in bone homeostasis. *J. Biomed. Sci.* 19(1), 36.

Yang, J., Guo, J., & Yuan, J. (2008). In vitro antioxidant properties of rutin. *LWT-Food. Sci. Technol.* 41(6), 1060-1066.

Yongmanitchai, W. (1989). Omega-3 fatty acids: alternative sources of production. *Process Biochem.* 24, 117-125.

Yuan, C., Jin, Z., Xu, X., Zhuang, H., & Shen, W. (2008). Preparation and stability of the inclusion complex of astaxanthin with hydroxypropyl-β-cyclodextrin. *Food. Chem.* 109(2), 264-268.

Yuan, J. P., Chen, F., Liu, X., & Li, X. Z. (2002). Carotenoid composition in the green microalga *Chlorococcum*. *Food. Chem.* 76(3), 319-325.

Yuan, J. P., Peng, J., Yin, K., & Wang, J. H. (2011). Potential health-promoting effects of astaxanthin: A high-value carotenoid mostly from microalgae. *Mol. Nutr. Food. Res.* 55(1), 150-165.

Zeng, X., Yin, B., Hu, Z., Liao, C., Liu, J., Li, S., & Jiang, S. (2010). Total synthesis and biological evaluation of largazole and derivatives with promising selectivity for cancers cells. *Org. Lett.*, 12(6), 1368-1371.

In: Applied Algal Biotechnology
Editors: M. Arumugam, S. Kathiresan et al.
ISBN: 978-1-53617-524-0

Chapter 6

LIPIDS BIOSYNTHESIS IN MICROALGAE

Anila Narayanan[1,*], K. A. Wafha[2] and Shibin Mohanan[2]
[1]Department of Botany, St. Xavier's College, Aluva, Kerala
[2]Department of Botany, Nirmala College, Muvattupuzha, Kerala

ABSTRACT

Microalgae are photosynthetic microorganisms which are extremely diverse in its morphology, physiology and habitat. The microalgae are valued for its potential to produce wide varieties of value added products. There has been an intense interest in microalgal lipids due to its applications on food, chemical, pharmaceutical and cosmetic applications.PUFAs is one of the most important lipid with immense pharmaceutical potential. With the increasing interest in the field of lipid science especially those involved in the production and PUFAs and other valuable products, this field requires a comprehensive study which provides information on the various basic aspects of lipid biosynthesis that can evidently promote lipid induction and should also be industrially feasible. The present chapter intends to provide a comprehensive review of the current knowledge of lipid biosynthesis and it's regulation in microalgae, pathways involved in lipid biosynthesis will also be discussed along with methods for augmentation of lipid production. Effect of nutrient stress on the growth, biomass and lipid production and accumulation is being discussed. Genetic engineering approaches used for enhancing lipid production either by forward or reverse genetics, random or site directed is being discussed. Use of latest gene editing tools in strain development and pathway modification for improving lipid synthesis and accumulation is discussed. It is reckoned that high potential of microalgal lipid will force research to find effective and efficient ways to manipulate bio synthetic pathway towards cost effective algal cultivation and production. Besides lipids other valuable compounds produced by algae will increase the financial viability of micro algae cultivation.

Keywords: microalgae, lipids, biosynthetic pathways, strain development

[*] Corresponding Author's Email: anilanarayanan@gmail.com.

Abbreviations

PUFAs	Polyunsaturated fatty acids
TAGs	Triglycerides
FFAs	Free fatty acids
ACC	acetyl-CoA carboxylase
ACS	Acetyl CoA Synthase
ME	Malic Enzyme
PFK2	Phosphofructo-2-kinase
GK	Glycerol Kinase
G3PDH	Glycerol-3-phosphate Dehydrogenase
FDX	Ferredoxins

6. Introduction

Lipids are organic molecules with biological origin that are insoluble in water and soluble in organic solvents such as, alcohols and ethers. They are the major components in the cell responsible for energy storage, structural maintenance and participating in signalling pathways (Cooper, 2000; Fahy et al., 2011). Lipids are diverse group of organic compounds available in plants, animal and micro-organisms (Antonio et al., 2000). The lipid group comprises of different compounds, including fatty acids, fat-soluble vitamins, triglycerides, wax, phospholipids, ceramides, cholesterol and other steroids. Due to its higher demand in food and feed supplements, pharmaceutical application, biofuel, etc. the industrial production of lipids is of much significance. Among the several sources which can produce lipids, microalgal cultivation get more advantages. Alternative source such as, bacteria, fungi and plants are currently being explored for commercial production. Fungi need an organic carbon source for its production and typically have long growth periods which decreases its preference as a source organism (Barclay et al., 1994). Where, plants require arable land, have longer growth times compared to algae and have no enzymatic activity for producing long chain PUFAs (Polyunsaturated fatty acid), unless genetically modified (Ursin, 2003). Microalgal lipid producers can naturally grow fast under a variety of autotrophic, mixotrophic and heterotrophic conditions with high fatty acid production potential (Liang et al., 2009). These products are being used as food additives, animal feed, vitamins, pigments, pharmaceutical compounds, cosmetics and potentially as a biofuel source (Schenk et al., 2008, Van et al., 2009, Spolaore et al., 2006).

The microalgae received considerable interest in lipid production primarily because of its rapid growth potential, area requirement and low cost substrate (Stephens et al., 2010). Furthermore, they are capable of all year round production using the non-agricultural land. It can also reduce the use of freshwater sources (Hill et al., 2006). Considering the yield of microalgae which could reach 55tons ha^{-1} $year^{-1}$ and along with the effective lipid productivity under stress conditions and its great flexibility and adaptability towards diverse environments (Zhu et al., 2007; Faried et al., 2017; Tandon and Liu, 2017; Hasnain et al., 2018; Lee et al., 2015; Choo et al., 2017; Tan et al., 2018; Zullaikah et al., 2019) encourage industrial microalgal-lipid production (Zhu et al., 2007; Chen et al., 2012; Zhou et al., 2017).

Additionally, following advantages of microalgae make it more suitable for lipid production.

Algae are the fast growers when compared to other sources and they can double their numbers in every few hours. Daily harvesting is another beneficial property and its yield is many times greater than other lipid resources. Algae mainly store their lipids in the form of energy such as, oils and carbohydrates leads about 1000 to as many as 5,000 gallons of biofuels per acre per year (Chisti et al., 2007, Haag et al., 2007, Jhones and Mayfield, 2012). Algae Consume CO_2 like plants when grown under sunlight. They fix carbon dioxide (CO_2) and release oxygen (O_2). Their yield highly depends on greater availability of CO_2. Therefore, its cultivation reduces the CO_2 emitted from power plants, ethanol facilities, and other sources to a particular extent. Algal cultivation, do not compete with agriculture. In many cases, unsuitable traditional agriculture sources are used. These include water sources, which cannot be used by other crops, such as sea-, brackish- and wastewater (Susmita et al., 2017).

Microalgal classification can be grouped into Chlorophyceae (green algae), Pyrrophyceae (dinoflagellates), Rhodophyceae (red algae), Phaeophyceae (brown algae), Pyrrophyceae (dinoflagellates) and Chrysophyceae (yellow-green algae) (Susmita et al., 2017). The most important microalgae are green algae (*Chlamydomonas*) (Lili et al., 2018), colonial (*Volvox*) (Moseley et al., 1980) or filamentous (*Spirogyra*) (Zhang et al., 2016), red algae (*Polysiphonia*) (Saravanakumar et al., 2008), brown algae (*Macrocystis*, *Sargassum*). The algal lipids are different from other sources because of the following properties listed below:

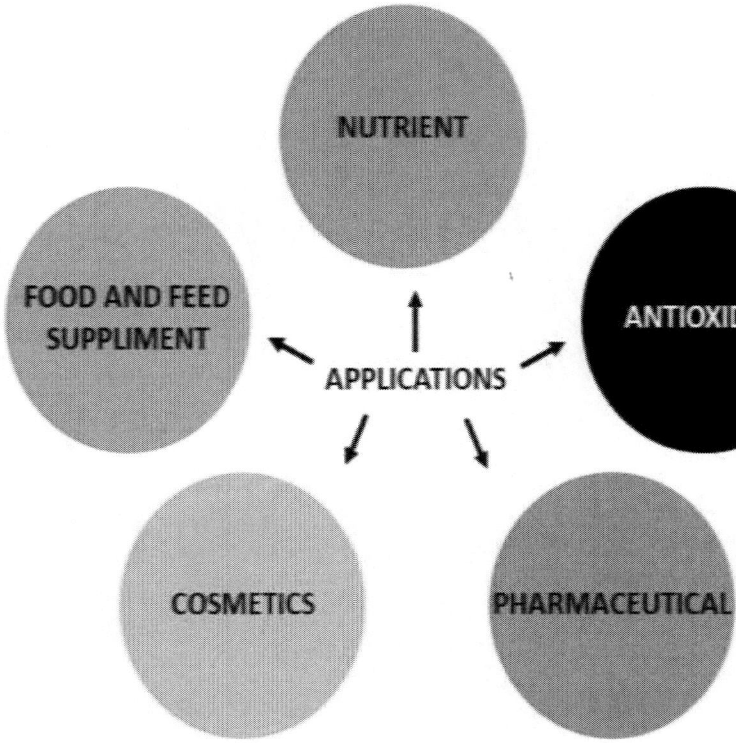

Figure 1. Application of microalgal lipids.

6.1. Microalgal Lipids

The significance of microalgae applications initiated in the 20th century. Now there are several species which are currently cultivated in large scale and producing tons of various metabolites per year. There has been an increased attention in the use of microalgal lipids in numerous commercial applications, such as in food, chemical and pharmaceutical industries and cosmetology. Revealing high interest in microalgal lipids is the noteworthy research that has performed in the last decade on all aspects concerning microalgal lipid production. These include fundamental research on the mechanisms to enhance the productivity.

Microalgae cells are principally composed of four major components such as, proteins, carbohydrates, nucleic acids and lipids. Among these, lipids are the energy-rich compound in the cell and important in signalling process. It acts as building blocks for the structural and functional maintenance of living cells (Figure 1). They are not appreciably soluble in water that makes them different from all of the other molecules. It is generally categorised in two major groups such as, polar-structural lipids and nonpolar-neutral lipids. Each of these lipids has specific roles in microalgae metabolism and growth period. Some lipids are imperative structural components of biological membranes, while others may act as key intermediates in the cell signalling pathways (Aitken et al., 2014). The amount of distribution of these microalgal lipids vary with the type of species, growth conditions, metabolic rate and ambient environments.

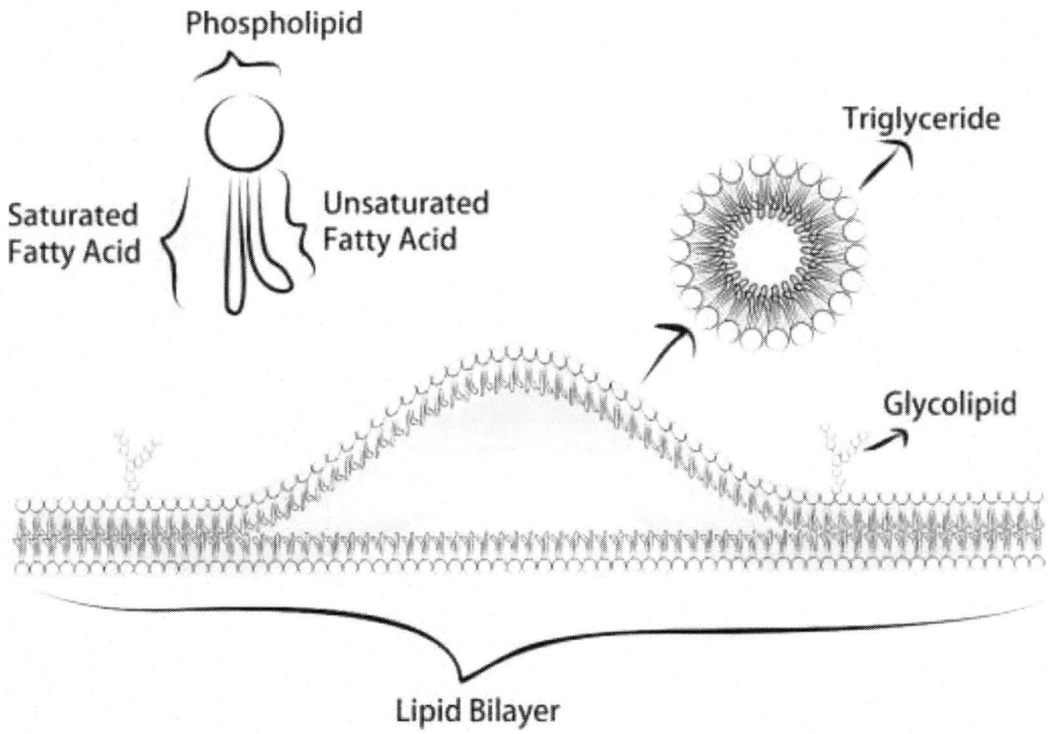

Figure 2. Type of lipids in Microalgae.

6.1.1. Polar-Structural Lipids

The polar-structural lipids is mainly comprised of glycolipids and phospholipids. They are the fundamental compounds bound to the organelle membranes (Hu et al., 2008, Sukenik et al., 1988) to protect the cell.They are simply called as lipid bilayer and consists of polar hydrophilic head and two hydrophobic fatty acid tails (Olofsson et al., 2012). Moreover, some polar lipids may also behave as important intermediates in cell signalling pathways (e.g., sphingolipids, inositol lipids, oxidative products) and contribute in reacting to changes in the environmental parameters. Sterols also belongs to polar lipids.

6.1.2. Glycolipids

Glycolipid is a polar unsaturated lipid with carbohydrate attached by glycosidic bond. Its main role is to maintain the stability of cell membrane. Compare with other lipids, glycolipids are quite abundant in microalgae and are considered an important source of fatty acids (FAs). They are located in the membrane of chloroplasts and thylakoids, and are important signal and regulatory molecules (Harwood et al., 2009; Siegenthaler and Murata, 1998; Hölz *et*, al 2007). Glycolipids in microalgae exist as glyceroglycolipids, it poses a glycerol backbone that anchors on acyl chains (Petroutsos et al., 2014).Mostly glycolipids in microalgae are found in form of Monogalactosyl Diacylglycerols (MGD), Digalactosyl Diacylglycerols (DGD) and Sulfoquinovosyl Diacylglycerols (SQGD), depending on the composition of the glycosidic head (Harwood et al., 2009; He et al., 2011; Kim et al., 2013) (Table 1). They are the major components of the plastid lipids (Wang and Benning, 2012). The roles and functions of glycolipids depend on their structure and composition.They are mainly synthesized in the chloroplast, within the envelope membranes of plastids and likewise in endoplasmic reticulum (Petroutsos et al., 2014).

Table 1. Types, properties and functions of gylycolipids such as, Monogalactosyl Diacylglycerols(MGD), Digalactosyl Diacylglycerols (DGD) and Sulfoquinovosyl Diacylglycerols (SQGD)

	Digalactosyl Diacylglycerols	Monogalactosyl Diacylglycerols	Sulfoquinovosyl Diacylglycerols
Structure	Terminal α-galactose moiety (1→6) linked to the inner β-galactose residue	16- and 18-carbon *n*-3 fatty acids, with more than 20 carbon atoms and more than three double bonds (Petroutsos et al., 2014).	Composed of a monoglycosyldiacylglycerol with a sulfonic acid linked in the position 6 of the monosaccharide moiety (Gustafson et al., 1989).
Shape	Cylindrical shape	Unsaturated fatty acid- conical shape Saturated fatty acids- cylindrical form (Lee, 2000).	Cylindrical shape
Function	Substituting for phosphoglycerolipids to various extraplastidial membranes (Yang et al., 2015).	Structural stabilization and function of membranes and trafficking (Boudière et al., 2013),	"Membrane mosaic" (Babu and Wu, 2003).
Distribution lipid bilayer	15%–35%	(40%–55%)	2%–40% (Siegenthaler and Murata, 1998).

6.1.3. Phospholipids

Phospholipids, also called phosphatides. They are the key polar components that makes the cell membranes, which act as a selective permeable barrier for organelles and cells and participate directly in membrane fusion events. Furthermore, they are also known to play a vital role in algal cellular signaling and cell-cell interactions (Lu et al., 2012).The quantity and composition of phospholipids varies from species to species, in spite of their developmental and environmental changes (Eastmond et al., 2010). Phospholipids are similar to triglycerides and are usually found with backbone of phosphatidic acid (PA), formed from L-glycerol 3-phosphate with two fatty acids esterified. Its structure and composition strongly depends on physicochemical and dynamical properties of membranes (Yang et al., 2012). The algae contains three major phospholipids, phosphatidylglycerol (PG), phosphatidylethanolamine (PE), and phosphatidylcholine (PC) (Carballeira et al., 2002). The precursors of phospholipid biosynthesis are DAG and FFA; they are derived from TAG, as well as membrane phospholipids (Beal et al., 2013). These phospholipids also play a significant role biological processes, such as stress response, photosynthesis, etc. (Arisz and Munnik, 2011).

6.1.4. Polyunsaturated Fatty Acids (PUFAs)

Microalgae are rich source of polyunsaturated fatty acids.PUFAs have more than one double bond in their long carbon chain structure. Eukaryotic microalgae are the primary producers of long chain PUFA (Khozin-Goldberg et al., 2016). PUFA biosynthesis is one of the central metabolic pathways in microalgae which act as building blocks for acyl glycerolipids (Guschina and Harwood, 2006), and play significant role in the generation of lipid signalling molecules (Mosblech et al., 2009). There are two classes of PUFAs ω-6 and ω -3, distinguished by the position of the terminal double bond from the methyl end of the fatty acid. In microalgae, its biosynthesis is compartmentalized and initiated from endoplasmic reticulum. They are found in both chloroplast and extraplastidial compartments, therefore it could assumed that they are shuffled between them. The long-chain PUFAs fatty acid in microalgae include linolenic acid, arachidonic acid, eicosapentaenoic acid (EPA) and docosahexaenoic acid (DHA) (Borowitzka, 2013; Patil and Patil, 2015). EPA and DHA are significant in bio membranes and cellular functions (Valentine and Valentine, 2004).

6.2. Nonpolar-Neutral Lipids

Non-polar lipids are also called as storage lipids i.e., Triglycerides (TAGs), Diglycerides (DAGs), Monoglycerides (MAGs), Free Fatty Acids (FFAs), Hydrocarbons, Prenyl derivatives, and Pyrrol derivatives. Their storage form varies depends on the algal species, growth phase and environmental conditions.

6.2.1. Triglycerides (TAGs)

Storage lipids are mainly in the form of triglycerides (TAGs). Its accumulation in cell is low in exponential growth phase and higher in stationary phase (Brown et al., 1997; Tonon et al., 2002). They poses high content of saturated and some unsaturated FAs.Its major function is to supply metabolic energy.

In the presence of light at times when the energy input, through carbon assimilation, exceeds the immediate metabolic needs of the cell, TAGs synthesized are accumulated in cytosolic lipid bodies, and then reprocessed for polar lipid synthesis in the absence of light (Sharma et al., 2013). Rather than as an energy storage product, PUFA rich TAGs may endue specific acyl groups to Mono GalactosylDiacyl Glycerol (MGDG) and other polar lipids (Sharma et al., 2013). They do not perform any structural role in micro algae and are commonly deposited in cytoplasm of microalgal cells (Damiani et al., 2010). Monoglycerides (MAGs) and Diglycerides (DAGs) also belong to the category in which triglycerides are reacted in methanol in the presence of a catalyst in Trans-esterification process. Chemically it neutralizes the free fatty acids in a triglyceride and also may serve as a base for free fatty acids (FFAs), to avoiding lipotoxicity (Kurat et al. 2006).

Storage lipids are mainly in the form of triglycerides (TAGs) and they are packaged in simple structures termed as lipid droplets also referred to as oil bodies. Oil bodies comprise of a central core of TAG within a monolayer of membrane lipids through budding from the ER. They remain in the cell as oil bodies and unable to integrate into membrane bilayers due to lack of polar head groups. Its synthesis is initiated after a sequence of biochemical pathways and results in swelling of ER membrane which leads to budding of growing lipid droplets into the cytosol. Rather than serving as oil bodies that play vital roles in lipid metabolism, they play a mojor role in homeostasis and trafficking (Walther and Farese 2009; Wilfling et al. 2014).

6.2.2. Free Fatty Acids (FFAs)

Fatty acid composition varies with algae depends on the cell requirement. Rather than storage function they also plays key role in structural maintenance of the cell.The Fatty acid predominantly stored in the form of monounsaturated fatty acids (MUFA) and saturated fatty acids (SAFA). Polyunsaturated fatty acids (PUFA) comprise the structural lipid fraction and belongs to the category of polar lipids (Hu et al., 2008; Olofsson et al., 2012). Sometimes the concentrations of saturated fatty acids (SFA) is higher in cells because ofthe presence of palmitic acid (16:0). Levels of monounsaturated fatty acid (MUFA) also found to be higher in many algae. Both of them comprise the lipid fraction of storage for microalgae (Olofsson et al., 2012). The fatty acid composition in microalgae is usually a mixture of unsaturated fatty acids, such as palmitoleic (16:1), oleic (18:1), linoleic (18:2), and linolenic acid (18:3). Saturated fatty acids, palmitic (16:0) and stearic (18:0) are also present to some extent (Jiang et al., 2010).Commercially saturated fatty acids (SAFA) 14:0, 16:0, and 18:0 are of limited taxonomic value as biochemical markers. This is because of the availability at various concentrations in all algal groups. However, monounsaturated fatty acids (MUFA), 16:1, 18:1 are potential biomarkers (palmitoleic acid).

During the cultivation of microalgae at higher temperature, MUFAs in various microalgae decreased, while SAFAs increased due to the fluidity of cell membranes (Renaud et al., 2002). Fatty acid can also act as a precursor for the synthesis of nonpolar hydrocarbons using light energy (Wijffels et al., 2013), which varies from species to species.

6.3. Lipid Accumulation

The importance for microalgae lipid begins principally from the fact that these organisms are able to synthesize considerable quantities of lipids with a wide range of application. There are widespread analytical methodologies associated with the reported values, ensuring its potential for greater variability. From species to species, environmental and cultural conditions such as temperature, irradiance and, mostly, nutrient availability greatly influence the lipid content and composition in algal cells. The lipid accumulation varies in different algal classes. Several micro algal species are capable of accumulating considerable quantities of lipid. The total lipid content in microalgae constitute up to 80% in dry biomass, but even in normal conditions the lipid productivity is actually low. Cyanobacteria usually have the least amount of cellular lipid content due to their high protein demand for the phycobilisomes and in contrast, heterokonts and haptophytes exhibit the highest lipid content (Finkel et al., 2016). The general distribution of most important polar and nonpolar lipids in microalgae showed in the Table 1.

The accumulation of lipids within a cell differs with types. The polar lipids are mainly accountable for structural assembly, where non-polar lipids act as storage. Different types of lipids were discussed in the previous section. The major function of Polar glycol lipids is to maintain the stability of the cell. They are also important signal and regulatory molecules. Phospholipids are key polar components of the cell membrane which directly participate in membrane fusion. As a result, glycolipids and phospholipids can be found in all the organelle membranes. However, PUFAs are only found in chloroplast and extra-plastidial compartments. The non-polar structural lipid especially TAGs are mainly found in the form of free oil droplets in the cytoplasm of microalgae for performing various functions such as metabolism, homeostasis and trafficking.

Table 2. The general distribution of polar and nonpolar lipids in microalgae

Lipids		Concentration (%)
Glycolipid	Total	0.07-11.9
	MGD	0.02-4.28
	DGD	0.02-9.09
	SQGD	0-1.19
Phospholipid	PG	0.11-2.89
	PE	0-0.34
	PC	0.01-5.10
Fatty acid	TFA	99
	PUFA	01-40.7
	MUFA	0.5-48.8
	SFA	1.7-25.9
TAG		80-90

MGD – Monogalactosyl Diacylglycerols, DGD – Digalactosyl Diacylglycerols, SQGD – Sulfoquinovosyl Diacylglycerols, PG - phosphatidylglycerol, PE - phosphatidylethanolamine, PC -phosphatidylcholine (PC), TFA – Total fatty acids, PUFA - Polyunsaturated fatty acids, MUFA - monounsaturated fatty acids, SFA - saturated fatty acids, TAG - triglycerides

Table 3. Accumulation of lipids in the cell and their functions

Lipids	Site	Function
	Polar lipids	
Glycolipids	Organelle membrane	Structural maintenance Stability Signalling Membrane mosaic Trafficking
Phospholipids		Structural maintenance Membrane fusion
PUFAs		Chloroplast and extraplastidial membrane
	Non polar lipids	
TAGs	Cytoplasm	Catabolism Homeostasis Trafficking
FFAs		Provide energy Energy transport

The thylakoids of chloroplasts are the major intra-cellular membranes of algae, so their lipid composition depends on its distribution. Predominantly the lipid in chloroplasts is MGD, with smaller amounts of DGDG and the two negatively charged lipids, SQDG and PG. In fact, negligible amounts of other lipids have been found in isolated chloroplasts. It denotes that microalgae lipid composition have a similar percentage lipid composition to those from higher plants.

6.4. Biosynthesis of Lipids

In general, lipid biosynthesis is an energy intensive process. It is very important to understand the intracellular reaction networks and machinery involved in the lipid biosynthesis. It involves a series of pathways with several enzymes have been reported. The most important pathway for the production of lipids in microalgae involves de-novo synthesis of fatty acids in the chloroplasts. Mostly microalgae can synthesize polysaccharides, proteins, lipids and hydrocarbons via photosynthesis. In microalgae, lipid biosynthesis starts with photosynthesis and followed by several pathways such as, Calvin cycle, citric acid cycle and glycolysis, finally to synthesise acetyl-CoA (Ac-CoA). It is the building block for fatty acid synthesis which ultimately leads to the synthesis of TAGs.

6.4.1. Photosynthesis (Calvin Cycle)

Generally, Calvin cycle is the initial pathway for lipid synthesis. The regulation of this cycle is important in substantially to convert CO_2 to glycerate-3-phosphate (G3P). This cycle plays a central role in algal metabolism, which takes place in chloroplast. Here a series of 11 enzymatic reactions occur for carbon fixation. This cycle can divided into three major stages such as, carboxylation, reduction, and regeneration. It is initiated by ribulose 1,5-bisphosphate carboxylase/oxygenase (Rubisco), that catalyzes the carboxylation of the CO_2 acceptor molecule by converting in to ribulose 1,5-bisphosphate (RuBP). Followed by aldolase enzyme responsible for the reversible conversion of dihydroxyacetone phosphate (DHAP) and glyceraldehyde 3-phosphate to fructose 1,6-bisphosphate (FBP) and son on (Nakahara et al., 2003). This glyceraldehyde 3-phosphate is the primary compound for glycolysis.

6.4.2. Glycolysis

This is a central metabolic pathway characterized by a series of reactions that results in the conversion of glyceraldehyde 3-phosphate into pyruvate. Glycerol-3-phosphate are located both cytosol and plastids. So this cycle can found in both cytosol and plastids in microalgae. Calvin cycle is characterized by the production of a variety of intermediates and function as substrates in additional pathways. Additional pathways includes gluconeogenesis, lipid metabolism, the pentose phosphate pathway, and the TCA.

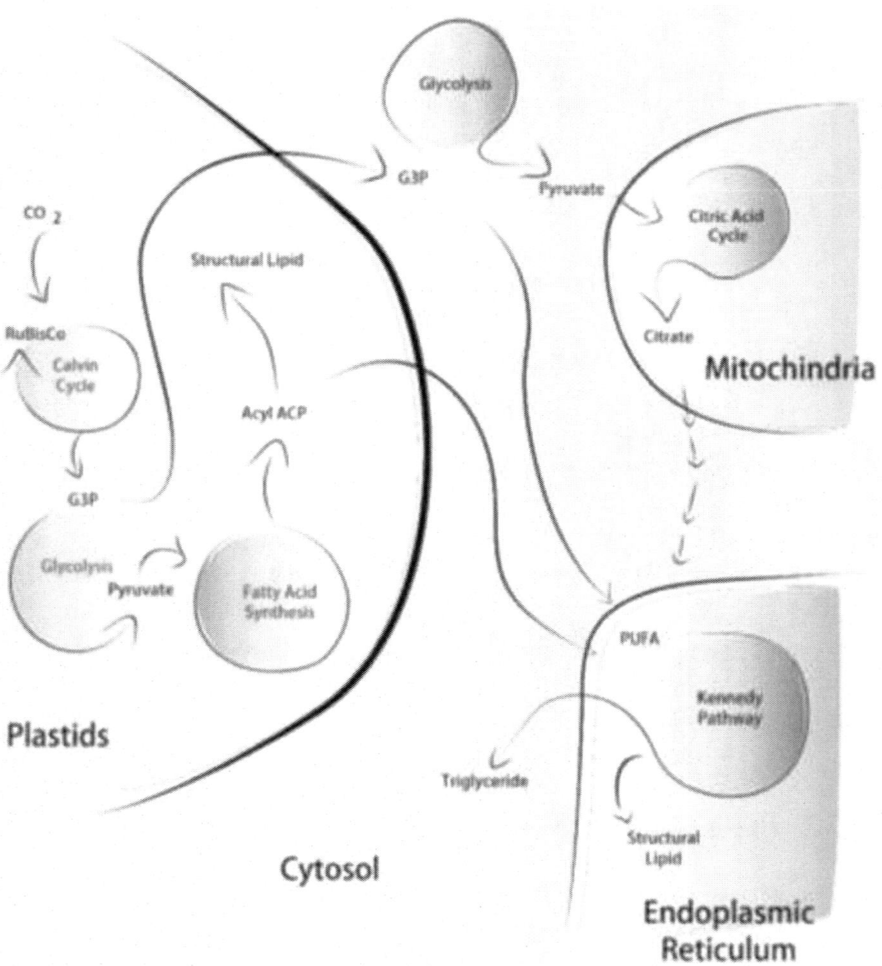

Figure 3. Pathway of biosynthesis of structural lipids and storage lipids.

6.4.3. Citric Acid Cycle

The citric acid cycle (TCA) or Krebs cycle, in microalgae is carried out in mitochondria when glyceraldehyde 3 Phosphate enter to mitochondrial space with the conversion to acetyl Co A. This process is to generate energy via the oxidation of acetate and plays a critical role in the PUFA synthesis. The reverse of TCA cycle is a series of chemical reactions to produce carbon compounds from carbon dioxide and water. It is an alternative to photosynthesis which produces organic molecules and ultimately to generate an acetyl CoA.This process requires a number of reduction reactions with a series of enzymes such as ATP citrate lyase, 2-oxoglutarate: ferredoxin oxido-reductase, and pyruvate: ferredoxin oxidoreductase. ATP citrate lyase is the one of the important enzymes to perform reverse TCA. Once the citrate is synthesised by this cycle it is then moved to cytosolic space.

6.4.4. Fatty Acid Synthesis

Once the acetyl-CoA accumulates in the algal cell, they begin to produce fatty acid with fatty acid synthesis pathway. Acetyl-CoA serves as the precursor in support of fatty acid synthesis in the plastids. Followed by the Biotin-dependent carboxylation, acetyl-CoA give rise to malonyl-CoA (Davis et al. 2000). It is the central carbon donor of fatty acid synthesis in the presence of acetyl-CoA carboxylase (ACCase) (Cronan, 2000). This reaction is catalysed in both plastid and cytosol (Baba and Shiraiwa, 2013; Davis et al., 2000; Khozin-Goldberg and Cohen, 2011; Lei et al., 2012). In plastid, Acetyl-CoA is synthesised from pyruvate by pyruvate decarboxylase and in cytosol, by ATP citrate lyase from citrate. In algae, ACCs exist in two different forms, the heteromeric and homomeric found in plastid or cytosol. Afterwards the malonyl-CoA in plastids is transferred to the acyl-carrier protein (ACP), which is one of the subunits of the fatty acid synthase (FAS) complex, with the help of malonyl-CoA:ACP transacetylase (Blatti et al., 2012; Greenwell et al., 2010; Hu et al., 2008). ACP contains a thiol group which is responsible for its transformation to malonyl-ACP via forming thioesters with malonyl-CoA. Generally FAS is responsible for the further elongation of the fatty acids. It is carried out with the extension of the acyl chain with malonyl-ACP(Harwood, 1996).

The FAS pathway initiated through the ketoacyl-ACP synthase (KAS) catalyzes the condensation of malonyl-ACP or the growing acyl chain to form ketobutyryl-ACP or ketoacetyl-ACP. The resulting compound then undergo three subsequent steps such as, reduction, dehydration and reduction to butyryl-ACP, palmitoyl-ACP, stearoyl-ACP, acetyl-CoA and Oleoyl-ACP respecteivly (Mühlroth et al., 2013;Yu W.L. et al., 2011). At last, fatty acyl-ACP formed in the plastid act as the precursors for the synthesis of both polar and nonpolar lipids.The elongation of fatty acids is predominantly completed in plastids by two methods. First one, is the removal of acyl group by the enzyme glycerol-3-phosphate acyltransferase and second one hydrolyzes the thioester bonds by acyl-ACP thioesterase located in the chloroplast envelopeto release free fatty acids (FFAs) (Ohlrogge Browse, 1995). The acyl chains directly involved in structural lipids (mostly glycolipids) synthesis are processed in the plastid. These FFAs are transported to the outer membrane of the plastids to the cytosol (Koo et al., 2004; Li et al., 2013) to produce acyl-CoA pool in the cytosol (Chapman and Ohlrogge, 2012; Bellou et al., 2014).

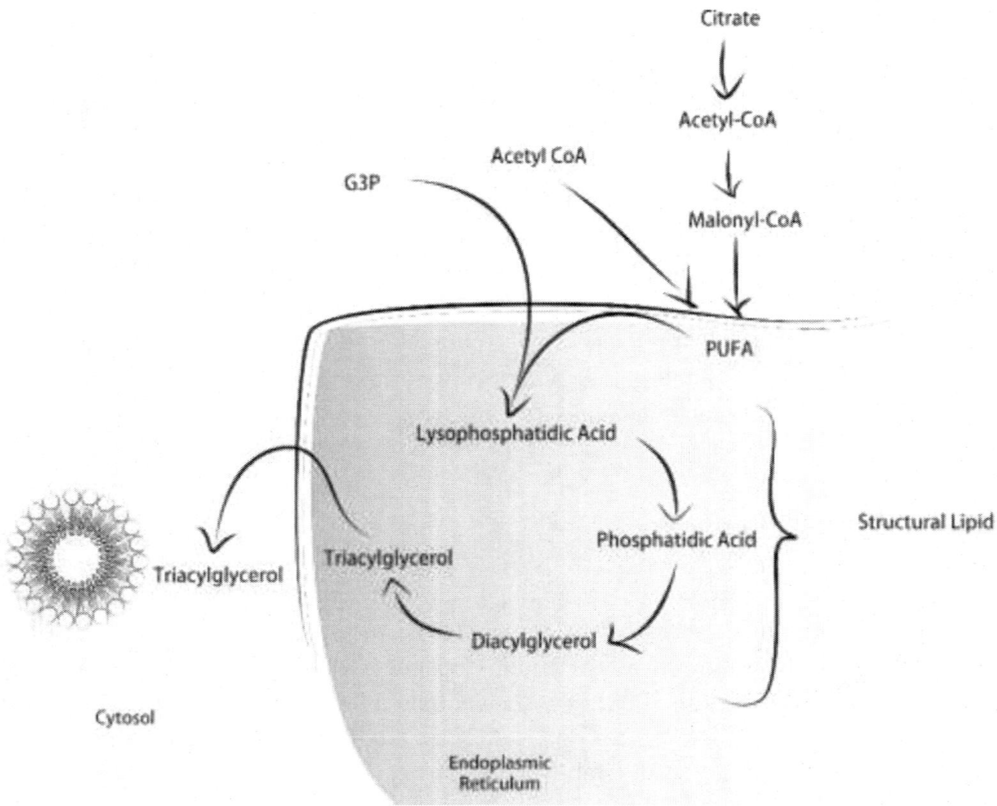

Figure 4. Synthesis of triglyceride in microalgae by Kennedy pathway.

6.4.5. PUFA Synthesis

The PUFA synthesis take place in the ER membrane by the condensation of acyl-CoA with Malonyl-Co A. The acetyl CoA comes from plastids and Malonyl-Co A from cytosolic acetyl CoA by acetyl CoA carboxylase enzyme. This process requires the presence of specific elongases and desaturases, which act primarily on palmitic, stearic and oleic acids. Fatty acid elongation occurs in both plastids and ER (Kunst and Samuels, 2009; Ohlrogge and Browse, 1995). Fatty acid elongase is a complex that comprises four subunits namely β-ketoacyl-CoA synthase (KCS), β-ketoacyl-CoA reductase, β-hydroxyacyl-CoA dehydratase and enoyl-CoA reductase (Baba and Shiraiwa, 2013; Yu W.L. et al., 2011). KCSs enzymes are involved in the synthesis of "fatty acid elongase" (FAE) and "elongase of very long-chain fatty acid" (ELOVL). ELOVL responsible for sphingolipid biosynthesis where FAE for TAGs or wax biosynthesis (Khozin-Goldberg and Cohen, 2011; Venegas-Calerón et al., 2010).

6.4.6. Kennedy Pathway- Triacylglycerol Formation (TAG)

TAG formation takes place in specialized organelles, i.e., the mitochondria located in the endoplasmic reticulum. This process results in the formation neutral lipids, to store fatty acids and structural lipids. The TAG synthesis initiated with the condensation of glycerol-3-

phosphate (G3P) with PUFAs, to form lysophosphatidate (LPA). This condensation process is catalysed by acyl-CoA phosphate acyl-transfererase (GPAT). This process involves cytosolic glycerol-3-phosphate (G3P) and PUFAs from ER membrane. This enzymatic activity of this enzyme is low in TAG synthesis pathyway, so it was potentially the rate-limiting step (Coleman et al. 2004). When the LPA further condensed with another acyl-CoA,it can produce phosphatidate (PA) which is followed by dephosphorylation by phosphatidic acid phosphates (PAP) to synthesize diacylglycerol.PA and DAG may act as precursors for the synthesis of plastidial membrane structural lipids (Fan et al., 2011; Ohlrogge and Browse, 1995). Finally, TAG is catalyzed by diacyglycerol acyl-transferase (DGAT). This enzyme is the important regulator for this pathway. In algae DGAT exist in two isoforms such as DGAT1 and DGAT2, both catalyzing the same reaction but have significant variations in sequence. In the final step of Kennedy pathway TAGs is then stored in oil bodies (Scott et al. 2010).

6.5. Gene Expression Profile of Lipid

The lipid synthesis in microalgae occurs under a series of biosynthesis pathway, all of these cycles are processed by key regulator of genes that control the production of lipids in microalgae. Over expression of the genes leads to accumulation of lipids in higher amounts. The most important genes that regulate lipid biosynthesis are, ACC(acetyl-CoA carboxylase), ACS (Acetyl CoA Synthase), Malic Enzyme (ME), PFK2 (Phosphofructo-2-kinase), Glycerol Kinase (GK), glycerol-3-phosphate Dehydrogenase (G3PDH), NAD(H) Kinase, Ferredoxins (FDX) and TAG synthesis coding genes.

6.5.1. ACC (Acetyl-CoA Carboxylase)

Acetyl-CoA carboxylase (ACC) gene, which codes for the enzyme that carboxylates acetyl-CoA to malonyl-CoA was first isolated from the microalga *Cyclotella cryptica*. This gene was overexpressed which enhanced the lipid productivity to 2–3-folds. It also suggests that over expression of ACC enzyme alone might not be sufficient to enhance the whole lipid biosynthesis pathway (Sheehan et al. 1998). This clearly indicated that ACC had no direct impact on lipid biosynthesis. However, the simultaneous over expression along with malic enzyme (ME), is only responsible for the conversion of malate to pyruvate to elevate the total lipid content (Talebi et al., 2014). However, increase in the level of expression of ACCase in *Chlorella sorokiniana*, also reported high lipid accumulation (Wan et al., 2011). This enzyme has been characterized as multi-domain protein with four different polypeptide subunits, including biotin carboxylase, biotin carboxyl carrier protein, and carboxyl transferase (α-CT and βCT) (Thelen and Ohlrogge 2002). Previous studies showed that the elevation of lipid content and altered fatty acid composition depends on the expression of ACCase (Amid et al. 2012; Andre et al. 2012; Chandrakar et al. 2013; Fukuda et al. 2013; Hunkeler et al. 2016).

Table 4. List of genes involved in lipid biosynthesis and their function

Gene	Function	Organism	References
ACC (acetyl-CoA carboxylase)	↑ Total lipid content	*D.salina*	Talebi et al., 2014
ACS (Acetyl CoA Synthase)	↑ Biomass productivity and FA profile ↑ TAG content ↑ TAG accumulation	*Schizochytriumsp, C.reinhardtii*	Yan et al., 2013; Rengel et al., 2018
Malic Enzyme (ME)	↑ Lipid content	*P.tricornutum*	Xue et al., 2015
PFK2 (Phosphofructo-2-kinase)	↑ Glycolytic activity	*T.pseudonana*	Abbriano et al., 2018
Glycerol Kinase (GK)	↑ Lipid productivity ↑ G3P content	*F.solaris;* *S.qudricauda*	Muto et al., 2015; Gomma et al., 2015
glycerol-3-phosphate Dehydrogenase (G3PDH)	↑ intracellular G3P level	*S.quadricauda* *P.tricornutum,* *C.minutissima*	Gomma et al., 2015; Yao et al., 2014; Hsieh et al., 2015
NAD(H) Kinase	↑ Lipid content	*C.pyrenoidosa*	Fan et al., 2015
Ferredoxins (FDX)	↑ Lipid level	*C.reinhardtii*	Huang et al., 2015
GPAT (glycerol-3-phosphate acyltransferase)	↑ TAG content	*C.reinhardtii,* *C.minutissima*	Boyle et al., 2012; Hsieh et al., 2012
Lysophosphatidic acid acyltransferase (LPAAT)	↑ TAG content	*C.reinhardtii,* *C.minutissima*	Yamaoka et al., 2016; Hsieh et al., 2012
Phosphatidic acid phosphatase (PAP)	↑ Lipid content ↑ TAG content	*C.reinhardtii,* *C.minutissima*	Deng et al., 2013; Hsieh et al., 2012
Diacylglycerol acyltransferase (DGAT)	↑ Lipid content ↑ Tag content	*N.oceanica,* *C.reinhardtii,* *T.pseudonana,* *C.minutissima,* *S.obliquus,* *C.tricornutum, T.chu*	Ubeda-Minguez et al., 2017; Niu et al., 2014; Iwai et al., 2014; La Russa et al., 2012; Chen et al., 2016; Hsieh et al., 2015; Ahmad et al., 2015; Li et al., 2016
Phospholipid:diacylglycerol acyltransferase (PDAT)	↑ TAG content	*C.reinhardtii,* *C.minutissima*	Hsieh et al., 2012; Zhu et al., 2018

"↑"increase the level, "TAG" triglyceride, "G3P" glycerol-3-phosphate

6.5.2. ACS (Acetyl CoA Synthase)

In microalgae, ACS is involved in metabolism of carbon sugars. Without directly participating in lipid production they enhance fatty acid profile. The reaction is carried out by two enzymatic steps that converts acetate to acetyl-CoA (Kumari et al. 1995). Once the Acetyl CoA is synthesized, it is then moved to TCA cycle and the pathways continues(Yan et al., 2012). This Acetyl CoA flux from TCA cycle to lipid synthesis enhanced TAG accumulation in *Chlamydomonas reinhardtii* (Deng et al., 2013; Plancke et al., 2014). The overexpression of ACS gene in microalgae *Schizochytrium* sp. and *C. Reinhardtii* lead to enhanced acetyl-CoA levels and an increase in lipid production (Rengel et al., 2018; Yan et al., 2013).

6.5.3. Malic Enzyme (ME)

Malic enzyme (ME) was responsible for the oxidative decarboxylation of malate to pyruvate along with NADH, and CO_2 (Vongsangnak et al., 2012; Chang and Tong, 2003). NADH plays a vital in fatty acid biosynthesis, which provides energy for cell metabolism (Wynn and Ratledge, 1997; Wynn and Ratledge, 2000). Malic Enzyme is also involved in metabolic pathways, including lipogenesis, energy metabolism and photosynthesis (Tsuchida et al., 2001; Moreadith and Lehninger, 1984; Gibala et al., 2000; Drincovich et al., 2001; Xue et al., 2015).

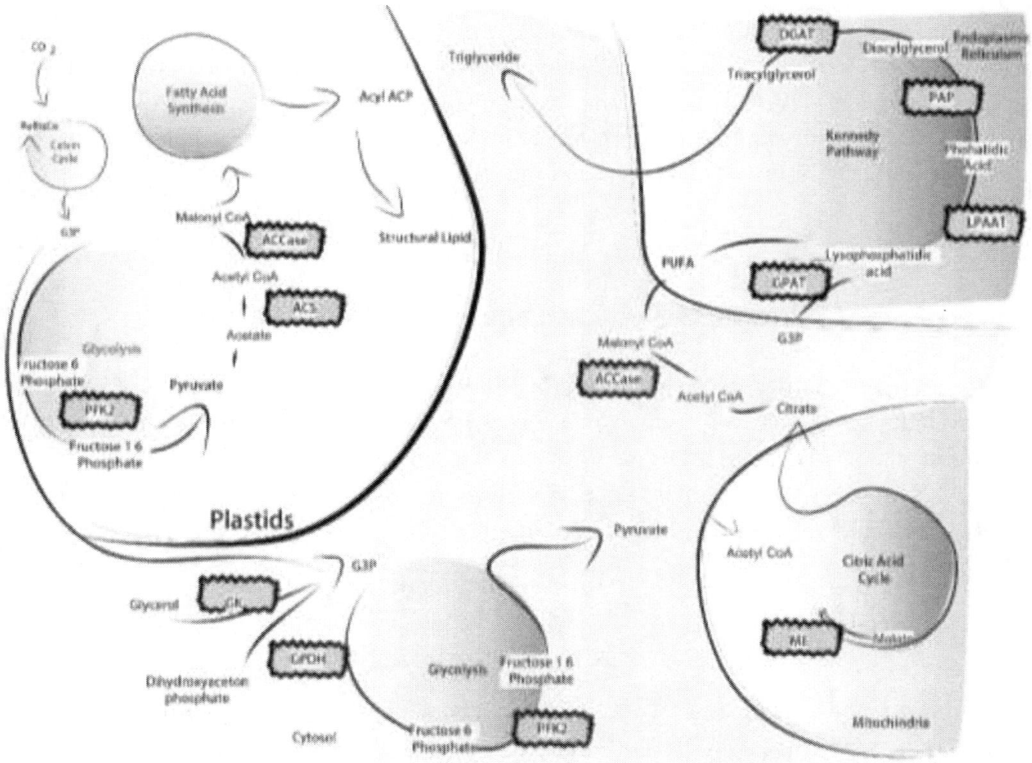

Figure 5. Genes involved in lipid biosynthesis (ACC (acetyl-CoA carboxylase), ACS (Acetyl CoA Synthase), Malic Enzyme (ME), PFK2 (Phosphofructo-2-kinase), Glycerol Kinase (GK), glycerol-3-phosphate Dehydrogenase (G3PDH), NAD(H) Kinase, Ferredoxins (FDX), GPAT (glycerol-3-phosphate acyltransferase), lysophosphatidic acid acyltransferase (LPAAT), phosphatidic acid phosphatase (PAP), diacylglycerol acyltransferase (DGAT), and Phospholipid:diacylglycerol acyltransferase (PDAT).).

6.5.4. PFK2 (Phosphofructo-2-Kinase)

Glycolysis is one of the pathways in lipid biosynthesis and can be found in plastid and cytoplasm. In glycolysis the enzyme that regulate fructose-6-bisphosphate to fructose-1,6-bisphosphate was catalysed by phosphor fructokinase (PFK) (Okar and Lange, 1999). This is an important allosteric regulator with bifunctional enzyme property phosphofructo-2-kinase/ fructose-2,6-bisphosphatase (PFK2/F2BP) (Rider et al., 2004). The overexpression of PFK2 increased glycolytic flux into lipids (Abbriano et al., 2018)

6.5.5. Glycerol Kinase (GK)

Glycerol-3-phosphate (G3P) is a precursor for TAG biosynthesis through the Kennedy pathway (Griffiths et al., 1985). The overexpression of glycerol kinase (GK) can improve lipid productivity (Yu et al., 2013). It can convert glycerol to glycerol-3-phosphate (G3P). Therefore, the overexpression of the *GK* gene leads to accumulation of intracellular G3P, which then stimulate TAG synthesis and accumulation. Studies showed that overexpression of the endogenous glycerol kinase (*GK*) gene in oleaginous marine diatom, *Fistulifera solaris*, accelerates glycerol metabolism and improves lipid and biomass productivities (Muto et al., 2015).

6.5.6. Glycerol-3-Phosphate Dehydrogenase (G3PDH)

Two different pathways can synthesize G3P, either via glycerol kinase (GK)-mediated phosphorylation or glycerol-3- phosphate dehydrogenase (GPDH)-mediated reduction (Harding et al., 1975). The GPDH gene overexpression can increase lipid content by 40% (Vigeolas et al., 2017). The microalgae, *Scenedesmus quadricauda* by the stimulation of G3PDH gene shunted the lipid production (Gomma et al., 2015).

6.5.7. NAD(H) Kinase

Lipid biosynthesis is an energy requiring process, which is dominated by ATP production and the reducing agent NADH.The transcriptome analysis indicated the switch from heterotrophy to light environment results in the generation of reducing energy equivalents (NADPH) that drive biosynthesis reactions (lipid synthesis) (Chai et al., 2006). Thus, increasing the level of NADPH may eventually contribute to lipid accumulation. The gene that code for NADH kinase is FACHB-9 in *C. pyrenoidosa* (Fan et al., 2015).

6.5.8. Ferredoxins (FDX)

Similar to NADH kinase, Ferredoxin (FDX) is an intracellular reducing equalent. They are the final electron carrier proteins; function as major electron donors in metabolic pathways. These FDX genes have different expression modes in response to various environmental conditions. Their expression also varied according to aerobic and anaerobic conditions, where the FDX gene are highly expressed in anaerobic conditions (Terauchi et al., 2009; Winkler et al., 2012). Studies on *Chlamydomonas* showed that they could increase the lipid productivity with FDX gene expression (Huang et al., 2015).

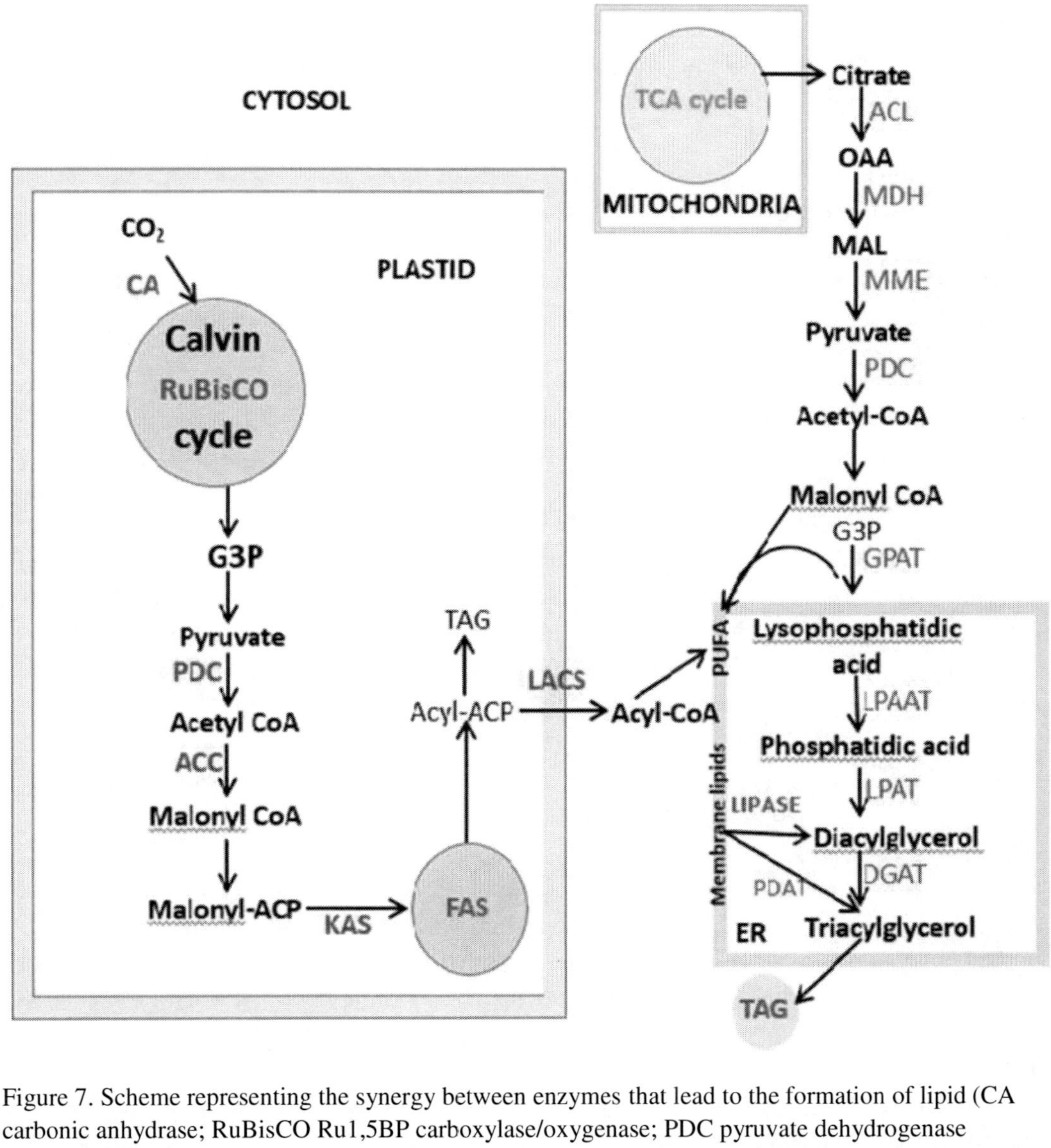

Figure 7. Scheme representing the synergy between enzymes that lead to the formation of lipid (CA carbonic anhydrase; RuBisCO Ru1,5BP carboxylase/oxygenase; PDC pyruvate dehydrogenase complex; ACC acetyl-CoA carboxylase; KAS 3-ketoacyl-ACP synthase; ACL ATP-citrate lyase; MDH malate dehydrogenase; MME NADP-malic enzyme; PDC pyruvate dehydrogenase complex; G PAT glycerol-3-phosphate acyltransferase; LPAAT lyso-phosphatidic acid acyltransferase; L PAT lyso-phosphatidylcholine acyltransferase; DGAT diacylglycerol acyltransferase; PDAT phospholipid diacylglycerol acyltransferase(adapted from Jagadevan etal., 2018)

6.5.9. TAG Synthesis Coding Genes

Kennedy pathway consist a series of five steps with actively participating four enzymes. Each of them coded by different genes such as, GPAT (glycerol-3-phosphate acyltransferase) (Boyle et al., 2012), lysophosphatidic acid acyltransferase (LPAAT) (Yamaoka et al., 2016), phosphatidic acid phosphatase (PAP) (Deng et al., 2013), diacylglycerol acyltransferase (DGAT) (Ahmad et al., 2015) and Phospholipid:diacylglycerol acyltransferase (PDAT)(Hsieh et al., 2012). Each of them have individual role in glycerol production. Their expression also have a significant role in lipid metabolism.

6.5.10. Growth Promoter Gene

The growth promoting agents synthesis such as, Auxin, IAA (Indole Acetic acid), etc have critical role in lipid. They can improve the lipid accumulation within the cell which has been reported in *C. vulgaris* (Piotrowska and Bajguz, 2014), *C. pyrenoidosa* and *Scenedesmus quadricauda*. Different types of genes regulated them, so its expression plays a key role in lipid synthesis. rbcL (ribulose 1, 5-bisphosphate carboxylase/oxygenase) is the key gene involved in photosynthesis of algae. It is the initiative pathway of lipid biosynthesis. Therefore, the expression of rbcL (ribulose 1, 5-bisphosphate carboxylase/oxygenase) genes suggested that IAA could regulate the certain metabolic pathways related to carbon fixation and lipid biosynthesis. ACCD (acetyl-CoA carboxylase) genes similar to rbcL gene involved in the regulation of IAA results in carbon fixation and lipid biosynthesis (Liu et al., 2016). 1-naphthaleneacetic acid (NAA) is a promoter involved in lipid productivity in of *C. vulgaris*. Further analysis showed that NAA could also up-regulate the KAS1 and SAD genes,. which regulated fatty acids biosynthesis in microalgae (Liu et al., 2017).

CONCLUSION

Microalgae are an important source of economically and medicinally important lipids. Over the years there is an increasing interest in microalgal lipids due to its high medicinal properties, industrial application, etc. The very fact that microalgae can be stimulated to produce and accumulate lipids has drawn great interest towards microalgae as a sustainable source of lipids. It is anticipated that with the ever increasing population and energy demand, microalgae would play an important role as feed stock for lipid production. With its ability to fix CO_2 from different sources both gaseous and fixed forms, numerous biotechnological applications for environmental protection such as bioremediation, bioassay and bio-monitoring has been devised using microalgae.

Commercial production of lipids from microalgae especially PUFAs seem to be much attainable and economically feasible than biodiesel. With limited sources of PUFAs and other lipids, microalgae have a great industrial potential. With ever increasing demand, applications of PUFAs will compel research towards seeking for effective solutions on crucial problems related to biochemical restrictions affecting microalgal biomass, lipid production and harvesting techniques. More research is needed in all aspects of lipid production especially

PUFAs including biochemical, molecular aspects, genome characterization and isolation and characterization of new oleaginous strains.

Along with lipids, several valuable products like carotenoids, astaxanthin, phenols, phycocyanin, acetogenins, terpenes, etc. pf pharmacological interest can also be extracted from microalga. Production of these valuable compounds along with lipds will increase the economic viability of microalgal production of metabolites.

REFERENCES

Abbriano, R., Vardar, N., Yee, D. & Hildebrand, M. (2018). Manipulation of a glycolytic regulator alters growth and carbon partitioning in the marine diatom *Thalassiosira pseudonana*. *Algal Res.*, *32*, 250-258.

Abida, H., Ruchaud, S., Rios, L., Humeau, A., Probert, I., De Vargas, C., Bach, S. & Bowler, C. (2013). Bioprospecting marine plankton. *Mar Drugs.*, *11*(11), 4594-4611.

Ahmad, I., Sharma, A.K., Daniell, H. & Kumar, S. (2015). Altered lipid composition and enhanced lipid production in green microalga by introduction of brassica diacylglycerol acyltransferase 2. *Plant Biotechnol. J.*, *13*(4), 540-550.

Aitken, D., Bulboa, C., Godoy-Faundez, A., Turrion-Gomez, J.L. & Antizar-Ladislao, B. (2014). Life cycle assessment of macroalgae cultivation and processing for biofuel production. *J. Clean. Prod.*, *75*, 45-56.

Amid, A., Lytovchenko, A., Fernie, A.R., Warren, G. & Thorlby, G.J. (2012). The sensitive to freezing3 mutation of *Arabidopsis thaliana* is a cold-sensitive allele of homomeric acetyl-CoA carboxylase that results in cold-induced cuticle deficiencies. *J. Exp. Bot.*, *63*(14), 5289-5299.

Andre, C., Haslam, R.P. & Shanklin, J. (2012). Feedback regulation of plastidic acetyl-CoA carboxylase by 18: 1-acyl carrier protein in *Brassica napus*. *Proc. Natl. Acad. Sci.*, *109*(25), 10107-10112.

Antonio, M., Rabasco, Alvarez. & María, Luisa González Rodríguez. (2000). Lipids in pharmaceutical and cosmetic preparations, *GRASAS ACEITES.*, *51*(1), 74-96.

Arisz, S.A. & Munnik, T. (2011). The salt stress-induced LPA response in Chlamydomonas is produced via PLA2 hydrolysis of DGK-generated phosphatidic acid. *J Lipid Res.*, *52*(11), 2012-2020.

Armbrust, E.V., Berges, J.A., Bowler, C., Green, B.R., Martinez, D., Putnam, N.H., Zhou, S., Allen, A.E., Apt, K.E., Bechner, M. & Brzezinski, M.A. (2004). The genome of the diatom *Thalassiosira pseudonana*: ecology, evolution, and metabolism. *Science.*, *306*(5693), 79-86.

Baba, M. & Shiraiwa, Y. (2013). Biosynthesis of lipids and hydrocarbons in algae. In Photosynthesis. *IntechOpen.*

Babu, A.G., Wu, X., Abrol, K., Ahmad, A.N. & Kim, D.P. (2017). Cultivation of an indigenous *Chlorella sorokiniana* with phytohormones for biomass and lipid production under N-limitation. *Algal Res.*, *23*, 178-185.

Balamurugan, S., Wang, X., Wang, H.L., An, C.J., Li, H., Li, D.W., Yang, W.D., Liu, J.S. & Li, H.Y. (2017). Occurrence of plastidial triacylglycerol synthesis and the potential regulatory role of AGPAT in the model diatom *Phaeodactylum tricornutum. Biotechnol Biofuels.*, *10*(1), 97.

Barclay, W.R., Meager, K.M. & Abril, J.R. (1994). Heterotrophic production of long chain omega-3 fatty acids utilizing algae and algae-like microorganisms. *J. Appl. Phycol.*, *6*(2), 123-129.

Beacham, T.A. & Ali, S. T. (2016). Growth dependent silencing and resetting of DGA1 transgene in *Nannochloropsis salina*, *Algal. Res.*, *14*, 65-71.

Beal, C.M., Hebner, R.E., Romanovicz, D., Mayer, C.C. & Connelly, R. (2013). Progression of lipid profile and cell structure in a research-scale production pathway for algal biocrude. *Renew.*, *50*, 86-93.

Bellou, S., Baeshen, M.N., Elazzazy, A.M., Aggeli, D., Sayegh, F. & Aggelis, G. (2014). Microalgal lipids biochemistry and biotechnological perspectives. *Biotechnol. Adv.*,*32*(8), 1476-1493.

Benavente-Valdés, J.R., Aguilar, C., Contreras-Esquivel, J.C., Méndez-Zavala, A. & Montañez, J. (2016). Strategies to enhance the production of photosynthetic pigments and lipids in chlorophycae species. *Biotechnol. Rep.*, *10*, 117-125.

Berges, J.A., Charlebois, D.O., Mauzerall, D.C. & Falkowski, P.G. (1996). Differential effects of nitrogen limitation on photosynthetic efficiency of photosystems I and II in microalgae. *Plant Physiol.*, *110*(2), 689-696.

Blaby, I.K., Glaesener, A.G., Mettler, T., Fitz-Gibbon, S.T., Gallaher, S.D., Liu, B., Boyle, N.R., Kropat, J., Stitt, M., Johnson, S. & Benning, C. (2013). Systems-level analysis of nitrogen starvation–induced modifications of carbon metabolism in a *Chlamydomonas reinhardtii* starchless mutant. *The Plant Cell.*, *25*(11), 4305-4323.

Blatti, J.L., Beld, J., Behnke, C.A., Mendez, M., Mayfield, S.P. & Burkart, M.D. (2012). Manipulating fatty acid biosynthesis in microalgae for biofuel through protein-protein interactions. *PloS one.*, *7*(9), 42949.

Bona, F., Capuzzo, A., Franchino, M. & Maffei, M.E. (2014). Semicontinuous nitrogen limitation as convenient operation strategy to maximize fatty acid production in Neochloris oleoabundans. *Algal Res.*, *5*, 1-6.

Borowitzka, M.A. (2013). High-value products from microalgae - their development and commercialisation. *J Appl Phycol.*, *25*(3), 743-756.

Boudière, L., Michaud, M., Petroutsos, D., Rébeillé, F., Falconet, D., Bastien, O., Roy, S., Finazzi, G., Rolland, N., Jouhet, J. & Block, M.A. (2014). Glycerolipids in photosynthesis: composition, synthesis and trafficking. *Biophysica Acta-Bioenergetics.*, *1837*(4), 470-480.

Bowler, C., Allen, A.E., Badger, J.H., Grimwood, J., Jabbari, K., Kuo, A., Maheswari, U., Martens, C., Maumus, F., Otillar, R.P. & Rayko, E. (2008). The *Phaeodactylum* genome reveals the evolutionary history of diatom genomes. *Nature.*, *456*(7219), 239.

Boyle, N.R., Page, M.D., Liu, B., Blaby, I.K., Casero, D., Kropat, J., Cokus, S.J., Hong-Hermesdorf, A., Shaw, J., Karpowicz, S.J. & Gallaher, S.D. (2012). Three acyltransferases and nitrogen-responsive regulator are implicated in nitrogen starvation-induced triacylglycerol accumulation in *Chlamydomonas. J. Biol. Chem.*, *287*(19), 15811-15825.

Breuer, G., Lamers, P.P., Martens, D.E., Draaisma, R.B. & Wijffels, R.H. (2012). The impact of nitrogen starvation on the dynamics of triacylglycerol accumulation in nine microalgae strains. *Bioresour. Technol.*, *124*, 217-226.

Brown, M.R., Jeffrey, S.W., Volkman, J.K. & Dunstan, G.A. (1997). Nutritional properties of microalgae for mariculture. *Aquaculture.*, *151*(1-4), 315-331.

Carballeira, N.M., Miranda, C. & Rodríguez, A.D. (2002). Phospholipid fatty acid composition of *Gorgonia mariae* and *Gorgonia ventalina.* Comparative biochemistry and physiology part B: *Biochem. Mol. Biol.*, *131*(1), 83-87.

Cases, S., Smith, S.J., Zheng, Y.W., Myers, H.M., Lear, S.R., Sande, E., Novak, S., Collins, C., Welch, C.B., Lusis, A.J. & Erickson, S.K. (1998). Identification of a gene encoding an acyl CoA: diacylglycerol acyltransferase, a key enzyme in triacylglycerol synthesis. *Proc. Natl. Acad. Sci.*, *95*(22), 13018-13023.

Chai, M. F., Wei, P. C., Chen, Q. J., An, R., Chen, J., Yang, S. & Wang, X. C. (2006). NADK3, a novel cytoplasmic source of NADPH, is required under conditions of oxidative stress and modulates abscisic acid responses in *Arabidopsis*, *Plant J.*, *47*, 665–674.

Chandrakar, B., Jain, A., Roy, S., Gutlapalli, V. R., Saraf, S., Suppahia, A., Verma, A., Tiwari, A., Yadav, M. & Nayarisseri, A. (2013). Molecular modeling of acetyl-CoA carboxylase (ACC) from *Jatropha curcas* and virtual screening for identification of inhibitors, *J Pharm Res.*, *6*, 913–918.

Chang, G. G. & Tong, L. (2003). Structure and function of malic enzymes, a new class of oxidative decarboxylases *Biochemistry.*, *42*, 12721–12733.

Chang, W.C., Zheng, H.Q. & Chen, C.N.N. (2016). Comparative transcriptome analysis reveals a potential photosynthate partitioning mechanism between lipid and starch biosynthetic pathways in green microalgae. *Algal Res.*, *16*, 54-62.

Chapman, K.D. & Ohlrogge, J.B. (2012). Compartmentation of triacylglycerol accumulation in plants. *J. Biol. Chem.*, *287*(4), 2288-2294.

Chaturvedi, R. & Fujita, Y. (2006). Isolation of enhanced eicosapentaenoic acid producing mutants of *Nannochloropsis oculata* ST-6 using ethyl methane sulfonate induced mutagenesis techniques and their characterization at mRNA transcript level. *Phycol. Res.*, *54*(3), 208-219.

Chen, C.Y., Kao, A.L., Tsai, Z.C., Chow, T.J., Chang, H.Y., Zhao, X.Q., Chen, P.T., Su, H.Y. & Chang, J.S. (2016). Expression of type 2 diacylglycerol acyltransferse gene DGTT1 from *Chlamydomonas reinhardtii* enhances lipid production in *Scenedesmus obliquus*. *Biotechnol. J.*, *11*(3), .336-344.

Chen, G., Qu, S., Wang, Q., Bian, F., Peng, Z., Zhang, Y., Ge, H., Yu, J., Xuan, N., Bi, Y. & He, Q. (2014). Transgenic expression of delta-6 and delta-15 fatty acid desaturases enhances omega-3 polyunsaturated fatty acid accumulation in *Synechocystis sp.* PCC6803. *Biotechnol Biofuels.*, *7*(1), 32.

Chen, G.Q. & Chen, F. (2006). Growing phototrophic cells without light. *Biotechnol. Lett.*, *28*(9), 607-616.

Chen, J.W., Liu, W.J., Hu, D.X., Wang, X., Balamurugan, S., Alimujiang, A., Yang, W.D., Liu, J.S. & Li, H.Y. (2017). Identification of a malonyl CoA-acyl carrier protein transacylase and its regulatory role in fatty acid biosynthesis in oleaginous microalga *Nannochloropsis oceanica*. *Biotechnol Appl Bioc.*, *64*(5), 620-626.

Chen, L., Liu, T., Zhang, W., Chen, X. & Wang, J. (2012). Biodiesel production from algae oil high in free fatty acids by two-step catalytic conversion, *Bioresour Technol.*, *111*, 208–14.

Chisti, Y. (2007). Biodiesel from microalgae. *Biotechnol. Adv.*, *25*(3), 294-306.

Choo, M.Y., Oi, L.E., Show, P.L., Chang, J.S., Ling, T.C., Ng, E.P., Phang, S.M. & Juan, J.C. (2017). Recent progress in catalytic conversion of microalgae oil to green hydrocarbon: A review. *J Taiwan Inst Chem Eng*., *79*, 116-124.

Coleman, R.A. & Lee, D.P. (2004). Enzymes of triacylglycerol synthesis and their regulation. *Prog Lipid Res*., *43*(2), 134-176.

Cooper, G.M., Hausman, R.E. & Hausman, R.E. (2000). *The cell: a molecular approach*, (Vol. *10*). Washington, DC, ASM press.

Cronan, Jr. J.E. & Waldrop, G.L. (2002). Multi-subunit acetyl-CoA carboxylases. *PROG Lipid Res.*, *41*(5), 407-435.

Damiani, M. C., Popovich, C. A., Constenla, D. & Leonardi, P. I.(2010). Lipid analysis in *Haematococcus pluvialis* to assess its potential use as a biodiesel feedstock. *Bioresour. Technol.*, *101* (11), 3801-3807.

Davis, M. S., Solbiati, J. & Cronan, J. E. (2000). Overproduction of acetyl-CoA carboxylase activity increases the rate of fatty acid biosynthesis in *Escherichia coli*, *J Biol Chem*., *275*(37), 28593–8.

de Jaeger, L., Verbeek, R.E., Draaisma, R.B., Martens, D.E., Springer, J., Eggink, G. & Wijffels, R.H. (2014). Superior triacylglycerol (TAG) accumulation in starchless mutants of *Scenedesmus obliquus*: (I) mutant generation and characterization. *Biotechnol Biofuels*., *7*(1), 69.

de Oliveira Dal'Molin, C.G., Quek, L.E., Palfreyman, R.W. & Nielsen, L.K. (2011). December. AlgaGEM–a genome-scale metabolic reconstruction of algae based on the *Chlamydomonas reinhardtii* genome. In *BMC Genomics*, (Vol. *12*, No. 4, p. S5). BioMed Central.

Dean, A.P., Sigee, D.C., Estrada, B. & Pittman, J.K. (2010). Using FTIR spectroscopy for rapid determination of lipid accumulation in response to nitrogen limitation in freshwater microalgae. *Bioresour. Technol.*, *101*(12), 4499-4507.

Deng, X., Cai, J. & Fei, X. (2013). Effect of the expression and knockdown of citrate synthase gene on carbon flux during triacylglycerol biosynthesis by green algae *Chlamydomonas reinhardtii*. *BMC Biochem*., *14*(1), 38.

Deng, X.D., Cai, J.J. & Fei, X.W. (2013). Involvement of phosphatidate phosphatase in the biosynthesis of triacylglycerols in *Chlamydomonas reinhardtii*. *J Zhejiang Univ Sci B*., *14*(12), 1121-1131.

Derelle, E., Ferraz, C., Rombauts, S., Rouzé, P., Worden, A.Z., Robbens, S., Partensky, F., Degroeve, S., Echeynié, S., Cooke, R. & Saeys, Y. (2006). Genome analysis of the smallest free-living eukaryote *Ostreococcus tauri* unveils many unique features. *Proc. Natl. Acad. Sci.*, *103*(31), 11647-11652.

Domergue, F., Lerchl, J., Zähringer, U. & Heinz, E. (2002). Cloning and functional characterization of *Phaeodactylum tricornutum* front-end desaturases involved in eicosapentaenoic acid biosynthesis. *Eur J Biochem*., *269*(16), 4105-4113.

Drincovich, M.F., Casati, P. & Andreo, C.S. (2001). NADP-malic enzyme from plants: a ubiquitous enzyme involved in different metabolic pathways. *FEBS Lett.*, *490*(1-2), 1-6.

Dunahay, T.G., Jarvis, E.E. & Roessler, P.G. (1995). Genetic transformation of the diatoms *Cyclotella cryptica* and *Navicula saprophila*. *J. Phycol*., *31*(6), 1004-1012.

Dunahay, T.G., Jarvis, E.E., Dais, S.S. & Roessler, P.G. (1996). Manipulation of microalgal lipid production using genetic engineering. *Appl Biochem Biotech*., *57*(1), 223.

Eastmond, P. J., Quettier, A. L., Kroon, J. T. M., Craddock, C., Adams, N.& Slabas, A. R. (2010). Phosphatidic Acid Phosphohydrolase1 and 2 Regulate Phospholipid Synthesis at the Endoplasmic Reticulum in *Arabidopsis*, *The Plant Cell.*, *22*(8), 2796–2811.

El Baky, H.H.A., El-Baroty, G.S., Bouaid, A., Martinez, M. & Aracil, J. (2012). Enhancement of lipid accumulation in *Scenedesmus obliquus* by optimizing CO2 and Fe3+ levels for biodiesel production. *Bioresour. Technol.*, *119*, 429-432.

El-Sheek, M.M. & Rady, A.A. (1995). Effect of phosphorus starvation on growth, photosynthesis and some metabolic processes in the unicellular green alga *Chlorella kessleri.Phyton.*

Fahy, E., Cotter, D., Sud, M. & Subramaniam, S. (2011). Lipid classification, structures and tools. *Biochim. Biophys. Acta, Mol. Cell. Biol. Lipids.*, *1811*(11), 637-647.

Fakhry, E.M. & El Maghraby, D.M. (2015). Lipid accumulation in response to nitrogen limitation and variation of temperature in *Nannochloropsis salina. Bot. Stud.*,*56*(1), 6.

Fan, J., Andre, C. & Xu, C. (2011). A chloroplast pathway for the de novo biosynthesis of triacylglycerol in *Chlamydomonas reinhardtii. FEBS Lett.*, *585*(12), 1985-1991.

Fan, J., Ning, K., Zeng, X., Luo, Y., Wang, D., Hu, J., Li, J., Xu, H., Huang, J., Wan, M. & Wang, W. (2015). Genomic foundation of starch-to-lipid switch in oleaginous *Chlorella sp.Plant Physiol.*, *169*(4), 2444-2461.

Fan, Y., Yuan, C., Jin, Y., Hu, G.R. & Li, F.L. (2018). Characterization of 3-ketoacyl-coA synthase in a nervonic acid producing oleaginous microalgae Mychonastes afer. *Algal Res.*, *31*, 225-231.

Faried, M., Samer, M., Abdelsalam, E., Yousef, R.S., Attia, Y.A. & Ali, A.S. (2017). Biodiesel production from microalgae: Processes, technologies and recent advancements. *RENEW SUST ENERG REV.*, *79*, 893-913.

Finkel, Z.V., Follows, M.J., Liefer, J.D., Brown, C.M., Benner, I. & Irwin, A.J. (2016). Phylogenetic diversity in the macromolecular composition of microalgae. *PLoS One.*, *11*(5), 0155977.

Fukuda, N., Ikawa, Y., Aoyagi, T. & Kozaki, A. (2013). Expression of the genes coding for plastidic acetyl-CoA carboxylase subunits is regulated by a location-sensitive transcription factor binding site, *Plant Mol. Biol.*, *82*(4-5), 473-483.

Gibala, M.J., Young, M.E. & Taegtmeyer, H. (2000). Anaplerosis of the citric acid cycle: role in energy metabolism of heart and skeletal muscle. *Acta Physiol. Scand.*, *168*(4), 657-665.

Gomma, A.E., Lee, S.K., Sun, S.M., Yang, S.H. & Chung, G. (2015). Improvement in oil production by increasing Malonyl-CoA and Glycerol-3-Phosphate pools in *Scenedesmus quadricauda. Indian J Microbiol.*, *55*(4), 447-455.

Gong, Y., Hu, H., Gao, Y., Xu, X. & Gao, H. (2011). Microalgae as platforms for production of recombinant proteins and valuable compounds: progress and prospects. *J Ind Microbiol Biotechnol.*, *38*(12), 1879-1890.

Gorain, P.C., Bagchi, S.K. & Mallick, N. (2013). Effects of calcium, magnesium and sodium chloride in enhancing lipid accumulation in two green microalgae. *Environ. Technol.*, *34*(13-14), 1887-1894.

Greenwell, H.C., Laurens, L.M.L., Shields, R.J., Lovitt, R.W. & Flynn, K.J. (2009). Placing microalgae on the biofuels priority list: a review of the technological challenges. *J R Soc Interface.*, *7*(46), 703-726.

Griffiths, G., Stobart, A.K. & Stymne, S. (1985). The acylation of sn-glycerol 3-phosphate and the metabolism of phosphatidate in microsomal preparations from the developing cotyledons of safflower (*Carthamus tinctorius L.*) seed. *Biochem. J.*, *230*(2), 379.

Griffiths, M.J., van Hille, R.P. & Harrison, S.T. (2012). Lipid productivity, settling potential and fatty acid profile of 11 microalgal species grown under nitrogen replete and limited conditions. *J. Appl. Phycol.*, *24*(5), 989-1001.

Guschina, I.A. & Harwood, J.L. (2006). Lipids and lipid metabolism in eukaryotic algae. *Prog. Lipid Res.*, *45*(2), 160-186.

Gustafson, K.R., Cardellina, J.H., Fuller, R.W., Weislow, O.S., Kiser, R.F., Snader, K.M., Patterson, G.M. & Boyd, M.R. (1989). AIDS-antiviral sulfolipids from cyanobacteria (blue-green algae). *JNCI*: *J Natl Cancer Inst.*, *81*(16), 1254-1258.

Haag, A.L. (2007). Algae bloom again. News @ *Nature.*, *447.*

Hamilton, M.L., Haslam, R.P., Napier, J.A. & Sayanova, O. (2014). Metabolic engineering of *Phaeodactylum tricornutum* for the enhanced accumulation of omega-3 long chain polyunsaturated fatty acids. *Metab. Eng.*, *22*, 3-9.

Hardie, L.P., Balkwill, D.L. & Stevens, S.E. (1983). Effects of iron starvation on the physiology of the cyanobacterium *Agmenellum quadruplicatum. Appl. Environ. Microbiol.*, *45*(3), 999-1006.

Harding, Jr. J.W., Pyeritz, E.A., Copeland, E.S. & White, 3rd. H.B. (1975). Role of glycerol 3-phosphate dehydrogenase in glyceride metabolism. Effect of diet on enzyme activities in chicken liver. *Biochem J.*, *146*(1), 223.

Harwood, J.L. & Guschina, I.A. (2009). The versatility of algae and their lipid metabolism. *Biochimie.*, *91*(6), 679-684.

Harwood, J. L. (1996). Recent advances in the biosynthesis of plant fatty acids. *Biochim. Biophys. Acta, Lipids Lipid Metab.*, *1301*, 7–56.

He, H., Rodgers, R.P., Marshall, A.G. & Hsu, C.S. (2011). Algae polar lipids characterized by online liquid chromatography coupled with hybrid linear quadrupole ion trap/fourier transform ion cyclotron resonance mass spectrometry. *Energ Fuel.*, *25*(10), 4770-4775.

Hemaiswarya, S., Raja, R., Kumar, R.R., Ganesan, V. & Anbazhagan, C. (2011). Microalgae: a sustainable feed source for aquaculture. *J Microb Biot.*, *27*(8), pp.1737-1746.

Hill, J., Nelson, E., Tilman, D., Polasky, S. & Tiffany, D. (2006). Environmental, economic, and energetic costs and benefits of biodiesel and ethanol biofuels. *Proc Natl Acad Sci.*, *103*(30), 11206-11210.

Hlavova, M., Turoczy, Z. & Bisova, K. (2015). Improving microalgae for biotechnology—From genetics to synthetic biology. *Biotechnol. Adv.*, *33*(6), 1194-1203.

Ho, S.H., Chen, C.Y. & Chang, J.S. (2012). Effect of light intensity and nitrogen starvation on CO2 fixation and lipid/carbohydrate production of an indigenous microalga *Scenedesmus obliquus* CNW-N. *Bioresour. Technol.*, *113*, 244-252.

Hölzl, G. & Dörmann, P. (2007). Structure and function of glycoglycerolipids in plants and bacteria. *Prog Lipid Res.*, *46*(5), 225-243.

Hsieh, H.J., Su, C.H. & Chien, L.J. (2012). Accumulation of lipid production in *Chlorella minutissima* by triacylglycerol biosynthesis-related genes cloned from *Saccharomyces cerevisiae* and *Yarrowia lipolytica*. *J Microbiol.*, *50*(3), 526-534.

Hu, Q., Sommerfeld, M., Jarvis, E., Ghirardi, M., Posewitz, M., Seibert, M. & Darzins, A. (2008). Microalgal triacylglycerols as feedstocks for biofuel production: perspectives and advances. *Plant J.*, *54*(4), 621-639.

Huang, L.F., Lin, J.Y., Pan, K.Y., Huang, C.K. & Chu, Y.K. (2015). Overexpressing ferredoxins in *Chlamydomonas reinhardtii* increase starch and oil yields and enhance electric power production in a photo microbial fuel cell. *Int. J. Mol. Sci.*, *16*(8), 19308-19325.

Huang, L.F., Lin, J.Y., Pan, K.Y., Huang, C.K. & Chu, Y.K. (2015). Overexpressing ferredoxins in *Chlamydomonas reinhardtii* increase starch and oil yields and enhance electric power production in a photo microbial fuel cell. *Int. J. Mol. Sci*, *16*(8), 19308-19325.

Hunkeler, M., Stuttfeld, E., Hagmann, A., Imseng, S. & Maier, T. (2016). The dynamic organization of fungal acetyl-CoA carboxylase. *Nat. Commun.*, *7*, 11196.

Iskandarov, U. S., Sitnik, N., Shtaida, S., Didi-Cohen, S., Leu, I., Khozin-Goldberg, Z., Cohen, S. & Boussiba. (2016). Cloning and characterization of a GPAT-like gene from the microalga *Lobosphaeraincisa* (Trebouxiophyceae): overexpression in *Chlamydomonasreinhardtii* enhances TAG production, *J. Appl. Phycol.*, *28*, 907-919.

Iwai, M., Ikeda, K., Shimojima, M. & Ohta, H. (2014). Enhancement of extraplastidic oil synthesis in *Chlamydomonas reinhardtii* using a type-2 diacylglycerol acyltransferase with a phosphorus starvation–inducible promoter. *Plant Biotechnol J.*, *12(*6), 808-819.

Jagadevan, S., Banerjee, A., Banerjee, C., Guria, C., Tiwari, R., Baweja, M. & Shukla, P. (2018). Recent developments in synthetic biology and metabolic engineering in microalgae towards biofuel production. *Biotechnol Biofuels.*, *11*(1), 185.

Jiang, Y., Yoshida, T. & Quigg, A. (2012). Photosynthetic performance, lipid production and biomass composition in response to nitrogen limitation in marine microalgae. *PLANT Physiol Bioch.*, *54*, 70-77.

Jiang, Z., Meng, D., Mu, H. & Yoshikawa, K. (2010). Study on the hydrothermal drying technology of sewage sludge. *Sci China Ser E.*, *53*(1), 160-163.

John, E.H. & Flynn, K.J. (2000). Modelling phosphate transport and assimilation in microalgae; how much complexity is warranted?*Ecol. Model.*,*125*(2-3), 145-157.

Jones, C.S. & Mayfield, S.P. (2012). Algae biofuels: versatility for the future of bioenergy. *Curr Opin Biotech.*, *23*(3), 346-351.

Kalla, N. & Khan, S. (2016). Effect of nitrogen, phosphorus concentrations, pH and salinity ranges on growth, biomass and lipid accumulation of *chlorella vulgaris*. *Int J Pharm Sci Res.*, *7*(1), 397-05.doi: 10.13040/IJPSR.0975-8232.7 (1).397-05.

Kaye, Y., Grundman, O., Leu, S., Zarka, A., Zorin, B., Didi-Cohen, S., Khozin-Goldberg, I. & Boussiba, S. (2015). Metabolic engineering toward enhanced LC-PUFA biosynthesis in *Nannochloropsis oceanica*: Overexpression of endogenous Δ12 desaturase driven by stress-inducible promoter leads to enhanced deposition of polyunsaturated fatty acids in TAG. *Algal Res.*, *11*, 387-398.

Khan, S.A., Khan, A., Khan, S.A., Beg, M.A., Ali, A. & Damanhouri, G. (2017). Comparative study of fatty-acid composition of table eggs from the Jeddah food market and effect of value addition in omega-3 bio-fortified eggs. *Saudi J. Biol. Sci.*, *24*(4), 929-935.

Khozin-Goldberg, I. & Cohen, Z. (2011). Unraveling algal lipid metabolism: recent advances in gene identification. *Biochimie.*, *93*(1), 91-100.

Khozin-Goldberg, I., Leu, S. & Boussiba, S. (2016). Microalgae as a source for VLC-PUFA production. *Lipidsplant & Algae Develop.*, 471-510.

Kim, S.H., Liu, K.H., Lee, S.Y., Hong, S.J., Cho, B.K., Lee, H., Lee, C.G. & Choi, H.K. (2013). Effects of light intensity and nitrogen starvation on glycerolipid, glycerophospholipid, and carotenoid composition in *Dunaliella tertiolecta* culture. *PLoS One.*, *8*(9), 72415.

Klaitong, P., Faaroonsawat, S. & Chungjatupornchai, W. (2017). Accelerated triacylglycerol production and altered fatty acid composition in oleaginous microalga *Neochlorisoleoabundans* by overexpression of diacylglycerol acyltransferase 2. *Microb. Cell. Fact.*, *16*.

Kobayashi, T., Sakaguchi, K., Matsuda, T., Abe, E., Hama, Y., Hayashi, M., Honda, D., Okita, Y., Sugimoto, S., Okino, N. & Ito, M. (2011). Increase of eicosapentaenoic acid in thraustochytrids through thraustochytrid ubiquitin promoter-driven expression of a fatty acid Δ5 desaturase gene. *Appl. Environ. Microbiol.*, *77*(11), 3870-3876.

Kumari, S., Singh, P., Gupta, S.K. & Kumar, S. (2015). Genetic Engineering Tools for Enhancing Lipid Production in Microalgae. *Algae & Envt. Sust.*

Kumari, S., Tishel, R., Eisenbach, M. & Wolfe, A.J. (1995). Cloning, characterization, and functional expression of acs, the gene which encodes acetyl coenzyme A synthetase in Escherichia coli. *J Bacteriol.*, *177*(10), 2878-2886.

Kunst, L. & Samuels, L. (2009). Plant cuticles shine: advances in wax biosynthesis and export. *Curr Opin Plant Biol.*, *12*(6), 721-727.

Kurat, C.F., Natter, K., Petschnigg, J., Wolinski, H., Scheuringer, K., Scholz, H., Zimmermann, R., Leber, R., Zechner, R. & Kohlwein, S.D. (2006). Obese yeast: triglyceride lipolysis is functionally conserved from mammals to yeast. *J. Biol. Chem.*, *281*(1), 491-500.

La Russa, M., Bogen, C., Uhmeyer, A., Doebbe, A., Filippone, E., Kruse, O. & Mussgnug, J.H. (2012). Functional analysis of three type-2 DGAT homologue genes for triacylglycerol production in the green microalga *Chlamydomonas reinhardtii. J. Biotechnol.*, *162*(1), pp.13-20.

Lee, A.G. (2000). Membrane lipids: it's only a phase. *Curr Biol.*, *10*(10), R377-R380.

Lee, O.K., Seong, D.H., Lee, C.G. & Lee, E.Y. (2015). Sustainable production of liquid biofuels from renewable microalgae biomass. *Ind. Eng. Chem. Res.*, *29*, 24-31.

Lei, A., Chen, H., Shen, G., Hu, Z., Chen, L. & Wang, J. (2012). Expression of fatty acid synthesis genes and fatty acid accumulation in *Haematococcus pluvialis* under different stressors. *Biotechnol Biofuels.*, *5*(1), 18.

Li, D.W., Cen, S.Y., Liu, Y.H., Balamurugan, S., Zheng, X.Y., Alimujiang, A., Yang, W.D., Liu, J.S. & Li, H.Y. (2016). A type 2 diacylglycerol acyltransferase accelerates the triacylglycerol biosynthesis in heterokont oleaginous microalga *Nannochloropsis oceanica. J. Biotechnol.*, *229*, 65-71.

Li, Y., Han, D., Yoon, K., Zhu, S., Sommerfeld, M. & Hu, Q. (2013). Molecular and cellular mechanisms for lipid synthesis and accumulation in microalgae: biotechnological implications. *Handbook of Microalgal Culture.*, 545-565.

Li, Y., Horsman, M., Wang, B., Wu, N. & Lan, C.Q. (2008). Effects of nitrogen sources on cell growth and lipid accumulation of green alga *Neochloris oleoabundans. Appl Microbiol Biotechnol.*, *81*(4), 629-636.

Li, Y., Qin, J., Ball, A. & Moore, R. (2009). Perspectives of marine phytoplankton as a source of nutrition and bioenergy.

Li, Z., Meng, T., Ling, X., Li, J., Zheng, C., Shi, Y., Chen, Z., Li, Z., Li, Q., Lu, Y. & He, N. (2018). Overexpression of malonyl-CoA: ACP transacylase in *Schizochytrium* sp. to improve polyunsaturated fatty acid production. *J. Agric. Food Chem.*, *66*(21), 5382-5391.

Liang, K., Zhang, Q., Gu, M. & Cong, W. (2013). Effect of phosphorus on lipid accumulation in freshwater microalga *Chlorella sp. J Appl Phycol.*, *25*(1), 311-318.

Liang, M.H. & Jiang, J.G. (2013). Advancing oleaginous microorganisms to produce lipid via metabolic engineering technology. *Prog Lipid Res.*, *52*(4), 395-408.

Lim, D.K., Schuhmann, H., Thomas-Hall, S.R., Chan, K.C., Wass, T.J., Aguilera, F., Adarme-Vega, T.C., Dal'Molin, C.G., Thorpe, G.J., Batley, J. & Edwards, D. (2017). RNA-Seq and metabolic flux analysis of *Tetraselmis* sp. M8 during nitrogen starvation reveals a two-stage lipid accumulation mechanism. *Bioresour. Technol.*, *244*, 1281-1293.

Liu, J., Huang, J., Sun, Z., Zhong, Y., Jiang, Y. & Chen, F. (2011). Differential lipid and fatty acid profiles of photoautotrophic and heterotrophic *Chlorella zofingiensis*: assessment of algal oils for biodiesel production. *Bioresour. Technol.*, *102*(1), 106-110.

Liu, J., Qiu, W. & Song, Y. (2016). Stimulatory effect of auxins on the growth and lipid productivity of *Chlorella pyrenoidosa* and *Scenedesmus quadricauda. Algal Res.*, *18*, 273-280.

Liu, L., Pohnert, G. & Wei, D. (2016). Extracellular metabolites from industrial microalgae and their biotechnological potential. *Mar Drugs.*, *14*(10), 191.

Liu, T., Liu, F., Wang, C., Wang, Z. & Li, Y. (2017). The boosted biomass and lipid accumulation in *Chlorella vulgaris* by supplementation of synthetic phytohormone analogs. *Bioresour. Technol.*, *232*, 44-52.

Liu, Z.Y., Wang, G.C. & Zhou, B.C. (2008). Effect of iron on growth and lipid accumulation in Chlorella vulgaris. *Bioresour. Technol.*, *99*(11), 4717-4722.

Lu, N., Wei, D., Chen, F. & Yang, S.T. (2012). Lipidomic profiling and discovery of lipid biomarkers in snow alga *Chlamydomonas nivalis* under salt stress. *Eur. J. Lipid Sci. Technol.*, *114*(3), 253-265.

Lv, H., Qu, G., Qi, X., Lu, L., Tian, C. & Ma, Y. (2013). Transcriptome analysis of *Chlamydomonas reinhardtii* during the process of lipid accumulation. *Genomics.*, *101*(4), 229-237.

Matsuzaki, M., Misumi, O., Shin-i, T., Maruyama, S., Takahara, M., Miyagishima, S.Y., Mori, T., Nishida, K., Yagisawa, F., Nishida, K. & Yoshida, Y. (2004). Genome sequence of the ultrasmall unicellular red alga *Cyanidioschyzon merolae* 10D. *Nature*, *428*(6983), 653.

Merchant, S.S., Prochnik, S.E., Vallon, O., Harris, E.H., Karpowicz, S.J., Witman, G.B., Terry, A., Salamov, A., Fritz-Laylin, L.K., Maréchal-Drouard, L. & Marshall, W.F. (2007). The *Chlamydomonas* genome reveals the evolution of key animal and plant functions. *Science.*, *318*(5848), 245-250.

Miller, R., Wu, G., Deshpande, R.R., Vieler, A., Gärtner, K., Li, X., Moellering, E.R., Zäuner, S., Cornish, A.J., Liu, B. & Bullard, B. (2010). Changes in transcript abundance in *Chlamydomonas reinhardtii* following nitrogen deprivation predict diversion of metabolism. *Plant Physiol.*, *154*(4), 1737-1752.

Moreadith, R.A. & Lehninger, A.L. (1984). The pathways of glutamate and glutamine oxidation by tumor cell mitochondria. Role of mitochondrial NAD (P)+-dependent malic enzyme. *J Biol Chem.*, *259*(10), 6215-6221.

Mosblech, A., Feussner, I. & Heilmann, I. (2009). Oxylipins: structurally diverse metabolites from fatty acid oxidation. *Plant Physiol Biochem.*, *47*(6), 511-517.

Moseley, K.R. & Thompson, G.A. (1980). Lipid composition and metabolism of *Volvox carteri.Plant Physiol.*, *65*(2), 260-265.

Mühlroth, A., Li, K., Røkke, G., Winge, P., Olsen, Y., Hohmann-Marriott, M.F., Vadstein, O. & Bones, A.M. (2013). Pathways of lipid metabolism in marine algae, co-expression network, bottlenecks and candidate genes for enhanced production of EPA and DHA in species of *Chromista*. *Mar Drugs.*,*11*(11), 4662-4697.

Muto, M, Tanaka, M, Liang, Y, Yoshino, T, Matsumoto, M.& Tanaka, T. (2015). Enhancement of glycerol metabolism in the oleaginous marine diatom *Fistulifera solaris* JPCC DA0580 to improve triacylglycerol productivity, *Biotechnol Biofuels.*, *8*(1), 4.

Nakahara, Y., Kanamori, Y., Kiuchi, M. & Kamimura, M. (2003). *In vitro* studies of hematopoiesis in the silkworm: cell proliferation in and hemocyte discharge from the hematopoietic organ. *J. Insect Physiol.*, *49*(10), 907-916.

Nguyen, H.M., Baudet, M., Cuine, S., Adriano, J.M., Barthe, D., Billon, E., Bruley, C., Beisson, F., Peltier, G., Ferro, M. & Li-Beisson, Y. (2011). Proteomic profiling of oil bodies isolated from the unicellular green microalga *Chlamydomonas reinhardtii*: with focus on proteins involved in lipid metabolism. *Proteomics*, *11*(21), 4266-4273.

Niu, Y. F., Zhang, M. H., Li, D. W., Yang, W. D., Liu, J. S., Bai, W. B.& Li, H. Y. (2013). Improvement of Neutral Lipid and Polyunsaturated Fatty Acid Biosynthesis by Overexpressing a Type 2 Diacylglycerol Acyltransferase in Marine Diatom *Phaeodactylum tricornutum*. *Mar Drugs.*, *11*(11), 4558–4569.

Nymark, M., Sharma, A.K., Sparstad, T., Bones, A.M. & Winge, P. (2016). A CRISPR/Cas9 system adapted for gene editing in marine algae. *Sci Rep.*, *6*, 24951.

Ohlrogge, J. & Browse, J. (1995). Lipid biosynthesis. *Plant Cell.*, *7*(7), 957.

Okar, D. & Lange, A. (1999). Fructose-2,6-bisphosphate and control of carbohydrate metabolism in eukaryotes. *Biofactors*, *10*, 1–14.

Olofsson, M., Lamela, T., Nilsson, E., Bergé, J.P., Del Pino, V., Uronen, P. & Legrand, C. (2012). Seasonal variation of lipids and fatty acids of the microalgae *Nannochloropsis oculata* grown in outdoor large-scale photobioreactors. *Energies*, *5*(5), 1577-1592.

Ördög, V., Stirk, W.A., Bálint, P., van Staden, J. & Lovász, C. (2012). Changes in lipid, protein and pigment concentrations in nitrogen-stressed *Chlorella minutissima* cultures. *J Appl Phycol.*, *24*(4), 907-914.

Osorio-Fierros, A., Cronin, K., Ring, D., Méndez-Zavala, A., Morales-Oyervides, L. & Montañez, J.C. (2017). Influence of granulation process parameters on food tablet properties formulated using natural powders (*Opuntia ficus* and *Chlorella sp.*). *Powder Technol.*, *317*, 281-286.

Palenik, B., Grimwood, J., Aerts, A., Rouzé, P., Salamov, A., Putnam, N., Dupont, C., Jorgensen, R., Derelle, E., Rombauts, S. & Zhou, K. (2007). The tiny eukaryote *Ostreococcus* provides genomic insights into the paradox of plankton speciation. *Proc. Natl. Acad. Sci.*, *104*(18), 7705-7710.

Patil, A.D. & Patil, N. (2015). Mitigation of Global Warming Through Biological Carbon Sequestration Using Micro Algae. *Int J Sci Res.*, *4*, 1559-1564.

Petrie, J.R., Liu, Q., Mackenzie, A.M., Shrestha, P., Mansour, M.P., Robert, S.S., Frampton, D.F., Blackburn, S.I., Nichols, P.D. & Singh, S.P. (2010). Isolation and characterization of a high-efficiency desaturase and elongases from microalgae for transgenic LC-PUFA production. *Mar Biotechnol.*, *12*(4), 430-438.

Petroutsos, D., Amiar, S., Abida, H., Dolch, L.J., Bastien, O., Rébeillé, F., Jouhet, J., Falconet, D., Block, M.A., McFadden, G.I. & Bowler, C. (2014). Evolution of galactoglycerolipid biosynthetic pathways–from cyanobacteria to primary plastids and from primary to secondary plastids. *Progress in Lipid Research.*, *54*, 68-85.

Pignolet, O., Jubeau, S., Vaca-Garcia, C. & Michaud, P. (2013). Highly valuable microalgae: biochemical and topological aspects *J Ind Microbiol Biotechnol.*, *40*(8), 781-796.

Piotrowska-Niczyporuk, A. & Bajguz, A. (2014). The effect of natural and synthetic auxins on the growth, metabolite content and antioxidant response of green alga *Chlorella vulgaris* (Trebouxiophyceae). *Plant Growth Regul.*, *73*(1), 57-66.

Plancke, C., Vigeolas, H., Höhner, R., Roberty, S., Emonds-Alt, B., Larosa, V., Willamme, R., Duby, F., Onga Dhali, D., Thonart, P. & Hiligsmann, S. (2014). Lack of isocitrate lyase in C hlamydomonas leads to changes in carbon metabolism and in the response to oxidative stress under mixotrophic growth. *Plant J.*, *77*(3), 404-417.

Posewitz, M. C., Smolinski, S. L., Kanakagiri, S., Melis, A., Seibert, M. & Ghirardi, M. L. (2004). Hydrogen photoproduction is attenuated by disruption of an isoamylase gene in *Chlamydomonas reinhardtii. Plant Cell*, *16*, 2151–2163.

Potvin, G. & Zhang, Z. (2010). Strategies for high-level recombinant protein expression in transgenic microalgae: a review. *Biotechnology advances.*, *28*(6), 910-918.

Qu, C.B., Wu, Z.Y. & Shi, X.M. (2008). Phosphate assimilation by Chlorella and adjustment of phosphate concentration in basal medium for its cultivation. *Biotechnol Lett.*, *30*(10), 1735.

Radakovits, R., Eduafo, P.M. & Posewitz, M.C. (2011). Genetic engineering of fatty acid chain length in *Phaeodactylum tricornutum. Metab Eng.*, *13*(1), 89-95.

Radakovits, R., Jinkerson, R.E., Darzins, A. & Posewitz, M.C. (2010). Genetic engineering of algae for enhanced biofuel production. *Eukaryotic cell.*, *9*(4), 486-501.

Raja, R., Coelho, A., Hemaiswarya, S., Kumar, P., Carvalho, I.S. & Alagarsamy, A. (2018). Applications of microalgal paste and powder as food and feed: An update using text mining tool. *Beni-Suef univ. J. Basic Appl. Sci.*

Ren, H.Y., Liu, B.F., Kong, F., Zhao, L., Xie, G.J. & Ren, N.Q. (2014). Enhanced lipid accumulation of green microalga Scenedesmus sp. by metal ions and EDTA addition. *Bioresour. Technol.*, *169*, 763-767.

Renaud, S.M., Thinh, L.V., Lambrinidis, G. & Parry, D.L. (2002). Effect of temperature on growth, chemical composition and fatty acid composition of tropical Australian microalgae grown in batch cultures. *Aquaculture. 211*(1-4), 195-214.

Rengel, R., Smith, R.T., Haslam, R.P., Sayanova, O., Vila, M. & León, R. (2018). Overexpression of acetyl-CoA synthetase (ACS) enhances the biosynthesis of neutral lipids and starch in the green microalga *Chlamydomonas reinhardtii. Algal Res.*, *31*, 183-193.

Rider, M.H., Bertrand, L., Vertommen, D., Michels, P.A., Rousseau, G.G. & Hue, L. (2004). 6-phosphofructo-2-kinase/fructose-2, 6-bisphosphatase: head-to-head with a bifunctional enzyme that controls glycolysis. *Biochem. J.*, *381*(3), 561-579.

Roessler, P.G. & Ohlrogge, J.B. (1993). Cloning and characterization of the gene that encodes acetyl-coenzyme A carboxylase in the alga *Cyclotella cryptica. J. Biol. Chem.*, *268*(26), 19254-19259.

Roessler, P.G., Bleibaum, J.L., Thompson, G.A. & Ohlrogge, J.B. (1994). Characteristics of the gene that encodes acetyl-CoA carboxylase in the diatom *Cyclotella cryptica*. *Ann N Y Acad Sci.*, *721*, 250-256.

Ruangsomboon, S. (2012). Effect of light, nutrient, cultivation time and salinity on lipid production of newly isolated strain of the green microalga, *Botryococcus braunii* KMITL 2. *Bioresour. Technol.*, *109*, 261-265.

Sajjadi, B., Chen, W.Y., Raman, A.A.A. & Ibrahim, S. (2018). Microalgae lipid and biomass for biofuel production: A comprehensive review on lipid enhancement strategies and their effects on fatty acid composition. *Renew Sust Energ Rev.*, *97*, 200-232.

Saravanakumar, D.E.M., Folb, P.I., Campbell, B.W. & Smith, P. (2008). Antimycobacterial Activity of the Red Alga *Polysiphonia virgata*. *Pharm Biol.*, *46*(4), 254-260.

Sathasivam, R., Radhakrishnan, R., Hashem, A. & Abd_Allah, E.F. (2017). Microalgae metabolites: A rich source for food and medicine. *Saudi J. Biol. Sci.*

Schenk, P.M., Thomas-Hall, S.R., Stephens, E., Marx, U.C., Mussgnug, J.H., Posten, C., Kruse, O. & Hankamer, B. (2008). Second generation biofuels: high-efficiency microalgae for biodiesel production. *Bioenergy Res.*, *1*(1), 20-43.

Scott, S.A., Davey, M.P., Dennis, J.S., Horst, I., Howe, C.J., Lea-Smith, D.J. & Smith, A.G. (2010). Biodiesel from algae: challenges and prospects. *Curr Opin Biotech*, *21*(3), 277-286.

Scranton, M.A., Ostrand, J.T., Georgianna, D.R., Lofgren, S.M., Li, D., Ellis, R.C., Carruthers, D.N., Dräger, A., Masica, D.L. & Mayfield, S.P. (2016). Synthetic promoters capable of driving robust nuclear gene expression in the green alga *Chlamydomonas reinhardtii*. *Algal Res*, *15*, 135-142.

Shah, M., Mahfuzur, R., Liang, Y., Cheng, J.J. & Daroch, M. (2016). Astaxanthin-producing green microalga *Haematococcus pluvialis*: from single cell to high value commercial products. *Front. Plant Sci.*, *7*, 531.

Shah, S.H., Raja, I.A., Rizwan, M., Rashid, N., Mahmood, Q., Shah, F.A. & Pervez, A. (2018). Potential of microalgal biodiesel production and its sustainability perspectives in Pakistan. *Renew Sust Energ Rev.*, *81*, 76-92.

Sharma, K.K., Garg, S., Li, Y., Malekizadeh, A. & Schenk, P.M. (2013). Critical analysis of current microalgae dewatering techniques. *Biofuels*. *4*(4), 397-407.

Sharma, P.K., Saharia, M., Srivstava, R., Kumar, S. & Sahoo, L. (2018). Tailoring microalgae for efficient biofuel production. *Front Mar Sci.*, *5*, 382.

Sheehan, J., Dunahay, T., Benemann, J. & Roessler, P. (1998). Look back at the US department of energy's aquatic species program: biodiesel from algae; close-out report (No. NREL/TP-580-24190). *National Renewable Energy Lab.* Golden, CO. (US).

Shi, H., Chen, H., Gu, Z., Zhang, H., Chen, W. & Chen, Y.Q. (2016). Application of a delta-6 desaturase with α-linolenic acid preference on eicosapentaenoic acid production in *Mortierella alpina*. *Microb. Cell. Fact.*, *15*(1), 117.

Siegenthaler, P.A. & Murata, N. eds. (2006). *Lipids in photosynthesis: structure, function and genetics* (Vol. *6*). Springer Science & Business Media.

Singh, R., Mattam, A.J., Jutur, P. & Yazdani, S.S. (2006). Synthetic biology in biofuels production. *J Cell Mol Med.*, *2*(1), 144-176.

Singh, R., Parihar, P., Singh, M., Bajguz, A., Kumar, J., Singh, S., Singh, V.P. & Prasad, S.M. (2017). Uncovering potential applications of cyanobacteria and algal metabolites in biology, agriculture and medicine: current status and future prospects. *Front. Microbiol.*, *8*, 515.

Spijkerman, E. & Wacker, A. (2011). Interactions between P-limitation and different C conditions on the fatty acid composition of an extremophile microalga. *Extremophiles.*, *15*(5), 597.

Spolaore, P., Joannis-Cassan, C., Duran, E. & Isambert, A. (2006). Commercial Hölzland Dörmann Stephens, E., Ross, I.L., Mussgnug, J.H., Wagner, L.D., Borowitzka, M.A., Posten, C., Kruse, O. & Hankamer, B. (2010). Future prospects of microalgal biofuel production systems. *Trends Plant Sci.*, *15*(10), 554-564.

Sukenik, A., Bilanovic, D. & Shelef, G. (1988). Flocculation of microalgae in brackish and sea waters. *Biomass*, *15*(3), 187-199.

Sun, X.M., Ren, L.J., Zhao, Q.Y., Ji, X.J. & Huang, H. (2018). Enhancement of lipid accumulation in microalgae by metabolic engineering. *BBA-Mol Cell Biol L.*

Susmita, Das., Ajitabh, Bora., Hemanta, K., Gogoi, Sushil, Kumar., Manderia, Vanlalhmuaka.& Raju, P. S. (2017). Lipid induction in microalgae, *J. Algal Biomass Utln*, *8*(4), 39-50.

Sydney, E.B., Sturm, W., de Carvalho, J.C., Thomaz-Soccol, V., Larroche, C., Pandey, A. & Soccol, C.R. (2010). Potential carbon dioxide fixation by industrially important microalgae. *Bioresour Technol.*, *101*(15), 5892-5896.

Takagi, M., Watanabe, K., Yamaberi, K. & Yoshida, T. (2000). Limited feeding of potassium nitrate for intracellular lipid and triglyceride accumulation of *Nannochloris* sp. UTEX LB1999. *Appl Microbiol Biotechnol.*, *54*(1), 112-117.

Talebi, A.F., Tohidfar, M., Bagheri, A., Lyon, S.R., Salehi-Ashtiani, K. & Tabatabaei, M. (2014). Manipulation of carbon flux into fatty acid biosynthesis pathway in *Dunaliella salina* using AccD and ME genes to enhance lipid content and to improve produced biodiesel quality. *J, Biofuel Res.*, *1*(3), 91-97.

Tan, X.B., Lam, M.K., Uemura, Y., Lim, J.W., Wong, C.Y. & Lee, K.T. (2018). Cultivation of microalgae for biodiesel production: a review on upstream and downstream processing. *Chinese J Chem Eng.*, *26*(1), 17-30.

Tandon, P. & Jin, Q. (2017). Microalgae culture enhancement through key microbial approaches. *Renew Sust Energ Rev.*, *80*, 1089-1099.

Terauchi, A.M., Lu, S.F., Zaffagnini, M., Tappa, S., Hirasawa, M., Tripathy, J.N., Knaff, D.B., Farmer, P.J., Lemaire, S.D., Hase, T. & Merchant, S.S. (2009). Pattern of expression and substrate specificity of chloroplast ferredoxins from *Chlamydomonas reinhardtii*. *J. Biol. Chem.*, *284*(38), 25867-25878.

Thelen, J.J. & Ohlrogge, J.B. (2002). The multisubunit acetyl-CoA carboxylase is strongly associated with the chloroplast envelope through non-ionic interactions to the carboxyltransferase subunits. *Arch. Biochem. Biophys.*, *400*(2), 245-257.

Tonon, T., Harvey, D., Larson, T.R. & Graham, I.A. (2002). Long chain polyunsaturated fatty acid production and partitioning to triacylglycerols in four microalgae. *Phytochemistry.*, *61*(1), 15-24.

Torzillo, G. & Masojidek, J. (2008). Mass Cultivation of Freshwater Microalgae. vol. *Encyclopedia of Ecology*, 2226-2235. Sven Erik Jorgensen and Brian Fath.

Trentacoste, E.M., Shrestha, R.P., Smith, S.R., Glé, C., Hartmann, A.C., Hildebrand, M. & Gerwick, W.H. (2013). Metabolic engineering of lipid catabolism increases microalgal lipid accumulation without compromising growth. *Proc. Natl. Acad. Sci.*, *110*(49), 19748-19753.

Tsuchida, H., Tamai, T., Fukayama, H., Agarie, S., Nomura, M., Onodera, H., Ono, K., Nishizawa, Y., Lee, B.H., Hirose, S. & Toki, S. (2001). High level expression of C4-specific NADP-malic enzyme in leaves and impairment of photoautotrophic growth in a C3 plant, rice. *Plant Cell Physiol.*, *42*(2), 138-145.

Úbeda-Mínguez, P., García-Maroto, F. & Alonso, D.L. (2017). Heterologous expression of DGAT genes in the marine microalga *Tetraselmis chui* leads to an increase in TAG content. *J Appl Phycol.*, *29*(4), 1913-1926.

Ursin, V.M. (2003). Modification of plant lipids for human health: development of functional land-based omega-3 fatty acids. *Nutr. J.*, *133*(12), 4271-4274.

Valentine, R.C. & Valentine, D.L. (2004). Omega-3 fatty acids in cellular membranes: a unified concept. *Prog Lipid Res.*, *43*(5), 383-402.

Venegas-Caler6n, M., Sayanova, O. & Napier, J.A. (2010). An alternative to fish oils: metabolic engineering of oil-seed crops to produce omega-3 long chain polyunsaturated fatty acids. *Prog Lipid Res.*, *49*(2), 108-119.

Vigeolas, H., Waldeck, P., Zank.& Geigenberger, P. (2007). Increasing seed oil content in oil-seed rape (*Brassica napusL*) by over-expression of a yeast glycerol-3-phosphate dehydrogenase under the control of a seed-specific promoter, *Plant Biotechnol J.*, *5*, 431–441.

Vongsangnak, W., Zhang, Y., Chen, W., Ratledge, C. & Song, Y. (2012). Annotation and analysis of malic enzyme genes encoding for multiple isoforms in the fungus *Mucor circinelloides* CBS 277.49. *Biotechnol Lett.*, *34*(5), 941-947.

Vymazal, J. (1995). *Algae and element cycling in wetlands.* Lewis Publishers Inc.

Walther, T.C. & Farese, Jr. R.V. (2009). The life of lipid droplets. *BBA-Mol Cell Biol L.*, *1791*(6), 459-466.

Wan, M., Liu, P., Xia, J., Rosenberg, J.N., Oyler, G.A., Betenbaugh, M.J., Nie, Z. & Qiu, G. (2011). The effect of mixotrophy on microalgal growth, lipid content, and expression levels of three pathway genes in *Chlorella sorokiniana. Appl Microbiol Biot.*, *91*(3), 835-844.

Wang, C., Li, Y., Lu, J., Deng, X., Li, H. & Hu, Z. (2018). Effect of overexpression of LPAAT and GPD1 on lipid synthesis and composition in green microalga *Chlamydomonas reinhardtii. J Appl Phycol*, *30*(3), 1711-1719.

Wang, Z. & Benning, C. (2012). *Chloroplast lipid synthesis and lipid trafficking through ER–plastid membrane contact sites.*

Wang, Z.T., Ullrich, N., Joo, S., Waffenschmidt, S. & Goodenough, U. (2009). Algal lipid bodies: stress induction, purification, and biochemical characterization in wild-type and starchless *Chlamydomonas reinhardtii. Eukaryotic cell.*, *8*(12), 1856-1868.

Wei, H., Shi, Y., Ma, X., Pan, Y., Hu, H., Li, Y., Luo, M., Gerken, H. & Liu, J. (2017). A type-I diacylglycerol acyltransferase modulates triacylglycerol biosynthesis and fatty acid composition in the oleaginous microalga, *Nannochloropsis oceanica. Biotechnol Biofuels.*, *10*(1), p.174.

Wijffels, R.H., Kruse, O. & Hellingwerf, K.J. (2013). Potential of industrial biotechnology with cyanobacteria and eukaryotic microalgae. *Curr Opin Biotech.*, *24*(3), 405-413.

Wilfling, F., Haas, J.T., Walther, T.C. & Farese, Jr. R.V. (2014). Lipid droplet biogenesis. *Curr Opin Cell Biol.*, *29*, 39-45.

Winkler, M., Hemschemeier, A., Jacobs, J., Stripp, S. & Happe, T. (2010). Multiple ferredoxin isoforms in *Chlamydomonas reinhardtii*–their role under stress conditions and biotechnological implications. *Eur. J. Cell Biol.*, *89*(12), 998-1004.

Worden, A. Z., Lee, J. H., Mock, T., Rouzé, P., Simmons, M. P., Aerts, A. L., Allen, A. E., Cuvelier, M. L., Derelle, E., Everett, M. V. & Foulon, E. (2009). Green evolution and dynamic adaptations revealed by genomes of the marine picoeukaryotes Micromonas. *Science.*, *324*(5924), 268-272.

Wu, Y. H., Yu, Y., Li, X., Hu, H. Y. & Su, Z. F. (2012). Biomass production of a *Scenedesmus* sp. under phosphorous-starvation cultivation condition. *Bioresour. Technol.*, *112*, 193-198.

Wymer, P. E. O. & Thake, B. (1980). The importance of phosphorus in microalgal growth and species composition in mixed populations: experiments and simulations. *Philos Trans R Soc Lond B Biol Sci.*, *209*(1176), 333-353.

Wynn, J. P. & Ratledge, C. (1997). Malic enzyme is a major source of NADPH for lipid accumulation by *Aspergillus nidulans*. *Microbiology.*, *143*(1), 253-257.

Wynn, J. P. & Ratledge, C. (2000). Evidence that the rate-limiting step for the biosynthesis of arachidonic acid in *Mortierella alpina* is at the level of the 18: 3 to 20: 3 elongase. *Microbiology.*, *146*(9), 2325-2331.

Xin, L., Hong-Ying, H., Ke, G. & Ying-Xue, S. (2010). Effects of different nitrogen and phosphorus concentrations on the growth, nutrient uptake, and lipid accumulation of a freshwater microalga *Scenedesmus* sp. *Bioresource Technol.*, *101*(14),

Xu, L., Cheng, X. & Wang, Q. (2018). Enhanced lipid production in *Chlamydomonas reinhardtii* by co-culturing with *Azotobacter chroococcum*. *Front Plant Sci.*, *9*, 741.

Xue, J., Niu, Y. F., Huang, T., Yang, W. D., Liu, J. S. & Li, H. Y. (2015). Genetic improvement of the microalga *Phaeodactylum tricornutum* for boosting neutral lipid accumulation. *Metab Eng.*, *27*, 1-9.

Xue, J., Niu, Y. F., Huang, T., Yang, W. D., Liu, J. S. & Li, H. Y. (2015). Genetic improvement of the microalga *Phaeodactylum tricornutum* for boosting neutral lipid accumulation. *Metab Eng.*, *27*, 1-9.

Yamaoka, Y., Achard, D., Jang, S., Legéret, B., Kamisuki, S., Ko, D., Schulz-Raffelt, M., Kim, Y., Song, W. Y., Nishida, I. & Li-Beisson, Y. (2016). Identification of a Chlamydomonas plastidial 2-lysophosphatidic acid acyltransferase and its use to engineer microalgae with increased oil content. *Plant Biotechnol J.*, *14*(11), 2158-2167.

Yan, J., Cheng, R., Lin, X., You, S., Li, K., Rong, H. & Ma, Y. (2013). Overexpression of acetyl-CoA synthetase increased the biomass and fatty acid proportion in microalga *Schizochytrium*. *Appl Microbiol Biotechnol.* *97*(5), pp.1933-1939.

Yang, D., Song, D., Kind, T., Ma, Y., Hoefkens, J. & Fiehn, O. (2015). Lipidomic analysis of *Chlamydomonas reinhardtii* under nitrogen and sulfur deprivation. *PLoS One.*, *10*(9), 0137948.

Yang, J., Yang, S. & Yuan, Y. J. (2012). Integrated investigation of lipidome and related signaling pathways uncovers molecular mechanisms of tetramethylpyrazine and butylidenephthalide protecting endothelial cells under oxidative stress. *Mol Biosyst.*, *8*(6), 1789-1797.

Yao, Y., Lu, Y., Peng, K. T., Huang, T., Niu, Y. F., Xie, W. H., Yang, W. D., Liu, J. S. & Li, H. Y. (2014). Glycerol and neutral lipid production in the oleaginous marine diatom

Phaeodactylum tricornutum promoted by overexpression of glycerol-3-phosphate dehydrogenase. *Biotechnol Biofuels*., *7*(1), 110.

Yu, K. O., Jung, J., Ramzi, A. B., Choe, S. H., Kim, S. W., Park, C. & Han, S. O. (2013). Development of a Saccharomyces cerevisiae strain for increasing the accumulation of triacylglycerol as a microbial oil feedstock for biodiesel production using glycerol as a substrate. *Biotechnol Bioeng*., *110* (1), 343-347.

Yu, W. L., Ansari, W., Schoepp, N. G., Hannon, M. J., Mayfield, S. P. & Burkart, M. D. (2011). Modifications of the metabolic pathways of lipid and triacylglycerol production in microalgae. *Microb Cell Fact.*, *10*(1), 91.

Zäuner, S., Jochum, W., Bigorowski, T. & Benning, C. (2012). A cytochrome b5-containing plastid-located fatty acid desaturase from *Chlamydomonas reinhardtii*. *Eukaryotic cell.*, *11*(7), 856-863.

Zhang, J., Hao, Q., Bai, L., Xu, J., Yin, W., Song, L., Xu, L., Guo, X., Fan, C., Chen, Y. & Ruan, J. (2014). Overexpression of the soybean transcription factor GmDof4 significantly enhances the lipid content of *Chlorella ellipsoidea*. *Biotechnol Biofuels.*, *7*(1), 128.

Zhang, W., Zhao, Y., Cui, B., Wang, H. & Liu, T. (2016). Evaluation of filamentous green algae as feedstocks for biofuel production. *Bioresour. Technol.*, *220*, 407-413.

Zhou, D., Qiao, B., Li, G., Xue, S. & Yin, J. (2017). Continuous production of biodiesel from microalgae by extraction coupling with transesterification under supercritical conditions. *Bioresour. Technol.*, *238*, 609-615.

Zhu, L., Nugroho, Y. K., Shakeel, S. R., Li, Z., Martinkauppi, B. & Hiltunen, E. (2017). Using microalgae to produce liquid transportation biodiesel: what is next? *Renew Sust Energ Rev.*, *78*, 391-400.

Zhu, Z., Yuan, G., Fan, X., Fan, Y., Yang, M., Yin, Y., Liu, J., Liu, Y., Cao, X., Tian, J. & Xue, S. (2018). The synchronous TAG production with the growth by the expression of chloroplast transit peptide-fused ScPDAT in *Chlamydomonas reinhardtii*. *Biotechnol Biofuels.*, *11*(1), 156.

Zou, L. G., Chen, J. W., Zheng, D. L., Balamurugan, S., Li, D. W., Yang, W. D., Liu, J. S. & Li, H. Y. (2018). High-efficiency promoter-driven coordinated regulation of multiple metabolic nodes elevates lipid accumulation in the model microalga *Phaeodactylum tricornutum*. *Microb. Cell. Fact*, *17*(1), 54.

Zullaikah, S., Utomo, A. T., Yasmin, M., Ong, L. K. & Ju, Y. H. (2019). Ecofuel conversion technology of inedible lipid feedstocks to renewable fuel. *Adv. Eco-FuelsSust. Envt.*, 237-276.

In: Applied Algal Biotechnology
Editors: M. Arumugam, S. Kathiresan et al.
ISBN: 978-1-53617-524-0

Chapter 7

STRATEGIES FOR LIPID ENHANCEMENT IN MICROALGAE

***B. Vanavil*[1,3,*], *B. S. Sujitha*[1,2,*] *and Muthu Arumugam*[1,2,†]**
[1]Microbial Processes and Technology Division,
National Institute for Interdisciplinary Science and Technology (NIIST),
Council of Scientific and Industrial Research (CSIR), Trivandrum, Kerala, India
[2]Academy of Scientific and Innovative Research (AcSIR), CSIR, India
[3]Department of Biotechnology,
[3]Kalasalingam Academy of Research and Education, Krishnankoil, Tamilnadu, India

ABSTRACT

Increasing energy crisis due to alarming use and depletion of fossil fuels leads to increased interest in search of biofuel alternatives like bioethanol and biodiesel. Much attention is paid to the production of biofuel using various biomasses. Biofuel production from oilseed and food crops faces ethical food security concerns and nurtures issues like agricultural land availability and water. Microalgae are a valuable source of bioactive metabolites such as polysaccharides, lipids, pigments, proteins, vitamins and antioxidants that have a wide range of applications as biofuels, health supplements, pharmaceuticals, cosmetics and can be employed for bioremediation, wastewater treatment, and CO_2 mitigation in the atmosphere. The lipid-rich microalgal oil can be exploited for biodiesel production. Hence, microalgae are currently viewed as one of the promising alternatives for renewable, sustainable and eco-friendly feedstock for biofuel production due to its advantages of higher photosynthetic efficiency, biomass production and growth rate compared to the terrestrial crops. In spite of these potential advantages and applications, algae-based fuel has not been used commercially owing to its high production cost attributed to low lipid content in the cells and the high cost of downstream processing. The future scope for commercialization of algal-based biodiesel lies in the enhancement of lipid content of the cells leading to improvement in process economics. In this perspective, this chapter reviews various strategies or approaches for lipid enhancement

[*] Joint first author, contributed equally.
[†] Corresponding Author'S Email: arumugam@niist.res.in; aasaimugam@gmail.com; https://orcid.org/0000-0003-4697-6925.

in microalgae. These lipid inducing tactics can be combined in a suitable way to optimize the over-accumulation. Further, the chapter also reviews the prospects of lipid enhancement in microalgae.

Keywords: microalgae, biodiesel, strain improvement, nitrogen stress-mediated lipid accumulation, triacylglycerol, bioreactor

7.1. Algae-Based Biofuel - An Introduction

An increase in the world's population, rapid industrialization and increasing demand for transportation lead to the alarming use of fossil fuels. Hence, to face the challenge of the energy crisis and to address the issue of global warming due to the release of carbon dioxide from fossil fuels, the interest is increasing in search of new renewable, sustainable and clean energy source (Faried et al., 2017). Biofuel is one such alternative to overcome the demand for renewable fuel resources. Biofuels can be produced via several renewable feedstocks such as edible, non-edible oils and lignocellulosic biomass (Milano et al., 2016).

Microalgae are photosynthetically driven biofactories that utilize sunlight, CO_2 and water (Adeniyi et al., 2018) to produce a variety of metabolites such as polysaccharides, lipids, pigments, proteins, vitamins and anti-oxidants which can be used as source of biofuels, health supplements, pharmaceuticals, cosmetics and can be potentially employed for bioremediation, wastewater treatment, and CO_2 mitigation in atmosphere (Rizwan et al., 2018). Biomass can be used as food, feed, and fertilizer (Rizwan et al., 2018). Microalgae have got the potential to yield biofuels without compromising food supplies, forest or arable land (Milano et al., 2016) that are commonly encountered by other sources of biofuels.

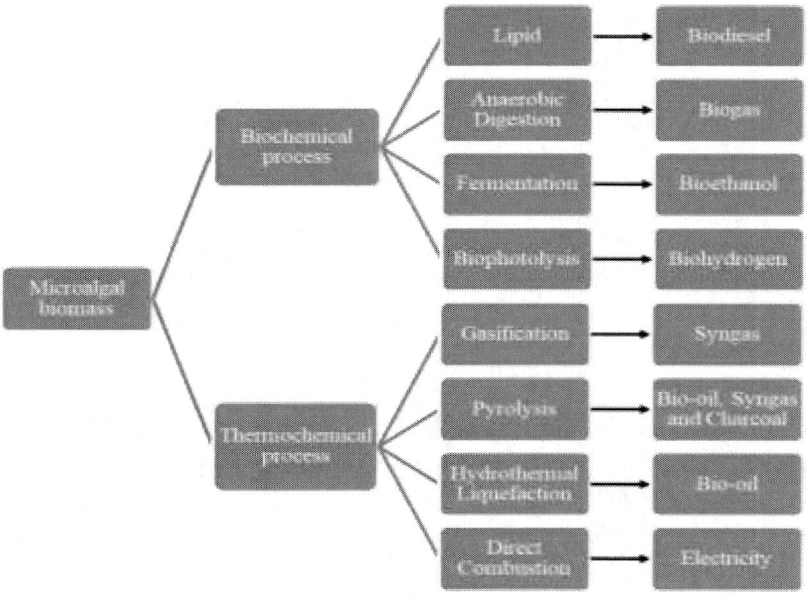

Figure 7.1. Different routes for the production of bioenergy sources from microalgae.

Microalgae can be employed to produce different types of renewable biofuels. The strategies for energy conversion from microalgal biomass can be classified as biochemical conversion and thermochemical conversion. After extraction of lipids from the algal biomass, the residual biomass can be processed into other biofuels by biochemical or thermochemical processes. The residual biomass can be further used as a biofertilizer. The strategies are classified as given in Figure 7.1. (Chisti, 2007; Schneider et al., 2012; Prabandono & Amin, 2015).

7.2. Biodiesel from Microalgae - Pros and Cons

Monoalkyl esters of long-chain fatty acids (FAME) are defined as biodiesel. Biodiesel is generally considered as clean fuel and offers advantages like reduction of exhaust emissions when compared with fossil fuels, high cetane number, biodegradability, non-toxicity, lack of sulfur, inherent lubricity, positive energy balance, higher flash point, renewability, and origin from domestic sources (Tabatabaei et al., 2015; Faried et al., 2017). Microalgae capture solar energy and use water and atmospheric CO_2 to accumulate organic elements such as lipids under appropriate conditions. They are neutral lipids that are accumulated and stored as triacylglycerols (TAGs) in the thylakoid membrane of microalgae. These lipids serve as a hopeful feedstock for biodiesel production through the process of transesterification of TAGs into several types of fatty acid methyl esters (Zhu, Li, & Hiltunen, 2016; Anand et al., 2019).

The advantages of using microalgae for biodiesel production include rapid doubling time and high oil content compared to oil crops (Chisti, 2007). Also, algal biomass productivity is several folds higher when compared to terrestrial crops (Faried et al., 2017). They do not compete with food crops for water and arable land as they can grow in a wide range of aquatic atmospheres ranging from freshwater to saline and even wastewater with lower land area requirements for the production of a comparable amount of fuel (Faried et al., 2017). However, currently, the usage of microalgal biodiesel has not come up to commercial scale due to high production costs and technical barriers such as slow growth rate, low biomass production, and lipid content. These issues need to be overcome in support of using microalgae as an economically viable feedstock for biofuels (Chisti, 2007).

7.3. Strategies for Lipid Enhancement in Microalgae

Maximizing the overall lipid productivity can be an effective solution for the problem of process economics (Sibi et al., 2016) and lipid content and biomass productivity are two most crucial associated parameters (Shin et al., 2018). During favorable conditions, the fatty acids synthesized by microalgae are primarily used for membrane glycolipids and phospholipids production. Under the stressful or unfavorable situation, they divert their lipid biosynthetic pathway to accumulate more amounts of triacylglycerol (TAG) as neutral lipids at an expense of carbon uptake for cell proliferation (Sibi et al., 2016). This chapter deals with discussing various conventional and recent advances in lipid enhancement strategies with application to biodiesel production (Figure 7.3a).

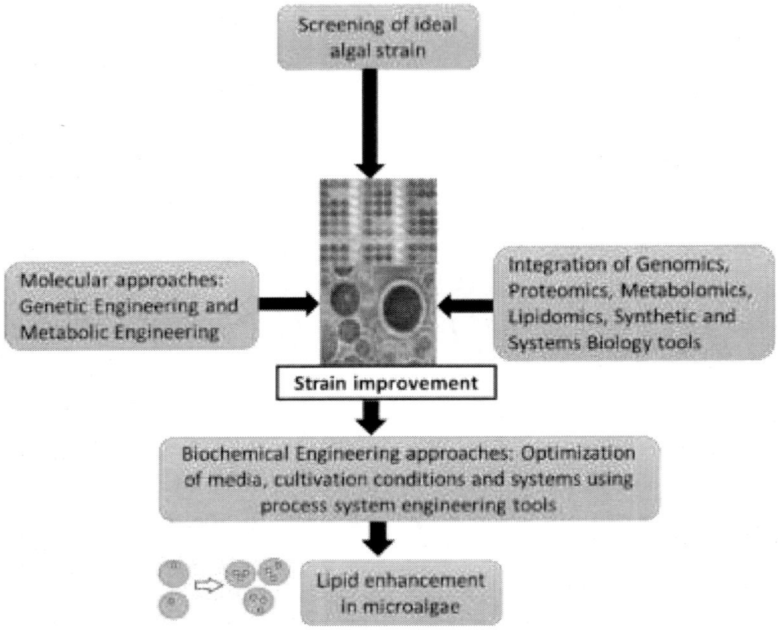

Figure 7.3a. Strategies of strain improvement for lipid enhancement in microalgae.

7.3.1. Screening and Selection of Suitable Microalgal Species

The first stage of lipid enhancement in algae is to explore their immense natural biodiversity and select the suitable species for biodiesel production. It is preferred to isolate the native strains that are likely to be acclimatized to local environmental conditions (Lim & Schenk, 2017). Species of microalgae and the cultivation conditions play a vital role in the lipid content of microalgae (Milano et al., 2016) and the content varies between 10-50% with different species of microalgae. As the total lipid content of a microalgal strain reflects in the quantity of biodiesel, it is essential to identify and select strains with the following characteristics: high lipid content, fast growth rate, appropriate lipid composition i.e., fatty acid profile (Lim et al., 2012), ease of harvesting (Shin et al., 2018), stress resistance (Nascimento et al., 2013). Fast growth rate not only results in rapid accumulation of biomass density leading to a reduction in production cost with increase in the yield per harvest volume but also reduces the risk of contamination (Shin et al., 2018).

Strain selection has to be done specifically on the fatty acid profiles with a right blend of saturated and monounsaturated short-chain fatty acids suitable for biodiesel requirements (Schlagermann et al., 2012) such as very low oxidative potential whilst retaining a good CFPP rating and cetane number (Lim et al., 2012). Under stress conditions, the growth rate of microalgae may get reduced leading to lower biomass production. Hence, the strain has to be tolerant of various extreme stress conditions including thermotolerance and the ability to grow at a varied range of salinities (Chu, 2017). Besides, it is highly appreciable if the strains possess the ability to grow at higher cell density, overexpress and excrete TAG. Isolation, preservation, and characterization of algal species are now automated by the advent of technologies like flow cytometry and robotics, deep sequencing and microfluidics (Larkum et al., 2012).

7.3.2. Strain Improvement

It is extremely difficult to screen and isolate a microalgal species satisfying all of the above characteristics. However, it can be satisfied to a certain extent through strain improvement techniques.

7.3.2.1. Traditional Approaches

Breeding is a traditional approach for strain improvement in microalgae. Following breeding technologies have been applied in the screening of microalgae with high lipid content and biomass. Selective Breeding is the basis of all methods and is widely used. Construction of mutants using chemical and physical mutagenesis is an effective method of strain improvement in microalgae. Chemical mutagens such as sodium azide, ethyl methanesulfonate, N-methyl-N- nitro-N-nitrosoguanidine and pingyangmycin are used to induce the mutations. The common radiation sources for physical mutation are ultraviolet, laser and gamma radiations. Cell Fusion refers to the protoplast fusion of two or more cells to form hybrid cells possessing new biological and genetic characteristics. As the genetic traits are unstable, only a few studies report the use of cell fusion for strain improvement in algae (Ma et al., 2014).

7.3.2.2. Modern Approaches

7.3.2.2.1. Genetic Engineering

Recently, molecular approaches are being employed to enhance biomass and lipid accumulation in microalgae for biodiesel production. Molecular approaches involve the development of transgenic microalgae which requires information of genomics, transcriptomics and the metabolic pathways for identification of the target genes for engineering. To engineer microalgae with high lipid content, molecular level identification of key genes associated with lipid metabolism including lipid biosynthesis and catabolism, as well as pathways involved in the modification of the length and saturation of fatty acids is crucial (Chu, 2017).

Particle bombardment, glass bead agitation, microinjection, electroporation, and *Agrobacterium tumefaciens*-mediated transformation are different techniques used for the manipulation of microalgae (Chu, 2017). The latest genetic engineering tools such as CRISPR-cas9 and RNAi technology has huge space for strain improvement of algal species (Sharma et al., 2018). An outline of the lipid biosynthesis pathway is depicted in Figure 7.3b. The TAG synthesis pathway is comprised of three main steps: (1) carboxylation of acetyl-CoA to form malonyl-CoA (2) acyl chain elongation; and (3) TAG formation.

Overexpression of genes involved in TAG biosynthetic pathway enzymes such as Acetyl-CoA carboxylase (ACC), Fatty acid synthetase (FAS), Lysophosphatidate acyl-transferase (LPAT) and Acyl-CoA: diacylglycerol acyl-transferase (DGAT) can lead to enhancement of lipid production (Courchesne et al., 2009; Lim & Schenk, 2017). Overexpression of enzymes such as Acetyl-CoA synthase (ACS), Malic enzyme (ME), ATP: citrate lyase (ACL), CiS (citrate synthase) (Lim & Schenk, 2017) that are not directly involved in lipid metabolism will also influence the lipid accumulation rate by increasing the pool of vital metabolites for lipid biosynthesis. Lipid catabolism can be suppressed for enhancing lipid accumulation in microalgae. There are also attempts to genetically modify microalgae to increase the quality and composition of the lipids by engineering the carbon chain length and degree of unsaturation of the microalgal fatty acids through manipulation of the expression of thioesterase (Rasala et al., 2013; Chu, 2017; Sharma et al., 2018).

7.3.2.2.2. Metabolic Engineering

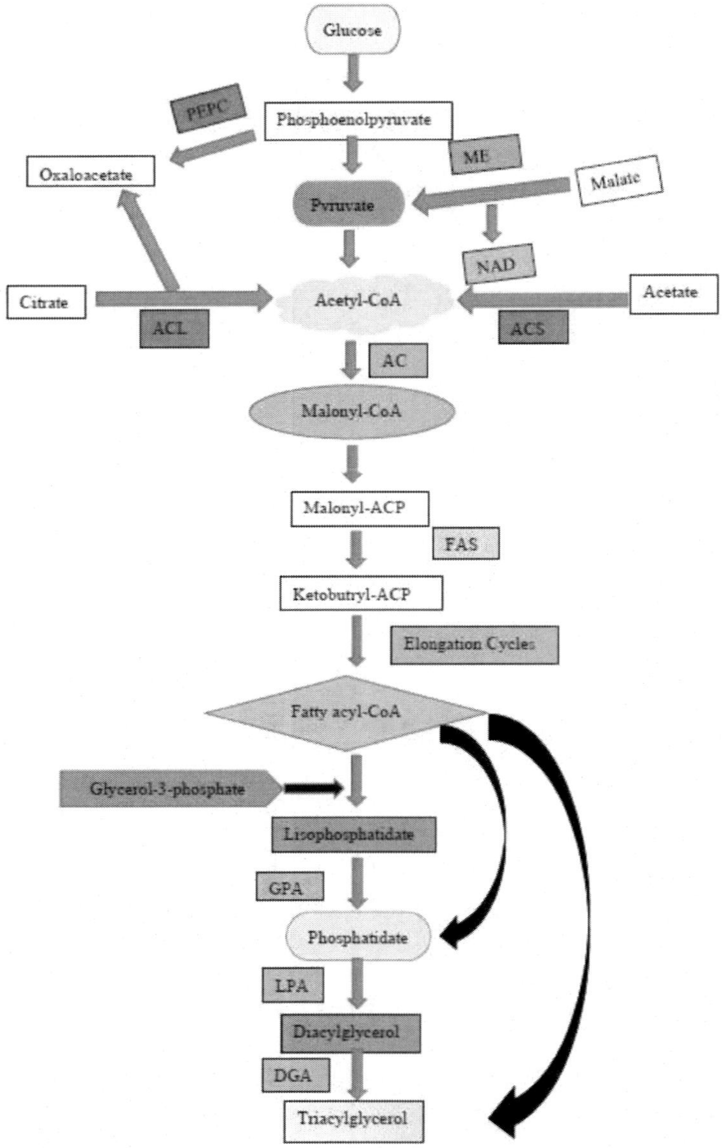

PEPC - Phosphoenolpyruvate carboxylase, ME - Malic Enzyme, NADH - Nicotinamide Adenine Dinucleotide Hydrogen, ACL - ATP Citrate Lyase, ACS- Acetyl-CoA Synthetase, ACC - Acetyl-Coenzyme A Carboxylase, FAS - Fatty-Acyl-CoA Synthase, GPAT - Glycerol-3-Phosphate Acyl Transferase, LPAT - Lyso Phosphatidate Acyl Transferase, DGAT - Diacyl Glycerol Acyl Transferase.

Figure 7.3b. TAG synthesis pathway (modified from Courchesne et al., 2009).

As far as metabolic engineering is considered, stalling off the competing pathways possibly will also augment the TAG biosynthesis. This can be achieved by (i) blocking the metabolic pathways that lead to the accumulation of other energy-rich storage compounds, such as starch. (ii) blocking β-oxidation, principal metabolic pathway responsible for the degradation of fatty acids, (iii) blocking phospholipid biosynthesis, (iv) blocking the conversion of phosphoenolpyruvate to oxaloacetate, which is catalyzed by the phosphoenolpyruvate carboxylase (PEPC). Recently, the use of regulatory factors such as transcription factors (TFs) to control the activity of multiple enzymes related to the production of desired metabolites has triggered extensive interests. This approach affects a large number of genes involving in multiple metabolic pathways that result in integrated up- or down-regulation of these pathways simultaneously. This emerging metabolic engineering approach can be employed for improving the production of lipids in microalgae (Courchesne et al., 2009).

7.4. MANIPULATION OF NUTRIENT FACTORS

The neutral lipid accumulation in oleaginous microalgae depends on strain and culture conditions. A potent strain can be devised to modify the neutral lipid composition by altering the nutrient composition of growth media. Nutrient limitation or deprivation is the commonly practiced strategy for increasing lipid accumulation in several microalgal species (Lima et al., 2018). The advantage of nutrient manipulation is that it can be practically feasible for large scale cultivation.

7.4.1. Optimization of Culture Media Composition and Carbon Dioxide

The composition of nutritional media for microalgae is simple compared to other microorganisms. The nutritional media of microalgae comprises macronutrients (N, P, Mg, S, K), micronutrients or trace metals like Cl, Fe, Mn, Zn, Cu, Mo and vitamins (Mata, Martins, & Caetano, 2010). These major elements in algal culture media are crucial for different cellular activities like photosynthesis, respiration, cell division, protein synthesis, etc. (Klok et al., 2013; Zhang et al., 2014; Singh et al., 2016). Thus optimization of the macro and micronutrients in the growth media will enhance neutral lipid accumulation. Limitation of nutrients such as nitrogen, phosphorous, iron, sulfur, silicon, etc. especially activates the neutral lipid accumulation in microalgae. In the mass cultivation of microalgae, the nutrients are to be optimized along with minimal nutrients for increased biomass and neutral lipid (Blair, Kokabian, & Gude, 2014). Under nutritional stress conditions, the metabolism and biochemical composition of microalgae get altered. Thus an enhanced production of biofuels can be attained by tuning the nutrient composition (Figure 7.4).

CO_2 is essential for the phototrophic cultivation of microalgae. Adequate supply in an optimal level of CO_2 will increase biomass and lipid content in microalgae. A decrease in biomass and lipid production is observed when the level of CO_2 is below optimal as low CO_2 concentration reduces the photosynthetic rate (Morales, Sanchez, & Revah, 2017). Several researchers have reported the concentration of CO_2 varies between species to species with altered neutral lipid content in microalgae. (Ortiz Montoya et al., 2014), reported that CO_2 sparging of 8% (v/v) in *Chlorella vulgaris* enhanced lipid accumulation as 29.5 mg/L. Similarly, *Chlamydomonas sp.* JSC4 when sparged with 4% (v/v) CO_2 showed lipid productivity as 169.1 mg/L (Nakanishi et al., 2014). The maximum lipid content was observed in *S. armatus* (Ho, Chen, & Chang, 2012), *Nannochloropsis* sp. (Jiang et al., 2011) and *Chlorococcum littorale* (Ota et al., 2009) as 22.4%, 59.9%, and 34% respectively.

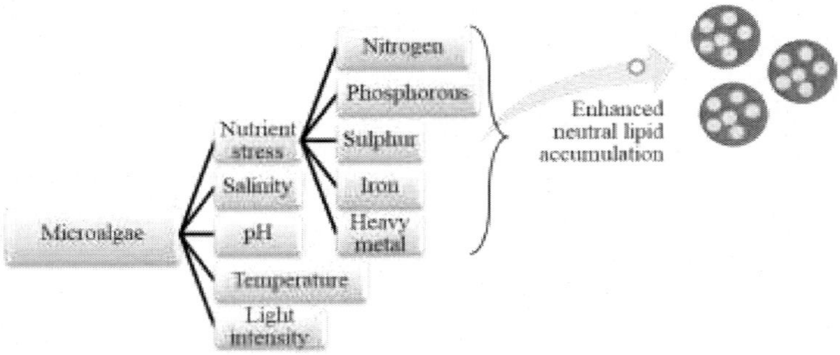

Figure 7.4. Strategies for lipid enhancement in microalgae.

7.4.2. Nitrogen Starvation

The nitrogen in growth medium directly facilitates the growth of cells as it forms the major backbone for the synthesis of proteins and nucleic acids (Li et al., 2008; Arumugam et al., 2013; Sulochana & Arumugam, 2016). Due to these peculiarities, nitrogen stress induction is widely practiced for enhanced lipid accumulation in microalgae. The nitrogen stress induction is performed in two-stage cultivation processes; in the first stage, the microalgae are grown in nitrogen sufficient media to obtain maximum cell density. Further, in the second stage, the harvested biomass is subjected to nitrogen-deficient or nitrogen-limited media to induce nitrogen stress (Anand & Arumugam, 2015; Minhas et al., 2016; Sulochana & Arumugam, 2016; Zhu, Li, & Hiltunen, 2016). Nitrogen stress can be induced by either complete removal (nitrogen deprived) or by limiting the supply of nitrogen to the medium (nitrogen-limited). Either nitrogen limitation or deprivation proved to enhance the lipid accumulation in microalgae. Among the nutritional stress, the most extensively followed strategy is nitrogen deprivation because it is directly related to the lipid metabolism of microalgae and can yield more lipid within hours when compared to other stresses (Table 7.4). The lipids are stored in a highly specialized structure called lipid body, which made up of major lipid droplet protein (MLDP). MLDP is the principal structural component and it recruits other lipid components involved in lipid body maturation. The presence of MLDP protein was detected in *S. quadricauda* using immunostaining in salt stress and nitrogen-starved stress condition (Javee et al., 2016).

The assimilable form of nitrogen sources like ammonium, nitrate, peptone, and urea altered the biomass and neutral lipid accumulation in microalgae (Chen & Chen, 2006). The nitrogen stress-induced *Nannochloropsis sp.*, *Haematococcus pluvialis* shows lipid accumulation after two days with a sharp decline in chlorophyll content (Li et al., 2008; Rodolfi et al., 2009). Nitrogen limitation also leads to the enhanced biosynthesis of carbohydrates in several microalgal species, such as a fourfold increase in carbohydrate content in *Tetraselmis subcordiformis*, and a 29% increase of carbohydrate content in *S. obliquus* CNW-N (Ji et al., 2011; Ho, Chen, & Chang, 2012).

Table 7.4. Nitrogen stress-induced lipid accumulation in microalgae

Microalgae	Lipid content (% dcw)	N stress Lipid (% dcw)	Reference
Neochloris oleoabundans	13	29	Adams et al., 2013
Scenedesmus dinmorphus	9	20	Adams et al., 2013
Chlorella vulgaris	10	40	Adams et al., 2013
Monoraphidium sp.	36.68	44..4	Zhao et al., 2016
Ankistrodesmus falcatus	36.54	34.4	Alvarez-Diaz et al., 2014
Nannochloris sp.	31	51	Takagi et al., 2000
Chlorella sorokiniana	15	21	Adams et al., 2013
Chlorella oleofaciens	12	35	Adams et al., 2013
Scenedesmus naegleii	10	21	Adams et al., 2013

7.4.3. Phosphorous Limitation

Phosphorous is an important microalgal growth component that actively takes part in cellular metabolism such as signal transduction, energy generation, nucleic acid synthesis, respiration and photosynthesis (Raghothama, 2000; El-Kassas, 2013; Singh et al., 2016). Phosphorous limitation results in enhanced lipid accumulation as evidenced in *Scenedesmus sp.* LX1 where a 53% increase in lipid content was observed (Xin et al., 2010). During phosphorous starvation, the freshwater microalga *Scenedesmus obliquus* showed an increment in lipid content from 10 to 29.5% (Mandal & Mallick, 2009). Likewise, the lipid accumulation was observed in *Pavlova lutheri*, *Nannochloropsis sp.* and *Phaeodactylumtric ornutum* but there was a decrease in lipid content in *Tetraselmis sp.* and *Nannochloris atomus* during phosphorus limitation (Reitan, Rainuzzo, & Olsen, 1994; Rodolfi et al., 2009). The polyphosphate forms stored in nutrient sufficient condition was utilized in the phosphorous limited conditions as phosphorous reserves (Liang et al., 2013; Qi et al., 2013; Chu et al., 2014). Also, the sequential effect of phosphorous deficiency-induced reduction in membrane lipid was compromised by non-phosphorous glycolipids and sulpholipids. Liang et al., 2013 reported that maximum lipid content was found to be 23.60% in *Chlorella sp.* at a concentration of 32 μM phosphorous.

7.4.4. Metal Stress

7.4.4.1. Iron

Among the trace metals, iron is important for the normal metabolism and functioning of microalgae. Iron has a major role in photosynthetic electron transport chain and impairment in iron metabolism that leads to a drastic reduction in CO_2 fixation, nitrogen assimilation and alteration of lipid content in microalgae (van-Oijen et al., 2004; Liu et al., 2008; Buitenhuis & Geider, 2010; Cheng & He, 2014). Low iron concentration resulted in decreased chlorophyll concentration, which leads to low biomass and lipid content. When *C. vulgaris* and *S.*

obliquus were grown under the iron-deficient condition, the lipid content increased up to 19.6 and 24.4% (dry cell weight) with a reduction in biomass yield (Singh & Mallick, 2014).

7.4.4.2. Sulfur

Sulfur limitation altered fatty acid content of *Chlamydomonas reinharditi* into twice (Matthew et al., 2009). Under sulfur and phosphorous depletion in *C. reinhardtii*, a decrease in sulfoquinovosyl diacylglycerol (SQDG) and an increase of two-fold in phosphor glycerides was observed (Sato et al., 2000). Furthermore, Branyikova et al. (2011) reported that *Chlorella vulgaris* synthesized 50% more starch under sulfur-limiting conditions than under sulfur-replete conditions.

7.4.4.3. Silicon

Silicon is an important trace metal for diatoms as it influences cell metabolism and it is a major component of diatom cell walls. According to Griffiths & Harrison, 2009, silicon limitation increased cellular lipid content in *Chaetoceros muelleri*, *Cyclotella cryptica*, and *Naviculas aprophila* as 89, 110 and 104% dry cell weight respectively.

7.4.5. Growth Hormones

Phytohormones are chemical messengers that actively take part in the growth, metabolic processes, senescence and resistance to biotic and abiotic stresses of plants. They mainly comprise five different classes like auxin, gibberellins, cytokinins, abscisic acid and ethylene (Han et al., 2018). The synthesis and metabolism of phytohormones are not yet described in microalgae rather the presence of phytohormone was detected in various microalgal species (Tarakhovskaya et al., 2007; Lu & Xu, 2015). According to several researchers, the phytohormones particularly shoots up during stress conditions to induce cell defense mechanism for adapting the harsh environmental conditions. The research groups also evidenced that the exogenous supplementation of these phytohormones improves the growth and lipid pattern of microalgae.

Wu et al., 2018 reported that Abscisic acid (ABA- 1mg/L), Salicylic acid (SA- 10mg/L) and Jasmonic acid (JA- 0.5mg/L) induced 1.5-2 fold elevated lipid content in *Chlorella vulgaris* ZF strain than the controls. Similarly, the supplementation of auxins, cytokinins, ABA, polyamines, brassinosteroids, JA and SA improved the adaptability to environmental stress tolerance to *Chlorella vulgaris* (Bhola et al., 2011; Piotrowska-Niczyporuk et al., 2012). Enhanced lipid accumulation was observed in *Monoraphidium sp.* by the addition of melatonin (Li et al., 2017) and fluvic acid (Che et al., 2017). The increased level of phytohormones may induce cellular ROS production, gene expression, and stress tolerance eventually leads to lipid accumulation (Che et al., 2017; Chu, 2017). Individually IAA (Indole Acetic Acid) promoted growth in *C. vulgaris* (50 ppm), *Nannochoropsis oceanica* (Udayan & Arumugam, 2017) and *Scenedesmus obliquus* (10^{-5} M) (Salama et al., 2014) as 11-19 times and 1.9 fold than the control respectively. Cytokinins enhance the accumulation of photosynthetic pigments in *C. vulgaris* (Piotrowska & Czerpak, 2009) and *Dunaliella salina* (Mousavi et al., 2016). Also, exogenous supplementation of zeatin (0.1 mg/L) increased the growth rate in *Acutodesmus obliquus*. With the increased growth, zeatin and kinetin promoted lipid accumulation of about 64.95 and 63.06% respectively in *A. obliquus* (Renuka et al., 2017). GA3 (Gibberellic Acid) induced the rapid growth of *Microcystis aeruginosa* by uptake of more nitrogen and utilization of carbohydrates (Pan et al., 2008). Other findings observed in *Chlorella pyrenoidosa* increased growth and lipid accumulation (292.3 mg/g) (Du et al., 2015). ABA, stress-responsive hormone suppressed the growth in *Haematococcus pluvialis* with elevated carotenoids by the exogenous supplementation (Kobayashi et al., 1997). A similar growth pattern was observed in other microalgae like *Coscinodiscus granii* (Kentzer & Mazur, 1991) and *Nannochloropsis oceanica* (Lu et al., 2014). When comparing this negative impact of ABA to freshwater microalga *Scenedesmus quadricauda*, the biomass concentration sustained in nitrogen stressed condition rather a rapid decline in biomass (Sulochana & Arumugam, 2016). Ethylene well studied as a growth inhibitor but promoted growth depending upon the concentration supplemented. Exogenous ethylene in *H. pluvialis* increased astaxanthin production (Gao & Meng, 2007).

7.4.6. Utilization of Low-Cost Substrates

To reduce the microalgal oil production cost and more suited for commercial diesel blend, low-cost substrates were utilized. The low-cost substrates include industrial effluents, lignocellulosic materials, industrial wastes (glycerol), residues and by-products (Dias et al., 2019). The oleaginous microalga *Schizochytrium limacinum* produces docosahexanoic acid (DHA) insignificant amount when grown in different carbon sources such as glucose, glycerol or fructose (Yokochi et al., 1998; Chi et al., 2007). Qin et al., 2019 reported that the *Yarrowia lipolytica* and *C. vulgaris* use 250 mM glycerol and 40 mM of dairy wastewater as substrate. Also, the mixotrophic cultivation of *Rhodotorula glutinisa* and *C. vulgaris* using effluent from the seafood processing plant and molasses from sugar cane plant as substrates (Cheirsilp, Suwannarat, & Niyomdecha, 2011) would reduce the substrate cost with large scale cultivation.

7.4.7. Addition of Anti-Oxidants

Oxidative stress induces the Reactive oxygen species (ROS) and eventually leads to cellular damage. To suppress the enormous ROS generation, the cell antioxidant scavenging system will get turned on. Several studies showed that antioxidant has a direct effect on growth and lipid production in different microalgae. According to Ren et al., 2014, DHA and biomass yield has improved by 30.44% and 16.16% respectively with the addition of ascorbic acid (9 g/L) for *Schizochytrium sp.* Similarly in *Crypthecodinium cohnii,* an increase in cell number and fatty acid production was observed by the addition of sesamol (Liu et al., 2015). In addition to the antioxidant property, the ascorbic acid also enhances the expression of enzyme glucose-6-phosphate dehydrogenase.

7.5. Manipulation of Growth Parameters

The growth factors other than nutrients such as light intensity, temperature, pH and salinity, changes in these parameters also lead to enhanced lipid accumulation in microalgae.

7.5.1. Light Intensity

Being a photoautotrophic microorganism, microalgae requires a light source for normal growth and metabolism. Also, the light intensity is a known parameter for the growth of microalgae. The light source and intensity obtained by microalgae will affect the photosynthetic activity. Moreover, different light intensities and wavelengths change the lipid metabolism in microalgae altering the lipid content. Typically, the amount of polyunsaturated fatty acids (structural lipids) increases under low light conditions, whereas high light promotes the accumulation of saturated and mono-unsaturated fatty acids (storage lipids).

High light intensity leads to oxidative damage of PUFA, and is also required for the synthesis of C16:1 (3 *trans*) and alters the level of this fatty acid in microalgae. Pal et al., 2011 evidenced that maximum lipid content was observed in *Nannochloropsis sp.* under 700 µmol photons $s^{-1}m^{-2}$ light intensity. Under higher light intensities in *Pavlova lutheri,* increased lipid content was associated with an increase in both cell population and weight per cell profile (Carvalho & Malcata, 2005). The effect of different light intensities on algal lipid composition was successfully demonstrated in a detailed study on *Chlorella vulgaris* (Hassan, Aljobory, & Al-Jumaily, 2013). Thus, the lipid productivity of different microalgae can be attenuated with high light stress. Similarly, there is increased lipid productivity in *C. vulgaris*, *Raphidocelis subcapitata*, *Mychonastes homosphaera* and *Scenedesmus sp.* under high light irradiance (Tang et al., 2011; Liu et al., 2012; Gonçalves, Pires, & Simões, 2013). Therefore saturated and monounsaturated fatty acids are preferred sources for biodiesel and it may be feasible to use relatively high light intensities to improve biodiesel yield.

7.5.2. Temperature

The temperature has an essential role in the growth and lipid composition of microalgae. The optimum temperature for growth and lipid content in microalgae varies between species to species (Sibi et al., 2016). For example, the optimum temperature for *Scenedesmus sp.* was found to be 20°C (Xin et al., 2011). Converti et al. (2009) have reported that the lipid content increased from 7.9 to 14.9% between 20-25°C. According to Renaud et al., 2002, there is a shift in fatty acid composition with temperature i.e, at lower temperatures, the unsaturated fatty acid found to be increased and saturated fatty acid content will be increased at a higher temperature. As lipid profile changes at different temperatures, the properties of algal-derived biodiesel would also change for different climates and seasons and hence seasonal cultivation can be practiced for biodiesel production.

7.5.3. Salinity

Microalgal cultivation using freshwater is a major concern when practicing large scale cultivation. Therefore, microalgae that can adapt to the marine environment will be a good choice as an industrially potent strain. Salinity got attention as it created osmotic stress and eventually led to ROS generation with lipid accumulation in microalgae (Asulabh, Supriya, & Ramachandra, 2012). Increase in salinity may result in a significant increase in the lipid content of several microalgal species, including *Chlorococcum sp.* (Harwati, Willke, & Vorlop, 2012), *Botryococcus braunii* (Zhila, Kalacheva, & Volova, 2011), *Chlorella protothecoides* (Campenni et al., 2013) and *Nannochloropsis sp.* (Pal et al., 2011). *Dunaliella* cells accumulated about 70% of lipid content under high salt concentration (Mutsumi Takagi & Yoshida, 2006). When diatom *Nitzschialaevis* was subjected to high salt concentrations, 71.3% of total EPA (Eicosa Pentanoic Acid) was obtained (G. Chen, Jiang, & Chen, 2008).

7.6. Maximizing Biomass Production

As discussed previously, nutrient stress strategies are successful for enhancing lipid accumulation in microalgae while resulting in low biomass yield thereby affecting the overall lipid productivity. Hence, maximizing the biomass yield is very important as the total lipid output relies on the total biomass content. Stress, CO_2 fixation rate, and light utilization efficiency play a vital role in governing biomass productivity. To improve the biomass yield, conventional approaches focussed on the optimization of nutrients and culture conditions, usage of mixed microalgal-microbial consortia, genetic modification, and photobioreactor (PBR) design improved photosynthetic efficiency is essential. Transgenic microalgal strains tolerant to abiotic stress through enhanced reactive oxygen species (ROS) scavenging (Ramanna, Rawat, & Bux, 2017) can be employed.

7.7. Bioreactor Strategies for Maximizing Lipid Production

To achieve maximal biomass and lipid productivity, the cultivation system plays an important role. Two types of bioreactors are usually employed for the cultivation of microalgae (i) open or raceway or high rate algal ponds (HRAP) (ii) photobioreactors. While the former is cost-effective and has low energy requirements the latter offers a closed environment with resourceful capture and utilization of light (Benavente-Valdés et al., 2016; Chu, 2017; Shin et al., 2018).

7.7.1 Two-Stage Cultivation

In this strategy, microalgae are cultured under two different culture conditions in a separate bioreactor (by changing either the carbon source or type of reactor) to increase biomass productivity and lipid accumulation. A hybrid system with a combination of open ponds and photobioreactors is an attractive strategy to maximize the production of algal biomass as well as lipids for biofuel feedstock. The microalgal growth will be favored by first stage PBR under nutrient-sufficient conditions to attain high biomass concentration whereas the second stage open pond favors the production of the lipids when subjected to nutrient deprivation (a stress condition). In this way, two-stage cultures can overcome the limitation of nutrient stress-induced reduction in biomass productivity (Figure 7.8). The use of vertical tubular-type photobioreactors has been reported for enhancing lipid production (Yeh & Chang, 2011; Chen et al., 2014).

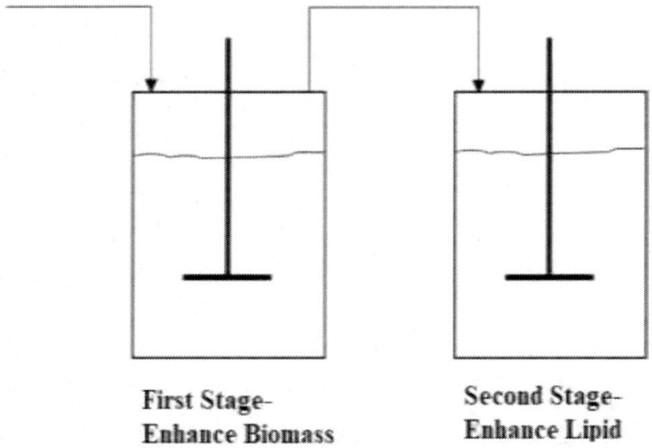

Figure 7.8. Two-stage cultivation process for lipid enhancement.

7.8. Co-Cultivation of Microalgae with Other Microalgae, Yeasts, and Bacteria

Co-culturing of microalgae with other microorganisms, especially bacteria and yeasts will induce biomass and lipid production. Using co-cultivation, the production cost can also be decreased by coupling wastewater treatment with lipid production by microalgae. Further, the synergistic effects by the co-cultivation of two different microalgae and the nutrient-starvation are directly related to the increased lipid yield.

7.9. Application of Process System Engineering Tools

Process system engineering tools such as artificial intelligence (AI), particularly neural learning, statistical and evolutionary learning-based techniques can be used as an innovative approach for optimizing the lipid productivity in microalgae. Bayesian and naïve Bayes models, clustering, hidden Markov and nearest neighbor models are statistical learning techniques that are used in AI. Genetic algorithm (GA), particle swarm optimization (PSO), ant colony optimization (ACO), bees algorithms are widely used in evolutionary learning methods. Hybrid AI methods like Neuro-fuzzy, Neuro-genetic, and Fuzzy-genetic can be exploited in the future (Jha et al., 2017).

7.10. Strategies concerned with Downstream Processing

Downstream processing of algal lipids contributes to the major cost of biodiesel production using microalgae (Ranjith Kumar et al., 2015). The production costs can be cut down considerably by manipulating the microalgal cells to secrete TAG directly into the growth medium (Chu, 2017). Bio-based solvents such as ethyl acetate, ethyl lactate, cyclopentyl methyl ether, and 2-methyl tetrahydrofuran can serve as an alternative for conventional petroleum-based solvents by providing enhanced extraction yields and fatty acid composition suitable for practical applications (Mahmood et al., 2017).

Conclusion and Future Prospectives

With the growing need for renewable energy, microalgal lipid serves as a boon for biodiesel production. At the same time, for commercialization of algal biofuels the rate and total accumulation of microalgal lipids have to be improved. This chapter discussed various approaches or strategies for enhancing lipid productivity. These strategies can be employed as a stand-alone approach or integration of one or more strategies will lead to appreciable enhancement of lipid production in microalgae. Advancements in genomics, transcriptomics, metabolomics proteomics, and lipidomics and emerging strategies such as synthetic and systems biology provide scope for enhanced lipid production.

Omics technology can serve as a transformation tool for achieving higher lipid productivity. Genomics, Transcriptomics, Proteomics, Metabolomics helps in understanding the biological functions in microalgae and contribute to the development of strains with superior productivity. Genomics involves identification of functional gene using next-generation sequencing methods, pyrosequencing, sequencing by ligation, real-time sequencing that can be exploited to increase the rate of lipid biosynthesis. Gene expression analysis can be done from the transcriptome produced during lipid biosynthesis. Proteomics involves the analysis of various biological processes at the protein level. Metabolomics assists in the study of metabolic pathways of microalgae leading to improved production of lipids through pathway engineering of lipid metabolism in microalgae which are crucial before metabolic engineering can be applied to enhance lipid productivity in microalgae. The lipidomic analysis helps in understanding the fundamentals of lipid metabolism and TAG accumulation in microalgae (Anand et al., 2019). The integration of omics techniques in microalgae will accelerate the efforts in the elucidation of mechanisms and potential engineering targets of lipid accumulation (Figure 7.12).

Systems biology combines the high throughput experimental data generated from omics technology with mathematical models and computational tools for the development of comprehensive cellular metabolic models. The development of appropriate mathematical models can also describe the microalgal system along with the prediction of the system's response to perturbations enabling dynamic modeling of algal cellular metabolism.

Application of synthetic biology microalgal technology is an emerging field wherein the artificial regulatory circuits are built to control the cellular behavior based on user-defined inputs/signals/stimuli to generate the desired output such as biochemicals by making required changes in metabolism. Synthetic biology paves the way in the construction of artificial biological systems by coupling the engineering strategy with genetic, metabolic and regulatory data from experiments.

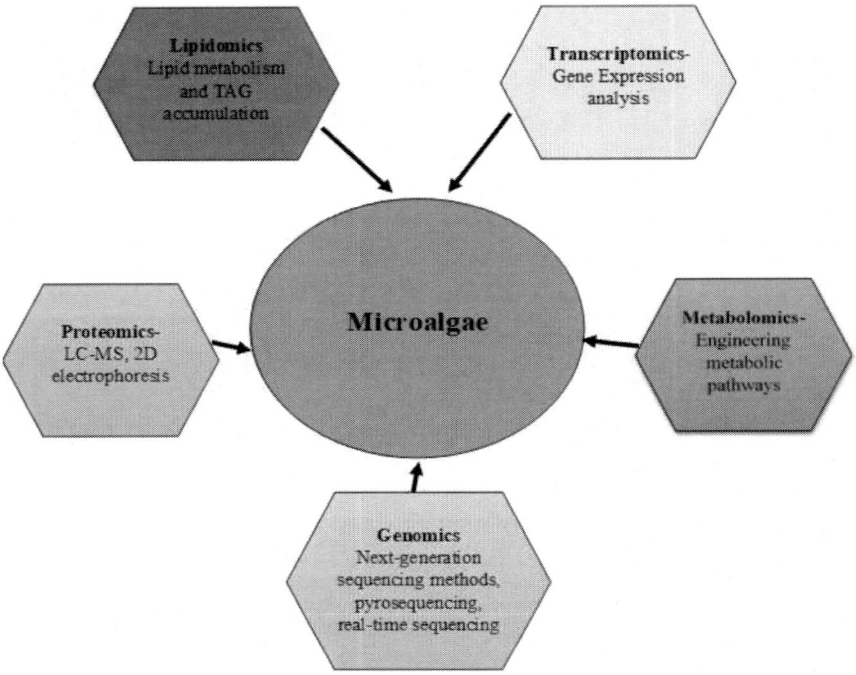

Figure 7.12. Integration of Omics technology towards enhancement of TAG accumulation in microalgae.

REFERENCES

Adams, C., Godfrey, V., Wahlen, B., Seefeldt, L., & Bugbee, B. (2013). Understanding precision nitrogen stress to optimize the growth and lipid content tradeoff in oleaginous green microalgae. *Bioresour. Technol.* 131, 188–194.

Adeniyi, O. M., Azimov, U., & Burluka, A. (2018). Algae biofuel: Current status and future applications. *Renew. Sustain. Energy Rev.* 90, 316–335.

Álvarez-Díaz, P. D., Ruiz, J., Arbib, Z., Barragán, J., Garrido-Pérez, C., & Perales, J. A. (2014). Lipid production of microalga *Ankistrodesmus falcatus* increased by nutrient and light starvation in a two-stage cultivation process. *Appl. Biochem. Biotechnol.* 174, 1471–1483.

Anand, J., & Arumugam, M. (2015). Enhanced lipid accumulation and biomass yield of *Scenedesmus quadricauda* under nitrogen starved condition. *Bioresour. Technol.* 188, 190–194.

Anand, V., Kashyap, M., Samadhiya, K., & Kiran, B. (2019). Strategies to unlock lipid production improvement in algae. *Int. J. Environ. Sci. Technol.* 16, 1829–1838.

Arumugam, M., Agarwal, A., Arya, M. C., & Ahmed, Z. (2013). Influence of nitrogen sources on biomass productivity of microalgae *Scenedesmus bijugatus*. *Bioresour. Technol.* 131, 246–249.

Asulabh, K. S., Supriya, G., & Ramachandra, T. V. (2012). Effect of salinity concentrations on growth rate and lipid concentration in Microcystis sp., *Chlorococcum sp.* and *Chaetoceros sp.* In *Natl. Conf. Conserv. Manag. Wetl. Ecosyst. Sch. Environ. Sci. Mahatma Gandhi Univ. Kottayam, Kerala.*

Benavente-Valdés, J. R., Aguilar, C., Contreras-Esquivel, J. C., Méndez-Zavala, A., & Montañez, J. (2016). Strategies to enhance the production of photosynthetic pigments and lipids in chlorophycae species. *Biotechnol. Reports* 10, 117–125.

Bhola, V., Desikan, R., Santosh, S. K., Subburamu, K., Sanniyasi, E., & Bux, F. (2011). Effects of parameters affecting biomass yield and thermal behaviour of *Chlorella vulgaris*. *J. Biosci. Bioeng.* 111, 377–382.

Blair, M. F., Kokabian, B., & Gude, V. G. (2014). Light and growth medium effect on *Chlorella vulgaris* biomass production. *J. Environ. Chem. Eng.* 2, 665–674.

Brányiková, I., Maršálková, B., Doucha, J., Brányik, T., Bišová, K., Zachleder, V., & Vítová, M. (2011). Microalgae—novel highly efficient starch producers. *Biotechnol. Bioeng.* 108, 766–776.

Buitenhuis, E. T., & Geider, R. J. (2010). A model of phytoplankton acclimation to iron–light colimitation. *Limnol. Oceanogr.* 55, 714–724.

Campenni, L., Nobre, B. P., Santos, C. A., Oliveira, A. C., Aires-Barros, M. R., Palavra, A. M. F., & Gouveia, L. (2013). Carotenoid and lipid production by the autotrophic microalga *Chlorella protothecoides* under nutritional, salinity, and luminosity stress conditions. *Appl. Microbiol. Biotechnol.* 97, 1383–1393.

Carvalho, A. P., & Malcata, F. X. (2005). Optimization of ω-3 fatty acid production by microalgae: crossover effects of CO_2 and light intensity under batch and continuous cultivation modes. *Mar. Biotechnol.* 7, 381–388.

Che, R., Huang, L., Xu, J. W., Zhao, P., Li, T., Ma, H., & Yu, X. (2017). Effect of fulvic acid induction on the physiology, metabolism, and lipid biosynthesis-related gene transcription of *Monoraphidium sp.* FXY-10. *Bioresour. Technol.* 227, 324–334.

Cheirsilp, B., Suwannarat, W., & Niyomdecha, R. (2011). Mixed culture of oleaginous yeast *Rhodotorula glutinis* and microalga *Chlorella vulgaris* for lipid production from industrial wastes and its use as biodiesel feedstock. *N. Biotechnol.* 28, 362–368.

Chen, C. Y., Yeh, K. L., Chang, H. Y., & Chang, J. S. (2014). Strategies to improve oil/lipid production of microalgae in outdoor cultivation using vertical tubular-type photobioreactors. *Energy Procedia* 61, 2755–2758.

Chen, G. Q., & Chen, F. (2006). Growing phototrophic cells without light. *Biotechnol. Lett.* 28, 607–616.

Chen, G., Jiang, Y., & Chen, F. (2008). Salt-induced alterations in lipid composition of diatom *Nitzschia laevis* (Bacillariophyceae) under heterotrophic culture condition (1). *J. Phycol.* 44, 1309–1314.

Cheng, D., & He, Q. (2014). Assessment of environmental stresses for enhanced microalgal biofuel production–an overview. *Front. energy Res.* 2, 26.

Chi, Z., Pyle, D., Wen, Z., Frear, C., & Chen, S. (2007). A laboratory study of producing docosahexaenoic acid from biodiesel-waste glycerol by microalgal fermentation. *Process Biochem.* 42, 1537–1545.

Chisti, Y. (2007). Biodiesel from microalgae. *Biotechnology Advances* 25, 294–306.

Chu, F. F., Chu, P. N., Shen, X. F., Lam, P. K. S., & Zeng, R. J. (2014). Effect of phosphorus on biodiesel production from *Scenedesmus obliquus* under nitrogen-deficiency stress. *Bioresour. Technol.* 152, 241–246.

Chu, W. L. (2017). Strategies to enhance production of microalgal biomass and lipids for biofuel feedstock. *Eur. J. Phycol.* 52, 419–437.

Converti, A., Casazza, A. A., Ortiz, E. Y., Perego, P., & Del Borghi, M. (2009). Effect of temperature and nitrogen concentration on the growth and lipid content of *Nannochloropsis oculata* and *Chlorella vulgaris* for biodiesel production. *Chem. Eng. Process. Process Intensif.* 48, 1146–1151.

Courchesne, N. M. D., Parisien, A., Wang, B., & Lan, C. Q. (2009). Enhancement of lipid production using biochemical, genetic and transcription factor engineering approaches. *J. Biotechnol.* 141, 31–41.

Dias, C., Santos, J., Reis, A., & da Silva, T. L. (2019). Yeast and microalgal symbiotic cultures using low-cost substrates for lipid production. *Bioresour. Technol. Reports* 100261.

El-Kassas, H. Y. (2013). Growth and fatty acid profile of the marine microalga *Picochlorum sp.* grown under nutrient stress conditions. *Egypt. J. Aquat. Res.* 39, 233–239.

Faried, M., Samer, M., Abdelsalam, E., Yousef, R. S., Attia, Y. A., & Ali, A. S. (2017). Biodiesel production from microalgae: Processes, technologies and recent advancements. *Renew. Sustain. Energy Rev.* 79, 893–913.

Gao, Z. Q., & Meng, C. X. (2007). Impact of extraneous ethylene concentrations to astaxanthin accumulation of *Haematoccus pluvialis*. *Food Sci.* 10, 93.

Gonçalves, A. L., Pires, J. C. M., & Simões, M. (2013). Lipid production of *Chlorella vulgaris* and *Pseudokirchneriella subcapitata*. *Int. J. Energy Environ. Eng.* 4, 14.

Griffiths, M. J., & Harrison, S. T. L. (2009). Lipid productivity as a key characteristic for choosing algal species for biodiesel production. *J. Appl. Phycol.* 21, 493–507.

Han, X., Zeng, H., Bartocci, P., Fantozzi, F., & Yan, Y. (2018). Phytohormones and effects on growth and metabolites of microalgae: a review. *Fermentation* 4, 25.

Harwati, T. U., Willke, T., & Vorlop, K. D. (2012). Characterization of the lipid accumulation in a tropical freshwater microalgae *Chlorococcum sp. Bioresour. Technol.* 121, 54–60.

Hassan, F. M., Aljobory, I. F., & Al-Jumaily, E. F. (2013). Stimulation of biodiesel production from two algae: *Chlorella vulgaris* Berjerinck and Nitzschiapalea (Kütz) Smith, and study their some growth parameters under different light intensity. *J. Env. Sci. Toxicol. Food Technol* 6, 31–42.

Ho, S. H., Chen, C. Y., & Chang, J. S. (2012). Effect of light intensity and nitrogen starvation on CO2 fixation and lipid/carbohydrate production of an indigenous microalga *Scenedesmus obliquus* CNW-N. *Bioresour. Technol.* 113, 244–252.

Javee, A., Sulochana, S. B., Pallissery, S. J., & Arumugam, M. (2016). Major lipid body protein: a conserved structural component of lipid body accumulated during abiotic stress in *S. quadricauda* CASA-CC202. *Frontiers in Energy Research*, 4, 37.

Jha, S. K., Bilalovic, J., Jha, A., Patel, N., & Zhang, H. (2017). Renewable energy: Present research and future scope of Artificial Intelligence. *Renew. Sustain. Energy Rev.* 77, 297–317.

Ji, C. F., Yu, X. J., Chen, Z. A., Xue, S., Legrand, J., & Zhang, W. (2011). Effects of nutrient deprivation on biochemical compositions and photo-hydrogen production of *Tetraselmis subcordiformis*. *Int. J. Hydrogen Energy* 36, 5817–5821.

Jiang, L., Luo, S., Fan, X., Yang, Z., & Guo, R. (2011). Biomass and lipid production of marine microalgae using municipal wastewater and high concentration of CO_2. *Appl. Energy* 88, 3336–3341.

Kentzer, T., & Mazur, H. (1991). Abscisic acid as endogenous inhibitor of the marine diatom *Coscinodiscus granii*. *Acta Physiol. Plant.* 13, 153–157.

Klok, A. J., Martens, D. E., Wijffels, R. H., & Lamers, P. P. (2013). Simultaneous growth and neutral lipid accumulation in microalgae. *Bioresour. Technol.* 134, 233–243.

Kobayashi, M., Hirai, N., Kurimura, Y., Ohigashi, H., & Tsuji, Y. (1997). Abscisic acid-dependent algal morphogenesis in the unicellular green alga *Haematococcus pluvialis*. *Plant Growth Regul.* 22, 79–85.

Larkum, A. W. D., Ross, I. L., Kruse, O., & Hankamer, B. (2012). Selection, breeding and engineering of microalgae for bioenergy and biofuel production. *Trends Biotechnol.* 30, 198–205.

Li, D., Zhao, Y., Ding, W., Zhao, P., Xu, J. W., Li, T., Ma, H., & Yu, X. (2017). A strategy for promoting lipid production in green microalgae *Monoraphidium sp.* QLY-1 by combined melatonin and photoinduction. *Bioresour. Technol.* 235, 104–112.

Li, Y., Horsman, M., Wang, B., Wu, N., & Lan, C. Q. (2008). Effects of nitrogen sources on cell growth and lipid accumulation of green alga *Neochloris oleoabundans*. *Appl. Microbiol. Biotechnol.* 81, 629–636.

Liang, K., Zhang, Q., Gu, M., & Cong, W. (2013). Effect of phosphorus on lipid accumulation in freshwater microalga *Chlorella sp*. *J. Appl. Phycol.* 25, 311–318.

Lim, D. K. Y., Garg, S., Timmins, M., Zhang, E. S. B., Thomas-Hall, S. R., Schuhmann, H., Li, Y., & Schenk, P. M. (2012). Isolation and evaluation of oil-producing microalgae from subtropical coastal and brackish waters. *PLoS One* 7, e40751.

Lim, D. K. Y., & Schenk, P. M. (2017). Microalgae selection and improvement as oil crops: GM vs non-GM strain engineering. *AIMS Bioeng.* 4, 151–161.

Lima, J. F. De, José, A., Queiroz, D. M., Maria, R., Figueirêdo, F. De, Hérica, R., Rocha, C., Souza, W. R. De, & Agroenergy, E. (2018). Enhancement of Lipid Accumulation in *Chlorella sp . Farming Culture Held in Open Photobioreactors* 88–99.

Liu, B., Liu, J., Sun, P., Ma, X., Jiang, Y., & Chen, F. (2015). Sesamol enhances cell growth and the biosynthesis and accumulation of docosahexaenoic acid in the microalga *Crypthecodinium cohnii*. *J. Agric. Food Chem.* 63, 5640–5645.

Liu, J., Yuan, C., Hu, G., & Li, F. (2012). Effects of light intensity on the growth and lipid accumulation of microalga *Scenedesmus sp.* 11-1 under nitrogen limitation. *Appl. Biochem. Biotechnol.* 166, 2127–2137.

Liu, Z. Y., Wang, G. C., & Zhou, B. C. (2008). Effect of iron on growth and lipid accumulation in *Chlorella vulgaris*. *Bioresour. Technol.* 99, 4717–4722.

Lu, Y., Tarkowská, D., Turečková, V., Luo, T., Xin, Y., Li, J., Wang, Q., Jiao, N., Strnad, M., & Xu, J. (2014). Antagonistic roles of abscisic acid and cytokinin during response to nitrogen depletion in oleaginous microalga *Nannochloropsis oceanica* expand the evolutionary breadth of phytohormone function. *Plant J.* 80, 52–68.

Lu, Y., & Xu, J. (2015). Phytohormones in microalgae: a new opportunity for microalgal biotechnology? *Trends Plant Sci.* 20, 273–282.

Ma, C., Liu, B. F., Ren, H. Y., & Ren, N. Q. (2014). A review on biodiesel production: breeding technologies of microalgae containing rich lipid. In *Appl. Mech. Mater.* Vol. 472, pp. 759–763. Trans Tech Publ.

Mahmood, W. M. A. W., Theodoropoulos, C., & Gonzalez-Miquel, M. (2017). Enhanced microalgal lipid extraction using bio-based solvents for sustainable biofuel production. *Green Chem.* 19, 5723–5733.

Mandal, S., & Mallick, N. (2009). Microalga *Scenedesmus obliquus* as a potential source for biodiesel production. *Appl. Microbiol. Biotechnol.* 84, 281–291.

Mata, T. M., Martins, A. A., & Caetano, N. S. (2010). Microalgae for biodiesel production and other applications: A review. *Renew. Sustain. Energy Rev.* 14, 217–232.

Matthew, T., Zhou, W., Rupprecht, J., Lim, L., Thomas-Hall, S. R., Doebbe, A., Kruse, O., Hankamer, B., Marx, U. C., & Smith, S. M. (2009). The metabolome of *Chlamydomonas reinhardtii* following induction of anaerobic H_2 production by sulfur depletion. *J. Biol. Chem.* 284, 23415–23425.

Milano, J., Ong, H. C., Masjuki, H. H., Chong, W. T., Lam, M. K., Loh, P. K., & Vellayan, V. (2016). Microalgae biofuels as an alternative to fossil fuel for power generation. *Renew. Sustain. Energy Rev.* 58, 180–197.

Minhas, A. K., Hodgson, P., Barrow, C. J., & Adholeya, A. (2016). A review on the assessment of stress conditions for simultaneous production of microalgal lipids and carotenoids. *Front. Microbiol.* 7, 546.

Morales, M., Sánchez, L., & Revah, S. (2017). The impact of environmental factors on carbon dioxide fixation by microalgae. *FEMS Microbiol. Lett.* 365, 262.

Mousavi, P., Morowvat, M. H., Montazeri-Najafabady, N., Abolhassanzadeh, Z., Mohagheghzadeh, A., Hamidi, M., Niazi, A., & Ghasemi, Y. (2016). Investigating the effects of phytohormones on growth and β-carotene production in a naturally isolates stain of *Dunaliella salina*. *J. Appl. Pharm. Sci.* 6, 164–171.

Nakanishi, A., Aikawa, S., Ho, S. H., Chen, C. Y., Chang, J. S., Hasunuma, T., & Kondo, A. (2014). Development of lipid productivities under different CO_2 conditions of marine microalgae *Chlamydomonas sp.* JSC4. *Bioresour. Technol.* 152, 247–252.

Nascimento, I. A., Marques, S. S. I., Cabanelas, I. T. D., Pereira, S. A., Druzian, J. I., de Souza, C. O., Vich, D. V., de Carvalho, G. C., & Nascimento, M. A. (2013). Screening microalgae strains for biodiesel production: lipid productivity and estimation of fuel quality based on fatty acids profiles as selective criteria. *Bioenergy Res.* 6, 1–13.

Ortiz Montoya, E. Y., Casazza, A. A., Aliakbarian, B., Perego, P., Converti, A., & de Carvalho, J. C. M. (2014). Production of *Chlorella vulgaris* as a source of essential fatty acids in a tubular photobioreactor continuously fed with air enriched with CO_2 at different concentrations. *Biotechnol. Prog.* 30, 916–922.

Ota, M., Kato, Y., Watanabe, H., Watanabe, M., Sato, Y., Smith Jr, R. L., & Inomata, H. (2009). Fatty acid production from a highly CO_2 tolerant alga, *Chlorocuccum littorale*, in the presence of inorganic carbon and nitrate. *Bioresour. Technol.* 100, 5237–5242.

Pal, D., Khozin-Goldberg, I., Cohen, Z., & Boussiba, S. (2011). The effect of light, salinity, and nitrogen availability on lipid production by *Nannochloropsis sp. Appl. Microbiol. Biotechnol.* 90, 1429–1441.

Pan, X., Chang, F., Kang, L., Liu, Y., Li, G., & Li, D. (2008). Effects of gibberellin A3 on growth and microcystin production in *Microcystis aeruginosa* (Cyanophyta). *J. Plant Physiol.* 165, 1691–1697.

Piotrowska-Niczyporuk, A., Bajguz, A., Zambrzycka, E., & Godlewska-Żyłkiewicz, B. (2012). Phytohormones as regulators of heavy metal biosorption and toxicity in green alga *Chlorella vulgaris* (Chlorophyceae). *Plant Physiol. Biochem.* 52, 52–65.

Piotrowska, A., & Czerpak, R. (2009). Cellular response of light/dark-grown green alga *Chlorella vulgaris* Beijerinck (Chlorophyceae) to exogenous adenine-and phenylurea-type cytokinins. *Acta Physiol. Plant.* 31, 573–585.

Prabandono, K., & Amin, S. (2015). Biofuel production from microalgae. Handbook of marine microalgae. *Biotechnol. Adv. Elsevier Inc* 145–158.

Qi, H., Wang, J., & Wang, Z. (2013). A comparative study of maximal quantum yield of photosystem II to determine nitrogen and phosphorus limitation on two marine algae. *J. sea Res.* 80, 1–11.

Qin, L., Wei, D., Wang, Z., & Alam, M. A. (2019). Advantage assessment of mixed culture of *Chlorella vulgaris* and *Yarrowia lipolytica* for treatment of liquid digestate of yeast industry and cogeneration of biofuel feedstock. *Appl. Biochem. Biotechnol.* 187, 856–869.

Raghothama, K. G. (2000). Phosphate transport and signaling. *Curr. Opin. Plant Biol.* 3, 182–187.

Ramanna, L., Rawat, I., & Bux, F. (2017). Light enhancement strategies improve microalgal biomass productivity. *Renew. Sustain. Energy Rev.* 80, 765–773.

Ranjith Kumar, R., Hanumantha Rao, P., & Arumugam, M. (2015). Lipid extraction methods from microalgae: a comprehensive review. *Front. Energy Res.* 2, 61.

Rasala, B. A., Gimpel, J. A., Tran, M., Hannon, M. J., Miyake-Stoner, S. J., Specht, E. A., & Mayfield, S. P. (2013). Genetic engineering to improve algal biofuels production. In *Algae for biofuels and energy*. pp. 99–113. Springer.

Reitan, K. I., Rainuzzo, J. R., & Olsen, Y. (1994). Effect of nutrient limitation on fatty acid and lipid content of marine microalgae 1. *J. Phycol.* 30, 972–979.

Ren, L. J., Sun, G., Ji, X. J., Hu, X., & Huang, H. (2014). Compositional shift in lipid fractions during lipid accumulation and turnover in *Schizochytrium sp. Bioresour. Technol.* 157, 107–113.

Renaud, S. M., Thinh, L. V., Lambrinidis, G., & Parry, D. L. (2002). Effect of temperature on growth, chemical composition and fatty acid composition of tropical Australian microalgae grown in batch cultures. *Aquaculture* 211, 195–214.

Renuka, N., Guldhe, A., Singh, P., Ansari, F. A., Rawat, I., & Bux, F. (2017). Evaluating the potential of cytokinins for biomass and lipid enhancement in microalga *Acutodesmus obliquus* under nitrogen stress. *Energy Convers. Manag.* 140, 14–23.

Rizwan, M., Mujtaba, G., Memon, S. A., Lee, K., & Rashid, N. (2018). Exploring the potential of microalgae for new biotechnology applications and beyond: A review. *Renew. Sustain. Energy Rev.* 92, 394–404.

Rodolfi, L., Chini Zittelli, G., Bassi, N., Padovani, G., Biondi, N., Bonini, G., & Tredici, M. R. (2009). Microalgae for oil: Strain selection, induction of lipid synthesis and outdoor mass cultivation in a low-cost photobioreactor. *Biotechnol. Bioeng.* 102, 100–112.

Salama, E. S., Kabra, A. N., Ji, M. K., Kim, J. R., Min, B., & Jeon, B. H. (2014). Enhancement of microalgae growth and fatty acid content under the influence of phytohormones. *Bioresour. Technol.* 172, 97–103.

Sato, N., Hagio, M., Wada, H., & Tsuzuki, M. (2000). *Environmental effects on acidic lipids of thylakoid membranes.* Portland Press Limited.

Schlagermann, P., Göttlicher, G., Dillschneider, R., Rosello-Sastre, R., & Posten, C. (2012). Composition of algal oil and its potential as biofuel. *J. Combust.* 2012.

Schneider, R. C. S., Bjerk, T. R., Gressler, P. D., Souza, M. P., Corbellini, V. A., & Lobo, E. A. (2012). Potential production of biofuel from microalgae biomass produced in wastewater. In *Biodiesel-Feedstocks, Prod. Appl.* IntechOpen.

Sharma, P. K., Saharia, M., Srivstava, R., Kumar, S., & Sahoo, L. (2018). Tailoring microalgae for efficient biofuel production. *Front. Mar. Sci.* 5, 382.

Shin, Y. S., Choi, H. Il, Choi, J. W., Lee, J. S., Sung, Y. J., & Sim, S. J. (2018). Multilateral approach on enhancing economic viability of lipid production from microalgae: a review. *Bioresour. Technol.* 258, 335–344.

Sibi, G., Shetty, V., & Mokashi, K. (2016). Enhanced lipid productivity approaches in microalgae as an alternate for fossil fuels–A review. *J. Energy Inst.* 89, 330–334.

Singh, D. K., & Mallick, N. (2014). Accumulation potential of lipids and analysis of fatty acid profile of few microalgal species for biodiesel feedstock. *J Microbiol Biotech Res* 4, 37–44.

Singh, P., Kumari, S., Guldhe, A., Misra, R., Rawat, I., & Bux, F. (2016). Trends and novel strategies for enhancing lipid accumulation and quality in microalgae. *Renew. Sustain. Energy Rev.* 55, 1–16.

Sulochana, S. B., & Arumugam, M. (2016). Influence of abscisic acid on growth, biomass and lipid yield of *Scenedesmus quadricauda* under nitrogen starved condition. *Bioresour. Technol.* 213, 198–203.

Tabatabaei, M., Karimi, K., Sárvári Horváth, I., & Kumar, R. (2015). Recent trends in biodiesel production. *Biofuel Res. J.* 2, 258–267.

Takagi, M., Watanabe, K., Yamaberi, K., & Yoshida, T. (2000). Limited feeding of potassium nitrate for intracellular lipid and triglyceride accumulation of *Nannochloris sp.* UTEX LB1999. *Appl. Microbiol. Biotechnol.* 54, 112–117.

Takagi, M., & Yoshida, T. (2006). Effect of salt concentration on intracellular accumulation of lipids and triacylglyceride in marine microalgae *Dunaliella* cells. *J. Biosci. Bioeng.* 101, 223–226.

Tang, H., Chen, M., Garcia, M. E. D., Abunasser, N., Ng, K. Y. S., & Salley, S. O. (2011). Culture of microalgae *Chlorella minutissima* for biodiesel feedstock production. *Biotechnol. Bioeng.* 108, 2280–2287.

Tarakhovskaya, E. R., Maslov, Y. I., & Shishova, M. F. (2007). Phytohormones in algae. *Russ. J. Plant Physiol.* 54, 163–170.

Udayan, A., & Arumugam, M. (2017). Selective enrichment of Eicosapentaenoic acid (20: 5n-3) in *N. oceanica* CASA CC201 by natural auxin supplementation. *Bioresour. Technol.* 242, 329–333.

van Oijen, T., van Leeuwe, M. A., Gieskes, W. W. C., & de Baar, H. J. W. (2004). Effects of iron limitation on photosynthesis and carbohydrate metabolism in the Antarctic diatom *Chaetoceros brevis* (Bacillariophyceae). *Eur. J. Phycol.* 39, 161–171.

Wu, G., Gao, Z., Du, H., Lin, B., Yan, Y., Li, G., Guo, Y., Fu, S., Wei, G., & Wang, M. (2018). The effects of abscisic acid, salicylic acid and jasmonic acid on lipid accumulation in two freshwater *Chlorella* strains. *J. Gen. Appl. Microbiol.* 64, 42–49.

Xin, L., Hong-Ying, H., Ke, G., & Ying-Xue, S. (2010). Effects of different nitrogen and phosphorus concentrations on the growth, nutrient uptake, and lipid accumulation of a freshwater microalga *Scenedesmus sp. Bioresour. Technol.* 101, 5494–5500.

Xin, L., Hong-Ying, H., & Yu-Ping, Z. (2011). Growth and lipid accumulation properties of a freshwater microalga *Scenedesmus sp.* under different cultivation temperature. *Bioresour. Technol.* 102, 3098–3102.

Yeh, K., & Chang, J. (2011). Nitrogen starvation strategies and photobioreactor design for enhancing lipid content and lipid production of a newly isolated microalga *Chlorella vulgaris* ESP-31: Implications for biofuels. *Biotechnol. J.* 6, 1358–1366.

Yokochi, T., Honda, D., Higashihara, T., & Nakahara, T. (1998). Optimization of docosahexaenoic acid production by *Schizochytrium limacinum* SR21. *Appl. Microbiol. Biotechnol.* 49, 72–76.

Zhang, C., Zhang, Y., Zhuang, B., & Zhou, X. (2014). Strategic enhancement of algal biomass, nutrient uptake and lipid through statistical optimization of nutrient supplementation in coupling *Scenedesmus obliquus*-like microalgae cultivation and municipal wastewater treatment. *Bioresour. Technol.* 171, 71–79.

Zhao, Y., Li, D., Ding, K., Che, R., Xu, J. W., Zhao, P., Li, T., Ma, H., & Yu, X. (2016). Production of biomass and lipids by the oleaginous microalgae *Monoraphidium sp.* QLY-1 through heterotrophic cultivation and photo-chemical modulator induction. *Bioresour. Technol.* 211, 669–676.

Zhila, N. O., Kalacheva, G. S., & Volova, T. G. (2011). Effect of salinity on the biochemical composition of the alga *Botryococcus braunii* Kütz IPPAS H-252. *J. Appl. Phycol.* 23, 47–52.

Zhu, L. D., Li, Z. H., & Hiltunen, E. (2016). Strategies for Lipid Production Improvement in Microalgae as a Biodiesel Feedstock. *Biomed. Res. Int.* 2016, 1–8.

In: Applied Algal Biotechnology
Editors: M. Arumugam, S. Kathiresan et al.
ISBN: 978-1-53617-524-0

Chapter 8

GENETIC TRANSFORMATION AND METABOLIC ENGINEERING IN ALGAE

Shanmugam Kathiresan* and M. Ram Kumar
Department of Molecular Biology, School of Biological Sciences,
Madurai Kamaraj University, Madurai, Tamilnadu, India

ABSTRACT

Microalgae are fascinating group of a photosynthetic organisms having many biotechnological applications. They produce numerous natural products, many novel or unique which are considered to be a rich resource of functional foods. Strain development in algae for the production of high value compounds through the conventional methods is a tedious process and hence this can be easily done through a biotechnological approach. Genetic transformation is one of the biotechnological methods generally carried out to improve the production of existing compounds, to mediate the degradation of compounds, or to produce one or more new compounds by redirecting the enzymatic reaction. The primary aim of genetic transformation in algae is to regulate the metabolic pathways existing in the algal system by overexpression of a single gene, multiple gene combinations or a transcription factor, or use of RNAi/antisense to knockout the specific genes in their biosynthetic pathway in order to increase the content or change the composition of high-value metabolite/compounds. The fast growth rate and relatively low production cost of algae system makes the transgenic production of value metabolite and recombinant products on an industrial scale very simple. To date, many methods of gene transfer techniques have been attempted in these diverse groups of algae that ultimately end up with many positive outcomes. Here, some of the genetic transformation strategies attempted in different groups of algae and its impact on the biosynthetic pathways and their possibility of commercialization for the benefit of human welfare have been discussed.

Keywords: microalgae, genetic transformation, metabolic engineering, *Agrobacterium*

*Corresponding Author's Email: kskathir@gmail.com.

8.0. Introduction

All the organisms in nature have a unique character at the species level but there are some variation in characters from individuals to individuals. Algae are the one among such organism that possess many unique characters at the genus level and even at the species level. The variations in characters are mainly grouped into genotypic and phenotypic. The genotypic character is the one influenced by the whole genetic constitution of an organism. The phenotypic character is the one which normally observed by the morphological features (like appearance, anatomy, etc.) of an organism. Variation is the cause of differences between individuals. Generally, genetic variation in an organism or in algae is due to sexual reproduction i.e hybridization with superior phenotype. But the possibility of getting the superior progeny through the natural process is difficult. As algae are a photosynthetic group of organisms, it possesses many selectable features which are not present in many other species. The metabolic pathways present in the algal system are very much specialized. In recent years microalgae are realized for their potential for the production of high-value compounds such as carotenoids, polysaccharides, high value polyunsaturated fatty acids and other secondary metabolites of nutraceutical and pharmaceutical importance. When compare to cyanobacteria, eukaryotic microalgae possess several unique metabolic properties which are similar to those found in higher plants. Even though the application of genetic engineering to improve the quality or quantity character (whether phenotypic or genotypic) in eukaryotic microalgae is in its early stage, the necessary advancements to manipulate the microalgal system is essential. As of plant system, it is not a difficult process to culture the algae for the genetic transformation process. It is also known that the microalgae are adopted as a useful model organism for studying various physiological systems. Its ability to grow under both photosynthetic and heterotrophic conditions allows these groups as a useful genetic tool to study many metabolic engineering strategies. Since the smallest nature of cell size, fast growth rate, and minimum requirement of nutrients in the culture medium make the microalgal system to adapt for the easy transformation process.

8.1. Genetic Transformation – Is It Needed for Algae

Synthesizing the complex metabolic pathways in the plant system is difficult through genetic transformation strategies and it will be easy if the same can be carried out in the algal system due to the nature of the growth parameter and response within the cell in a short time. As microalgae comprise a collection of species that offer variable advantages for biotech and biomed industries, gene transfer technologies were further developed form any microalgae to allow the expression of foreign genes/proteins of interest. Generally, genetic transformation is carried out to improve the production of existing compounds, to mediate the degradation of compounds, or to produce new compounds by redirecting one or more enzymatic reactions. Approaches for achieving genetic/metabolic engineering include overexpression of a specific gene in the biosynthetic pathway, or use of RNAi/antisense technology to knock out the genes in their biosynthetic pathway in order to increase the content or change the composition of high-value compounds which is mainly used as food or feed. Due to the fast growth rate of microalgae and relatively less production cost makes the transgenic production of high-value metabolite and recombinant products on industrial scale very simple in the algal system and hence genetic transformation is possible in algae and it is essentially needed.

8.2. Possibility of Gene Transformation in Algae

Unlike that of plant cells, algal cells are also highly efficient and responsive for the genetic transformation/manipulation. So far, many successful genetic transformations have been demonstrated in more than a dozen algal species. Most of them were achieved by nuclear transformation. The early transformation study was made in the Cyanobacterium *Synechocystis* in 1970 (Shestakov, 1970). Many species of green microalgae have been subjected for transformation and stable transformation has been shown for the unicellular model organism *Chlamydomonas reinhardtii* (Debuchy et al. 1989; Kindle et al. 1989), and the transient transformation was also demonstrated. The stable genetic transformation was achieved in other green microalgae like *Volvox carteri* (Schiedlmeier et al. 1994), and the transient transformation was also observed in *Ulva lactuca* (Huang et al. 1996). The unicellular red algae *Cyanidioschyzon merolae* also attempted for genetic transformation (Minoda et al. 2004). Diatoms are the main producers in aquatic environments and they are responsible for the majority of the photosynthesis that occurs in freshwater and particularly in the oceans. Transformation of two dinoflagellates, *Amphidinium* and *Symbiodinium* (Ten Lohuis and Miller 1998), and one euglenoid, *Euglena gracilis*, has also to be considered in this aspect. Also, many species of Cyanobacteria, e.g.,*Spirulina*, *Anabaena*, or *Synechocystis*, can be transformed by electroporation or conjugation (Koksharova and Wolk 2002). Even *Agrobacterium*-mediated transformation methods also attempted for a viable and stable method for the T-DNA integration (Kumar et al. 2004; Coll 2006; Kathiresan et al. 2009, 2010; Anila et al. 2011). And now, many such transformation attempts are being carried out in algal species to improve the genotype or phenotype of the algal species. By having all these studies it is proven that the algal system will serve as a good candidate species for genetic transformation strategies.

Genetic transformation in microalgae mainly concentrating to improve the nature of the algal species to achieve new and economically viable production systems (Barbosa 2010; Greenwell and Laurens et al. 2010; Michael Hannon 2010; Scott, Sharma et al., 2012). In the present scenario, research in transgenic microalgae mainly focused on the utilization of carotenoids, high-value lipids, for biofuel and also for other secondary metabolites.The increasing demand for these compounds from natural sources has promoted large scale cultivation of microalgae. Genes encoding enzymes that are directly involved in specific metabolite synthesis that is being investigated and further development of transformation techniques will permit a considerable increase of secondary metabolites.

8.3. Genetic Transformation Methods in Algae

The basis of almost all algal transformation methods is to cause temporal permeabilization of the cell membrane, enabling DNA molecules to enter the cell. The entrance of the DNA into the nucleus and integration into the genome occurs without any external help. DNA integration mainly occurs by illegitimate recombination events, resulting in the integration of the introduced DNA and thus, culminates in stable genetic transformation. There are a couple of working transformation methods for algal systems that enable the recovery of viable transformants. The basic mechanism of these methods is to somehow disturb the cell wall and cell membrane of the algal cell and allow the gene to reach the nucleus via nuclear pore to integrate into the nucleus. Using the above-mentioned methods genes responsible for many metabolites such as lipids, terpenoids, carotenoids, and other different therapeutic recombinant proteins have been overexpressed.

Different strategies are being followed for the genetic transformation studies in microalgae. The methods include micro-particle bombardment (or biolistic), cell agitation with micro- or macroparticles (e.g., glass beads), protoplast fusion, protoplast transformation with polyethylene glycol or whole-cell transformation by means of electroporation and *Agrobacterium*-mediated transformation. Among these different methods, *Agrobacterium tumefaciens* mediated transformation produce is the efficient, viable and stable method for the T-DNA integration (Kumar et al. 2004; Coll 2006; Kathiresan et al. 2009; Anila et al. 2011). Cyanobacteria are bacteria, they can be transformed by established techniques, e.g., by means of electroporation, by conjugative transfer of vectors from *E. coli* (Wolk 1984), and by a natural DNA uptake system which is present in *Synechocystis* sp. PCC 6803 and in *Thermosynechococcus elongatus* (Iwai, Katoh et al. 2004). A successful and convenient gene expression method for diatoms was achieved by Xie et al., (2014).

The nuclear transformation of algae generally results in random integration of transgenes in plants. The same kind of integration has been found in microalgal transgenics. Recently in oil-producing marine microalgae, *Nannochloropsis* sp, high-frequency homologous recombination was observed after the transformation and selection (Kilian, Benemann et al. 2011). When compared to the eukaryotic microalgae, homologous recombination is easy to achieve in cyanobacteria (Xiaonan Zang 2007).

8.3.1. Electroporation

Naked cells, protoplasts, cell wall reduced mutants and other cells with thin walls can also be transformed by this method. Here electronic pulse temporarily disturbs the phospholipid bilayer of the cell membrane, allowing molecules like DNA to pass. Cells of *Chlamydomonas reinhardtii* (Brown et al. 1991), *Cyanidioschyzon merolae* (Minoda et al. 2004), *Dunaliella salina* (Geng et al. 2003), and *Chlorella vulgaris* (Chow and Tung 1999) have been transformed in this way. Such types of attempts were also made using the gene phytoene desaturase (*pds*) by Steinbrenner and Sandmann (2006). Results revealed that nitrate reductase gene from microalga *Phaeodactylum tricornutum* cloned into the *Chlorella vulguris* genome by electroporation method. Overall transformation efficiency observed that, 0.03% which is relatively high compared with other available *Chlorella* transformations systems. Guo et al., (2013) reported that the efficient genetic transformation system was achieved in *Scenedesmus obliquus*. Experimental results showed that high transformation efficiency was obtained through the electroporation method using a plasmid containing chloramphenicol resistance gene (CAT) as a selectable marker and the green fluorescent protein gene (*gfp*) as a reporter.Stable transformants maintained in the plasmid for more than 6 months and showed its strain improvement in biotechnology applications. Zhang et al., (2014) reported that high-efficiency nuclear transformation in diatom *Phaeodactylum tricornutum* using electroporation. Diatom cells could be transformed and maximum transformation frequency of 2.8 × 10^{-5} cells was obtained. The protocol of genetic transformation in diatom species will be helpful for other eukaryotic microalgae. Similarly, Li et al., (2014) developed high-efficiency nuclear transformation in oleaginous marine *Nannochloropsis* species using PCR product. The work revealed that β-tubulin promoter regions of all recognized species of *Nannochloropsis* were successfully transformed in all five marine species by electroporation. PCR fragment's based transformation system was established in this *Nannochloropsis* species. In this study, an easy and much higher efficient transformation protocol was reported. Jaegera et al., (2017) successfully transformed the oleaginous microalga *Monoraphidium neglectum* using the electroporation method. Results showed that pretreatment steps were applied to break the rigid natural cell wall of microalga. Muñoz et al. (2018) developed a simple and reliable method of improved DNA/protein delivery in microalgae. The optimum electroporation conditions were revealed with the help of sytox green and propidium iodide by monitoring cell permeability and viability respectively. Here, voltage settings determined for the microalgae *Chlamydomonas reinhardtii, Chlorella vulgaris, Neochloris oleoabundans* and *Acutodesmus obliquus,* electroporation transformation was successfully predicted for four microalgae species. A rapid and high-efficiency transformation of *Chlamydomonas reinhardtii* by square-wave electroporation was developed by Wang et al., (2019). The study revealed that the rapid and high transformation frequency (2–6 × 10^3 transformants per μg exogenous DNA) in *Chlamydomonas reinhardtii* through a square wave electroporation system. An overall, this method is an easy and fast technique to develop transgenic algae.

8.3.2. Particle Bombardment

The most popular method is micro-particle bombardment, also referred to as micro-projectile bombardment, particle gun transformation, gene gun transformation, or simply biolistic method. This method makes use of DNA-coated heavy-metal (mostly gold) micro-projectiles and allows the transformation of almost any type of cell, regardless of the thickness or rigidity of the cell wall, and it also allows the transformation of organelles. This course of action was successfully applied in green microalgae like *Chlamydomonas reinhardtii, Haematococcus pluvaialis* and other algae groups (Table 8.1) like red algae, brown algae, diatoms, euglenids, dinoflagellates, and cyanobacteria. Khatiwada et al., (2019) reported that *Agrobacterium*-mediated genetic transformation, using biolistic bombardment and electroporation to obtain stable nuclear transformants in *Euglena gracilis.* In this study, two 3′UTR fragments of the *gapC* gene from *Euglena gracilis* were added at the 5′and 3′ ends of the pCambia1302 to promote homologous integration of the transforming DNA into the genome and results in 1 transformant per 10,000 cells plated. This method is quite simple, but technically the frequency of transformation is low.

8.3.3. Glass Bead Method

Since this method is becoming familiar from the last decade, Feng et al., (2009) reported that efficient genetic transformation in *Dunaliella salina* was developed by this method. Among the three methods studied, the glass beads method was found to be efficient for genetic engineering in *D. salina*. An efficient glass-bead agitation method has been developed by Cui et al., (2012) in the genetic transformation of *Platymonas subcordiformis*. This was the first report of stable transformation in *Platymonas subcordiformis* via glass bead methods. Furthermore, Irshad Ahmad et al., (2015) enhanced the lipid content, both lysophosphatidic acyltransferase gene (*c-lpaat*) & glycerol-3-phosphate dehydrogenase gene (*c-gpd1*), optimized according to the codon bias of *Chlamydomonas reinhardtii*, were inserted into the genomic DNA of model microalga *C. reinhardtii* by the glass bead method. From these studies, it is proved that the glass bead method can also successfully suitable for genetic transformation studies in different algal species.

8.3.4. Cell Suspension Method

Another less complex and less expensive transformation procedure involve preparation of a suspension of microalgae that is then agitated in the presence of micro- or macro-particles, polyethylene glycol, and DNA. Several investigators have used silicon carbide (SiC) whiskers (~0.3- 0.6 μm thick and ~5-15 μm long) as micro-particles. These hard and rigid micro-particles allowed the transformation of cells with intact cell walls including *Chlamydomonas reinhardtii* (Dunahay 1993), *Symbiodinium microadriaticum* and *Amphidinium* sp. (Ten Lohuis and Miller 1998); however, cell wall reduced algae seem to be more appropriate when applying this method. Cell-wall free protoplasts of the green alga *'Chlorella' ellipsoidea* can be transformed without any micro- or macro-particles (Jarvis and Brown 1991); agitation of the protoplast in the presence of polyethylene glycol and DNA is sufficient. As this method require more technical efforts this method is not that much famous as other methods.

8.4. *AGROBACTERIUM*-MEDIATED GENE TRANSFER

This method is becoming more popular since many attempts ended up with positive results. The *Agrobacterium*-mediated genetic transformation method was initially attempted in dicot plants and then to monocot plants. By knowing the knowledge of*Agrobacterium*-mediated genetic transformation, now it is followed in different algal species also. The *Agrobacterium* infection causes tumors ("crown galls") in dicots and some monocots, and, astonishingly, some algae become infected, but they do not develop tumors. *Agrobacterium*-mediated transformation was initially demonstrated to work in the multicellular red alga *Porphyra yezoensis* (Cheney et al. 2001) and, most surprisingly, in the unicellular green alga *Chlamydomonas reinhardtii* (Kumar et al. 2004).

Kumar et al. (2004) made the first successful *Agrobacterium tumefaciens* mediated genetic transformation in the model organism *Chlamydomonas reinhardtii* which is stable and reliable even up to 20 generations. As followed by, Kathiresan et al. (2009) showed the first

Agrobacterium tumefaciens mediated genetic transformation in another freshwater green microalgae *Haematococcus pluvialis* and found the transformants were stable for more than two years with 15 days period of subculture. And also the regulation of the carotenoid biosynthetic pathway using the beta carotene ketolase (*bkt*) gene isolated from the same *H. pluvialis* for the enhanced production of astaxanthin also observed by Kathiresan (2009); Kathiresan et al. (2015). Anila et al. (2011) reported the possibility of *Agrobacterium tumefaciens* transformation in the halotolerant green microalgae *Dunaliella bardawil.* Using the *bkt* gene isolated from *Haematococcus pluvialis* the β-carotene pathway in *Dunaliella bardawil* was extended for the production of astaxanthin (Anila, 2011). Pratheesh, et al., (2014) reported an efficient protocol for the *Agrobacterium*-mediated genetic transformation of microalga *Chlamydomonas reinhardtii.* Here, an efficient genetic transformation of microalga *Chlamydomonas reinhardtii* was observed in pre-treatment of *Agrobacterium* in TAP induction medium (pH 5.2) containing 100 μM acetosyringone and 1 mM glycine betaine and infection of *Chlamydomonas* with the induced *Agrobacterium* greatly improved transformation frequency. This optimized protocol was double the number of transgenic events on selection media compared to that of previous reports. Further, Prasad et al., (2014) reported that stable genetic transformation was achieved in haptophyte *Isochrysis* sp. using *Agrobacterium tumefaciens*. Co-cultivation of *Isochrysis sp.* with *Agrobacterium tumefaciens* strain LBA 4404 harboring the binary vector pCAMBIA 1380-pds-L504R yielded norflurazon-resistant (NR) colonies visible on selective plates after 20–30 days. Cocultivation was observed in *Isochrysis* sp. with *A. tumefaciens* in ASW medium containing 200 μM of acetosyringone for 72 h produced the highest number of NR cells. Héctor Berrios et al., (2015) successfully transformed *Botryococcus braunii* using a cellulase pretreatment method. It was revealed that a simple, effective and reliable genetic transformation attempts made in cellulase pretreatment results in enhanced penetration of *Agrobacterium* harboring the recombinant construct showed the successful stable transformation of *B. braunii* cell with the *PSI103.* Simon et al., (2015) reported that phenolic compounds like vanillin, vanillic acid, gallic acid, coumarin, caffeic acid, and acetosyringone, extracted and identified from the spent medium of *Dunaliella salina* has shown to facilitate the virulence (vir) gene induction in *Agrobacterium* strain A348. Phenolic compounds, vanillin showed the highest concentration of 38.12 ± 0.88 μg L^{-1}in spent medium and significantly activates vir gene induction. Identified phenolic compounds from the spent medium are suitable for facilitating efficient *Agrobacterium*-mediated transformation in *D. salina.* Recently, Norzagarayet al., (2018) successfully established an efficient genetic transformation method in *Dunaliella tertiolecta* by *Agrobacterium tumefaciens.* With these many attempts, the *Agrobacterium* method is becoming more popular in the algal system which is simple, easy and reliable for many generations.

8.5. Other Successful Gene Transformation Attempts

In the last few years, successful genetic transformation of ~25 algal species has been demonstrated (Table 8.1); most of these were achieved by nuclear transformation. Ten species of green algae have been transformed, the stable transformation has been shown for seven of them, and one of which was the unicellular model organism *Chlamydomonas reinhardtii*

(Debuchy et al. 1989; Kindle et al. 1989), and the transient transformation was demonstrated in the other three. All of these green algae are unicellular species except for *Volvox carteri*, for which stable transformation has been shown (Schiedlmeier et al. 1994), and *Ulva lactuca*, which was transiently transformed (Huang et al. 1996). Species of red algae have also transformed so far, two of which are unicellular, and one of which is the ultra-small unicellular *Cyanidioschyzon merolae* (Minoda et al. 2004). All four of the multicellular species are macroalgae. Two multicellular species come from the genus *Porphyra*; the stable transformation was demonstrated in one of these species, *Porphyra yezoensis* (Cheney et al. 2001).

Table 8.1. Transformable algal species.

Groups	Species	Transformation	Reference
Green algae	*Chlamydomonas reinhardtii*	Stable	Debuchy et al. 1989; Kindle et al. 1989
	Haematococcus pluvialis	Stable	Steinbrenner and Sandmann 2006
	Dunaliella salina	Stable	Geng et al. 2003, 2004; Tan et al. 2005
	Dunaliella viridis	Stable	Sun et al. 2006
	Volvox carteri	Stable	Schiedlmeier et al. 1994
	Chlorella sorokiniana	Stable	Dawson et al. 1997
	Chlorella' ellipsoidea	Transient	Jarvis and Brown 1991
	Chlorella' kessleri	Stable	El-Sheekh 1999
	Chlorella vulgaris	Transient	Chow and Tung 1999
	Ulva lactuca	Transient	Huang et al. 1996
Red algae	*Porphyra yezoensis*	Stable	Cheney et al. 2001
	Porphyra miniata	Transient	Kübler et al. 1993
	Kappaphycus alvarezii	Transient	Mikami et al. 1991
	Gracilaria changii	Transient	Gan et al. 2003
	Porphyridium sp.	stable (chloroplast)	Lapidot et al. 2002
Brown algae	*Laminaria japonica*	Stable	Qin et al. 1999
	Undaria pinnatifida	Stable	Qin et al. 2003
Diatoms	*Phaeodactylum tricornutum*	Stable	Apt et al. 1996; Falciatore et al. 1999, 2000; Zaslavskaia et al. 2000, 2001
	Navicula saprophila	Stable	Dunahay et al. 1995
	Cylindrotheca fusiformis	Stable	Fischer et al. 1999
	Cyclotella cryptica	Stable	Dunahay et al. 1995
	Thalassiosira weissflogii	Transient	Falciatore et al. 1999
Euglenids	*Euglena gracilis*	stable (chloroplast)	Doetsch et al. 2001
Dinoflagellates	*Amphidinium sp.*	Stable	Ten Lohuis and Miller 1998
	Symbiodinium microadriaticum	Stable	Ten Lohuis and Miller 1998
Cyanobacteria	*Spirulina platensis*	Stable	Kawata et al. 2004
	Anabaena sp.	Stable	Thiel and Poo 1989
	Synechocystis sp.	Stable	Dzelzkalns and Bogorad 1986

8.6. Metabolic Pathway Engineering

Yet another major development in the field of microalgae is Metabolic Pathway Engineering. The progress in the algal genetic engineering was extremely slow until recently little work has been done by adopting a genetic engineering approach to improve the algae. The methods successfully used for transformation in other systems failed when applied to algae. Techniques to introduce DNA into algal cells with suitable promoters, new selectable marker genes, and expression vectors have to be standardized. Currently, all these requirements have been fulfilled for the diatom *Phaeodactylum*, the green alga, *Chlamydomonas, Haematococcus pluvialis, Botryococcus,* and the blue-green algae, *Synechococcus* and *Synechocystis* (Boussiba2000). The success of genetic engineering lies in the improvement of nutritional value, product yield with optimal production parameters.

8.7. Identification and Isolation of Algal Genes

So far, many genes from algal species have been identified and isolated for biotechnological purposes. Not only the genes but also the other elements such as promoters, transcription factors are being in the process of identification and isolation. Díaz-Santos et al., (2013) reported that the efficiency of heterologous promoters in microalgae and the use of different transformation methods. The efficiency was evaluated using the heterologous promoters of cauliflower mosaic virus 35S (CaMV 35S) and *Agrobacterium* nopaline synthase (NOS) genes. These promoters were fused to the paromomycin conferring-resistance aminoglycoside 3′-phosphotransferase encoding gene (APHVIII), and *C. reinhardtii* was transformed by the glass beads agitation method. Here, to identify the significantly higher transformation efficiencies and a higher level of APHVIII expression in those transformants harboring the NOS promoter than in those transformed with CaMV 35S promoter. And hence, the NOS promoter is widely used for genetic manipulation of higher plants that has been very rarely used for the transformation of microalgae.

When observing the suitability of the promoter sequences for the transgenic production in microalgae, the endogenous promoter like Rubisco small subunit (RbcS2) or the ubiquitin (Ubi1) promoter or the heterologous promoters CaMV35S and SV40 have been successfully used for microalgal genetic transformation.` The CaMV35S, being a constitutive promoter,has been studied universally for higher plants which also works well in several algal stains and showed strong expression (Kumar et al. 2004; Kathiresan et al. 2009; Kathiresan 2009,2015; Anila et al. 2011). Also, the SV40, a polyomavirus promoter, has been shown to work in *H. pluvialis* and in *C. reinhardtii* (Coll 2006). Cheng et al., (2006) characterized a heat-shock gene (hsp70) from green alga *Volvox carteri*. The study showed that the multiple heat shock elements (HSEs) and is highly inducible by heat shock and it is orthologous to HSP70A of the related alga *Chlamydomonas reinhardtii.*

8.8. Cloning and Functional Characterization of Algal Genes

Thousands of algal genes are being used for the expression studies by cloning and functional characterization. Zhong et al. (2011) functionally characterized various algal carotenoid ketolases that ketolating zeaxanthin efficiently to produce a high amount of astaxanthin in transgenic *Arabidopsis."* This work reveals that extending the carotenoid pathway to astaxanthin in plants is of scientific and industrial interest. β-carotene ketolase (BKT) that catalyzes the formation of ketocarotenoids in transgenic plants typically results in low levels of astaxanthin. The result observed that in transgenic *Arabidopsis thaliana* expression of CrBKT developed orange leaves which accumulated astaxanthin up to 2 mg /g dry weight with a 1.8-fold increase in total carotenoids. This study advances a decisive step towards the utilization of plants for the production of high-value astaxanthin. Liu et al. (2013) cloned a phytoene desaturase (PDS) gene and characterized by the unicellular green microalga *Chlamydomonas reinhardtii*. Functional complementation analysis revealed *C. reinhardtii*PDS (CrPDS) catalyzes the conversion of phytoene to the colored carotenoid -carotene. A single amino acid substitution, L505F, enhanced its desaturation activity by 29%, as indicated by an in vitro enzymatic assay. CrPDS-L505F exhibited a 27.7-fold higher resistance to the herbicide norflurazon. Recently Thiyagarajan et al. (2018) demonstrated the isolation and functional characterization of delta 6 desaturase from the marine microalga *Isochrysis* sp which convert the essential fatty acid to high-value polyunsaturated fatty acids like stearidonic acid and gamma-linolenic acid.

8.9. Marker Gene Transformation for Antibiotic Selection

Cui et al., (2014) reported that the *bar* gene as a selectable marker on chloroplast transformation of *Platymonas (Tetraselmis) subcordiformis*. The *gfp* reporter gene and *bar* gene as a selectable marker, transformation vectors of *P. subcordiformis* chloroplast were constructed with endogenous fragments rrn16S–trnI (left) and trnA–rrn23S (right) as a recombination site of the chloroplast genome. Plasmids were transferred into *P. subcordiformis* via particle bombardment. GFP (green fluorescence protein) was localized in the chloroplast of *P. subcordiformis*, confirming the activity of *Chlamydomonas reinhardtii* promoter. This result was indicated that the bar gene was successfully integrated into the chloroplast genome via homologous recombination. A simple and non-invasive method of nuclear transformation of intact-walled *Chlamydomonas reinhardtii* was established by Kim et al., (2014). An efficient transformation using DNA and aminoclay results in a transformation efficiency of 5.0×10^2 transformants/μg DNA**.** Díaz-Santoset al., (2016) reported the new approach to express transgenes in microalgae and increased the flocculation ability of *Chlamydomonas reinhardtii*. The results showed that the cotransformation with two naked promoterless genes, selectable antibiotic-resistant gene (bleomycin-resistant gene) and gene of interest (flocculin gene from *Saccharomyces bayanus*) for the flocculation process in yeast. The transformation percentage was significantly high and effectively transcribed and translated. Run et al., (2016) developed a stable nuclear transformation method for industrially important alga *Chlorella pyrenoidosa*. Results showed that the pGreen 0029 vector containing the eGFP gene under the control of a ubiquitin promoter and NptII, which is a selectable marker gene was constructed and explored and optimization was done by electroporation methods. Optimal transformation efficiency was obtained and successful transformants were screened and stably integrated and expressed in transgenic *Chlorella pyrenoidosa*. Mini et al., (2018) reported that a new selection marker paromomycin (Paro) in *Agrobacterium*-mediated genetic transformation of *Chlamydomonas reinhardtii.* Results suggested that *Agrobacterium*-mediated transformation carrying within the T-DNA borders and an expression cassette containing the Luciferase (Luc) reporter gene. Next-generation sequencing was confirmed the nuclear insertion points region, which is occurred in exons or untranslated regions (UTRs) for 10 out of 10 *Agrobacterium*-mediated and 9 out of 13 of electroporation-mediated insertions. Buck et al., (2018) reported that a new selection marker blasticidin-S deaminase for genetic transformation in the diatom *Phaeodactylum tricornutum.* This work revealed that blasticidin-S and tunicamycin turned out to be lethal to wild-type cells at low concentrations. The successful double resistant (against blasticidin-S and nourseothricin) *P. tricornutum* cell lines were observed. Muñoz et al., (2019) reported that stable genetic transformation of the green algae *Acutodesmus obliquus* and *Neochloris oleoabundans* was accomplished based on *E. coli* conjugation to improve TAG levels. The first time achieved the successful transformation of green oleaginous microalgal strains by transferring an expression vector via conjugation from *E. coli.* Successful expression of selectable marker and reporter genes was achieved in both microalgal species. Stable gene expression and unsuccessful plasmid recovery suggest integration into microalgal genomes.

8.10. Metabolic Engineering - Carotenoid Production

Carotenoid production through metabolic engineering was first reported by Harker et al., (1997) that the gene, crtO, encoding P-C-4-oxygenase which converts P-carotene to canthaxanthin from *Haematococcus pluvialis* was successfully cloned in green alga cyanobacterium *Synechococcus* PCC7942. The transferred crtO gene activates the carotenoid pathway and to produce some astaxanthin. Specifically, P-carotene hydroxylase, which normally converts p-carotene to zeaxanthin, can also function in the biosynthesis of astaxanthin. Furthermore, Zhong et al., (2011) reported that ketolating zeaxanthin of various algal carotenoid ketolases reveals that it is essential for the high production of astaxanthin in transgenic *Arabidopsis.* This work revealed that extending the carotenoid pathway to astaxanthin in plants for industrial interest. B-carotene ketolase (BKT) that catalyzes the formation of keto carotenoids in transgenic plants typically results in low levels of astaxanthin. In transgenic *Arabidopsis thaliana* expressing the CrBKT developed orange leaves which accumulated astaxanthin up to 2 mg/g dry weight with a 1.8-fold increase in total carotenoids. This study advances a decisive step towards the utilization of plants for the production of high-value astaxanthin. On the other hand, Li et al., (2011) cloned and characterized the carotene desaturase gene from *Chlorella protothecoides* CS-41. The work revealed that the carotene desaturase gene (zds) was isolated from *Chlorella protothecoides.* The full-length cDNA sequence was 2031 bp and contained 1755 bp putative open reading frame which encodes a 584 amino acid and molecular weight was 63.7 kDa. Nucleotide and putative protein had sequence identities of 72.5% and 69.5% with those of the green alga *Chlamydomonas reinhardtii*, respectively. Baldo et al., (2011) reported that enhanced carotenoids biosynthesis in *Chlamydomonas reinhardtii* was achieved by nuclear transformation using a phytoene synthase gene isolated from *Chlorella zofingiensis.*Results showed that isolation and characterization of the phytoene synthase gene from the green microalga *Chlorella zofingiensis* (CzPSY), involved in the first step of the carotenoids biosynthetic pathway. The overexpression of CzPSY in *C. reinhardtii,* by nuclear transformation, has led to an increase in the corresponding CzPSY transcript level as well as in the content of the carotenoids violaxanthin and lutein which were 2.0 and 2.2-fold higher than in untransformed cells. Harker et al., (1997) reported that biosynthesis of ketocarotenoids in transgenic cyanobacteria expressing the algal gene for β-C-4-oxygenase, crtO. The results revealed that phytoene desaturase (PDS) gene from the unicellular green microalga *Chlamydomonas reinhardtii*was cloned and characterized. Functional complementation analysis revealed *C. reinhardtii* PDS (CrPDS) catalyzes the conversion of phytoene to the colored carotenoid carotene. Glass bead-mediated delivery displayed a high transformation efficiency of *C. reinhardtii* with CrPDS-L505F, demonstrating clearly that the engineered endogenous CrPDS is a dominant selectable marker for *C. reinhardtii* and possibly for other green algae. Simon et al., (2016) reported that heterologous expression of β-carotene hydroxylase in *Dunaliella salina* by *Agrobacterium*-mediated genetic transformation. The β-carotene hydroxylase from the chlorophycea microalga, *Chlamydomonas reinhardtii* was amplified. The amplified β-carotene hydroxylase gene with rubisco promoter and chloroplast transit peptide was introduced into the genome of the *D.salina* by *Agrobacterium*-mediated genetic transformation. The carotenoid analysis showed a 3-fold increase of violaxanthin at growth stage and a 2-fold increase of zeaxanthin under stress conditions in transformed *D. salina.* Galarza et al., (2018) achieved over-accumulation of astaxanthin in *Haematococcus pluvialis* through chloroplast genetic engineering. The work revealed that the carotenoid biosynthesis pathway of *Haematococcus pluvialis* has been genetically modified to overproduce astaxanthin. The endogenous phytoene desaturase coding sequence (pds) has been codon-optimized and overexpressed in the chloroplast of *H. pluvialis* under the control of the psbA Promoter/UTR. Astaxanthin accumulated up to 67% higher in transformed strains than in wild-type induction with high light intensity and nitrogen depleted conditions. Sarnaik et al., (2018) successfully improved zeaxanthin production in recombinant *Synechococcus* elongatus PCC 7942 under natural light conditions. In this study, CrtR (β-carotene oxygenase) gene from *Synechococcus elongatus* PCC 7002 was successfully cloned to *Synechococcus elongatus* PCC 7942 through homologous recombination to improve β-carotene flux towards zeaxanthin synthesis. This attempt successfully resulted in enhanced production of lutein-free zeaxanthin in the cyanobacterial system.

8.11. Metabolic Engineering for PUFA Production

Modification of lipid biosynthetic pathway also increases the quality of the lipids with regards to both high-value PUFA for health benefits and also suitability as diesel fuel feedstock and other industrial applications (Radakovitset al. 2010). Genes for this purpose include the group of fatty acid modifying enzymes, such as desaturases, elongases, thioesterases and carboxylases which have been studied in genetically modified plants in detail. But in algal species, especially in marine microalgae attempt is being made to improve the nature of the strain with respect to the production of high value polyunsaturated fatty acid (Napier 2007). Microalgal genetic engineering for enhanced lipid production has mainly focused on manipulating the key genes associated with lipid synthesis through forward or reverse genetics (Radakovits et al,. 2010; Hlavová et al., 2015). Gene manipulation of genes involved in the lipid biosynthetic pathway by random as well as site-directed mutagenesis to improve lipid production has been reported by various workers. Chemical mutagens such as EMS have been employed by Chaturvedi and Fujita (2006) to improve the eicosapentaenoic acid accumulation in *Nannocholoropsis oculata.* Lipid metabolism is highly complex and intricate and is coupled with many other regulatory pathways in plants and microalgae. Metabolic engineering of the lipid biosynthetic pathway in algae is mainly aimed at the overexpression of enzymes involved in lipid synthesis and storage, sometimes along with other enzymes that are involved in the reduction of lipid catabolism mechanisms (Figure 1). Many available reports are aimed at manipulating the genes involved in TAG synthesis, thereby increasing the TAG content in microalgae which are suitable to be used as biofuel. TAG with better cold flow properties are optimum for the use as biofuel and so the algal strains producing lipids with less unsaturation are preferred over saturated ones. As many of the pathways share the carbon pool, a piece of extensive knowledge on carbon pooling is highly required before any gene manipulation. Scranton *et. al.,* (2016) has investigated carbon partitioning between starch and lipid in *C. reinhardtii* and to identify the key genes involved in TAG biosynthesis. The Heat shock method successfully attempted and transcription level was increased and fatty acids also significantly increase of 44.5 and 67.5% lipid content in *C. reinhardti*. Especially, C18 (Stearic acid and C18:1 Oleic acid), notably increased, while, unsaturated fatty acids dramatically decreased. This study reveals that genetic manipulation and intermittent heat shock to enhance lipid production, especially the production of long-chain saturated fatty acids, using *C. reinhardti* (Wang et al. 2018). *Phaeodactylum tricornutum* can accumulate up to 30% of the omega-3 long-chain polyunsaturated fatty acid (LC-PUFA) eicosapentaenoic acid (EPA) and, as such, is considered a good source for the industrial production of EPA. Here, transgenic strain Pt_Elo5 has been investigated for the scalable production of EPA and DHA (Hamilton et al. 2015). The first time, higher DHA production over EPA, generating an average EPA/DPA/DHA profile of 8.7%, 4.5%, and 12.3%, with omega-3 content up to 25.5% in genetically modified diatom *Phaeodactylum tricornutum*. Zorin et al., (2014) reported that oleaginous green alga *Lobosphaera parietochlorisincisa* a mutant strain; deficient in arachidonic acid biosynthesis was genetically transformed to synthesize long-chain PUFAs. This work successfully transformed LCPUFA gene desaturases and elongases for the production of high-value ω-3 LC-PUFA of the commercial interest. In a study, successful large-scale production of the LC-PUFA dihomo-g-linolenic acid (DGLA, 20:3 n-6) was achieved by Abu Ghosh et al., (2015). Results showed that the high-density cultures of the D5 desaturase-mutant microalga *Lobosphaera incise* produced a significant amount of LC-PUFA in dihomo-g-linolenic acid (DGLA, 20:3 n-6). Zhu et al., (2017) studied the overexpression of endogenous delta-6 fatty acid desaturase gene in *Phaeodactylum tricornutum*. Results revealed that overexpression of Δ6 desaturase genein *P. tricornutum significantly increased the total lipid content of 47.66% and EPA content of 38.101.mg/g DW in algal strain.* Overexpression of endogenous D6 desaturase gene resulted in increased EPA production was observed. Norashikin et al., (2018) improved the protocol of *Agrobacterium*-mediated genetic transformation of *C. vulgaris* using an endogenous omega-3 fatty acid desaturase gene with its promoter. The work revealed that overexpression of ω-3 FAD gene increased the C18:3n3 content in transgenic line which leads to higher α-linolenic acid (C18:3n3) content.

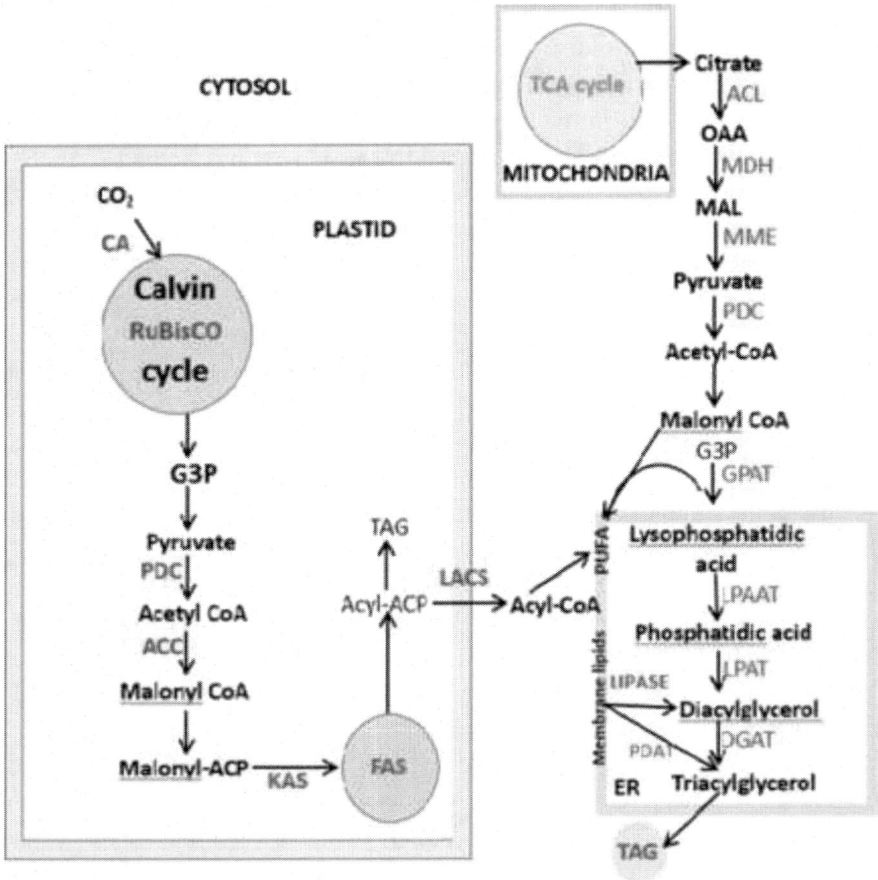

Figure 8.1. Scheme representing the synergy between enzymes that lead to the formation of lipid (CA carbonic anhydrase; RuBisCO Ru1,5BP carboxylase/oxygenase; PDC pyruvate dehydrogenase complex; ACC acetyl-CoA carboxylase; KAS 3-ketoacyl-ACP synthase; ACL ATP-citrate lyase; MDH malate dehydrogenase; MME NADP-malic enzyme; PDC pyruvate dehydrogenase complex; G PAT glycerol-3-phosphate acyltransferase; LPAAT lysophosphatidic acid acyltransferase; L PAT lysophosphatidylcholine acyltransferase; DGAT diacylglycerol acyltransferase; PDAT phospholipid diacylglycerol acyltransferase (adapted from Jagadevanet al., 2018).

8.12. Genetic Engineering to Increase the Length and Unsaturation of Fatty Acids

Many marine microalgae synthesize PUFAs, especially EPA and DHA, via a recently discovered anaerobic polyketide synthase (PKS) pathway. PUFA come under two main category viz ω-6 and ω-3 fatty acids based on the position of the first double bond. Due to their importance as neutraceuticals, many studies have been carried out to increase PUFA content in microalgae. Microalgae use mainly two pathways for the synthesis of PUFS; aerobic pathway and anaerobic pathway (PKS pathway). In *Nannochloropsis oceanica*, the endogenous Δ12-desaturase was cloned and overexpressed resulting in a 50–75% increase of ARA content (Kaye et al., 2015). Overexpression of endogenous Δ6 desaturase from *Mortierella alpina* and Δ15 desaturase from *Gibberella fujsikuroi* in the cyanobacterium

Synechocystis sp greatly enhanced ω-3 PUFAs accumulation, especially the yield of ALA and SDA (Chen et al., 2014).

However, as the catalytic activity of Δ6 desaturase varies based on the substrate (LA or ALA), its overexpression produced varied results in different microalgae spp. the conversion efficiency of marine microalga *Thalassiosira pseudonana* possesses a Δ6 desaturase with 80% efficiency for ALA in the ω-3 pool and 68% for LA in the ω-6 pool (Petrie et al., 2012). Genetically engineered *Mortierella alpine* with overexpression of Δ6 desaturase has a 26.2-fold increase in EPA content over wild-type *Mortierella alpine* (Shi et al., 2016) whereas a heterologous expression of the gene in *Phaeodactylum tricornutum* had little effect in EPA content (Hamilton et al., 2014). Kobayash et al., (2011) has reported the overexpression of the Δ5 desaturase gene from *Thraustochytrium aureum* in *Aurantiochytrium limacinum* driven by the thraustochytrid ubiquitin promoter. The engineered strains so produced had 4.6- and 13.2-fold increases of the EPA and ARA levels, respectively. *Phaeodactylum tricornutum* is reported as an efficient producer of EPA. As the EPA accumulation mechanism of these algae is highly efficient. The enzyme Δ5-elongase is responsible for the production of EPA in microalgae. Cloning and characterization of the Δ5-elongase gene from *Phaeodactylum tricornutum* were first reported by Domergue et al. (2002). *Phaeodactylum tricornutum* genetically engineered with Δ5-elongase from the picoalga *Ostreococcus tauri* resulted in an 8-fold increase of DHA accumulation, and its -expression with Δ6-desaturase further increased the DHA levels (Hamilton et al., 2014). DHA synthesis is mediated by Δ4-desaturases in microalgae and the same has been characterized in many microalgal spp like *Chlamydomonas reinhardtii* (Zäuneret al., 2012) and *Ostreococcus lucimarinus* (Ahmann et al., 2011).

8.13. Metabolic Engineering for Biodiesel Production

Improvement of neutral lipids in microalgae is an important strategy for biodiesel production. Trentacoste et al., (2013) successfully engineered *Thalassiosira pseudonana* for improved microalgal lipid accumulation without affecting the growth. These results indicated that the targeted knockdown of a multifunctional lipase/phospholipase/acyltransferase increased lipid yields without affecting growth in the diatom *Thalassiosira pseudonana*. Furthermore, Ahmad et al., (2015) reported that overexpression of *(BnDGT2)* key enzyme in *Chlamydomonas reinhardtii* enhanced the lipid and fuel content. The study successfully enhanced the lipid content, by transferring the lysophosphatidic acyltransferase gene (c-lpaat) & glycerol-3-phosphate dehydrogenase gene (c-gpd1) through codon optimization by the glass bead method. The method successfully increased the fatty acids stearic acid and Oleic acid by 44.5 and 67.5% respectively which is feasible for biodiesel. Antisense-expressing knockdown strains 1A6 and 1B1 exhibited wild-type–like growth and increased lipid content under both continuous light and alternating light/dark conditions. In a recent study, genetic improvement of the microalga *Phaeodactylum tricornutum* resulted in neutral lipid accumulation. It was showed that malic enzyme plays a major role in pyruvate metabolism and carbon fixation and further overexpression of *(PTME)* gene significantly increased the total lipid content up to 2.5-fold and 57.8% of dry cell weight with a similar growth rate to wild type (Xue et al., 2015). Furthermore, Shemesh, et al., (2016) reported that the inducible expression of *Haematococcus* oil globule protein in the diatom *Phaeodactylum tricornutum* enhanced lipid droplets and of TAG accumulation under nitrogen starvation. Experimental results revealed that the heterologous expression of a major LD protein fused with EGFP and expressed under the control of the strong light-inducible fcpA gene promoter and the novel nitrogen-starvation-inducible DGAT1 gene promoter of the green microalga *Haematococcus pluvialis* in the diatom *Phaeodactylum tricornutum* significantly enhanced oil productivity. Similarly, Chen et al., (2016) reported that overexpression of type 2 diacylglycerol acyltransferase gene DGTT1 from *Chlamydomonas reinhardtii* enhances lipid production in *Scenedesmus obliquus*. The study revealed that genetically engineered microalga improved growth and lipid productivity. The lipid content, biomass concentration, and biomass productivity obtained from 40 L tubular PBR were 127.8% 20.0%, and 232.6% higher than the wild-type strain. Transferring the DGTT1 gene into *S. obliquus* CPC2 can stimulate both microalgae growth and lipid production.

8.14. Problems in Genetic Transformation in Algae

There are different challenges and existing problems for transformation and integration of genes in microalgae which need to be overcome to fully utilize the potential of this microbe for molecular cloning and overproduction of different value-added compounds. In the transformation experiments of algae, a number of selectable markers like *npt, hpt, bar* has been shown to be successful in obtaining genetically modified strains. The use of antibiotic-resistant marker genes and reporter genes had considerably reduced the effort involved in transformant screening. Efforts were also made to standardize the co-cultivation procedure for both microalga and *Agrobacterium* in the process of *Agrobacterium*-mediated transformation. Since the nutrients for both organisms differ from each other, the transformation process depends on the co-cultivation medium and the age of the cells (Kumar et al. 2004; Kathiresan et al. 2009; 2015 and Anila et al. 2011). For the efficient transformation, cells from the late logarithmic growth phase are commonly used for transformation. These are some of the points to be considered before starting the genetic transformation and metabolic engineering in the algal system.

Conclusion

Genetic transformation in algae is complex and fast-growing technology. In general, the commercial algal biotechnology produces food additives, cosmetics, animal feed additives, pigments, polysaccharides, fatty acids, and biomass. But, the recent progress in algal transgenics promises a much broader field of application: molecular farming, the production of proteins or metabolites that are valuable to medicine or industry, seems to be feasible with transgenic algal systems. Definitely, the ability of transgenic algae to produce recombinant proteins, metabolites, antibodies, vaccines, insecticidal proteins, bio-fuel, and many other compounds has already been demonstrated. Genetic engineering of microalgae is necessary to produce economically feasible strains for metabolite production. Current efforts are focused on the manipulation of individual metabolic genes, but the outcomes are not stable and/or efficient for large-scale production. Genetic modifications that enhance the physiological properties of algal strains and optimization of algal production systems are further needed to improve the potential of this auspicious technology in the future.

REFERENCES

Abu-Ghosh, S., Pal-Nath, D., Markovitch, D., Solovchenko, A., Didi-Cohen, S., Portugal, I., Khozin-Goldberg, I., Cohen, Z. and Boussiba, S. (2015). A novel source of dihomo-γ-linolenic acid: Possibilities and limitations of DGLA production in the high-density cultures of the Δ5 desaturase-mutant microalga *Lobosphaera incisa*. *Eur. J. Lipid. Sci. Tech.* 117(6), 760-766.

Ahmad, I., Sharma, A. K., Daniell, H. and Kumar, S. (2015). Altered lipid composition and enhanced lipid production in green microalga by introduction of brassica diacylglycerol acyltransferase 2.*Plant Biotechnol. J.* 13(4), 540-550.

Ahmann, K., Heilmann, M. and Feussner, I. (2011). Identification of a Δ4-desaturase from the microalga *Ostreococcus lucimarinus*. *Eur. J. Lipid. Sci. Tech.* 113(7), 832-840.

Anila, N., Chandrashekar, A., Ravishankar, G.A. and Sarada, R. (2011). Establishment of *Agrobacterium* tumefaciens-mediated genetic transformation in *Dunaliella bardawil*. *Eur. J. Phycol.* 46(1), 36-44.

Berrios, H., Zapata, M. and Rivas, M. (2016). A method for genetic transformation of *Botryococcus braunii* using a cellulase pretreatment. *J. Appl. Phycol.* 28(1), 201-208.

Brown, L. E., Sprecher, S. L. and Keller, L. R. (1991). Introduction of exogenous DNA into *Chlamydomonas reinhardtii* by electroporation. *Methods Mol. Cell. Biol.* 11(4), 2328-2332.

Chaturvedi, R. and Fujita, Y. (2006). Isolation of enhanced eicosapentaenoic acid producing mutants of *Nannochloropsis oculata* ST-6 using ethyl methane sulfonate induced mutagenesis techniques and their characterization at mRNA transcript level. *Phycol. Res.* 54(3), 208-219.

Chen, C. Y., Kao, A. L., Tsai, Z. C., Chow, T. J., Chang, H. Y., Zhao, X. Q., Chen, P. T., Su, H. Y. and Chang, J. S. (2016). Expression of type 2 diacylglycerol acyltransferse gene DGTT1 from *Chlamydomonas reinhardtii* enhances lipid production in *Scenedesmus obliquus*. *Biotechnol. J.* 11(3), 336-344.

Chen, G., Qu, S., Wang, Q., Bian, F., Peng, Z., Zhang, Y., Ge, H., Yu, J., Xuan, N., Bi, Y. and He, Q. (2014). Transgenic expression of delta-6 and delta-15 fatty acid desaturases enhances omega-3 polyunsaturated fatty acid accumulation in *Synechocystis sp.* PCC6803. *Biotechnol. Biofuels.* 7(1), 32.

Cheney, D., Metz, B. and Stiller, J. (2001). *Agrobacterium*-mediated genetic transformation in the macroscopic marine red alga *Porphyra yezoensis*. *J. Phycol.* 37(11).

Coll, J. M. (2006). Methodologies for transferring DNA into eukaryotic microalgae: a review. *Span J Agric Res.* 4(4), 316-330.

Cui, Y., Jiang, P., Wang, J., Li, F., Chen, Y., Zheng, G. and Qin, S. (2012). Genetic transformation of *Platymonas* (Tetraselmis) *subcordiformis* (Prasinophyceae, Chlorophyta) using particle bombardment and glass-bead agitation. *Chin J Oceanol Limn.* 30(3), 471-475.

De Jaeger, L., Verbeek, R. E., Draaisma, R. B., Martens, D. E., Springer, J., Eggink, G. and Wijffels, R. H. (2014). Superior triacylglycerol (TAG) accumulation in starchless mutants of *Scenedesmus obliquus*: (I) mutant generation and characterization. *Biotechnol Biofuels*. 7(1), 69.

Debuchy, R., Pup`Ton, S., & Rochaix, J. D. (1989). The argininosuccinate lyase gene of *Chlamydomonas reinhardtii*: an important tool for nuclear transformation and for correlating the genetic and molecular maps of the ARG7 locus. *Embo J.* 8 (10), 8203-2809.

Derelle, E., Ferraz, C., Rombauts, S., Rouzé, P., Worden, A. Z., Robbens, S., Partensky, F., Degroeve, S., Echeynié, S., Cooke, R. and Saeys, Y. (2006). Genome analysis of the smallest free-living eukaryote *Ostreococcus tauri* unveils many unique features. *Proc. Natl. Acad. Sci.*, 103(31), 11647-11652.

Domergue, F., Lerchl, J., Zähringer, U. and Heinz, E. (2002). Cloning and functional characterization of *Phaeodactylum tricornutum* front-end desaturases involved in eicosapentaenoic acid biosynthesis. *Eur. J. Biochem*. 269(16), 4105-4113.

Feng, S., Xue, L., Liu, H. and Lu, P. (2009). Improvement of efficiency of genetic transformation for *Dunaliella salina* by glass beads method. *Mol. Biol. Rep.*36 (6), 1433.

Geng, D., Wang, Y., Wang, P., Li, W. and Sun, Y. (2003). Stable expression of hepatitis B surface antigen gene in *Dunaliella salina* (Chlorophyta). *J. Appl. Phycol.* 15(6), 451-456.

Greenwell, H. C., Laurens, L. M. L., Shields, R. J., Lovitt, R. W. and Flynn, K. J. (2009). Placing microalgae on the biofuels priority list: a review of the technological challenges. *J R SocInterface*. 7(46), 703-726.

Guo, S. L., Zhao, X. Q., Tang, Y., Wan, C., Alam, M. A., Ho, S. H., Bai, F. W. and Chang, J. S. (2013). Establishment of an efficient genetic transformation system in *Scenedesmus obliquus*. *J. Biotechnol.* 163(1), 61-68.

Hamilton, M. L., Haslam, R. P., Napier, J. A. and Sayanova, O. (2014). Metabolic engineering of *Phaeodactylum tricornutum* for the enhanced accumulation of omega-3 long chain polyunsaturated fatty acids. *Metabol Eng*. 22, 3-9.

Hannon, M., Gimpel, J., Tran, M., Rasala, B. and Mayfield, S. (2010). Biofuels from algae: challenges and potential. *Biofuels*. 1(5), 763-784.

Harker, M. and Hirschberg, J. (1997). Biosynthesis of ketocarotenoids in transgenic cyanobacteria expressing the algal gene for β-C-4-oxygenase, crtO. *FEBS letters*. 404(2-3), 129-134.

Hlavova, M., Turoczy, Z. and Bisova, K. (2015). Improving microalgae for biotechnology - From genetics to synthetic biology.*Biotechnol. Adv.*, 33(6).1194-1203.

Huang, X., Weber, J.C., Hinson, T.K., Mathieson, A.C. and Minocha, S.C., (1996). Transient expression of the GUS reporter gene in the protoplasts and partially digested cells of *Ulva lactuca* L. (Chlorophyta). *Bot. Mar.* 39 (1-6), 467-474.

Iwai, M., Katoh, H., Katayama, M. and Ikeuchi, M. (2004). Improved genetic transformation of the thermophilic cyanobacterium, *Thermosynechococcus elongatus* BP-1. *Plant Cell Physiol.* 45(2), 171-175.

Jaeger, D., Hübner, W., Huser, T., Mussgnug, J.H. and Kruse, O. (2017). Nuclear transformation and functional gene expression in the oleaginous microalga *Monoraphidium neglectum*. *J. Biotechnol.* 249, 10-15.

Jagadevan, S., Banerjee, A., Banerjee, C., Guria, C., Tiwari, R., Baweja, M. and Shukla, P. (2018). Recent developments in synthetic biology and metabolic engineering in microalgae towards biofuel production.*Biotechnol Biofuels*. 11(1), 185.

Kathiresan (2009). *Genetic Transformation of Green Alga Haematococcuspluvialis for the Regulation of Carotenoid Biosynthesis*. Thesis p. 171.

Kathiresan, S., Chandrashekar, A., Ravishankar, G.A. and Sarada, R. (2009). *Agrobacterium*-mediated transformation in the green alga *Haematococcus pluvialis* (Chlorophyceae, Volvocales) 1. *J. Phycol*. 45(3), 642-649.

Kathiresan, S., Chandrashekar, A., Ravishankar, G.A. and Sarada, R. (2015). Regulation of astaxanthin and its intermediates through cloning andgenetic transformation of β-carotene ketolase in *Haematococcus pluvialis. Journal of Biotechnology* 196–197: 33–41

Kaye, Y., Grundman, O., Leu, S., Zarka, A., Zorin, B., Didi-Cohen, S., Khozin-Goldberg, I. and Boussiba, S. (2015). Metabolic engineering toward enhanced LC-PUFA biosynthesis in *Nannochloropsis oceanica*: Overexpression of endogenous Δ12 desaturase driven by stress-inducible promoter leads to enhanced deposition of polyunsaturated fatty acids in TAG. *Algal Res.*11, 387-398.

Khatiwada, B., Kautto, L., Sunna, A., Sun, A. and Nevalainen, H. (2019). Nuclear transformation of the versatile microalga *Euglena gracilis*. *Algal Res*.37, 178-185.

Kilian, O., Benemann, C.S., Niyogi, K.K. and Vick, B. (2011). High-efficiency homologous recombination in the oil-producing alga *Nannochloropsis* sp. *Proc. Natl. Acad. Sci.* 108(52), 21265-21269.

Kobayashi, T., Sakaguchi, K., Matsuda, T., Abe, E., Hama, Y., Hayashi, M., Honda, D., Okita, Y., Sugimoto, S., Okino, N. and Ito, M. (2011). Increase of eicosapentaenoic acid in thraustochytrids through thraustochytrid ubiquitin promoter-driven expression of a fatty acid Δ5 desaturase gene. *Appl. Environ. Microbiol.* 77(11), 3870-3876.

Koksharova, O., and Wolk, C. (2002). Genetic tools for cyanobacteria. *Appl. Microbiol. Biotechnol.*, 58(2), 23-137.

Kumar, S.V., Misquitta, R.W., Reddy, V.S., Rao, B.J. and Rajam, M.V. (2004). Genetic transformation of the green alga - *Chlamydomonas reinhardtii* by *Agrobacterium* tumefaciens. *Plant Sci. J.* 166(3), 731-738.

Kumari, S., Singh, P., Gupta, S.K. and Kumar, S. (2015). Genetic Engineering Tools for Enhancing Lipid Production in Microalgae. In *Algae and Environmental Sustainability* (pp. 119-127). Springer, New Delhi.

Li, F., Gao, D. and Hu, H. (2014). High-efficiency nuclear transformation of the oleaginous marine *Nannochloropsis* species using PCR product.*Biosci. Biotech. Biochem.*78 (5), 812-817.

Matsuzaki, M., Misumi, O., Shin-i, T., Maruyama, S., Takahara, M., Miyagishima, S.Y., Mori, T., Nishida, K., Yagisawa, F., Nishida, K. and Yoshida, Y. (2004). Genome sequence of the ultrasmall unicellular red alga Cyanidioschyzon merolae 10D. *Nature*. 428(6983), 653.

Minoda, A., Sakagami, R., Yagisawa, F., Kuroiwa, T., and Tanaka, K. (2004). Improvement of culture conditions and evidence for nuclear transformation by homologous recombination in a red alga, *Cyanidioschyzon merolae* 10D. *Plant Cell Physiol.* 45(6), 667-671.

Muñoz, C.F., de Jaeger, L., Sturme, M.H., Lip, K.Y., Olijslager, J.W., Springer, J., Wolbert, E.J., Martens, D.E., Eggink, G., Weusthuis, R.A. and Wijffels, R.H. (2018). Improved

DNA/protein delivery in microalgae–A simple and reliable method for the prediction of optimal electroporation settings. *Algal Research*. 33, 448-455.

Norzagaray-Valenzuela, C.D., Germán-Báez, L.J., Valdez-Flores, M.A., Hernández-Verdugo, S., Shelton, L.M. and Valdez-Ortiz, A. (2018). Establishment of an efficient genetic transformation method in *Dunaliella tertiolecta* mediated by *Agrobacterium* tumefaciens. *J. Microbiol. Methods*. 150, 9-17.

Petrie, J.R., Shrestha, P., Zhou, X.R., Mansour, M.P., Liu, Q., Belide, S., Nichols, P.D. and Singh, S.P. (2012). Metabolic engineering plant seeds with fish oil-like levels of DHA. *PLoS One*. 7(11), 49165.

Prasad, B., Vadakedath, N., Jeong, H.J., General, T., Cho, M.G. and Lein, W. (2014). *Agrobacterium* tumefaciens-mediated genetic transformation of haptophytes (*Isochrysis* species). *Appl. Microbiol. Biotechnol*. 98(20), 8629-8639.

Pratheesh, P.T., Vineetha, M. and Kurup, G.M. (2014). An efficient protocol for the *Agrobacterium*-mediated genetic transformation of microalga *Chlamydomonas reinhardtii*. *Mol. Biotechnol*. 56 (6), 507-515.

Radakovits, R., Jinkerson, R.E., Darzins, A. and Posewitz, M.C. (2010). Genetic engineering of algae for enhanced biofuel production. *Eukaryotic cell*. 9(4), 486-501.

Schiedlmeier, B., Schmitt, R.C.D.I.G.E.R., Müller, W., Kirk, M.M., Gruber, H., Mages, W. and Kirk, D.L. (1994). Nuclear transformation of *Volvox carteri*. *Proceedings of the Proc. Natl. Acad. Sci.* 91(11), 5080-5084.

Scott, S.A., Davey, M.P., Dennis, J.S., Horst, I., Howe, C.J., Lea-Smith, D.J. and Smith, A.G. (2010). Biodiesel from algae: challenges and prospects. *Curr. Opin. Biotechnol.* 21(3), 277-286.

Scranton, M.A., Ostrand, J.T., Georgianna, D.R., Lofgren, S.M., Li, D., Ellis, R.C., Carruthers, D.N., Dräger, A., Masica, D.L. and Mayfield, S.P. (2016). Synthetic promoters capable of driving robust nuclear gene expression in the green alga *Chlamydomonas reinhardtii*. *Algal Res*, 15, 135-142.

Sharma, K.K., Schuhmann, H. and Schenk, P.M. (2012). High lipid induction in microalgae for biodiesel production.*Energies*. 5(5), 1532-1553.

Shemesh, Z., Leu, S., Khozin-Goldberg, I., Didi-Cohen, S., Zarka, A. and Boussiba, S. (2016). Inducible expression of *Haematococcus* oil globule protein in the diatom *Phaeodactylum tricornutum*: association with lipid droplets and enhancement of TAG accumulation under nitrogen starvation. *Algal Res,* 18, 321-331.

Shestakov, S.V., and Khyen, N.T. (1970). Evidence for genetic transformation in blue-green alga *Anacystis nidulans*.*Mol. Gen. Genet.* 107 (4), 372-375.

Simon, D.P., Narayanan, A., Gouda, K.M. and Sarada, R. (2015). Vir gene inducers in *Dunaliella salina*; an insight in to the *Agrobacterium*-mediated genetic transformation of microalgae. *Algal Res*. 11, 121-124.

Steinbrenner, J. and Sandmann, G. (2006). Transformation of the green alga *Haematococcus pluvialis* with a phytoene desaturase for accelerated astaxanthin biosynthesis. *Appl. Environ. Microbiol.* 72(12), 7477-7484.

Ten Lohuis, M.R., and Miller, D.J. (1998). Hypermethylation at CpG-motifs in the dinoflagellates 587 *Amphidinium carterae* (Dinophyceae) and *Symbiodinium microadriaticum* 588 (Dinophyceae): evidence from restriction analyses, 5-azacytidine and ethionine 589 treatment. *J. Phycol*, 34(152-159), 590.

Thiyagarajan, S., Arumugam, M., Senthil, N., Vellaikumar, S., and Kathiresan, S. (2018). Functional characterization and substrate specificity analysis of Δ6-Desaturase from marine microalga *Isochrysis*sp CASA CC101577 *Biotechnology Letters* – 40:577-584.

Wang, L., Yang, L., Wen, X., Chen, Z., Liang, Q., Li J., and Wang W. (2019). Rapid and high efficiency transformation ofChlamydomonas reinhardtiiby square-wave electroporation. *Bioscience Reports,* 39, 1-8.

Wijffels, R.H. and Barbosa, M.J. (2010). An outlook on microalgal biofuels. *Science.* 329(5993), 796-799.

Wolk, C.P., Vonshak, A., Kehoe, P. and Elhai, J. (1984). Construction of shuttle vectors capable of conjugative transfer from *Escherichia coli* to nitrogen-fixing filamentous cyanobacteria. *Proc. Natl. Acad. Sci.* 81(5), 1561-1565.

Xie, W.H., Zhu, C.C., Zhang, N.S., Li, D.W., Yang, W.D., Liu, J.S., Sathishkumar, R. and Li, H.Y. (2014). Construction of novel chloroplast expression vector and development of an efficient transformation system for the diatom *Phaeodactylum tricornutum. J. Mar. Biotechnol.*16 (5), 538-546.

Xue, J., Niu, Y.F., Huang, T., Yang, W.D., Liu, J.S. and Li, H.Y. (2015). Genetic improvement of the microalga *Phaeodactylum tricornutum* for boosting neutral lipid accumulation. *Metab Eng*. 27, 1-9.

Zang, X., Liu, B., Liu, S., Arunakumara, K.K.I.U. and Zhang, X. (2007). Optimum conditions for transformation of *Synechocystis* sp. PCC 6803. *J Microbiol.* 45(3), 241-245.

Zäuner, S., Jochum, W., Bigorowski, T. and Benning, C. (2012). A cytochrome b5-containing plastid-located fatty acid desaturase from *Chlamydomonas reinhardtii. Eukaryot. Cell* 11(7), 856-863.

Zhu, B.H., Tu, C.C., Shi, H.P., Yang, G.P. and Pan, K.H. (2017). Overexpression of endogenous delta-6 fatty acid desaturase gene enhances eicosapentaenoic acid accumulation in *Phaeodactylum tricornutum*. *Process Biochem*. 57, 43-49.

Zorin, B., Grundman, O., Khozin-Goldberg, I., Leu, S., Shapira, M., Kaye, Y., Tourasse, N., Vallon, O. and Boussiba, S. (2014). Development of a nuclear transformation system for oleaginous green alga *Lobosphaera* (Parietochloris) *incisa* and genetic complementation of a mutant strain, deficient in arachidonic acid biosynthesis. *PloS one*. 9(8), 105223.

In: Applied Algal Biotechnology
Editors: M. Arumugam, S. Kathiresan et al.
ISBN: 978-1-53617-524-0

Chapter 9

SCALE UP METHODS MICROALGAE CULTURES

Andreas Isdepsky
Algae Biomass and Energy System R&D Center (ABES),
University of Tsukuba, Ibaraki, Japan

ABSTRACT

Microalgae constitute a large and diverse group of mainly photosynthetic micro-organisms, classified in seven divisions and comprising several thousand species. The history of microalgae research goes back more than 100 years, when the first uni-algal culture, today known as *Chlorellavulgaris* was established (1890) and more than 60 years later, commercial large scale cultivation began in the early 1960's in Japan with the culture of *Chlorella*, used for manufacturing *Chlorella* pills.Since then scientists together with engineers have developed and designed a wide range of different culture systems associated with the type of algae, volume and commercial application of algae culture. Light, temperature, effect of mixing, nutrient demand including inorganic carbon, pH and, contamination risk must be considered during scaling-up to find the balance between each parameter for maximum productivity. Contamination in algae mass cultivation refers to other algae, fungi and bacteria, predators and parasites that are considered as a major problem in large scale operations is considered.

This chapter summarises and highlights the limiting factors of algal growth and product yield in microalgae cultures which have fundamental importance in the development of a commercial large scale algal process. This section provides details of all steps of the subject of microalgae mass cultivation applied worldwide by incorporating own experiences for euryhaline *Tetraselmis* species from small scale to pre-commercial scale.

Keywords: bio-energy, biomass, aqua-farming, algal-biotechnology, productivity

9. INTRODUCTION

Human population growth and the rapid industrialisation of developing countries are the key contributors for the increase in demand for food and energy. Recently, as a result of the focus on "climate change" interest in producing energy from biomasses such as microalgae, for useas feedstock for transport fuel continues to rise (Martinot et al. 2007; Verbruggen and Al Marchohi 2010). A commitment to decreasing greenhouse gas emissions underlies the need for renewable energy as the atmospheric concentrations of CO_2, NO_x,SO_x, CH_4 and other greenhouse gases increase (Heinimö and Junginger 2009). CO_2 one of the greenhouse gases, is produced by burning fossil fuel as a transport fuel and stationary power plants for electricity generation (Bilanovic et al. 2009). The concentration of atmospheric CO_2 has increased from 280 ppmv in 1870 to 375 ppmv in 2006 and it is estimated that the concentration will rise to 550 ppmv by 2050 (Rittmann 2008).

Microalgae as a fast-growing feedstock, have the ability to fix CO_2 while capturing solar energy with a predicted efficiency 10 to 50 times greater than that of terrestrial plants based on the weight of biomass produced per unit of area (Usui and Ikenouchi 1997). Further, microalgae constitute a large and diverse group of mainly photosynthetic micro-organisms, classified in seven divisions and comprising several thousand species (Borowitzka 1999). This high diversity of phytoplankton forms the basis of the marine food chain and produces roughly 50% of the O_2 on Earth (Rahmstorf 2002).

The history of microalgae research goes back more than 100 years, when the first unialgal culture, today known as *Chlorella vulgaris*, was established by the Dutch microbiologist Beijerinck (1890). More than 50 years later (after World War II, in 1948), a global focus on cultivation for algae mass culture began. Commercial large-scale culture began in the early 1960's in Japan with the culture of *Chlorella*, used for manufacturing *Chlorella* pills, extracts and other items for which there is now a well established market. This was followed by *Spirulina*in the early 1970's in Lake Texcoco, Mexico by Sosa Texcoco S.A (Durand-Chastel 1980). A large number of microalgae have been suggested and evaluated for their suitability for commercial exploitation (Borowitzka et al. 1984; Borowitzka 2013a). Algae mass cultivation with a high variety of microalgae species was also recognised for the application in aquaculture (De Pauw et al. 1984). Interest in applied phycology has continued, especially with studies on the use of algae as a renewable feedstock for transport fuel and as a microbial protein source (Borowitzka 1999).

The commercial production of *Dunaliella salina*, as a source of β-carotene, became the third major microalgal industry with production facilities being established in the early 1980s in Israel, the United States (US) and Australia (Soeder 1980b; Borowitzka 1997a; Spolaore et al. 2005).Two large *Dunaliella* plants are located in Australia each with over 400 ha of unmixed ponds were established by Western Biotechnology (Hutt Lagoon, Western Australia) and Betatene (Whyalla, South Australia) (both now owned by BASF) in 1986 (Borowitzka 1999). More recently, several plants producing *Haematococcus pluvialis*as a source of astaxanthin have been established in Israel, the USA and India (Spolaore et al. 2005; Borowitzka 2013a).

9.1. Mass Culture of Microalgae

During the whole period of large scale mass culture of algae, scientists together with engineers have developed and designed a wide range of different culture systems (Borowitzka 1999). Depending on the type of algae, volume and commercial application of algae culture, the systems can be basically separated into *open* and *closed* culture systems.

Table 9.1. Underlying principles in algae mass cultivation

<table>
<tr><th colspan="2">Light</th><th>Temperature</th><th colspan="2">pH, CO_2, O_2</th></tr>
<tr><td rowspan="6">Algal Cell specific</td><td>Light Capture</td><td>Photosynthesis efficiency</td><td colspan="2">Species dominance by pH</td></tr>
<tr><td>Limitations (P-E curve)</td><td>Degree of tolerance</td><td colspan="2">CO_2 solubility at various pH</td></tr>
<tr><td>Conversion efficiency (α)</td><td>Enzymatic activity (Q_{10})</td><td colspan="2">Productivity improvement at CO_2 addition</td></tr>
<tr><td>Acclimation pigment concentration and type</td><td>Biochemical composition</td><td colspan="2">Sources of inorganic carbon</td></tr>
<tr><td colspan="2">Photo-inhibition ⇔ light / temperature interaction</td><td colspan="2">*CO_2 competition with O_2Rubisco</td></tr>
<tr><td colspan="4">Photo- / Dark – Respiration ⇔ Temperature ↑ ⇔ O_2↑ *</td></tr>
<tr><td rowspan="3">Reactor specific</td><td rowspan="2">Light path - density/light penetration</td><td rowspan="3">Selection of algae strains</td><td colspan="2">CO_2 delivery</td></tr>
<tr><td colspan="2">Cost of inorganic carbon</td></tr>
<tr><td>** Light/dark frequency</td><td colspan="2">pH maintenance</td></tr>
<tr><th colspan="5">Scale-ability</th></tr>
<tr><th colspan="2">Nutrients</th><th>Contaminants</th><th>** Mixing & turbulence</th><th>Salinity</th></tr>
<tr><td rowspan="2">Algal Cell specific</td><td>C:N:P ratio</td><td>Factors leading to contamination</td><td>Mass transfer</td><td>Tolerance</td></tr>
<tr><td>Type and source of nutrients</td><td>Form of contamination</td><td>Shear</td><td>Solubility of CO_2 and O_2</td></tr>
<tr><td colspan="3" rowspan="2">Reactor specific</td><td>Gas exchange***</td><td rowspan="2">Management***</td></tr>
<tr><td>Cost***</td></tr>
</table>

Stars indicate the degree of impact during scaling up (*low / *** high).

Apart from open raceway ponds and enclosed systems, a shallow cascade system was developed for the cultivation of *Scenedesmus sp.* by Setlik (1970) cited in Borowitzka (1999) and is still running in Trebon, Czech Republic. This open system provides high growth rates even by maintaining the system with biomass densities of 10 g L^{-1} at less than 10 cm culture depth which is comparable and common for enclosed photo-bioreactors (Borowitzka 1999). Similar inclined slope systems were used in Bulgaria and Western Australia and a modified raceway-type system with sloped meandering channel was established in Peru (Borowitzka and Moheimani 2013). A two-step cultivation concept of combined open and enclosed reactors are used as hybrid systems for the production of high value products such as aastaxanthin in *Haematococcus pluvialis* in Hawaii (Olaizola 2000). However this concept is considered more expensive in capital and operation costs and is therefore not suitable for a commodity product such as biofuel (Fon Sing et al. 2013). The success of algae mass cultivation is based on the understanding of interactions between environmental factors such as solar irradiance, temperature and salinity and operational parameters which include nutrient availability and pH (Table 9.1).

9.1.1. Open-Culture Systems

Several open systems with different designs exist; and these vary in size, shape, materials of construction, as well as manner of agitation and inclination (Tredici 2004). Borowitzka (1999) summarised the most predominant culture systems for commercial algae mass culture as follows (the list includes the limiting characteristics of these systems):

- large open ponds (extensive open ponds)
  - unstirred, very low growth rate
  - light limited

- circular ponds with rotating arm for mixing (centre-pivot pond)
  - light limited
  - expensive concrete construction
  - high energy consumption for stirring
  - turbulence in the central part is very low, especially in large scale units

- raceway ponds
  - light limited

- tanks
  - light limited
  - mixing is limited in large tanks

- cascade systems
  - higher capital cost compared to unlined raceway ponds.

A pertinent question to ask is: Why are most commercial large algae culture systems open air systems? Commercial open systems are easier and less expensive to build and operate per unit area, in terms of scalability, monitoring/controlling, maintenance, and also large open systems are more durable than large closed reactors would be an acceptable answer (Tredici 2004). The largest open production units are the extensive shallow ponds located at Hutt Lagoon in Western Australia, used for producing β-carotene from *Dunaliella salina* (Borowitzka et al. 1984). Apart from extensive cultivation systems, very large culture systems are originated from raceway ponds arranged side by side, covering many hectares with a maximum surface area of 10,000 m^2(1 hectare) for a single pond (Dodd 1986).

However, in designing and building successful large algae production systems a compromise must be reached between high performance and economical constraints. Regarding the demand for biofuel production, a very large scale algae cultivation plant is needed and therefore, simply the best pertains for raceway ponds. These ponds have various opportunities for construction materials, for different parts per unit combined with high performance in design for optimal mixing properties.

Many options in design for mixing and turbulence are available for raceway ponds which are very important for open systems. These raceway ponds are principally driven by paddle wheels and are operated at between 20 and 30 cm depth. The depth is a compromise between the minimum water depth required to provide adequate light to the algal cells (i.e., a lower level means more light is available for the cells) and the minimum depth needed to maintain adequate mixing and to avoid large changes in ionic composition due to evaporation (Borowitzka 1999).

Further, the choice of the cultivation system can vary by local conditions, algae species and utilisation of the algae biomass. According to several studies the main disadvantage of open-air pond systems is that the actual productivity achieved is less than the theoretical predicted productivity (Richmond and Becker 1986; Becker 1994b; Borowitzka 1999; Pulz 2001; Grobbelaar 2009).

9.1.2. Enclosed-Culture Systems

Enclosed culture systems for microalgae are usually termed 'enclosed photo-bioreactors' (PBR). The design of photo-bioreactors is different from the design of bioreactors for other micro-organisms, because of the requirement of light as the major energy source for photoautotrophic micro-organisms such as microalgae. Enclosed photo-bioreactors are applied rather than open-culture systems for the following reasons:

- to minimise light limitation and to increase biomass productivity
- to minimise the risk of contamination
- to minimise water losses/consumption
- to optimise hydrodynamic characteristics
- to optimise process monitoring
- to optimise mass transfer in terms of:
  - CO_2 supply while minimising losses
  - removal of photosynthetically generated O_2 (closed systems are particularly bad in this respect compared to open ponds).

A large number of photo-bioreactors ranging from laboratory to industrial scale have been proposed and reviewed (Tamiya 1957; Tredici and Materassi 1992; Borowitzka 1997b; Grima et al. 1999; Pulz 2001; Masojidek et al. 2003; Eriksen 2008). Most of the PBRs are either tubular devices or flat panels and they can be characterised further on the basis of the mechanism of circulation of the culture, the orientation of the tubes or panels, and the type of gas exchange system (Molina et al. 1991). These include vertical, horizontal or serpentine tube, flat-plate, bubble column and airlift tubular reactors. However, each reactor configuration has its limitations in scalability per unit on the basis of principles in physics and biology. For example, a vertical reactor design is limited in height and volume for static reasons in material and construction.

Several experiments have demonstrated that not every algae species is suitable for the growth in a PBR (or it may be suitable only up to certain size), as a result of the interaction between algal physiology and reactor design in terms of fluid dynamics and light transfer (Torzillo et al. 1993; Moheimani and Borowitzka 2006; Posten 2009).

The 'Biocoil' is a helical tubular photo-bioreactor that appears to have the most promising design within the enclosed photo bioreactors. The tubes of the Biocoil are constructed from glass, clear Teflon® or poly vinyl chloride (PVC) lines in helical configuration attached around a central support tower (Robinson et al. 1988). This helical tubular photo-bioreactor was modified by Watanabe (1995) into a cone-shaped tubular photo bioreactor, which gave the advantage of better sun-light capture from direct overhead solar irradiance (Watanabe and Saiki 1997). Mixing in tubular photo-bioreactors is provided by different types of pumps, including airlifts, centrifugal, diaphragm or lobe pumps achieving turbulent flow with Reynolds numbers above 3300 (Grobbelaar, 2009).

The effect of shear stress on the physiology of the algal cell induced by turbulence depends on the algal species, and particularly with flagellate algal species, it can lead to cell breakage (Janssen et al. 2003; Miron et al. 2003; Barbosa 2004; Grobbelaar 2010). Photobioreactors (PBR) continue to have the major disadvantage of high capital costs of construction and operation compared to open systems and this is an even greater problem for low value products such as biofuel. PBR's achieve higher volumetric biomass productivity which reduces cost for the harvesting process and requires a smaller land area compared to open systems. This will be a major advantage for countries or areas where the land cost is very high. The major challenge for photo-bioreactors was found in the design process due to the scaling-up, which includes criteria such as surface-to-volume ratio, orientation and inclination, mixing and degassing, temperature regulation and, transparency, as well as the durability of the construction material and ease of maintenance (cleaning) combined with operation (Tredici 2004).

Presently, the largest commercial PBR systems are the modular tubular photobioreactor plant with about 300 km long glass tubes, on 10 acres of arid desert land at Algatechnologies in Israel and the 600 m^3 plant at Klötze in Germany. The latter plant consists of a tubular PBR system of 20 modules of 35 m^3 each, installed in a 1.2 ha greenhouse (http://www.algomed.de, 28.12.2019). Other commercial PBR systems exist in the USA for example, with a capacity of three times 25000 L at Aquasearch Inc. and in the United Kingdom (UK) at Biosynthesis with modules up to 5000 L (Tredici 2004; Borowitzka 2013a). For large commercial mass production of microalgae a combination of closed PBRs and open raceway ponds has also been proposed, and addressed in the literature as multistage systems or hybrid systems (Fon Sing et al. 2013).

The advantage of this concept is to use the closed PBR to produce a high cell density inoculum for the open ponds, thereby offering the possibility of optimised biomass production in a shorter time. This concept has been applied for valuable bio-compounds with algae species such as *Haematococcus* sp. (Grobbelaar 2000).

9.2. Limits to Growth in Open Raceway Ponds

The understanding of limiting factors to algal growth, and product yield in microalgae cultures is of fundamental importance in the development of a commercial large scale algal process. Microalgae cultures in open systems at densities greater than 0.5 g L^{-1} are unattainable over long periods due to a number of physical, chemical and biological factors (Abeliovich 1980; Javanmardian and Palsson 1991; Zittelli et al. 2004). Algal growth in cultivation systems is influenced by various environmental factors (Borowitzka 1998). The principal factors are:

- Light
- Temperature
- Nutrients
- Contamination
- Mixing/turbulence

9.2.1. Light

Light energy that reaches the entire surface of Earth is never homogeneous with a difference in quantity while all high density algae cultures all over the world are light limited (Qiang et al. 1998). Therefore not all areas around the world (including areas with high cloud cover, high rainfall and /or low winter temperatures) are suited for successful algae mass culture.

9.2.2. Photosynthesis and Photon Utilisation Efficiency

Photosynthesis efficiency has been used as a tool for biomass yield comparison between phytoplankton and plant crops (Walker 2009). For the theoretical maximum of photosynthesis efficiency, optimal conditions are assumed, including temperature. However, it has to be considered that photon availability depends on the location of cultivation as in some areas on earth 1370W $m^{-2}d^{-1}$of light energy reaches the surface whereas only 240 W $m^{-2}d^{-1}$ is the world average.

The absorption of solar energy by chlorophyll is maximal at about 680 nm and the chemical endproduct of the Z-scheme and the Benson–Calvin cycle is a carbohydrate molecule (Leegood et al. 1985). The Z-scheme calls for four electrons to be propelled by four photons through each of two photo systems for every molecule of O_2 released from water and every molecule of CO_2 incorporated into 'CH_2O'(Walker 2002). Less than 50% of the full sunlight spectrum is photosynthetically active irradiance (PAR), which is defined as irradiance in the 400 - 700 nm wavelengths range. For example Watanabe and Hall (1995) calculated the input of PAR (IPAR), expressing the quantity of irradiance impacting on the reactor as the product of PAR ($kJm^{-2}d^{-1}$) and the illuminated surface (m^2). According to Watanabe and Saiki (1997), the photosynthetic efficiency (%) (PE) was calculated based on Equation 9.1.

$$\mathrm{PE} = \frac{r_G H_G \times 100}{\mathrm{IPAR}} \tag{9.1}$$

Where *r*G is the maximum daily growth ($g_{dry\ weight}\ d^{-1}$) and *H*G the enthalpy of the total dry biomass (21.01 kJ $g_{dry\ weight}{}^{-1}$) (Watanabe and Hall 1996). Weyer et al. (2010) who reported an energy content for protein of 16.7 kJ g^{-1}, carbohydrate 15.7 kJ g^{-1} and lipid 37.6 kJ g^{-1} give an estimated average energy value of 26.9 kJ g^{-1} for dry algal cells with 50% lipid and 25% for protein and carbohydrate content.

The energy content of an average photon mole of visible light in this range is about 50 kcal (209 kJ) and based on the Z-scheme, eight photons are required to produce one organic carbon unit of "CH_2O" which gives a maximum photosynthetic efficiency of 14%.

In contrast to the Z-scheme values, several studies found that the mean quantum requirement for C3 plants is 9.4 and this will reduce the maximum photosynthesis efficiency to 11.9% (Delieu and Walker 1981; Walker and Osmond 1986; Walker 1989; 2009).

Walker also states that a more realistic maximum in photosynthetic efficiency is approximately 4.5% for C3 plants or microalgae under field conditions (Walker 2009). At the present time, reported biomass productivity data of commercial algal mass cultures have achieved an average of 15 $g_{AFDW}\ m^{-2}\ d^{-1}$ over a period of 365 days, equal to 54 $t_{AFDW}ha^{-1}\ yr^{-1}$, which represents only 1% of photosynthesis efficiency (Belay 1997; Walker 2009). The dimensions of the cultivation systems can be seen as potential reason for limiting the amount of light energy delivered to the algae culture. In open raceway ponds, which represent the system most commonly used in commercial algae mass production, light can be captured only from one surface area resulting in less light availability and consequently in low volumetric biomass productivities. In high density algae culture, a steep light gradient is formed from the upper layer of the culture which leads to a low efficiency of light conversion (Grobbelaar 1982).

Another reason for light limitation in large-scale algal cultures is the presence of abiotic and biotic light absorption in competition with the target algal population or biomass (Huisman and Weissing 1994). The algae culture exposed to the natural environment also undergoes daily to seasonal changes in solar irradiance.

To withstand these variations in light intensity, algal cells have mechanisms such as photoacclimation, that regulate the amount of pigments and biochemical composition of the photosynthetic machinery in response to irradiance (Hu 2004; Vonshak and Torzillo 2004; Dubinsky and Stambler 2009). Three methods exist to reduce light limitation in high-density algae cultures; one is to increase the mixing, the second method is to decrease the depth of the culture in open raceway ponds (down to the minimum level that is required for mixing), and the third is to reduce the cell density.

However, it is commonly observed that there is a cellular response with the increase in light-harvesting pigments such as chlorophyll a, b and c, phycobiliproteins and primary carotenoids when irradiance is decreased. In contrast, high light intensities tend to decrease in chlorophyll a and other primary pigments accompanied in species such as *Dunaliella salina* and *Haematococcus* sp. with an increase in secondary carotenoids acting as photo-protective substances such as β-carotene, zeaxanthin and astaxanthin (Falkowski and Laroche 1991; Hu 2004; Dubinsky and Stambler 2009). However, light utilisation efficiency and response mechanisms are dependent on other parameters such as temperature, nutrients and pH (Davison 1991; Falkowski and Laroche 1991; Geider et al. 1998).

9.2.3. Temperature

Temperature represents another important potential limitation for high biomass production rates in the outdoor cultivation of microalgae. Temperature is an important factor for all biological reactions, including light-dependent and light-independent biological reactions in phototrophs (Goldman 1980). The temperature range in which photosynthetic algae can be found is broad, ranking from snow or Antarctic algae with photosynthetic activities down to -7°C, to the thermophilic cyanobacteria which are able to photosynthesise up to 74°C (Soeder and Stengel 1974; Davison 1991). Several authors have categorised algae species according to their optimum temperature range: below 10°C for polar algae, 10-25°C for temperate or mesophilic algae, and greater than 25°C for thermophilic or desert algae species (Soeder and Stengel 1974; Barasanti and Gualtieri 2006).

There are significant different temperature optima between the species of algae and even between algae strains. For example, several *Chlorella* species with temperature optima of 25°C, 30°C, 35°C and 39°C have been reported by Sorokin (1959) and Lorenzen (1963) cited in Soeder and Stengel (1974). Optimum temperatures for algal cultures can be achieved only under controlled conditions while outdoor cultures are exposed to daily fluctuations in temperature, which lead either to the limitation or inhibition of photosynthesis. One classic example of limiting levels was found during the morning in outdoor cultures, when solar irradiance is rising more rapidly than the temperature, which constitutes a sub-optimal condition (Vonshak et al. 2001).

Aside from diurnal fluctuations in temperature, in many habitats the same algal species or plant is subjected to very wide seasonal variations in temperature, which can require a mechanism for acclimation or adaptation. The adaptability of a strain is under genetic control, whereas the ability to photosynthetically acclimate is affected by its environment (Raines and Lloyd 2004). Similarities are reported for higher plants by Berry and Bjorkman (1980), where adaptations may be considered as a genotypic variation, which enable plants to adapt efficiently under the temperature regimes of their various native habitats.

The growth of a given genotype under a cool regime may result in an improved photosynthetic capacity at a low temperature, whereas the growth of another genotype under a warm regime may result in an improved photosynthetic performance at high temperatures. The potential for such photosynthetic acclimation or adaptation to temperature varies between species (Berry and Bjorkman 1980). Therefore the optimum temperature of a strain is equal to the highest temperature of the previous environment before isolation (Berry and Bjorkman 1980; Davison 1991). The temperature effect on algal growth can be categorised into (1) biological enzymatic processes and (2) physico-chemical processes.

Photosynthetic enzymes as members of the biological enzymatic process can be said to do the following: (1) represent the maximum rate indicated at the light saturation point I_k at a certain light intensity dependent on the ratio between the concentration of photosynthetic enzymes and the concentration of photosynthetic pigments (Jorgensen and Steemann 1965). Higher I_k values are found in algae grown at higher light intensities, rather than in algae grown at low light intensities due to the adaptation mechanism.

However, adaptation due to the enzymatic processes in which photosynthesis gradually increases with increasing temperature is commonly associated with a Q_{10} factor (rate of change as a consequence of increasing the temperature by 10°C), which can vary between species (Jorgensen and Steemann 1965; Davison 1991). That is, photosynthesis increases at constant light intensity up to an optimum temperature that occurs over a range of several degrees rather than a single value that indicates the maximum photosynthetic rate (Davison 1991). Conversely, a higher temperature is not the only factor that promotes enzyme activity in photosynthesis processes respiration processes have also been considered in the literature with higher Q_{10} values than in photosynthesis processes (Marre 1962).

Therefore, higher temperature can also be less favourable for net biomass productivity if respiration per unit time is not significant lower than the net-biomass accumulation (Marre 1962; Jorgensen and Steemann 1965). Inhibitory effects of temperature on algal growth are either intracellular, such as enzymatic processes or external in the medium, such as physical chemical processes. One common effect on physical, chemical processes of temperature has been recognised on the solubility of CO_2 and O_2 as competitive partners in the culture medium for the substrate RuDP-carboxylase/oxygenase. Marre (1962) reported that the solubility coefficients (α) decreased between 25°C and 60°C from 0.76 to 0.36 for CO_2 and from 0.028 to 0.019 for O_2, whereas the concentration of O_2 in seawater is more than twice that of CO_2 in gaseous form (Marre 1962; Capel 1995).

The potential response of the targeted species of microalgae to fluctuation in temperature is also an important consideration when choosing the location for cultivation, as well as the cultivation system in culturing these algae in closed or open cultivation systems (Torzillo et al. 1993; Borowitzka 1995b).

9.2.3. CO_2, pH and O_2

The major inorganic nutrient compound is carbon, and it plays a central role in the biological energy conversion process due to photosynthesis, whereas the C:N and C:P ratios are completely under biological control (Falkowski 1997). Light energy conversion into biochemical energy in the form of ATP and $NADPH_2$, aside from light and temperature, requires a suitable carbon source such as CO_2 for photo-autotrophic growth of algae.

Increases in the amount of atmospheric CO_2 concentration raise concerns because it is the principal factor that causes the phenomenon of global warming which has resulted in global aims to reduce this specific greenhouse gas (Wang et al. 2008; Florides and Christodoulides 2009). Atmospheric CO_2 has a much higher solubility than O_2 due to chemical reactions forming carbonic acid and its dissociation products such as bicarbonate and carbonate (Capel 1995). This indicates great potential for biological CO_2 mitigation by phytoplankton. Consequently inorganic carbon exists in three forms and the relationship between these forms is influenced by pH (Emerson and Green 1938), temperature and salinity (Figure 9.1).

The total available amount of CO_2 at pH 8.2 with approximately 40 mg L^{-1} support a biomass production of 12 g_{AFDW} m^{-2} d^{-1} at 30 cm depth at 50% carbon content (Capel 1995), indicating a demand of CO_2 addition of once or twice per day (Benemann 1993). The high carbon content of algal biomass necessitates an inorganic carbon supply with an average demand of approximately 45 gCO_2 m^{-2} d^{-1} dependent on biomass productivity and required for Goldman's estimated biomass yield of 60 g_{DW} m^{-2} d^{-1} up to 120 gCO_2 m^{-2} d^{-1} by 100% efficiency of the supplying technology (Goldman 1980; Soeder 1980a).

To date, there is no proven technology for CO_2 supply with 100% efficiency for cultivation systems of large-scale algae biomass production. Researchers have suggested that CO_2utilisation in "closed" cultures appears to be more efficient owing to the prevention of atmospheric losses (Robinson and Toerien 1982), which can also be reduced by adding the gas through a special apparatus (Su et al. 2008).

The disadvantage of using CO_2 as an inorganic carbon source can be seen in the lower diffusion rate in water by a factor of 10^4 compared to air (Beardall and Raven 2013) accompanied by factors such as temperature and salinity reducing the solubility of gases (Kelly 1989; Wang et al. 2008).

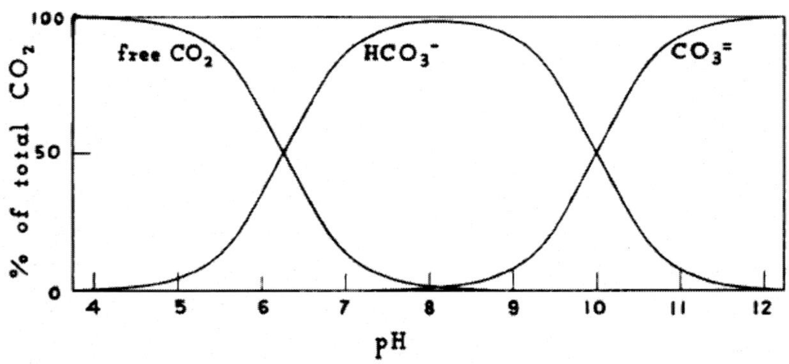

Figure 9.1. Proportion of total carbon dioxide in each of the three forms, free carbonic acid, bicarbonate ions, and carbonate ions as a function of hydrogen-ion concentration in freshwater from Emerson and Green(1938).

Currently, CO_2 in gaseous form is supplied to large scale open raceway ponds by using diffuser systems incorporated in 1 m deep carbonation channel to improve transfer efficiency due to longer resistance time. CO_2 supply for mass cultures, including transfer and storage, is still considered expensive, even by using industrial gases and waste gases from large CO_2 producers such as power stations, cement and fertilizer plants (Benemann 1993). Improvements could be made by co-location of the algae production facility with large CO_2 producers, where CO_2 can be supplied on demand by maintaining the optimum pH of the algae species (Benemann 2003; Fonseca et al. 2010). Soluble carbonates such as sodium bicarbonate and sodium carbonate are also potential CO_2 sources in microalgae culture systems (Wang et al. 2008).

Further attempts have been made by several scientists on the effect of pH and CO_2 on microalgae species selection for CO_2 sequestration (Goldman 1973; Shapiro 1973; Azov 1982; Richmond et al. 1982; Ono and Cuello 2003). Most of these investigations were carried out by comparing prokaryotic algae (cyanobacteria) with eukaryotic algae (green microalgae) grown under different pH conditions and consequently different forms of available inorganic carbon. Independent to each other, the results of these investigations indicated that cyanobacteria have favourable pH of >9 while green algae such as *Chlorella* sp. prefer pH conditions between pH 7 and 8. However, a major limitation for *Chlorella* sp. was found under high bicarbonate content and low gaseous CO_2 concentration (Richmond et al. 1982). Goldman (1973) concluded that the total inorganic carbon concentration (C_T) has a higher limiting impact than the particular availability of free CO_2 or carbonate and algal species difference occurs due to the difference in the half saturation coefficient (K_s) for C_T-limited growth. He also demonstrated in a pH range from 7.1 to 7.6 different K_s values for C_T limited growth of various green algae species (Goldman 1973).

The highest specific carbon uptake for *Scenedesmus obliquus*and *Chlorella vulgaris* (mg carbon per g of particular carbon per min) was observed at pH 7.2 compared to the specific carbon uptake at pH 9.2 (Azov 1982). In contrast, it has been considered that algae such as cyanobacteria with an optimum growth at pH>10, the capability of the direct use of HCO_3^- would result in their outcompeting other species that use only CO_2 (aqueous) (Goldman 1973).

A low pH close to neutral with artificially high concentrations of free CO_2 has been recommended for maximum biomass productivity in green algae (Shapiro 1973; Azov 1982). The downside of carbon assimilation in phytoplankton was found by an inverse increase of O_2 levels up to 500% saturation (30–33 mg O_2 L^{-1}) in the medium when in the CO_2 concentration decreased due to photosynthesis (Richmond et al. 1990), which changes the O_2/CO_2 ratio significantly (Ku and Edwards 1978).

Further, O_2 production in algae mass culture can be used as an indicator for biomass productivity if one knows the ratio between synthesised biomass and O_2 produced with respect to the interaction between temperature and irradiance (Richmond et al. 1990; Torzillo 1997). In contrast, O_2 levels exceeding 30 g L^{-1} demonstrated negative effects on specific growth and protein synthesis for the cyanobacterium *Spirulina*sp. (Vonshak et al. 1996; Torzillo 1997). High O_2 levels are considered more critical in enclosed photo-bioreactors due to poor degassing than in open raceway ponds and play an important role in determining the optimal tube length in tubular bioreactors with respect to the velocity of circulation and the number of degassing stations (Torzillo 1997). However, O_2 competes with CO_2 during photorespiration by using the same enzyme ribulose-1.5 bisphosphate carboxylase (Rubisco), to fix CO_2 as this enzyme is also an oxygenase (Geider and MacIntyre 2008).

A central role in photorespiration is played by P-glycolate as the product of the oxygenase inside the chloroplast and substrate of the photo-respiratory carbon oxidation cycle, where 75% of the carbon returning the photosynthetic carbon reduction cycle is to be oxidised outside the chloroplast (Badger 1985). High light intensities are favourable to glycolate formation and can be excreted in the medium, while the mechanism requires low CO_2 concentrations and high O_2 concentrations with half maximal activity of the oxidation reaction found at 68% O_2 equilibrium in water (Tolbert 1974).

Effects of the temperature on the intracellular solubilities of CO_2 and O_2 where the solubility of CO_2 decreases more than the solubility of O_2 at a higher temperature resulted in a change of CO_2/O_2 ratio (Ku and Edwards 1978; Brooks and Farquhar 1985). Therefore, the supply of CO_2 to the algae culture by maintaining a steady state of inorganic carbon concentration will stimulate the carbon concentrating mechanism and inhibit the oxygenase activity of Rubisco (Raven 1988).

Photosynthesis as a biochemical process includes enzyme activities that require optimal temperatures, nutrient concentrations and an optimal pH for enzyme kinetics. The pH optimum for ribulose-1.5 bisphosphate carboxylase activities is reported to be pH 7.8, with a rapid decline in activity beyond pH 8.3, accompanied by an increase in ribulose-bisphosphate-oxygenase activity. The enzyme ribulose-1.5 bisphosphate carboxylase including a series of reactions of the Calvin cycle are responsible for plant growth due to CO_2 reduction into carbohydrates (Kelly 1989). Further, the activity of the oxidation reaction was found with an optimum pH between 9 - 9.3 and required the presence of Mg^{2+}; at these ranges of pH, the Rubisco activity is completely inhibited (Tolbert 1974).

9.2.4. Nutrients

Despite from light and temperature, inorganic nutrients are the crucial limiting factors in phytoplankton growth and productivity. The elements found in organic matter produced by organisms including phototrophs vary across the entire periodic table of elements. However, the majority of the literature nominates sixteen elements, C, N, P, S, K, Mg, Ca, Sr, Fe, Mn, Zn, Cu, Co, Cd, B and Mo, as essential for growth (Goldman 1965; Capel 1995; Ho et al. 2003). In concentrated phytoplankton (e.g., algae mass cultivation), there is a classification of elements into bulk or macronutrients such as C, N, P (Table 9.2) and micronutrients or trace elements including Fe, Mn, Cu, Zn, Co and Mo, which are based on the required concentration due to growth limitation effects. Iron is occasionally considered as macronutrient due to the response in algal photosynthesis when added into natural lakes (Goldman 1965). However, a major limitation in microalgal growth is caused by macronutrients incorporated at high concentrations in C, N and P into cellular proteins, carbohydrates and lipids.

For example, when nutrients are not limited and light is the only growth limiting factor, the chemical composition of most algal species displays an elementary configuration of: 45 to 50% carbon, 8 to 10% nitrogen and 1% phosphorus equal to their biochemical composition, and 50% protein, 30% carbohydrates and 15% lipids (Goldman 1980). As discussed in the previous section, inorganic carbon for phototrophs plays a special role, where limitation occurs more in biological availability due to a low conversion rate between CO_2 and HCO_3^-

instead of the availability of the total amount of inorganic carbon (Riebesell and Wolf-Gladrow 2007).

Silicon (Si) as an element is sometimes considered a macronutrient in regards to diatoms, where the cell wall is distinct and composed of silicon dioxide (SiO_2) (Matthew and Edward 2010). For phytoplankton in general, the elemental composition is dependent on nutrient concentration and the light regime and can vary between species independently of the environmental conditions (Yentsch 1962; Burkhardt et al. 1999). Amounts of nitrogen and phosphorus supplied to autotrophic algae in mass cultivation are reported to range from 1.6 to 6.5 g (N) m^{-2} d^{-1} and 0.2 to 0.83 g (P) $m^{-2}d^{-1}$ (Toerien et al. 1987).

In fact, variations in nutrient concentration depend on biomass productivities or productivities of the target product. The best example of the effect of nutrient concentration is found in the two-stage cultivation of *Dunaliella salina* for the β-carotene production that requires in the first stage a nitrogen rich medium at low salinity for optimum biomass production, while the second stage requires low nitrogen concentration and high salinity for optimum β-carotene production (Ben-Amotz 1995). Many studies have been conducted to determine the optimal nutrient concentration for various algal species and their applications including their biochemical composition. Grobbelaar (2004) described the meaning of 'optimal nutrient concentration' as transition zone where algal growth is less affected by the addition of nutrients. The author also categorised nutrient concentrations as deficient zone at low concentration, adequate zone where increased nutrient concentration has no effect on growth, and toxic zone where increased nutrient concentration leads to a decline in growth (Grobbelaar 2004). Aside from concentrations and biological availability, it is also considered that the ratio between C: N: P first recognised by Redfield (1934) with a ratio of 106 C: 16 N: 1 P, plays an important role in avoiding the transition of elementary limitation for high yields (Falkowski 2000; Ho et al. 2003; Grobbelaar 2004). Research by Quigg et al. (2003) on yielded biomass in its elementary composition under identical growth conditions showed systematic variations in C:N:P ratios between the eukaryotic divisions remarkably similar to the stoichiometry after Redfield. However, Quigg et al. (2003) found a 30:1 N:P ratio of the Chlorophyceae compared to Prasinophyceae with 25:1 and diatoms with 10:1.

Table 9.2. Three major macro elements and their chemical forms used in microalgal biotechnology

Elements	Form of nutrition	Remarks	Reference
C	CO_2, HCO_3^-, CO_3^{2-}	– Low conversion rate between CO_2 and HCO_3^-	(Riebesell and Wolf-Gladrow 2007)
N	NO_3^-, NO_2^-, NH_4^+, NH_3, urea, N_2	– Likelihood of toxicity due to inhibitor effects of NH_4^+ and NH_3 by interrelation with pH	(Azov and Goldman 1982)
P	HPO_4^{2-}, PO_4^{3-}, $H_2PO_4^-$	– Increased pH due to photosynthesis causing precipitation of phosphate in seawater	(Olsen et al. 2006)

The differences for the effect of nutrient limitation between algae species can be seen as the result of evolution and diversity creating 'functional groups' in marine habitats including the diatoms, the coccolithophorids, nitrogen-fixing cyanobacteria and flagellates (Riebesell and Wolf-Gladrow 2007). From the ecological point of Riebesell and Wolf-Gladrow (2007), the classification for phytoplankton due to differences in cell composition is strongly based on seasonal elementary cycle and particular material in the ocean which is represented as the 'functional group' concept. Therefore, a major challenge is in maintaining a steady state of optimal nutrient concentration in outdoor mass cultivation with daily changes in solar irradiance and temperature accompanied by competition for limited resources.

Several attempts on the determination of nutrient uptake kinetics were made after Monod and Michaelis Menten indicated clearly that the uptake of limiting and non-limiting nutrients was found to be controlled by internal as well as external substrate concentrations. The interrelationship between external and internal nutrient concentration as a limiting factor was due to "luxury" (excess) consumption of one nutrient when growth was limited by another (Droop 1974). Nutrient uptake in phytoplankton is also influenced by cell size and form, because smaller spherical cells have a greater surface / volume ratio compared to larger cells. Greater surface / volume ratio promotes the diffusion across the boundary layer (Riebesell and Wolf-Gladrow 2007).

However, further aspects for nutrient consumption in phytoplankton must be considered; for example nutrient loss due to competitive organisms especially in open systems, or loss of P-nutrient due to precipitation during the pH rise in the medium due to photosynthesis. A mixture of different N-sources also can increase the rate of nutrient uptake, and it has been found that a ratio of nitrate to ammonium nitrogen of 3:1 yielded higher productivities than when supplying a single N-source (Mostert and Grobbelaar 1987). Conversely, light dependent photosynthesis influences nutrient uptake and is constant at light saturation (Grobbelaar 2004). Mesocosm studies on the effect of ultraviolet-B (UV-B) radiation on uptake of different nitrogen-sources such as NO_3^-, NH_4^+ and urea by phytoplankton showed that urea uptake was most sensitive. The uptake of urea was influenced by the level of UV-B radiation, but did not effect internal organic nitrogen composition (e.g., internal urea, free amino acids, and protein for all N-sources) (Fauchot et al. 2001; Grobbelaar 2004).

Temperature influenced biochemical reactions are measured as the temperature coefficient (Q_{10}) where the enzymatic reaction doubles with a 10°C rise, resulting in approximately twice as much nutrient uptake (Goldman and Carpenter 1974; Grobbelaar 2004). Movement in the form of water motion in natural habitats or velocity in mass culture reduces the diffusive boundary layer thickness resulting in increases in nutrient availability and uptake of a non-motile algal cell (Findenegg 1965a; Riebesell and Wolf-Gladrow 2007).

9.2.5. Salinity

Ecological differentiations of algae into halophilic or non-halophilic and stenohaline or euryhaline are made with based on their physiology and habitat distribution (Soeder and Stengel 1974). Although algae can adapt to a wide range of salinities, as recently discovered for a species of *Tetraselmis* (*Tetraselmis indica*) which was isolated from a salt pan with a salinity as high as 35% NaCl and as low as 3.5% (Arora et al. 2013). However, salinity tolerance in terms of survival differs from the maximum growth range within the tolerable range in salinity. For example, with an increase in salinity to 5%, the specific growth rate began to decline for *Tetraselmis chuii* (Ghezelbash et al. 2008). Adaptation to different NaCl concentrations is common for members of the genus *Tetraselmis* (Kirst 1975). However, limiting effects of salinity on algal growth and productivity are highly dependent on the microalgae species or strain (Cho et al. 2007).

The effect of salinity on microalgae growth is related to osmotic effects, which microalgae are able to regulate through diverse processes. The inability to balance the internal osmotic pressure with the external osmotic pressure can either lead to cell shrinking or result in elongation, ultimately leading to cell bursting (Kirst 1990; Fogg 2001). Uptake of ions and concentrations and solubility of Na^+ and K^- ions are very important in osmo-regulation in microalgae for surviving salinity changes (Kirst 1990; Blackwell and Gilmour 1991; Strizh et al. 2004). Apart from ionic compensation, synthesis of secondary metabolites (compatible solutes) such as glycerol, mannitol, sucrose, proline and, floridoside is important for salinity tolerance in microalgae. These organic solutes which are accumulated as a response to osmotic changes are compatible with the general metabolic processes of the cell (Borowitzka and Brown 1974; Kirst 1977; Kirst et al. 1988; Iwamoto and Shiraiwa 2005).

In macrophytes or species with inflexible cell walls, osmotic pressure is balanced by turgor pressure, as observed in the cyanobacterium *Scytonema* sp. isolated from desert soils which accumulates trehalose to create positive turgor pressure within the cells (Kirst and Bisson 1979; Edwards et al. 1988; Page-Sharp et al. 1999). The downside of the synthesis of secondary metabolites and solutes was found to bea high demand in energy for the synthesis of these compounds resulting in an enhancement in respiration and a transient repression in photosynthesis while anew steady state of growth is achieved (Hellebust 1976; Vonshak and Richmond 1981; Lu and Vonshak 1999).

Lu and Vonshak (1999) found that salt stressing *Spirulina platensis* was associated with a decrease in PS II activity, and a decrease in the overall activity of the electron transport chain accompanied by an increase in PS I activity and respiration. Biochemical changes in a high salinity environment are generally recognised by an increase in ash content (Ben-Amotz et al. 1985). The possibility of protein synthesis inhibition, can lead to an increase in carotenoids in *Dunaliella salina* and astaxanthin content in *Haematococcus pluvialis* (Ben-Amotz and Avron 1983b; Borowitzka and Borowitzka 1990; Borowitzka et al. 1991; Boussiba and Vonshak 1991; Ahmad and Hellebust 1993). Fast growing algae species with a great tolerance in salinity are most suitable for large scale mass cultivation using only saline water (Borowitzka 2013b).

9.2.6. Mixing

Mixing in algae mass cultivation is induced by paddle wheels, airlift pumps, screw or propeller pumps to develop turbulence in the cultivation unit to minimise cell sedimentation (Oswald 1988a). Turbulence is developed when the input force is high enough to shear the molecules from the upper layer of the water column and the ordered velocity gradient cannot capture the supplied energy resulting in the structure breaking into swirling and recoiling motion characteristics (Reynolds 2006). The purpose of turbulence is overcome by limitations in light and mass transfer, including gas exchange, or to optimise parameters in conjunction with each other in a liquid medium containing solids and gases regardless of the design and type of cultivation unit (Richmond and Grobbelaar 1986).

The effect of mixing in open ponds was found to increase the horizontal layer with higher photosynthesis saturation, while light limitation occurs at high cell densities when cells in the bottom layer are completely in darkness while the cells in the top layer receive excess light (Vonshak and Torzillo 2004). Even in shallow ponds or enclosed photo-bioreactors with a short light-path, light can be limiting due to self-shading, whereby turbulence increases the frequency between the light/dark and the light distribution to the individual algal cell (Richmond and Grobbelaar 1986).

Markedly higher productivities were achieved at short optical depth (<10 cm) combined with turbulence R_e>3000 (Grobbelaar 2009). Additionally, the photosynthetic rates increased on an average 2.1 times when the L:D frequencies were increased from 10 s to 10 ms equal to a range from 10 – 0.01 Hz and showed significant differences with an L:D ratio of 0.5–2.0 (Grobbelaar 2009). Furthermore, apart from influence on light availability to the algal cell, turbulence improved the gas exchange with respect to the air/water interface thus reducing the O_2 concentration (Chini Zittelli et al. 2006) by O_2 out-gassing and culture carbonation as pH regulatory effects particularly when air-lift systems are employed (Persoone et al. 1980). A further advantage of turbulence was found in the reduction of the boundary layer around the algal cell for more efficient nutrient uptake and metabolism rates (Grobbelaar 1994; Riebesell and Wolf-Gladrow 2007). Morphological characteristics such as form, size, viscosity and, chain or colony formation and density of the algal cell play an important role in association with nutrient uptake and the effect of turbulence particularly for large microalgae >50 µm with a smaller surface area to volume ratio, resulting in stronger nutrient limitation.

Conversely, small organisms with 1–10 µm diameters have faster diffusion of nutrients through their boundary layer, where most larger flagellates >10 µm, are able to swim more than 10 times their own body length per second to increase significantly their nutrient uptake by swimming (Mann and Lazier 2006b; Riebesell and Wolf-Gladrow 2007). In contrast to the benefits of turbulence, the energy input required increases exponentially when increasing turbulence in large open raceway ponds (Neenan et al. 1986; Oswald 1988a). Further, high Reynolds numbers as a quantitative indicator for turbulence mainly generated in airlift photobioreactors even at low gas -flow rates (Torzillo 1997) can lead to cell damage or cell mortality (Barbosa 2004; Moheimani et al. 2011). The relative sensitivities to shear force increase in the following order: green algae, blue-green algae, diatoms and dinoflagellates as the most sensitive algal group (Thomas and Gibson 1990). The difference in shear sensitivity can be found in associations between cell density, viscosity and cell wall structure (Mann and Lazier 2006a).

9.2.7. Contamination and Other Limitations/Inhibitions

The term contamination in algae mass cultivation refers to other algae, fungi and bacteria, predators and parasites that are considered as a major problem in large scale operations in which species such as rotifers and *Daphnia* are unpredictable and without chemical treatment, largely uncontrollable. Mechanical attempts for the removal of contaminating organisms by employing screens in open air ponds was found to be only successful for large contaminating organisms or foreign objects (Oswald 1980). Appearances of contamination indicate a drift from optimum or favourable growth conditions of the target culture. Environmental parameters such as irradiance and temperature, nutrient limitation, shift in salinity, pH or residence time of the culture are responsible for the decline in specific growth followed by overgrowth by contaminating organism and, in the worst case, the loss of the culture of the target algal population (Richmond 2004). Various scientists have pointed out for commercial microalgae such as *Spirulina* and *Dunaliella,* unique or extreme selective cultivation conditions favour the target algae strain as the dominant population in the cultivation system (Chaumont et al. 1988; Chaumont 1993; Richmond 2004). Saline water at high salinities, such as the >20% NaCl in which *Dunaliella salina* is growing as the dominant species, greatly reduces predators, especially the ciliate *Fabrea salina* and the brine shrimp *Artemia salina* (Post et al. 1983; Elloumi et al. 2009). Organisms capable of growing in high salinities under large scale open air operations, indicating harsh environmental conditions accompanied by extreme temperatures (>40°C) and very alkaline habitats require additional resistance to such conditions (Borowitzka and Borowitzka 1988a).

A major contamination problem in *Dunaliella salina* culture for β-carotene production was found in the overgrowth of non-carotenogenic *Dunaliella* species in saline water sources used for evaporation top up and medium replacement (Borowitzka 1992). Different strategies using high bicarbonate (e.g., 0.2 M) concentrations or additional ammonia (2 mM) markedly prevent contamination problems with *Chlorella* sp. in *Spirulina* mass cultures (Vonshak 1997a). Temporary high concentration of ammonia showed also success against rotifer contamination and chlorination treatment has been found successful for the elimination of grazers (Borowitzka and Moheimani 2013).

However, higher contamination risk is given to algal species such as *Haematococcus* or *Nannochloropsis* which grow in a more natural medium and normal environmental condition (Richmond 2004). Indeed, algae species such as *Nannochloropsis* in natural habitats with cell diameters of approximately 2 μm demonstrated their strengths in better nutrient uptake kinetics without diffusion limitation across the boundary layer especially at a low concentration of dissolved inorganic nitrogen due to better surface to volume ratio (Riebesell and Wolf-Gladrow 2007). Nutrient competition models have demonstrated, that if only one resource is limited, the organism that is able to reduce the demand to the lowest level can outcompete other organisms, whereby the competition for two resources is controlled by the ratio and rates of these supplied resources (Huisman and Weissing 1994). Cells that excrete substances in the medium, accumulated due to harvesting and medium recycling, demonstrated inhibitory growth characteristics for contaminating organism (De Freitas and Fredrickson 1978).

Further, the application of DCMU (3-(3,4-dichlorophenyl)-1,1-dimethylurea) as a selective herbicide has indicated different inhibitor effects on photosynthetic activity, and the growth rate was examined in several marine unicellular algae which has demonstrated an advantage for *Nannochloropsis sp.* (Gonen-Zurgil et al. 1996).

9.2.8. Scaling up of Microalgal Culture – Constraints and Requirements

Commercialising microalgae mass cultivation for biofuel production begins with fundamental research in the laboratory providing information about biological limitations. However, at this stage, there is no information in the literature available about technological reliability and long-term feasibility under environmental conditions for the final requirement of very large scale cultivation. The scaling up process is not only required for research and development, every microalgae production plant contains a series of ponds or PBR's at various scale to provide sufficient seeding culture for the production units. Therefore, it is essential to integrate scaling-up processes in the development of commercial plants regarding reactor type, size, and amount in association with site selection, nutrient sources and quantity, and water sources and amounts for more reliable economic modelling. In general, the series of ponds is following the inoculation ratio of 1:5 to 1:10 with a starting scale of 1-3 m^2 for outdoor cultivation. The pair wise configuration of ponds for each scale is recommended, while the number of smaller ponds should be higher for the purpose of onsite research and development. A typical layout of microalgae pilot plant and demonstration plant are shown in Figure 9.2.

The configuration of the ponds at different size varied significantly in material and construction method. Small ponds with a surface area of 1-3 m^2 are typically build out of fibre glass or coated plywood with epoxy paint. Timber sleeper covered with pond liner is possible for 20 m^2 or larger ponds, while earth wall build ponds are highly recommended for ponds with a surface area greater than 200 m^2 (Figure 9.3). Concrete build ponds are considered as a more expensive alternative to timber sleeper or earth walled build ponds which have been used at the pilot scale with a surface area between 20 and 100 m^2 (Figure 9.2B).

The design and configuration of the 4000 m^2 pond has to be more advanced compared to the 20 and 200 m^2 ponds. Oswald (1988a) and Dodd (1986) have proposed the integration of flow rectifier for curved raceway ponds and turning vanes for rectangular ponds such as shown in Figure 9.2B. The 1000 m^2 rectangular pond at the pilot plant in Minamisoma was chosen to demonstrate the feasibility of converting rice paddies into algae cultivation systems after the soil was contaminated by saline water due to the powerful tsunami in March 2011. The implementation of turning vanes is crucial for uniform mixing of rectangular ponds. Experience with curved earth walled ponds at the Muradel Demonstration Plant in Whyalla have led to the implementation of tear drops and flow rectifier for the 4000 m^2 pond (Figure 9.4).

Figure 9.2.Muradels pilot plant in Karratha, Northern Western Australia (A); Fukushima pilot plant in Minamisoma, Fukushima prefecture, Japan (B); Muradel's Demonstration Plant in Whyalla, South Australia (C).

Figure 9.3. Fibre glass build 2 x 2 m^2 ponds (A); 20 m^2 timber sleeper build and covered with HDPE liner at 1.5 mm thickness (B,C), 2 x 20 m^2 (D) and 200 m^2 (D) earth walled build ponds with 1.0 mm HDPE liner (C).

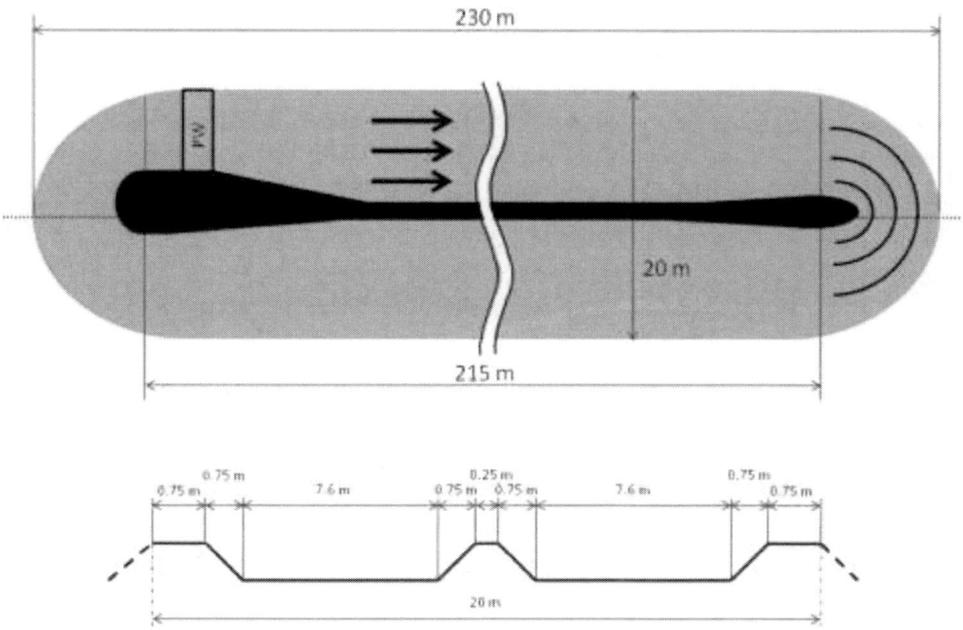

Figure 9.4. Configuration of the 4000 m^2 raceway pond with integrated tear-drop and flow rectifier.

The contribution of turbulence as a tool to overcome growth limitation has previously been described in this chapter and is one of the key parameters in the scaling-up process under engineering aspects for technological and physical barriers. Turbulence induced by velocity in raceway ponds is dependent on the dimensions of the culture system and the construction material in terms of friction (e.g., concrete, plastic liner, clay) and culture depth (Oswald 1988a; Borowitzka 2005; Fonseca et al. 2010). Oswald (1988a) estimated that a minimum velocity of 5 cm s^{-1} in all locations will prevent thermal stratification and keep algal cells in suspension, but 15 cm s^{-1} is required as the minimum channel velocity. Velocities higher than 30 cm s^{-1} create a trade-off in energy input and higher construction costs due to the need to prevent scouring. However, the experience at the Karratha pilot plant and later at the demonstration plant in Whyalla have demonstrated a significant loss in velocity by ponds with 200 m^2 surface area or greater. Simple observations on the culture surface of each pond size indicated the difference of velocity (Figure 9.5).

Figure 9.5. Microalgae velocity indicated by foam bubbles at the pond surface at the 20 m^2 pond - entire loop (A), at the 200 m^2 pond - first third of the second channel (B) and at the 4000 m^2 pond with flow rectifier showed almost no foam bubbles at the end of the first channel (C).

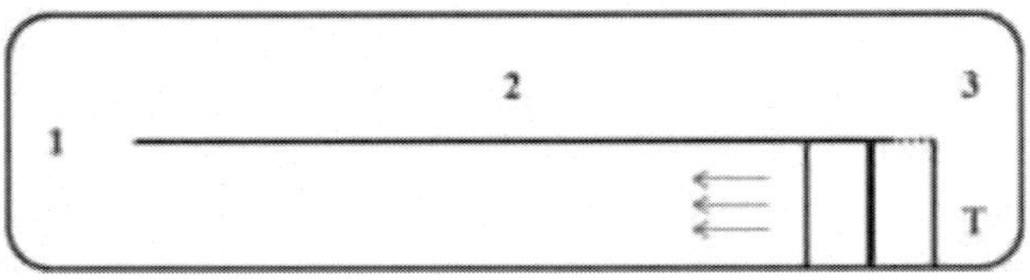

Figure 9.6. Bird's eye perspective of a paddle wheel driven raceway pond (T = tracer injection point and 1; 2; 3 are the pH -detection points). The arrows show the direction of flow.

Further investigations on the velocity for different ponds were conducted to confirm the visual observations. The flow rates in the raceway ponds were measured using the pH tracer method by introducing 1 M HCl and 1 M NaOH at the tracer injection point. The pH then was determined at three fixed points as demonstrated in (Figure 9.6) of increasing distance from the paddle wheel (1-3) in the open raceway pond, and the pH change was measured using the YSI 600XL multi-sensor electrode.

The culture velocity (*Ul*) was calculated using the distance from the tracer injection to the detection points (*Lt*) and the time (*t*) between injection and detection (Molina et al. 2001), as described by Equation 9.2.

$$Ul = \frac{Lt}{t} \tag{9.2}$$

Reynolds numbers were calculated from the specific flow rates and the dimensions of each cultivation systems.

The internal flow needed to be described by the hydraulic diameter, D_h, which is defined as four times the cross-sectional area (*A*), divided by the wetted perimeter (*P*) which for an open pond is the total perimeter of all channel walls that are in contact with the flow Equation 9.3.

$$D_h = \frac{4A}{P} \tag{9.3}$$

The Reynolds number was calculated by Equation 9.4

$$\mathrm{Re} = \frac{Ul.D_h.p}{\mu} \tag{9.4}$$

where:

p is the density of seawater at 20°C at 7% salinity (1051.8 kg m^{-3}) and μ is the dynamic viscosity of seawater at 20°C at 7% salinity (1.168x10^{-3} kg $m^{-1}s^{-1}$).

Investigations on hydrodynamics between the different scales demonstrated that the turbulence decreased with increasing distance from the paddle wheel of the 200 m^2 raceway pond according to the Reynolds numbers as shown in Table 9.4 with laminar flow occurring at the end of the first channel.

Turbulent flow at Reynolds number ≥2000 over the total distance from the paddle wheels was measured and estimated for the 2 m^2 and 20 m^2 raceway ponds at the pilot plant. High Reynolds numbers generated by the paddle wheels such as in 2 m^2 and 20 m^2 ponds resulted in a high biomass concentration and higher achievable cell densities during batch mode cultivation (Table 3). In comparison to the 2 m^2 and 20 m^2 ponds, the 200 m^2 pond with more than 50% laminar flow over the total channel length, demonstrated less achievable cell densities and less biomass concentration at similar salinity and pond depth.

Table 9.4. Cell densities, biomass concentration and Reynolds numbers for raceway ponds at different scales for batch -mode cultivation of MUR-233 with additional CO_2 and controlled pH over a period of eight days at the pilot plant in Karratha, where the distance from the paddle wheels increased from 1 to 3 as sampling locations

	2 m² raceway pond	**20 m² raceway pond**	**200 m² raceway pond**
Maximum cell density cells x 10^4 mL^{-1}	152 ± 8*	163 ± 11*	88 ± 3*
Maximum biomass concentration mg L^{-1}	720 ± 10.1*	768 ± 26.6*	469 ± 3.5*
Measurements points		*Reynolds number*	
PW / 0	-	-	5268
1	6022	8888	1708
2	4211	4223	955
3	2947	3005	723

Figure 9.7. Pre-cast concrete duct channel for 20 m2 ponds (Top panel); Pre-cast concrete foundations for 200 m2 ponds (middle panel); full poured concrete slab foundation for 4000 m2 pond (bottom panel).

Oswald (1988a) further concluded that for raceway ponds with an optimum velocity of 20-30 cm s^{-1} and culture depth 20 - 30 cm, a limitation in the channel width of up to 12 m is given, due to the paddle wheel shaft shear force. Considering the cost for the paddle wheels and the challenge for the installation of paddle wheels at 12 m in size, it is highly recommended to not exceed 6 meter. The size of the paddle wheel is determining the design of the paddle wheel station. For example, pre-cast concrete duct channels can be applied for ponds at 20 m^2 and pre-cast concrete foundations are suitable for 200 m^2 ponds while full poured concrete-slab foundations are required for the 4000 m^2 pond as illustrated in Figure 9.7.

Experience in large scale operations by various scientists has demonstrated several biological issues such as culture collapse after two months due to predators and algae contamination for *Nannochloropsis* sp. in 3000 m^2 raceway ponds (Sukenik 1999), as well as technical problems at the pilot scale such as rainwater running off into ponds, nutrients binding to clay walled ponds or leakage through the clay walls (Borowitzka 1991). Heavy tropical rainfall with >20 cm m^{-2} can exceed the operational pond level, leading and thus requiring where overflow protection and storm-water drainage as part of the facility design. Problems due to heavy rain have been recognised in culture dilution, nutrient washouts and parasite infection (Sinchumpasak 1980; Oswald 1988a). Therefore, it is essential to integrate scaling-up processes in the development of commercial plants regarding reactor type, size, and amount in association with site selection, nutrient sources and quantity, and water sources and amounts for more reliable economic modelling.

Soil erosion due to heavy rain had significant impact at the bottom level of the 4000 m^2 pond at the Muradel Demonstration Plant (Figure 9.8). The measurements have clearly indicated that soil erosion due to heavy rain occurred with significant changes at the pond level. Pond areas as deep as 120 mm were measured compared with high spots at low depth with 40 mm culture depth. These fluctuations in culture depth within the pond caused additional loss in velocity if the pond was operated at ≤ 20 cm culture depth. The most satisfied mixing effect was observed with approximately 30 cm culture depth at the 4000 m^2 pond.

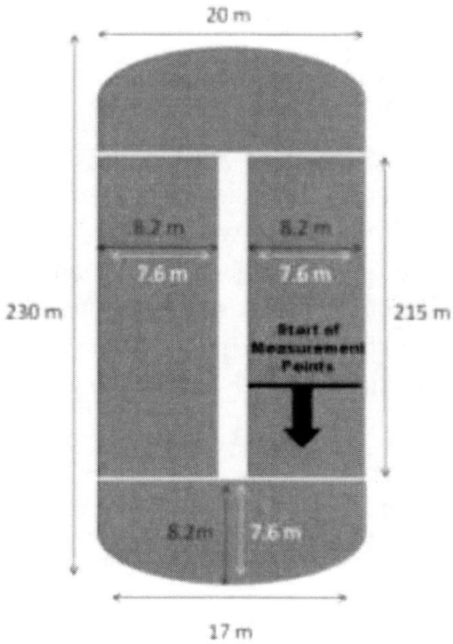

Measurement Points	Pond Level (mm)	1 Point = 16 m
1	75	0
2	80	16
3	90	32
4	90	48
5	55	64
6	40	80
7	65	96
8	50	112
9	60	128
10	60	144
11	50	160
12	60	176
13	70	192
14	40	208
15	50	224
16	75	240
17	85	256
18	10	272
19	60	288
20	120	304
21	120	320
22	85	336
23	50	352
24	90	368
25	45	384
26	60	400
27	60	416
28	40	432
29	60	448
Average	61.896552	

Figure 9.8. Measurements of the culture depth of the 4000 m^2 after heavy rain and soil erosion (8.2 m channel width at 20 cm culture depth while 7.6 m represents the bottom channel width.

Conclusion and Future Perspective

Regardless of the type of reactor, all the single issues discussed above, ranging from light, effect of mixing, temperature, nutrient demand including inorganic carbon, pH and, contamination risk must be considered during scaling-up to find the balance between each parameter for maximum productivity. Irrespective of the problem, it is a fact that the achieved productivity is less than the theoretical possible productivity due to poor light utilisation efficiency, temperature control, gas transfer and turbulence, and open air ponds are currently the systems that are used in commercial algal mass cultivation at a very large scale, (Borowitzka 1999).

However, apart from turbulence, scaling-up should consider further downstream processes such as integrated harvesting and medium recycling, dewatering, and product recovery. Challenges in scaling up have been also recognised with respect to carbonation and pH regulation, while the injection of CO_2 into the medium for pH regulation with current technologies has demonstrated that 60% is the maximum achievable efficiency in large raceway ponds (Becker 1994b). Commercial algae production plants contain multiple process units and many raceway ponds. There is a strong requirement for semi automated and fully automated control and monitoring systems for culture maintenance and process control (Olaizola 2000). Pulse amplitude modulated (PAM) fluorometers, devices for rapid and easy detection methods based on fluorescence techniques, are available to determine the physiological state of the algae cells, and flow cytometry can be applied to analyse dead and live cells, including bacterial detection (Kromkamp et al. 2009; Fonseca et al. 2010). Real time monitoring and control for system parameters such as pH, CO_2 addition, nutrient, light and temperature with integration into the harvest and post harvest process are essential in commercial algae production plants (Greenwell et al. 2010).

Acknowledgments

The author like to thank Professor Michael A. Borowitzka and Muradel Pty Ltd for enabling the great journey of experience combined with the financial support of the Australian State government, Australian Renewable Energy Agency (ARENA) and Aban Australia Pty Ltd. Further, the author wishes to acknowledge the invaluable support of colleagues at Muradel Pty Ltd as well as the teams at Murdoch University and Adelaide University. Muradel Pty Ltd retains the right to all intellectual property for *Tetraselmis* sp. strains MUR 167, 230, 232 and 233 including methods for reliable high productivity cultivation.

References

Abeliovich A (1980) Factors limiting algal growth in high-rate oxidation ponds. In: Shelef G, Soeder CJ (eds) *Algae biomass: production and use*. Elsevier/North-Holland Biomedical Press, Amsterdam, pp 205 - 216.

Ahmad I, Hellebust JA (1993) Protein biosynthesis in salt-shocked cells of *Stichococcusbacillaris* (Chlorophyceae). *Journal of Phycology* 29:294-300.

Arora M, Chandrashekar A, Leliaert F, Delany J, Mesbahi E (2013) *Tetraselmisindica* (Chlorodendrophyceae, Chlorophyta), a new species isolated from salt pans in Goa, India. *Eur J Phycol* 48:61 - 78.

Azov Y (1982) Effect of ph on inorganic carbon uptake in algal cultures. *Applied andEnvironmental Microbiology* 43:1300-1306.

Azov Y, Goldman JC (1982) Free ammonia inhibition of algal photosynthesis in intensive cultures. *Applied and Environmental Microbiology* 43:735-739.

Badger MR (1985) Photosynthetic Oxygen Exchange. *Annual Review of Plant Physiology* 36:27-53.

Barasanti L, Gualtieri P (2006) *Algae: Anatomy, Biochemistry and Biotechnology*. CRC Press, Boca Raton:320.

Barbosa MJ (2004) Overcoming shear stress of microalgae cultures in spargedphotobioreactors. *Biotechnology and Bioengineering* 85:78-85.

Beardall J, Raven JA (2013) Limits to Phototrophic Growth in Dense Culture: CO_2 Supply and Light. In: Borowitzka MA, Moheimani NR (eds) *Algae for Biofuels and Energy*. Springer Netherlands, Dordrecht, pp 91-97.

Becker EW (1994b) Large-scale cultivation. In: Becker EW (ed) *Microalgae: Biotechnology and Microbiology*. Cambridge University Press, New York, pp 63 - 171.

Beijerinck W (1890) Culturversuche mit *Zoochlorellen, Lichenengonidien* und anderen niederen Algen. *BotanischeZeitung*47:725 - 739.

Belay A (1997) Mass culture of *Spirulina* outdoors: the Earthrise Farms experience. In: Vonshak A (ed) *Spirulinaplatensis* (*Arthrospira*): *Physiology, cell-biology and biotechnology*. Taylor and Francis, London, pp 131-158.

Ben-Amotz A (1995) New mode of *Dunaliella* biotechnology: two-phase growth for beta-carotene production. *Journal of Applied Phycology* 7:65-68.

Ben-Amotz A, Avron M (1983b) On the factors which determine massive beta-carotene accumulation in the halotolerant alga *Dunaliellabardawil*. *Plant Physiology* 72:593-597.

Ben-Amotz A, Tornabene TG, Thomas WH (1985) Chemical profile of selected species of microalgae with emphasis on lipids. *Journal of Phycology* 21:72-81.

Benemann JR (1993) Utilization of carbon-dioxide from fossil-fuel - burning power-plants with biological-systems. *Energy Conversion and Management* 34:999-1004.

Benemann JR (2003) Professor William J. Oswald: An introduction. *Journal of Applied Phycology* 15:97-98.

Berry J, Bjorkman O (1980) Photosynthetic response and adaptation to temperature in higher plants. *Annual Review of Plant Physiology* 31:491-543.

Bilanovic D, Andargatchew A, Kroeger T, Shelef G (2009) Freshwater and marine microalgae sequestering of CO_2 at different C and N concentrations - Response surface methodology analysis. *Energy Conversion and Management* 50:262-267.

Blackwell JR, Gilmour DJ (1991) Physiological-response of the unicellular green-alga *Chlorococcum-submarinum* to rapid changes in salinity. *Archives of Microbiology* 157:86-91.

Borowitzka LJ (1991) Development of Western Biotechnology Algal Beta-Carotene Plant. *Bioresource Technology* 38:251-252.

Borowitzka LJ (1992) Beta-carotene production using algal biotechnology. *J NutrSciVitaminol*(Tokyo) Spec No:248-250.
Borowitzka LJ, Borowitzka MA (1990) Commercial production of beta-carotene by *Dunaliella-salina* in open ponds. *Bulletin of Marine Science* 47:244-252.
Borowitzka LJ, Brown AD (1974) The salt relations of marine and halophilic species of the unicellular green alga, *Dunaliella*. The role of glycerol as a compatible solute. *Archives of Microbiology* 96:37-52.
Borowitzka LJ, Moulton TP, Borowitzka MA (1984) The mass culture of *Dunaliellasalina* for fine chemicals: From laboratory to pilot plant. *Hydrobiologia* 116/117:115-134.
Borowitzka MA (1995b) Biotechnology and marine biodiversity. *Australasian Biotechnology* 5:213-216.
Borowitzka MA (1997a) Microalgae for aquaculture: opportunities and constraints. *Journal of Applied Phycology* 9:393-401.
Borowitzka MA (1997b) Tubular photobioreactors for large-scale algae production. *Phycologia* 36:10.
Borowitzka MA (1998) Limits to Growth. In: Wong Y-S, Tam NY (eds) *Wastewater Treatment with Algae*. Biotechnology Intelligence Unit. Springer Berlin, pp 203-226.
Borowitzka MA (1999) Commercial production of microalgae: ponds, tanks, tubes and fermenters. *Journal of Biotechnology* 70:313-321.
Borowitzka MA (2005) Culturing microalgae in outdoor ponds. In: Andersen RA (ed) *Algal culturing techniques*. Elsevier Academic Press, San Diego, pp 205-218.
Borowitzka MA (2013a) High-value products from microalgae-their development and commercialisation. *Journal of Applied Phycology* 25:743-756.
Borowitzka MA (2013b) Species and Strain Selection. In: Borowitzka MA, Moheimani NR (eds) *Algae for Biofuels and Energy*. Springer, Dordrecht, pp 77-89.
Borowitzka MA, Borowitzka LJ (1988a) *Dunaliella*. In: Borowitzka MA, Borowitzka LJ (eds) *Micro-algal biotechnology*. Cambridge University Press, Cambridge, pp 27-58.
Borowitzka MA, Huisman JM, Osborn A (1991) Culture of the astaxanthin-producing green alga *Haematococcuspluvialis* 1. Effects of nutrients on growth and cell type. *Journal of Applied Phycology* 3:295 - 304.
Borowitzka MA, Moheimani NR (2013) Open Pond Culture Systems. In: Borowitzka MA, Moheimani NR (eds) *Algae for Biofuels and Energy*, vol 5. Springer Netherlands, Dordrecht, pp 133-152.
Boussiba S, Vonshak A (1991) Astaxanthin accumulation in the green alga *Haematococcuspluvialis* 1. *Plant and Cell Physiology* 32:1077-1082.
Brooks A, Farquhar G (1985) Effect of temperature on the CO_2/O_2 specificity of ribulose-1, 5-bisphosphate carboxylase/oxygenase and the rate of respiration in the light. *Planta* 165:397-406.
Burkhardt S, Zondervan I, Riebesell U (1999) Effect of CO2 concentration on C: N: P ratio in marine phytoplankton: A species comparison. *Limnology and Oceanography*:683-690.
Capel V (1995) The seawater solution. In: Bearman G (ed) *Seawater: Its Composition, Properties and Behaviour*. Butterworth - Heinemann, Oxford, pp 85 - 127.
Chaumont D (1993) Biotechnology of algal biomass production: a review of systems for outdoor mass culture. *Journal of Applied Phycology* 5:593-604.

Chaumont D, Delepine R, Gudin C, Asensi A Energy bioconversion by *spirulina-maxima* in continuous culture. In: Delepine R, Gaillard J, Morand P (eds) *Valorisation Des Algues Et Autres Vegetaux Aquatiques*, 1988. pp 123-124.

ChiniZittelli G, Rodolfi L, Biondi N, Tredici MR (2006) Productivity and photosynthetic efficiency of outdoor cultures of *Tetraselmissuecica* in annular columns. *Aquaculture* 261:932-943.

Cho SH, JI SC, Hur SB, Bae J, Park IS, Song YC (2007) Optimum temperature and salinity conditions for growth of green algae *Chlorella ellipsoidea* and *Nannochlorisoculata*. *Fisheries Science* 73:1050-1056.

Davison IR (1991) Environmental effects on algal photosynthesis: temperature. *Journal of Phycology* 27:2-8.

De Freitas MJ, Fredrickson A (1978) Inhibition as a factor in the maintenance of the diversity of microbial ecosystems. *Journal of General Microbiology* 106:307-320.

De Pauw N, Morales J, Persoone G (1984) Mass culture of microalgae in aquaculture systems progress and constraints. *Hydrobiologia* 116 - 117:121-134.

Delieu T, Walker DA (1981) Polarographic measurement of photosynthetic oxygen evolution by leaf-disks. *New Phytologist* 89:165-178.

Dodd JD (1986) Elements of pond design and construction. In: Richmond A (ed) *Handbook of Microalgal Mass Culture*. Blackwell Science, Oxford, pp 265 - 283.

Droop M (1974) The nutrient status of algal cells in continuous culture. *Journal of the Marine Biological Association of the United Kingdom* 54:825-855.

Dubinsky Z, Stambler N (2009) Photoacclimation processes in phytoplankton: mechanisms, consequences, and applications. *Aquatic Microbial Ecology* 56:163-176.

Durand-Chastel H (1980) Production and use of *Spirulina* in Mexico. In: Shelef G, Soeder CJ (eds) *Algae biomass production and use*. Elsevier/North-Holland Biomedical Press, Amsterdam, pp 51-64.

Edwards D, Reed RH, Stewart W (1988) Osmoacclimation in *Enteromorphaintestinalis*: long-term effects of osmotic stress on organic solute accumulation. *Marine Biology* 98:467-476.

Elloumi J, Carrias J-F, Ayadi H, Sime-Ngando T, Bouain A (2009). Communities structure of the planktonic halophiles in the solar saltern of Sfax, Tunisia. *Estuarine, Coastal and Shelf Science* 81:19-26.

Emerson R, Green L (1938) Effect of Hydrogen-Ion Concentration on *Chlorella* Photosynthesis. *Plant Physiology* 13:157-168.

Eriksen NT (2008) The technology of microalgal culturing. *Biotechnology Letters* 30:1525-1536.

Falkowski PG (1997) Evolution of the nitrogen cycle and its influence on the biological sequestration of CO_2 in the ocean. *Nature* 387:272-275.

Falkowski PG (2000) Rationalizing elemental ratios in unicellular algae. *Journal of Phycology* 36:3-6.

Falkowski PG, Laroche J (1991) Acclimation to spectral irradiance in algae. *Journal of Phycology* 27:8-14.

Fauchot J, Gosselin M, Levasseur M, Mostajir B, Belzile C, Demers S, Roy S, Villegas PZ (2001) Influence of UV-B radiation on nitrogen utilization by a natural assemblage of phytoplankton. *Journal of Phycology* 36:484-496.

Findenegg I (1965a) Factors controlling primary productivity, especially with regard tp water replenishment, stratification, and mixing. In: Goldman CR (ed) *Primary productivity in aquatic environments*. University of California, Berkeley and Los Angeles, pp 106-119.

Florides GA, Christodoulides P (2009) Global warming and carbon dioxide through sciences. *Environment International* 35:390-401.

Fogg G (2001) Algal adaptation to stress - some general remarks. In: Rai LC, Gaur JP (eds) *Algal Adaptation to Environmental Stresses*. Springer Berlin Heidelberg, pp 1-19.

Fon Sing S, Isdepsky A, Borowitzka MA, Moheimani NR (2013) Production of biofuels from microalgae. *Mitigation and Adaptation Strategies for Global Change* 18:47-72.

Fonseca FAS, Vidal-Vieira JA, Ravagnani SP (2010) Transesterification of vegetable oils: Simulating the replacement of batch reactors with continuous reactors. *Bioresource Technology* 101:8151-8157.

Geider RJ, MacIntyre HL (2008) Physiology and biochemistry of photosynthesis and algal carbon aquisition. In: Williams PJB, Thomas DN, Reynolds CS (eds) *Phytoplankton Productivity: Carbon Assimilation inMarine and Freshwater Ecosystems*. Blackwell Science, Oxford, pp 44-77.

Geider RJ, MacIntyre HL, Kana TM (1998) A dynamic regulatory model of phytoplanktonic acclimation to light, nutrients, and temperature. *Limnology and Oceanography* 43:679-694.

Ghezelbash F, Farboodnia T, Heidari R, Agh N (2008) Effects of different salinities and luminance on growth rate of the green microalgae Tetraselmischuii. *Research Journal of Biological Sciences* 3:311-314.

Goldman CR (1965) Micronutrient limiting factors and their detection in natural phytoplankton populations. In: Goldman CR (ed) *Primary Productivity in Aquatic Environments*. University of California Press, Berkeley and Los Angeles.

Goldman JC (1973) Carbon-dioxide and pH - effect on species succession of algae. *Science* 182:306-307.

Goldman JC (1980) Physiological aspects in algal mass cultures. In: Soeder CJ, Shelef G (eds) *Algae Biomass production and use*. Elsevier/North-Holland Biomedical Press, Amsterdam, pp 343-360.

Goldman JC, Carpenter E (1974) A kinetic approach to effect of temperature on algal growth. *Limnology and Oceanography* 19:756-766.

Gonen-Zurgil Y, Carmeli-Schwartz Y, Sukenik A (1996) Selective effect of the herbicide DCMU on unicellular algae - a potential tool to maintain monoalgal mass culture of *Nannochloropsis*. *Journal of Applied Phycology* 8:415-419.

Greenwell H, Laurens L, Shields R, Lovitt R, Flynn K (2010) Placing microalgae on the biofuels priority list: a review of the technological challenges. *Journal of the Royal Society Interface* 7:703-726.

Grima EM, Fernández FGA, García Camacho F, Chisti Y (1999) Photobioreactors: light regime, mass transfer, and scale up. *Journal of Biotechnology* 70:231 - 247.

Grobbelaar JU (1982) *Potential of algal production*. Water SA 8:79-85.

Grobbelaar JU (1991) The influence of light/dark cycles in mixed algal cultures on their productivity. *Journal of AppliedPhycology* 1:333-340.

Grobbelaar JU (1994) Turbulence in mass algal cultures and the role of light-dark fluctuations. *Journal of Applied Phycology* 6:331-335.

Grobbelaar JU (2000) Physiological and technological considerations for optimising mass algal cultures. *Journal of Applied Phycology* 12:201-206.

Grobbelaar JU (2004) Algal Nutrition: Mineral Nutrition. In: Richmond A (ed) *Handbook of Microalgal Culture: Biotechnology and Applied Phycology*. Blackwell Science, Oxford, pp 97-115.

Grobbelaar JU (2009) Upper limits of photosynthetic productivity and problems of scaling. *Journal of Applied Phycology* 21:519-522.

Grobbelaar JU (2010) Microalgal biomass production: challenges and realities. *Photosynthesis Research* 106:135-144.

Heinimö J, Junginger M (2009) Production and trading of biomass for energy - An overview of the global status. *Biomass and Bioenergy* 33:1310-1320.

Hellebust JA (1976) Effect of salinity on photosynthesis and mannitol synthesis in the green flagellate *Platymonassuecica*. *Canadian Journal of Botany* 54:1735 - 1741.

Ho TY, Quigg A, Finkel ZV, Milligan AJ, Wyman K, Falkowski PG, Morel FMM (2003) The elemental composition of some marine phytoplankton. *Journal of Phycology* 39:1145-1159.

Hu Q (2004) Environmental Effects on Cell Composition. In: Richmond A (ed) *Handbook of Microalgal Culture*. Blackwell Science, Oxford, pp 83-94.

Huisman J, Weissing FJ (1994) Light-limited growth and competition for light in well-mixed aquatic environments - an elementary model. *Ecology* 75:507-520.

Iwamoto K, Shiraiwa Y (2005) Salt-regulated mannitol metabolism in algae. *Marine Biotechnology* 7:407-415.

Janssen M, Tramper J, Mur LR, Wijffels RH (2003) Enclosed outdoor photobioreactors: Light regime, photosynthetic efficiency, scale-up, and future prospects. *Biotechnology and Bioengineering* 81:193-210.

Javanmardian M, Palsson BO (1991) High-density photoautotrophic algal cultures: Design, construction, and operation of a novel photobioreactor system. *Biotechnology and Bioengineering* 38:1182-1189.

Jorgensen EG, Steemann N (1965) Adaption in Plankton Algae. In: Goldman CR (ed) *Primary Productivity in Aquatic Environments*. University of California Press, Berkeley and Los Angeles, pp 37 - 46.

Kelly GJ (1989) A comparison of marine photosynthesis with terrestrial photosynthesis: a biochemical perspective. In: Barnes H, Barnes M (eds) *Oceanography and Marine Biology: An Annual Review*, vol 27. Aberdeen University Press, Aberdeen, pp 11 - 44.

Kirst G (1977) Coordination of ionic relations and mannitol concentrations in the euryhaline unicellular alga, Platymonassubcordiformis (Hazen) after osmotic shocks. *Planta* 135:69-75.

Kirst G (1990) Salinity tolerance of eukaryotic marine algae. *Annual Review of Plant Biology* 41:21-53.

Kirst G, Bisson M (1979) Regulation of turgor pressure in marine algae: ions and low-molecular-weight organic compounds. *Functional Plant Biology* 6:539-556.

Kirst G, Janssen M, Winter U (1988) Ecophysiological investigations of *Chara vulgaris* L. grown in a brackish water lake: ionic changes and accumulation of sucrose in the vacuolar sap during sexual reproduction. *Plant, Cell & Environment* 11:55-61.

Kirst GO (1975) Wirkung unterschiedlicher Konzentrationen von NaCl und anderen osmotisch wirksamen Substanzen auf die CO_2-Fixierung der einzelligen Alge *Platymonas subcordiformis*. *Oecologia* 20:237-254.

Kromkamp JC, Beardall J, Sukenik A, Kopecky J, Masojidek J, BergeijkvS, Gabai S, Shaham E, Yamson A (2009) Short-term variations in photosynthetic parameters of *Nannochloropsis*cultures grown in two types of outdoor mass cultivation systems. *Aquatic Microbial Ecology* 56:309 - 322.

Ku SB, Edwards GE (1978) Oxygen inhibition of photosynthesis. *Planta*140:1-6.

Leegood RC, Walker DA, Foyer CH (1985) Regulation of the Benson-Calvin cycle In: Barber J, Baker NR (eds) *Photosynthetic Mechanisms and the Environment*. Elsevier Biomedical Press, Amsterdam, pp 189-258.

Lu C, Vonshak A (1999) Photoinhibition in outdoor Spirulinaplatensis cultures assessed by polyphasic chlorophyll fluorescence transients. *Journal of Applied Phycology* 11:355-359.

Mann K, Lazier J (2006a) Biology and boundary layers. In: Mann K, Lazier J (eds) *Dynamics of Marine Ecosystems: Biological-Physical Interactions in the Oceans*. 3 edn. Blackwell Publishing, Carlton, Australia, pp 9-67.

Mann K, Lazier J (2006b) *Dynamics of Marine Ecosystems: Biological-Physical Interactions in the Oceans*. Wiley.

Marre E (1962) Temperature. In: Lewin RA (ed) *Physiology and Biochemistry of Algae*. Academic Press, New York, pp 541 - 550.

Martinot E, Dienst C, Weiliang L, Qimin C (2007) Renewable energy futures: Targets, scenarios, and pathways. *Annu Rev Environ Resour*32:205-239.

Masojidek J, Papacek S, Sergejevova M, Jirka V, Cerveny J, Kunc J, Korecko J, Verbovikova O, Kopecky J, Stys D, Torzillo G (2003) A closed solar photobioreactor for cultivation of microalgae under supra-high irradiance: basic design and performance. *Journal of Applied Phycology* 15:239-248.

Matthew LJ, Edward CTT (2010) The diatoms: a primer. In: Smol JP, Stoermer EF (eds) *The Diatoms: Applications for the Environmental and Earth Sciences*. Cambridge University Press, pp 8-22.

Miron AS, Garcia M, Gomez AC, Camacho FG, Grima EM, Chisti Y (2003) Shear stress tolerance and biochemical characterization of *Phaeodactylumtricornutum* in quasi steady-state continuous culture in outdoor photobioreactors. *Biochemical Engineering Journal* 16:287-297.

Moheimani NR, Borowitzka MA (2006) The long-term culture of the coccolithophore*Pleurochrysiscarterae* (Haptophyta) in outdoor raceway ponds. *Journal of Applied Phycology* 18:703-712.

Moheimani NR, Isdepsky A, Lisec J, Raes E, Borowitzka MA (2011) *Coccolithophorid* algae culture in closed photobioreactors. *Biotechnology and Bioengineering* 108:2078-2087.

Molina E, Martinez E, Sanchez S, Garcia F, Contreras A (1991) The influence of temperature and the initial N: P ratio on the growth of microalgae *Tetraselmis* sp. *Process biochemistry* 26:183-187.

Mostert ES, Grobbelaar JU (1987) The influence of nitrogen and phosphorus on algal growth and quality in outdoor mass algal cultures. *Biomass* 13:219-233.

Neenan B, Feinberg D, Hill A, Mcintosh R, Terry K (1986) *Fuels from microalgae: technology status, potential, and research requirements. SERI.* Solar Energy Research Institute, Golden, Colorado, USA:149.

Olaizola M (2000) Commercial production of astaxanthin from *Haematococcuspluvialis* using 25,000-liter outdoor photobioreactors. *Journal of Applied Phycology* 12:499-506.

Olsen LM, Öztürk M, Sakshaug E, Johnsen G (2006) Photosynthesis-induced phosphate precipitation in seawater: ecological implications for phytoplankton. Mar*ine Ecology Progress Series* 319:103-110.

Ono E, Cuello JL Selection of optimal microalgae species for CO_2 sequestration. In: *Proceedings 2nd Annual Conference on Carbon Sequestration*, 2003. pp 1-7.

Oswald WJ (1980) Algal production - problems, achievements and potential. In: Shelef G, Soeder CJ (eds) *Algae biomass and use.* Elsevier / North-Holland Biomedical Press, Amsterdam, pp 1 - 8.

Oswald WJ (1988a) Large-scale algal culture systems (engineering aspects). In: Borowitzka MA, Borowitzka LJ (eds) *Micro-Algal Biotechnology.* Cambridge University Press, Cambridge, pp 357 - 394.

Page-Sharp M, Behm C, Smith GD (1999) Involvement of the compatible solutes trehalose and sucrose in the response to salt stress of a cyanobacterial*Scytonema* species isolated from desert soils. *BiochimicaetBiophysicaActa* 1472:519-528.

Persoone G, Morales J, Verlet L, De Pauw N (1980) Air-lift pumps and the effect of mixing on algal growth. In: Shelef G, Soeder CJ (eds) *Algae biomass: production and use.* Elsevier/North-Holland Biomedical Press, Amsterdam, pp 505 - 522.

Post FJ, Borowitzka LJ, Borowitzka MA, Mackay B, Moulton TP (1983) The protozoa of a Western Australian hypersaline lagoon. *Hydrobiologia* 105:95 - 113.

Posten C (2009) Design principles of photo-bioreactors for cultivation of microalgae. *Engineering in Life Sciences* 9:165-177.

Pulz O (2001) Photobioreactors: production systems for phototrophic microorganisms. *Applied Microbiology and Biotechnology* 57:287-293.

Qiang H, Zarmi Y, Richmond A (1998) Combined effects of light intensity, light-path and culture density on output rate of Spirulinaplatensis (Cyanobacteria). *European Journal of Phycology* 33:165-171.

Quigg A, Finkel ZV, Irwin AJ, Rosenthal Y, Ho TY, Reinfelder JR, Schofield O, Morel FMM, Falkowski PG (2003) The evolutionary inheritance of elemental stoichiometry in marine phytoplankton. *Nature* 425:291-294.

Rahmstorf S (2002) Ocean circulation and climate during the past 120,000 years. *Nature* 419:207-214.

Raines CA, Lloyd JC (2004) Molecular biological approaches to environmental effects on photosynthesis. In: Baker NR (ed) *Photosynthesis and the Environment.* Springer Netherlands, Dordrecht, pp 305-319.

Raven JA (1988) Limits to growth. In: Borowitzka MA, Borowitzka LJ (eds) *Micro-algal biotechnology.* Cambridge University Press, pp 331 - 356.

Redfield AC (1934) On the proportions of organic derivatives in sea water and their relation to the composition of plankton. In: (ed) DR (ed) *James Johnstone Memorial Volume.* Liverpool University Press, Liverpool, pp 176 -192.

Reynolds CS (2006) *The ecology of phytoplankton.* Cambridge University Press, New York:552.

Richmond A (2004) Biological principles of mass cultivation. *Handbook of Microalgal Culture: Biotechnology and Applied Phycology*:125-177.

Richmond A, Becker EW (1986) Technology aspects of mass cultivation: a general outline. In: Richmond A (ed) *Handbook of Microalgal Mass Culture*. Cambridge University Press, New York, pp 245-263.

Richmond A, Grobbelaar JU (1986) Factors affecting the output rate of *Spirulinaplatensis* with reference to mass cultivation. *Biomass* 10:253-264.

Richmond A, Karg S, Boussiba S (1982) Effects of bicarbonate and carbon-dioxide on the competition between *Chlorella-vulgaris* and *Spirulina-platensis*. *Plant and Cell Physiology* 23:1411-1417.

Richmond A, Lichtenberg E, Stahl B, Vonshak A (1990) Quantitative assessment of the major limitations on productivity of *Spirulinaplatensis* in open raceways. *Journal of Applied Phycology* 2:195-206.

Riebesell U, Wolf-Gladrow DA (2007) Supply and Uptake of Inorganic Nutrients. In: Williams PJIB, Thomas DN, Reynolds CS (eds) *Phytoplankton Productivity*. Blackwell Science Ltd, Oxford, pp 109-140.

Rittmann BE (2008) Opportunities for renewable bioenergy using microorganisms. *Biotechnology and Bioengineering* 100:203-212.

Robinson LF, Morrison AW, Bamforth MR (1988) Improvements relating to biosynthesis. *European Patent* 0 261 872,

Robinson R, Toerien D (1982) The algae-a future source of protein. *Developments in Food Proteins*:289-335.

Shapiro J (1973) Blue-Green Algae: Why They Become Dominant. *Science* 179:382-384.

Sinchumpasak O (1980) Microalgal biomass production in Thailand. In: Shelef G, Soeder CJ (eds) *Algae biomass: production and use*. Elsevier / North-Holland Biomedical Press, Amsterdam, pp 115-121.

Soeder C (1980a) Massive cultivation of microalgae: Results and prospects. *Hydrobiologia* 72:197-209.

Soeder C, Stengel E (1974) Physico-chemical factors affecting metabolism and growth rate. In: Stewart WDP (ed) *Algal physiology and biochemistry*. Blackwell Science, Oxford, pp 714 - 740.

Soeder CJ (1980b) The scope of microalgae for food and feed. In: Soeder CJ, Shelef G (eds) *Algae Biomass production and use*. Elsevier/ North-Holland Biomedical Press, Amsterdam, pp 9 - 20.

Spolaore P, Cassan CJ, Duran E, Isambert A (2005) Commercial Applications of Microalgae. *Journal of Bioscience and Bioengineering* 101:87-96.

Strizh I, Popova L, Balnokin YV (2004) Physiological aspects of adaptation of the marine microalga *Tetraselmis* (*Platymonas*) *viridis* to various medium salinity. *Russian Journal of Plant Physiology* 51:176-182.

Su ZF, Kang RJ, Shi SY, Cong W, Cai ZL (2008) An economical device for carbon supplement in large-scale micro-algae production. *Bioprocess and Biosystems Engineering* 31:641-645.

Sukenik A (1999) Production of eicosapentaenoic acid by the marine eustigmatophyte*Nannochloropsis*. In: Cohen Z (ed) *Chemicals from microalgae*. Taylor and Francis, London, pp 41-56.

Tamiya H (1957) Mass culture of algae. *Annual Review of Plant Physiology* 8:309-334.

Thomas WH, Gibson CH (1990) Effects of small-scale turbulence on microalgae. *Journal of Applied Phycology* 2:71-77.

Toerien DF, Grobbelaar JU, Walmsley RD (1987) Management of autotrophic mass cultures of micro-algae. In: Michael RG (ed) *Managed Aquatic Ecosystems*. Elsevier Science, Amsterdam, pp 111 - 123.

Tolbert NE (1974) Photorespiration. In: Stewart WDP (ed) *Algal Physiology and Biochemistry*. Blackwell Scientific Oxford, pp 474-504.

Torzillo G (1997) Turbular Bioreactors. In: Vonshak A (ed) *SpirulinaPlatensis* (*Arthrospira*): *Physiology, Cell Biology and Biotechnology*. Taylor & Francis Group, London, pp 101-115.

Torzillo G, Carlozzi P, Pushparaj B, Montaini E, Materassi R (1993) A 2-plane tubular photobioreactor for outdoor culture of *Spirulina*. *Biotechnology and Bioengineering* 42:891-898.

Tredici MR (2004) Mass Production of Microalgae: Photobioreactors. In: Richmond A (ed) *Handbook of Microalgal Culture*. Blackwell Science, Oxford, pp 178 - 214.

Tredici MR, Materassi R (1992) From open ponds to vertical alveolar panels - the Italian experience in the development of reactors for the mass cultivation of phototrophic microorganisms. *Journal of Applied Phycology* 4:221-231.

Usui N, Ikenouchi M (1997) Biological CO_2 fixation and utilization project by RITE .1. Highly-effective photobioreactor system. *Energy Conversion and Management* 38:487 - 492.

Verbruggen A, Al Marchohi M (2010) Views on peak oil and its relation to climate change policy. *Energy Policy* 38:5572-5581.

Vonshak A (1997a) Outdoor mass production of *Spirulina*: the basic consept. In: Vonshak A (ed) *Spirulinaplatensis* (*Arthrospira*): *Physiology, Cell Biology and Biotechnology*. Taylor & Francis Group, London, pp 79 - 100.

Vonshak A, Richmond A (1981) Photosynthetic and respiratory activity in *Anacystisnidulans* adapted to osmotic stress. *Plant Physiology* 68:504 - 505.

Vonshak A, Torzillo G (2004) Environmental stress physiology. In: Richmond A (ed) *Handbook of microalgal culture*. Blackwell Science, Oxford, pp 57-82.

Vonshak A, Torzillo G, Accolla P, Tomaselli L (1996) Light and oxygen stress in *Spirulinaplatensis* (cyanobacteria) grown outdoors in tubular reactors. *PhysiologiaPlantarum* 97:175-179.

Vonshak A, Torzillo G, Masojidek J, Boussiba S (2001) Sub-optimal morning temperature induces photoinhibition in dense outdoor cultures of the alga *Monodussubterraneus* (Eustigmatophyta). *Plant, Cell & Environment* 24:1113-1118.

Walker DA (1989) Automated measurement of leaf photosynthetic O_2 evolution as a function of photon flux-density. *Philosophical Transactions of the Royal Society of London SeriesB-Biological Sciences* 323:313-326.

Walker DA (2002) The Z-scheme - down hill all the way. *Trends in Plant Science* 7:183-185.

Walker DA (2009) Biofuels, facts, fantasy, and feasibility. *Journal of Applied Phycology* 21:509-517.

Walker DA, Osmond CB (1986) Measurement of photosynthesis invivo with a leaf disk electrode - correlations between light dependence of steady-state photosynthetic O_2 evolution and chlorophyll-a fluorescence transients. *Proceedings of the Royal Society of London Series B-Biological Sciences* 227:267-280.

Wang B, Li YQ, Wu N, Lan CQ (2008) CO_2 bio-mitigation using microalgae. *Applied Microbiology and Biotechnology* 79:707-718.

Watanabe Y, Hall DO (1995) Photosynthetic CO_2 fixation technologies using a helical tubular bioreactor incorporating the filamentous cyanobacterium*Spirulina-platensis*. *Energy Conversion and Management* 36:721-724.

Watanabe Y, Hall DO (1996) Photosynthetic CO_2 conversion technologies using a photobioreactor incorporating microalgae - Energy and material balances. *Energy Conversion and Management* 37:1321-1326.

Watanabe Y, Saiki H (1997) Development of a photobioreactor incorporating *Chlorella* sp. for removal of CO_2 in stack gas. *Energy Conversion and Management* 38:499-503.

Weyer KM, Bush DR, Darzins A, Willson BD (2010) Theoretical maximum algal oil production. *Bioenergy Research* 3:204-213.

Yentsch CS (1962) Marine plankton. In: Lewin RA (ed) *Physiology and biochemistry of algae.* Academic Press, New York, pp 771 - 797.

Zittelli GC, Rodolfi L, Tredici MR (2004) Industrial Production of Microalgal Cell-mass and Secondary Products - Species of High Potential: Mass Cultivation of *Nannochloropsis* in Closed Systems. In: Richmond A (ed) *Handbook of Microalgal Culture: Biotechnology and Applied Phycology*. Blackwell Science, Oxford, pp 298-303.

In: Applied Algal Biotechnology
Editors: M. Arumugam, S. Kathiresan et al.
ISBN: 978-1-53617-524-0

Chapter 10

HARVESTING OF MICROALGAE AND DOWNSTREAM PROCESSING

Raghu K. Moorthy[1,4] ***and Muthu Arumugam***[2,3,*]
[1]Department of Energy and Environment,
National Institute of Technology Tiruchirappalli,
Tiruchirappalli, India
[2]Microbial Process and Technology,
National Institute of Interdisciplinary Science and Technology (CSIR),
Trivandrum, India
[3]Academy of Scientific and Innovative Research,
Council of Scientific and Industrial Research (CSIR),
New Delhi, India
[4]Current affiliation: Department of Chemical Engineering,
Indian Institute of Technology Bombay,
Mumbai, India

ABSTRACT

Bioresources collectively provide significant contribution in renewable energy sector. Nearly one half of total renewable energy sources are met with useful biomass made available from thick forest and lush green vegetation spread over vast expanse of mainland. Over the past few decades, algae cultivation in large scale has been identified as a suitable alternative to meet the growing demand for biomass sources. Harvesting of microalgal biomass by control of abiotic conditions (light, temperature, pH) and biochemical composition of nutrient medium preferentially leads to high biomass yield and lipid content. Additional dewatering techniques in combination with microalgal harvesting are now widely reported to obtain pre-concentrated microalgal biomass with negligible moisture. Energy consumption could be reduced by more than half when a combination of harvesting techniques namely, batch sedimentation and flocculation is employed. To address industrial requirements such as high throughput and low harvesting

[*] Corresponsindg Author's Email/Website: arumugam@niist.res.in; https://orcid.org/0000-0003-4697-6925.

time, continuous flow systems could open up a whole new solution for existing scale-up challenges. This chapter covers the conventional harvesting techniques and discuss the importance of dewatering and lipid extraction in detail.

Keywords: bioenergy, omega-3 fatty acids, microalgal harvesting, sedimentation, adsorption, lipid extraction

10.1. Introduction

Bio-energy is the largest contributor among renewable energy sources available in the world. Technologies developed to utilize this form of energy are primarily focused towards bio-fuels, phytoremediation, food supplements, cosmetics and other value-added product applications (Milledge and Heaven, 2013). Non-vascular plants such as eukaryotic cyanobacteria and eukaryotes, green algae and diatoms are considered to be a rich source of biomass. Several abiotic factors including nutrient types and algal culture conditions is essential for their survival (Al Hattab et al., 2015). Solid-liquid unit operations are employed for harvesting of algal biomass. As an intermediate step in microalgal technologies to derive bio-energy, it is dependent on nature of microalgal species, aging of algal culture volume and more importantly, harvesting time. The end product applications, calorific content and energy costs during technology operation are other relevant factors. Using the output with lower moisture content from harvesting techniques such as sedimentation-flocculation, a dewatering step is to be included as a secondary technique (Schlesinger et al., 2012) to reduce moisture substantially before further downstream processing.

10.2. Significance of Microalgal Harvesting and Downstream Processing in Algal Biotechnology

Microalgal harvesting accounts for nearly 30% of production cost in a typical algal biotechnology scheme (Molina Grima et al., 2003). It is a necessary step to pre-concentrate the algal biomass solution well before dewatering. With suitable downstream unit operations, highly concentrated algal biomass as feed for several applications such as biofuel production and nutraceuticals Separation of solids from algal growth medium requires that proper downstream treatment is carried out. It is essential for larger scale-up processes that produce low volume, high value products. There is a need to build a sustainable model of harvesting with low energy requirements to ensure affordability without any compromise on overall efficiency of microalgal technology over time.

10.3. Harvesting Techniques

Main objective of microalgal harvesting is separation of solids from the liquid growth medium. It is currently performed using energy- and capital-intensive techniques like centrifugation and flotation (Medipally et al., 2015). For large scale processes with an objective of reduced energy requirements, sedimentation, flocculation or filtration techniques by different approaches are being considered and analyzed. At present, no single best method is available for cost-effective harvesting of algal biomass, and thus, various techniques are being adopted depending on the nature of microalgal species and specific use cases for the algal biomass.

Table 10.1. Overview of available microalgal biomass harvesting techniques

Harvesting technique	Algal species	Total suspended solids (TSS) as final concentration of biomass (%)	Energy consumption (kWh m^{-3})	Reference
Membrane-based filtration	*Chlorella vulgaris* *Scenedesmus sp.*	up to 22% when combined with centrifugation up to 16% with continuous flow	0.84 0.9 to 2.23	Bilad et al., 2012 Gerardo et al., 2014
Centrifugation (disc-type)	*Chlorella sp.* *Scenedesmus sp.*	Not reported 12-22	5.5 1.4	Sharma et al., 2013 Milledge and Heaven, 2013
Flotation	*Scenedesmus sp.* *Chlorella sp.*	10 1 to 6	Not reported 1.6	Chen et al., 1998 ASIO Inc.
Sedimentation	*Scenedesmus sp.*	2 to 7	0.41	Salim et al., 2012
Flocculation	*Chlorella sp.*	3 to 8	0.3 to 2	Milledge and Heaven, 2013

10.3.1. Centrifugation

The common harvesting technique followed in industrial production is centrifugation. It involves centrifugal action to impart shear stress on algal cells and accumulate them in a separate layer (Sharma et al., 2013). Large culture volumes could be processed with high throughput and low harvesting time. The process is very much energy-intensive and requires high production costs (excluding operation costs associated with downstream processing units).

10.3.2. Membrane Filtration and Flotation

Membrane acts as particle filters to entrap solids based on particle size. As a labor-intensive process with frequent maintenance requirements, formation of biomass filter-cake can be affected due to fluid transport properties (Gerardo et al., 2014). Flotation is highly dependent on air bubble flow rate for effective physical interactions between algae and collector materials (for froth formation) (Chen et al., 1998). The main disadvantage with flotation technique is the limited removal of algal biomass as it is largely dependent on the formation of air bubbles to which algal biomass gets attached.

10.3.3. Sedimentation-Flocculation

A simultaneous combination of batch sedimentation and flocculation is reported to be highly preferred among the microalgal harvesting techniques. It is advantageous in terms of being less energy-intensive, easy scale-up operation and low capital costs. If low harvesting time could be attained for a large set of different microalgal species, industries would rather move away from centrifugation towards batch sedimentation-flocculation techniques. To address this very problem, intensive research efforts are now devoted to developing biocompatible materials as flocculants and their physical interactions between microalgal cells (Schlesinger et al., 2012).

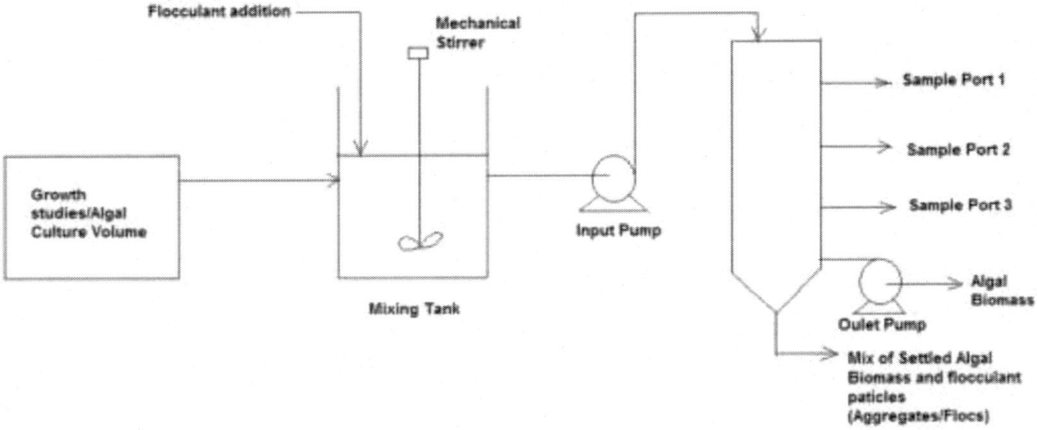

Figure 10.1. Typical process flow configuration for harvesting through sedimentation-flocculation techniques in microalgal technology.

Electrostatic attraction and bridging have been identified to play a major role in flocculation process. For example, Wang and co-workers used cationic polyacrylamide-iron oxide particles as magnetic flocculant to achieve a settling time of ten minutes (microalgal species, *B. braunii* and *C. ellipsoidea*) along with per cent recovery of more than 95% of total algal biomass available (Wang et al., 2013). In another case, pH of the liquid medium is reported to bring down the settling time by 50%. We have recently reported the usage of a novel concept of utilizing flocculant combination that is dependent on initial pH, algal biomass concentration, batch temperature and density gradient between algal biomass and organic flocculants (Moorthy et al., 2017). The low harvesting time (say, less than an hour) at relatively higher flocculation efficiency is an important aspect that needs to be factored in while one opts for a suitable flocculant.

Table 10.2. Different dewatering methods

Method	End-products	Reference
Solvent extraction	Biodiesel, docosahexaenoic acid (DHA), protein synthesis	Al Hattab et al., 2015
Adsorption or absorption	Oil spill sorption (as spill clean-up sorbent)	Mishra et al., 2012
Drying	Algal oil cake formation	Milledge and Heaven, 2013

10.4. Dewatering – An Essential Supplement to Harvesting

Production of high-value products require that moisture content is lowered to the maximum extent as much as possible. Biofuel applications mostly favor the dewatering step due to their higher probability of implementation in existing petroleum refineries and immediately gain rich dividends. For example, the extraction of algal oil from its biomass can be used as a blend to existing petroleum products and enhance the fuel efficiency (Mandotra et al., 2014). Different dewatering methods to obtain specific end-products after microalgal biomass harvesting are available in literature, with a select few of them listed down in Table 10.2.

10.5. Downstream Processing

Microalgae are a promising feedstock for the production of biofuels, value added bioactive molecules and nutraceuticals depending on its yield and fatty acids profile. Naturally, microalgae have the theoretical potential to meet the transportation fuel demand through the production of different forms of fuels such as bioethanol, biobutanol, biomethane, jet fuel, biohydrogen and thermochemical conversion products such as bio-oil, biocrude and syngas (Ranjithkumar et al., 2015). Although, microalgae have the potential for many different compounds, its chemical composition decides the suitability as a feedstock for specific utility. Every algae needs specific attention and each compounds requires specific downstream processing to extract and purify lipids. Describing all the downstream methods for all the compounds is beyond the scope of this book chapter. This subsection of the chapter describes the different methods used to extract the lipids for various applications including biodiesel. A comprehensive method on lipid extraction methods are presented in detail recently from our group (Ranjithkumar et al., 2015).

10.5.1. Total Lipid Extraction

10.5.1.1. Folch Method

This is the one of first method described for extraction of total lipid. In this method, Various organic solvents or its combination of different solvents used extract the total lipids. Folch method (Folch et al., 1957) uses chloroform–methanol (2:1 by volume) for extraction of lipids from complex endogenous cells. In Brief, the homogenised cells or biomasses were equilibrated with one-fourth volume of saline solution and mixed well. The resulting mixture was allowed for phase separation into two layers and lipids settle in the upper organic phase. This method is one of the oldest initiatives in lipid extraction, which formed the basis for development of many more new methods with specific tissues and samples. The above method with slight modifications is still being used for rapid estimation of algal lipids spectrophotometrically. It is rapid and easy processing of large number of samples; however, it is less sensitive.

10.5.1.2. Bligh and Dyer Method

The Bligh and Dyer (Bligh and Dyer, 1959) method is one of the widely practised methods for lipid extraction. It is similar to the Folch method, but differs in solvent/solvent and solvent/tissue ratios. This procedure is performed by extracting lipids from homogenized cell suspension using 1:2 (v/v) chloroform/methanol. The lipids from the chloroform phase are then extracted and processed by various procedures which are not described here. The above gravimetric method is still widely used for the estimation of lipids by algal technologists, and the same procedure is also followed for pilot-scale and large-scale extraction processes (for details, Ranjithkumar et al., 2015).

10.5.1.3. Extraction of All Classes of Lipids

Recently, by Matyash et al. (2008), described a modified Folch/Bligh and Dyer method, which can be used for the extraction of almost all major classes of lipids. In this method, a less water soluble organic solvent Methyl-tert-butyl ether (MTBE) was used for the extraction of lipids. This provides most accurate lipidome profile due to its low-density and easy phase separation facilitates complete extraction of lipids. A detailed procedure is described in our recent review article, the readers are advised to refer for further details (Ranjithkumar *et al.*, 2015).

10.5.1.4. Ecofriendly Solvent Extraction Methods

Although, chloroform is an effective organic solvent used for extraction of lipids, it is limited for lab level small scale extractions. If it is used for commercial scale extraction uses huge quantity of chloroform pose a serious environmental and health risks. Therefore, various less-toxic, but less effective, substitutes of green solvents such as such as ethanol, isopropanol, butanol, MTBE, acetic acid esters, hexane and various combinations of solvents have been investigated by many researchers for microalgal lipid extraction (Sheng et al., 2011). Detailed procedures are described elsewhere for large scale extraction (Ranjithkumar et al., 2015).

10.5.1.5. In Situ Lipid Hydrolysis and Supercritical in Situ Transesterification (SC-IST/E)

All the above mentioned methods are mostly for extracting lipids from dewatered, dried or lyophilized (freeze dried) algal biomass. Very recently, Levine *et al.* (2010) proposed a one-step in site lipid extraction and transesterification method for wet algal biomass. Details are described in recent review article Ranjithkumar et al. (2015), in brief, the wet algal biomass placed in stainless steel reactor, which is immersed in a pre-heated isothermal fluidized sand bath for required time and removed subsequently cooled with water. Then, the processed dried algal biomass of about one gram was mixed with 4 volumes of water in a relatively large reactor of 10 ml capacity, and the reaction was continued at increased temperature of 15, 30, 45, and 60 min at 250 °C. The main advantage of this method is simultaneous drying and dehydration converts the wet algal biomass into a dried solid biomass. This process facilitates precise loading of solids in hydrolysis reactions. After hydrolysis, the reaction mixture is allowed for cooling followed by phase separation, the top volume is crude biodiesel which can be separated using appropriate filter membrane.

In addition to the above methods, various other methods like Algal oil expellers, bead beating, microwave and ultra sonicwave assisted methods, electroporation, osmatic pressure, isotonic extraction and solvent free methods are reported by various groups (Ranjithkumar et al., 2015). In conclusion, a successful commercialisation of the algal biomass-based lipids for various applications depends on eco-friendly and cost-effective downstream processing method for lipid extraction. However, efficient lipid extraction and highest recovery remain the vital downstream processing difficulties in the algal biofuel industry.

10.6. Energy Efficient Strategies in Microalgal Biomass Harvesting

10.6.1. Algal Microfluidics

Continuous flow systems have been recognized as a solution to meet growing demand for alternate technologies to obtain high throughput at low energy costs (Jensen, 2001). Microalgal cell culture in confined volumes and subsequent pre-concentration using sedimentation is now successfully demonstrated using commercial microfluidic chips (Hønsvall et al., 2016). Size-based microalgal cell sorting could ensure cells of bigger size are flown into the harvesting units; leading to significant improvements in biomass yield and attain high recycle ratio for the microalgal culture media. High specific surface area made available for localized physical interactions could possibly reduce the overall equipment footprint relative to equivalent batch processing units.

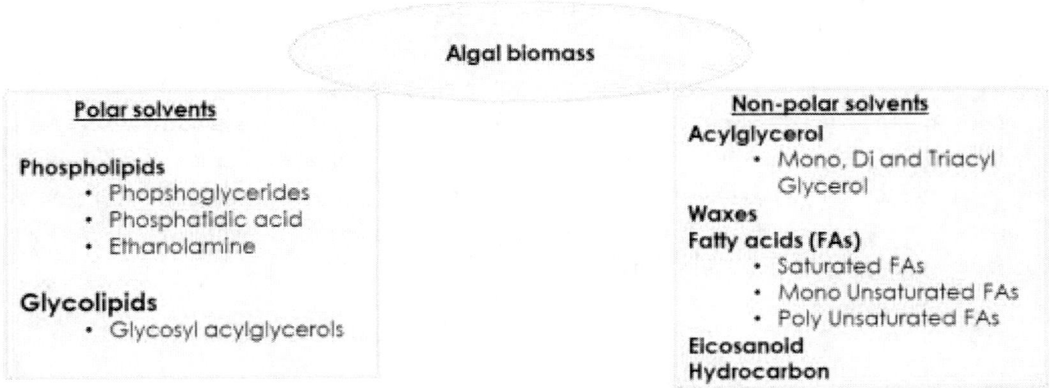

Figure 10.2. Different classes of microalgal lipids with example (Adapted from our review article, Ranjithkumar et al. 2016).

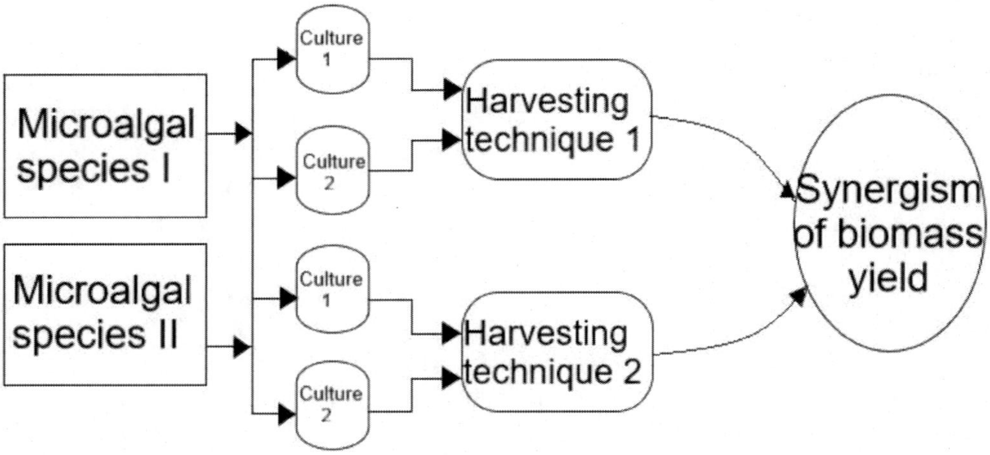

Figure 10.3. Process intensification in microalgal biomass harvesting through various microfluidic platforms.

The main disadvantages of this strategy are adaptability of its mechanical design towards wide range of microalgal species and high initial investment required to fabricate these chips at microfluidic scale.

10.6.2. Scale-Up Operation and Other Challenges in Biofuel Industry

Microalgal harvesting techniques at industrial scale (above 1 kL) demands an energy-efficient unit operation with negligible loss of biomass yield. Harvesting of biomass from a feed of dilute microalgal cell culture (biomass concentration of less than 5 g L^{-1}) should be accompanied with dewatering step necessarily (Molina Grima et al., 2003). The process economics is greatly affected by lack of a well-defined strategy to harvest microalgal biomass and subsequent number of downstream units required.

10.7. Summary and Future Perspective

Microalgal biomass harvesting through sedimentation-flocculation addresses major challenges in pre-concentration step in algal biotechnology. Scale-up operation of these methods heavily relies on three prominent factors, namely nature of microalgal species, energy requirements, capital and operational costs. Further, dewatering and lipid extraction as one of the important downstream processes may be considered as essential to ensure high value for various commercial end-products.

Acknowledgment

We are grateful to facilities and other useful resources provided at CSIR-National Institute of Interdisciplinary Science and Technology (CSIR-NIIST), Trivandrum, India and

Algal Research and Biotechnology Laboratory, Department of Energy and Environment, National Institute of Technology Tiruchirappalli, India. Financial support received from DBT through task force Project "No.BT/PR14219/AAQ/3/740/2015: Investigation on the synergistic effect of phytohormones on omega-3 fatty acid production from *Nannochloropsis sp*." is greatly acknowledged.

REFERENCES

Al Hattab, M., Ghaly, A., Hammouda, A. (2015). Microalgae Harvesting Methods for Industrial Production of Biodiesel: Critical review and comparative analysis, *J. Fundam. Renewable Energy Appl.*, 5: 2. doi: 10.4172/20904541.1000154.

ASIO Inc. (a company related to wastewater treatment technologies), Brno, Czech Republic. Company website: http://www.asio.cz/en.

Bilad, M. R., Vandamme, D., Foubert, I., Muylaert, K., Vankelecom, I. F. J. (2012). Harvesting microalgal biomass using submerged microfiltration membranes, *Bioresource Technol.*, 111, 343 - 352.

Bligh, E. G., Dyer, W. J. (1959). A rapid method of total lipid extraction and purification. *Can. J. Biochem. Phys.*, 37, 911 - 917. doi: 10.1139/o59-099.

Chen, Y. M., Liu, J. C., Ju, Y. H. (1998). Flotation removal of algae from water, *Colloid. Surface. B*, 12, 49 - 55.

Folch, J., Lees, M., Sloane Stanley, G. H. (1957). A simple method for the isolation and purification of total lipids from animal tissues. *J. Biol. Chem.*, 226, 497 - 509. PMID:13428781.

Gerardo, M. L., Radcliffe, D. L. O., Lovitt, R. W. (2014). Minimizing the energy requirement of dewatering *Scenedesmus sp.* by microfiltration: Performance, costs, and feasibility, *Environ. Sci. Technol.*, 48, 845 - 853.

Hønsvall, B. K., Altin, D., Robertson, L. J. (2016). Continuous harvesting of microalgae by new microfluidic technology for particle separation, *Bioresource Technol.*, 200, 360 - 365.

Jensen, K. F. (2001). Microreaction engineering - is small better?, *Chem. Eng. Sci.*, 56, 293 - 303.

Levine, R. B., Pinnarat, T., Savage, P. E. (2010). Biodiesel production from wet algal biomass through in situ lipid hydrolysis and supercritical transesterification. *Energy Fuels*, 24, 5235 - 5243. doi: 10.1021/ef1008314.

Mandotra, S. K., Kumar, P., Suseela, M. R., Ramteke, P. W. (2014). Fresh water green microalga *Scenedesmus abundans*: A potential feedstock for high quality biodiesel production, *Bioresource Technol.*, 156, 42 - 47.

Matyash, V., Liebisch, G., Kurzchalia, T. V., Shevchenko, A., Schwudke, D. (2008). Lipid extraction by methyl-tert-butyl ether for high-throughput lipidomics. *J. Lipid Res.,* 49, 1137 - 1146. doi: 10.1194/jlr.D700041-JLR200.

Medipally, S. R., Yusoff, F. M., Banerjee, S., Shariff. M. (2015). Microalgae as sustainable renewable energy feedstock for biofuel production, *BioMed Research International*, vol. 2015, Article ID 519513. doi: 10.1155/2015/519513.

Milledge, J. J., Heaven, S. (2013). A review of the harvesting of micro-algae for biofuel production, *Rev. Environ. Sci. Biotechnology*, 12, 165 - 178.

Mishra, P. K., Mukherji, S. (2012). Biosorption of diesel and lubricating oil on algal biomass, *3 Biotech.*, 2(4), 301 - 310.

Molina Grima, E., Belarbi, E. H., Acien-Fernandez, F. G., Robles-Medina, A. and Yusuf. C. (2003). Recovery of microalgal biomass and metabolites: process options and economics, *Biotechnol. Adv.*, 20(7 - 8), 491 - 515.

Moorthy, R. K., Premalatha, M., Arumugam, M. (2017). Batch sedimentation studies for freshwater green alga *Scenedesmus abundans* using combination of flocculants. Front. Chem., 5, 37.

Ranjith Kumar, R., Hanumantha Rao, P. and Arumugam, M. (2015). Lipid extraction methods from microalgae: a comprehensive review. *Front. Energy Res.*, 2, 61.

Salim, S., Vermuë, M. H., Wijffels, R. H. (2012). Ratio between autoflocculating and target microalgae affects the energy-efficient harvesting by bio-flocculation, *Bioresource Technol.*, 118, 49 - 55.

Schlesinger, A., Eisenstadt, D., Gil, A. B., Carmely, H., Einbinder, S., Gressel, J. (2012). Inexpensive non-toxic flocculation of microalgae contradicts theories; overcoming a major hurdle to bulk algal production, *Biotechnol. Adv.*, 30, 5, 1023 - 1030.

Sharma, K. K., Garg, S., Li, Y., Malekizadeh, A., Schenk, P. M. (2013). Critical analysis of microalgae dewatering techniques, *Biofuels*, 4: 4, 397 - 407.

Sheng, J., Vannela, R., Rittmann, B. E. (2011). Evaluation of methods to extract and quantify lipids from *Synechocystis* PCC 6803. *Bioresource Technol.*, 102, 1697 - 1703. doi: 10.1016/j.biortech.2010.08.007.

Wang, S. K., Wang, F., Hu, Y. R., Stiles, A. R., Guo, C., Liu, C. Z. (2014). Magnetic flocculant for high efficiency harvesting of microalgal cells, *ACS Appl. Mater. Interfaces,* 6, 1, 109 - 115.

In: Applied Algal Biotechnology
Editors: M. Arumugam, S. Kathiresan et al.
ISBN: 978-1-53617-524-0

Chapter 11

COMMERCIAL APPLICATIONS OF ALGAE IN THE FIELD OF BIOTECHNOLOGY

Jyothi Kaparapu, PhD
and Mohan Narasimha Rao Geddada*, PhD
Department of Botany, Andhra University, Visakhapatnam, India

ABSTRACT

Algae are important biological resources that have a wide range of biotechnological applications. Due to their high nutritional value, algae being mass cultured for health food. A variety of high-value products including polyunsaturated fatty acids (PUFA), pigments such as carotenoids and phycobiliproteins, and bioactive compounds are useful as nutraceuticals and pharmaceuticals, as well as for industrial applications. In terms of environmental biotechnology, microalgae are useful for bioremediation of agro-industrial wastewater, and as a biological tool for assessment and monitoring of environmental toxicants such as heavy metals, pesticides and, pharmaceuticals. In recent years, microalgae have attracted much interest due to their potential use as feedstock for biodiesel production. Algal biotechnology only really began to develop in the middle of the last century. Nowadays, there are numerous commercial applications of microalgae and Macroalgae. The use of microalgae in nanotechnology is a promising field of research with a green approach. The use of genetically modified algae for better production of different biotechnological compounds of interests is popular nowadays. Algal biomass production for sustainable biofuel production together with other high-value compounds is the major challenge of algal biotechnologists. Future research should focus on the improvement of production systems and the genetic modification of strains.

Keywords: algae, biofuel, biotechnology, nanotechnology, phycoremediation

* Corresponding Author's Email: gmnrao_algae@hotmail.com.

11.1. Introduction

More recently there has been a globally increasing search to meet demands in biomass and address the many challenges of an expanding human population, including climate change, aging populations, and food security resulting in the urgent need for sustainable biomass production from non-traditional sources. Consequently, there has been a surge of investment globally in marine biotechnology, with research primarily targeting sponges, invertebrates, microorganisms, and algae. Algae are of particular interest because of their apparent high productivity, unequaled diversity, and because of the many new discoveries in bioactivities and compound diversity, and technical developments in cultivation.

The term "algae" refers to a large diversity of unrelated phylogenetic entities, ranging from picoplanktonic cells to macroalgal kelps. Current systematic screening of algae is commonly based on systematic screening protocols, and disconnected from traditional phycological research, despite the high likelihood that the specific chemical adaptations of algal populations to extreme environments to yield discovery of interesting novel compounds. Algal biotechnology has gained significant importance in agri-horticulture (Khan et al., 2009, Craigie, 2011), marine functional foods (Freitas et al., 2012), cosmetics (Bedoux et al., 2014), and human health (Ioannou and Roussis, 2009) applications, including the provision of ingredients for functional foods.

The development of new products from algae involves many steps that are driven by new markets and consumer demands. The current dominant markets for high-value products of algae are in the field of agri-horticultural products, including animal feeds and plant stimulants, cosmeceuticals, and nutraceuticals. There is also a potential for new drug development although this requires longer development times and larger investments. Algal biomass can be derived from natural or cultivated resources, which may or may not be predefined by strain selection and characterization, or hybridization, or by targeted wild-harvesting of specific phenotypes or materials which are likely to vary according to their origin (Stengel et al., 2011).

The first use of microalgae by humans dates back 2,000 years to the Chinese, who used Nostoc to survive during famine. However, microalgal biotechnology only really began to develop in the middle of the last century, and the establishment of the Spirulina production plant in Mexico can be considered as the benchmark of microalgal biotechnology (Borowitzka, 1999). Particularly during the past two or three decades, algal biotechnology grew steadily into an important global industry with a diversified field of applications, and more and more new entrepreneurs began to realize the potentiality of microalgae in biotechnological exploitation. Currently, active researches are going on for the development of genetically modified algae and for biotechnological exploitation. Due to the marine and aquatic applications, algal biotechnology is sometimes also called blue biotechnology.

Commercial large-scale cultivation of microalgae (Chlorella) started in the early 1960s in Japan by the company Nihon Chlorella (Taipei, Taiwan) (Iwamoto, 2004). It was followed by Arthrospira culturing and harvesting facility in 1970 in Lake Texcoco (Mexico City, Mexico) (Borowitzka, 1999). The first use of algae in aquaculture appeared in the 1970s (Pulz and Scheibenbogen, 1998). The commercial production of *Dunaliella salina*, as a source of β-carotene, became the major microalgal industry established by Western Biotechnology (Hutt Lagoon, Australia) and Betatene in 1986. Other commercial plants were soon followed in

Israel and the USA. The large-scale production of cyanobacteria (blue-green algae) began in India at about the same time. Recently, several plants producing *Haematococcus pluvialis* as a source of astaxanthin have been established in the USA and India. Thus, in a short period of about 30 years, the algal biotechnology industry has grown and diversified significantly.

Based on the results from experimental studies it was interpreted as several compounds in algae can have amazing effects, providing a source of protein, amino acids, fatty acids, vitamins and minerals, and other biologically active phytochemicals (Becker, 2004).

11.2. Algae in Human and Animal Nutrition

Because of their high protein content and nutritive value *Chlorella* and *Spirulina* dominated the microalgal market as food (Becker, 2004; Pulz and Gross, 2004). These algal biomass is marketed as tablets, capsules, and liquids. Microalgae can act as a nutritional supplement or represent a source of natural food colorants (Soletto et al., 2005). *Arthrospira, Chlorella, Dunaliella salina, and Aphanizomenon flos - aquae* are commercially dominated by four strains. *Arthrospira* is used in human nutrition because of its high protein content and its excellent nutritive value (Soletto et al., 2005; Desmorieux and Decaen 2005). In addition, this cyanobacterium has various possible health-promoting effects, like alleviation of hyperlipidemia, suppression of hypertension, protection against renal failure, growth promotion of intestinal Lactobacillus, suppression of elevated serum glucose level, etc. (Liang et al., 2004). The world's largest producer of Arthrospira – the Hainan Simai Enterprising – is located in the Hainan province of China. This company has an annual production of 200 t of algal powder, which accounts for 25% of the total national output and almost 10% of the world output. Their Arthrospira-based products (tablets and powder) are distributed in over 20 countries around the world. A wide variety of nutraceuticals made from this microalga. For example, the Myanmar Spirulina Factory (Yangon, Myanmar) sells tablets, chips, pasta, and liquid extract, and Cyanotech (a plant in Kona, Hawaii, USA) produces products ranging from pure powder to packaged bottles under the name Spirulina Pacifica. For human consumption, Cognis Nutrition and Health, the world's largest producer of this strain, offers Dunaliella powder as an ingredient of dietary supplements and functional foods.

Many evaluations have shown the suitability of algal biomass as a feed supplement (Becker, 2004). Mainly, the microalgae *Spirulina* and, to some extent, *Chlorella* are used in this domain for many types of animals: cats, dogs, aquarium fish, ornamental birds, horses, poultry, cows, and breeding bulls (Spolaore et al. 2006). All these members are able to enhance the nutritional content of conventional feed preparations, and hence, they positively affect the physiology of these animals. Even very small amount of microalgal biomass can positively affect the physiology of animals by improving the immune response (Turner et al., 2002), resulting in growth promotion, disease resistance, enhancing stress resistance (Nath et al., 2012), antiviral and antibacterial action, improved gut function (Michalak et al., 2011), probiotic colonization stimulation, as well as by improved feed conversion, reproductive performance, and weight control (Jyothi, 2018; Harel and Clayton, 2004) and lipid metabolism. The external appearance of the animals may also be improved, resulting in healthy skin and a lustrous coat, for both farming animals (poultry, cows, breeding bulls) and pets (cats, dogs, rabbits, ornamental fishes, and birds) (Certik and Shimizu, 1999).

Seaweed is a valuable food source, rich in vitamins, minerals, proteins, polysaccharides, steroids, and dietary fibers. Phycocolloides such as agar-agar, carrageenan, and alginic acids are the main constituents of brown and red algal cell walls and are widely used in several food industries, especially in Asian countries such as China, Japan, and Korea (Cynthia et al., 2017). Agar-agar is a vegetarian substitute for gelatin and is frequently used as a soup thickener in ice cream and other desserts. Carrageenan has a similar application; it has gelling, thickening and stabilizing properties mainly in dairy and meat products. There are three main varieties of carrageenan, differing in their degree of sulphation. Kappa-carrageenan has one sulfate group per disaccharide, Lota-carrageenan has two sulfates and Lambda carrageenan has three sulfates per disaccharide. Porphyra, commonly known as nori, is the most widely consumed seaweed on earth. The edible seaweed belongs to the red algae and can be bought in supermarkets as dried, very thin sheets used to prepare sushi.

11.3. Aquaculture Feed

From an economic perspective, the importance of using algae in aquaculture has grown over the years as it can be utilized as an unconventional source of protein alternative to the much-used fish meal. Since the 1940s, microalgae became more and more important as live feed in aquaculture (shellfish or fish farming). Microalgal feeds are currently used mainly for the culture of larvae and juvenile shells and finfish as well as for raising the zooplankton required for feeding of juvenile animals (Chen, 2003). They are required for larval nutrition during a brief period, either for direct consumption in the case of mollusks and penaeid shrimp or indirectly as food for the live preys, mainly rotifers, copepods, and *Artemia nauplii*, which in turn are used for crustaceans and fish larvae feeding (Duerr et al., 1998).

The most frequently used microalgae in aquaculture are *Chlorella, Tetraselmis, Isochrysis, Pavlova, Phaeodactylum, Chaetoceros, Nannochloropsis, Skeletonema*, and *Thalassiosira* (Apt and Behrens., 1999). To support a better-balanced nutrition for animal growth, it is often advised to use mixed microalgae cultures, to have a good protein profile, adequate vitamin content, and high polyunsaturated fatty acids, mainly EPA, AA, and DHA, recognized as essential for survival and growth during the early stages of life of many marine animals (Volkman et al., 1989). One of the beneficial effects attributed to adding algae is an increase in ingestion rates of food by marine fish larvae which enhances growth and survival as well as the quality of the fry (Naas et al., 1992). Aquatic species, such as salmonids (salmon and trout), shrimp, lobster, sea bream, and koi carp, under intensive rearing conditions, need supplementation of carotenoids pigments in their diet to attain their characteristic muscle color. In addition to pigmenting effects, carotenoids, namely, astaxanthin and canthaxanthin, exert benefits on animal health and welfare, promote larval development, and provide growth and performance. A positive metabolic role of carotenoids in the nutrition of larval fish and survival of young fry was also discussed by Planas and Cunha (1999), and Lazo et al., (2000). The present group reported the use of nonconventional algal feed with improved results in growth, carcass composition, and pigmentation of goldfish (Khatoon et al., 2010a), rohu (Khatoon et al., 2010b), and prawn (Khatoon et al., 2009).

Silva-Neto et al., 2012 have suggested that kelp meal works as an excellent additive in pelleted feeds for penaeid shrimps and thus improved feed utilization efficiency in this slow feeding species. Valente et al., (2006) recommended that macroalgae such as *Gracilaria* and *Ulva* can be incorporated in sea bass feeds without affecting the performance of fish. Arieli et al. 1993 indicated that seaweed could be categorized as a low-energy high-nitrogen feedstuff in ruminants based on performance parameters of lambs when fed *Ulva lactuca*. Ruminants are unable to digest the carbohydrate fraction of *Chlorella* sp. and *Scenedesmus* sp., which may have been due to the specific characteristics of these algal strains (Hintz et al., 1966).

Taurine (the non-protein sulphonic acid) is usually an essential nutrient for carnivorous animals, including some fish, but it is not found in any land plants (Jyothi, 2017). However, although taurine has been less often investigated than amino acids, it has been reported in significant quantities in macroalgae such as *Laminaria, Undaria, and Porphyra* (Dawczynski et al., 2007) as well as certain microalgae, for example, the green flagellate *Tetraselmis*, the red unicellular alga *Porphyridium* (Flynn and Flynn, 1992), the dinoflagellate, *Oxyrrhis* (Opik and Flynn, 1989), and the diatom *Nitzschia* (Jackson et al., 1992).

11.4. Chemicals and Pharmaceuticals

A large number of studies have demonstrated the benefits and potential of algae as a source for biologically active compounds (Bansemir et al., 2006) and their antioxidant activity (Yuan et al., 2005). In the 1980s, the use of microalgae as a source of common and fine chemicals was the beginning of a new trend (de la Noue and de Pauw, 1988). Higher plants and animals lack the requisite enzymes to synthesize polyunsaturated fatty acids (PUFAs) of more than 18 carbons (Lee, 2001). Thus, they have to get them from their food.

Currently, algal docosahexaenoic acid (22:6 ω3, 6, 9, 12, 15, 18) produced from the dinoflagellate member, *Crypthecodinium cohnii*, is the only commercially available ω3-polyunsaturated fatty acid producer (Walker et al., 2005), but others like γ-linolenic acid from *Spirulina*, arachidonic acid from *Porphyridium*, and eicosapentaenoic acid from *Nannochloropsis, Phaeodactylum,* or *Nitzschia* have already demonstrated as potential for industrial production of their high-value compounds (Spolaore et al., 2006). Other fatty acids or lipids are isolated from *Phaeodactylum tricornutum* as a food additive; from *Odontella aurita* for pharmaceuticals, cosmetics, and baby food; and from *Isochrysis galbana* for animal nutrition (Pulz and Gross, 2004). Polysaccharides are isolated from Chlorella spp. for dietary supplements (Walker et al., 2005) and from *Porphyridium cruentum* for pharmaceuticals, cosmetics, and nutrition (Pulz and Gross 2004). Extracts from the cyanobacterium *Lyngbya majuscula* are used as immune modulators in pharmaceuticals and nutrition management (Pulz and Gross, 2004). Components of algae are frequently used in cosmetics as thickening agents, water-binding agents, and antioxidants. Cosmetics companies claim benefits on the skin or health in general from contents like carrageenan, other algal polysaccharides, algal proteins or lipids, vitamin A, vitamin B1, iron, phosphorus, sodium, copper, magnesium, calcium, or other elements.

Typical species that are used for cosmetics (have antiaging properties) are *Spirulina platensis, Nannochloropsis oculata, Chlorella vulgaris,* and *Dunaliella salina*. Carotenoids are utilized in pharmaceuticals, health food, and dietary supplements, cosmetics, and as a feed additive. Carotenoids act as provitamin A and as biological antioxidants, protecting cells and tissues from the damaging effects of free radicals and singlet oxygen. The carotenoids including β-carotene are produced in large amounts from the halophilic green microalga *Dunaliella salina and Dunaliella bardawil* (Walker et al. 2005). The most prominent xanthophyll is astaxanthin, which is extracted in a large scale from the green microalga *Haematococcus pluvialis* (Pulz and Gross, 2004). Further utilized xanthophylls are lutein, canthaxanthin, and zeaxanthin. The phycobiliproteins-phycoerythrin and phycocyanin are isolated from *Spirulina* and *Porphyridium* and are utilized for health food, pharmaceuticals, and cosmetics (Becker, 2004; Spolaore et al., 2006). Phycobiliproteins are not only used as pigments but have also been shown to have health-promoting properties. They are also used in research laboratories as labels for biomolecules (Spolaore et al., 2006). The first and most important application of phycocyanin is as food pigment, replacing current synthetic pigments (Becker, 2004). Dainippon Ink & Chemicals (Sakura) has developed a product called "Lina blue" which is used in chewing gum, ice sherberts, popsicles, candies, soft drinks, dairy products, and wasabi. They also sell another form of this pigment for natural cosmetics like lipstick and eyeliners (Viskari and Colyer, 2003).

11.5. Biofertilizers

Biofertilizer contains microorganisms that promote the adequate supply of nutrients to the host plants, ensure their proper development of growth and regulation in their physiology, and protect plants from soil-borne diseases to a certain degree. Among the microalgae, cyanobacteria have drawn much attention as prospective and rich sources of biologically active constituents that can be used as biofertilizers (Schlegel et al., 1999). Beneficial effects of cyanobacterial inoculation were reported, not for rice, wheat, soybean, oat, tomato, radish, cotton, sugarcane, maize, chili, bean, muskmelon, and lettuce (Saadatnia and Riahi, 2009; Maqubela et al., 2008). Efficient nitrogen- fixing blue-green algal strains are utilized for rice production. The capacity for enhanced biosynthesis of growth-promoting substances such as auxins, amino acids, sugars, and vitamins by algalization was reported that can enhance the growth of the plant (Misra and Kaushik, 1989).

Cyanobacteria play an important role in fixing atmospheric nitrogen due to the presence of heterocysts and buildup of soil fertility, consequently increasing rice growth and yield as a natural biofertilizer (Song et al., 2005). Blue-green algal inoculation with composite cultures was found to be more effective. Both free-living and symbiotic cyanobacteria have been harnessed in rice cultivation in India. A composite culture of BGA having heterocystous *Nostoc, Anabaena, Aulosira*, etc. is given as primary inoculum in trays, polyethene lined pots, and later mass multiplied in the field for application as soil-based flakes to the rice-growing field at the rate of 10 kg/ha. At many sites where algal inoculation was used for three to four consecutive cropping seasons, the inoculated algae establish well and the effect persisted over subsequent rice crop. The blue-green algal inoculum may be produced by several methods, viz., in tubs, galvanized trays, and small pits and also in field conditions. Nowadays, liquid fertilizers are in use. Liquid biofertilizers have greater potentials to fight with native population; cost saving on carrier material, pulverization, neutralization, sterilization, packing, and transport; easy and quick quality control protocols; better survival on seeds and soil; and very much easy to use by the farmer. The increasing demand for biofertilizers and awareness among farmers and planters in the use of biofertilizers have paved the way for fertilizer manufacturers and new entrepreneurs to get into biofertilizer production.

Marine macroalgae extracts also have been used in agricultural trends as soil conditioners to enhance crop productivity (Newton, 1951, Booth, 1969, Abdel-Raouf et al., 2012). For example, the brown macroalga *Ascophyllum nodosum* has been widely investigated and traditionally used as a soil biofertilizer and conditioning agent (Fan et al., 2011). In addition, methanolic extract of *A.nodosum* and to a lesser extent *Laminaria digitata*, *L. hyperborea*, and *Fucus serratus* have been applied for a large-scale production of biofertilizers due to their high content of betaines, an organic osmolyte compounds that can potentially play a crucial role in effective protection against salts, drought, and extreme temperature stress (Blunden et al., 2010). At present, it is important to improve the plant productivity using natural alternatives which should be safe to maintain ecological integrity (Nagy and Pinter, 2014).

11.6. Biofuel

Microalgae have long been recognized as potentially good sources for biofuel production because of their high oil content and rapid biomass production (Brennan and Owende, 2010). As with plant-derived feedstocks, algal feedstocks can be utilized directly or processed into liquid fuels and gas by a variety of biochemical conversion or thermochemical conversion processes (Amin, 2009; Brennan et al., 2010). Dried algal biomass may be used to generate energy by direct combustion (Kadam, 2002), but this is probably the least attractive use for algal biomass. Hydrogen can be produced from algae by bio- photolysis (Melis, 2002). The rise in world oil prices led to a sharp increase in biofuel production around the world. High oil species of microalgae cultured in growth-optimized conditions of photobioreactors have the potential to yield 5,000–15,000 gal of microalgal oil per acre per year. Microalgae are very efficient solar energy converters, and they can produce a great variety of metabolites (Chaumont, 2005). Biodiesel from microalgal oil is similar in properties to the standard biodiesel (Huang et al., 2010).

The algae that are used in biodiesel production are usually aquatic unicellular green algae. These types of algae are photosynthetic eukaryotes characterized by high growth rates and high population densities. Under good conditions, green algae can double its biomass in less than 24 h. Additionally, green algae can have huge lipid contents, frequently over 50%. This high yield, high-density biomass is ideal for excellent sources for biodiesel production. Algal oil contains saturated and monounsaturated fatty acids. The fatty acids were determined in the algal oil in the following proportions: 36% oleic (18:1), 15% palmitic (16:0), 11% stearic (18:0), 8.4% iso-linoleic (17:0), and 7.4% linoleic (18:2). The high proportion of saturated and monounsaturated fatty acids in this alga is considered optimal from a fuel quality standpoint, in that fuel polymerization during combustion would be substantially less than what would occur with polyunsaturated fatty acid-derived fuel (Gao et al., 2010). After oil extraction from algae, the remaining biomass fraction can be used as a high-protein feed for livestock (Schneider, 2006). This gives further value to the process and reduces waste.

Macroalgae possess plant-like characteristics, making its harvesting more easily compared to microalgae. The rhodophyte *Gracilaria vermiculophylla*, *G. multipartite*, chlorophyte *Ulva lactuca, U. intestinalis* and the phaeophyte *Padina boryana* were widely used in biofuel extraction due to high biomass production and relatively higher lipid content (Abomohra et al., 2017).

11.7. BIOREMEDIATION

Organic pollutants and heavy metals are considered to be a serious environmental problem for human health. Lead, cadmium, and mercury are the most frequent water- and soil-polluting heavy metals. The contamination of soils and aquatic systems by toxic metals and organic pollutants has recently increased due to anthropogenic activity. Phycoremediation is the use of microalgae for the removal or biotransformation of pollutants, including nutrients and xenobiotics from wastewater and CO_2 from waste air with concomitant biomass propagation (Moreno-Garrido, 2008). Recent studies have shown that microalgae can indeed support the aerobic degradation of various hazardous contaminants (Munoz et al., 2008). Many algal genera are found to have capabilities to accumulate heavy metals, thereby reducing their toxic effects (Schiewer and Volesky, 2000; Satıroglu et al., 2002; Arica et al., 2004). There has been a growing interest in using algae for biomonitoring eutrophication, organic pollutants, and inorganic pollutants. Many algae synthesize phytochelatins and metallothioneins that can form complexes with heavy metals and translocate them into vacuoles (Oswald and Golueke, 1960). The growth of microalgae is indicative of water pollution since they respond typically to too many ions and toxins. Blue-green algae are ideally suited to play a dual role in treating wastewater in the process of effective utilization of different constituents essential for growth leading to enhanced biomass production. Secondly, the role of microalgae is the accumulation and conversion of wastewater nutrients to biomass and lipids. Unicellular algae have shown great efficiency in the uptake of nutrients and have been found to show dominance in oxidation ponds (Pittman et al. 2010). *Chlorella, Ankistrodesmus*, and *Scenedesmus* species have been already successfully used for the treatment of olive oil, mill wastewaters, and paper industry wastewaters (Narro, 1987).

The algae have many features that make them ideal candidates for the selective removal and concentration of heavy metals, which include high tolerance to heavy metals, ability to grow both autotrophically and heterotrophically, large surface area/volume ratios, phototaxy, phytochelatin expression, and potential for genetic manipulation (Cai et al., 1995). *Caulerpa racemosa* is a low-cost biomaterial that could be used for the removal of boron species from aqueous solution (Bursali et al. 2009). *Dunaliella salina*, a green microalgae, have a high tendency for zinc accumulation followed by copper and cobalt, the lowest tendency was for cadmium, and this may be due to the importance of zinc as hydrogen transferring in photosynthesis (Liu et al. 2002). The present group also reported several algae-like *Oscillatoria, Anabaena, Rhizoclonium*, and *Chara* that accumulated lead, cadmium, and chromium from wastewater (Chakraborty et al., 2011). Arsenic-resistant genera recorded from the contaminated area were *Oscillatoria princeps*, *Oscillatoria limosa, Anabaena* sp., and *Phormidium laminosum* (Bhattacharya, 2011). Marinho et al., 2011 succeeded in using macroalgae *Gracilaria caudata* and the microcrustacean *Artemia franciscana* to remove nutrients from aquaculture effluents. Deniz and Ersanli, 2018 proved that a seaweed community composed of *Chaetomorpha* sp., *Polysiphonia* sp., *Ulva* sp. and *Cystoseira* sp. removed heavy metal ions from synthetic aqueous medium efficiently. The biomass resulting from the treatment of wastewaters can be applied for different aims, including the use as additives for animal feed, the extraction of added value products like carotenoids or other biomolecules, or the production of biofuel. The accumulation of heavy metals by algae provides an advantage for phytoremediation over other methods which are more costly and not environmental friendly.

11.8. PHYCONANOTECHNOLOGY

Nanotechnology and biotechnology are the two most fascinating technologies of the twenty-first century and unification of these two results in the development of nanobiotechnology which deals with the synthesis of nanostructures using biological organisms. Nanomaterials are gaining so much of interest due to their unique optical, chemical, photoelectron chemical, and electronic properties which are absent in the bulk material because of the quantum-size confinement imposed by nano-size (Kumar et al., 2003). Nanoparticles are synthesized by a number of chemical and physical procedures. Several manufacturing techniques are in use that employs atomistic, molecular, chemical, and particulate processing in a vacuum or a liquid medium (Daniel and Astruc, 2004). But most of these techniques are costly, as well as inefficient in materials and energy use. Green synthesis of nanoparticles is a significant process for the production of precious metals such as gold, silver, platinum, and palladium due to their various applications in material sciences. A number of microorganisms and higher plants have already been found to be competent to serve as eco-friendly nano factories for the synthesis of gold nanoparticles (GNP) and silver nanoparticles (SNP). Cyanobacteria and microalgae having the reducing power are already designated as potential bioreagents for nontoxic nanoparticle synthesis. They are effective in nanoparticle synthesis, as they grow rapidly in an eco-friendly atmosphere producing large amounts of biomass in a very short time and can accumulate a high quantity of metals followed by subsequent reduction. Their production costs are also negligible (Lin et al., 2005). The properties of nanogold change with varying sizes and shapes of the particles and are suitable for applying in diverse fields, starting from decorative coating paints to different fields of science like electronics, catalysis, pollution control, cancer therapy, drug delivery, biomedical assay, biosensor, bioimaging, etc. (Xu et al., 2004; Huang et al., 2006; Hauck et al., 2008). The "green" route for nanoparticle (NP) synthesis is therefore of great interest for economic prospects and applications. Moreover, nanoproducts, synthesized from algae, are expected to be biocompatible, therefore minimize environmental and human health risks. The use of SNPs and GNPs in drug delivery systems might be the future thrust in the field of medicine. Further research is needed to screen the most suitable algal taxa for large-scale controlled synthesis of gold and silver nanoparticles.

The synthesis of nanogold was first reported in 1995 in *Chlorella vulgaris* (Ting et al., 1995). Lengke et al., 2006, reported it in *Plectonema boryanum*. They reported cyanobacteria- and algae-based systemic process of gold and silver nanoparticle synthesis and biosynthesis of gold nanoparticle by cyanobacteria (*Lyngbya majuscula* and *Spirulina subsalsa*), green algae (*Rhizoclonium hieroglyphicum* and *R. riparium*), Spirogyra and diatoms (*Nitzschia obtusa* and *Navicula minima*) (Chakraborty et al., 2011, Nayak et al., 2006; Roychoudhury and Pal, 2014). Screening for different algae as bioreagent of GNP production and biosynthesis of gold nanorods by *Nostoc ellipsosporum* was done by Parial et al., (2012); Parial and Pal (2014) and of silver nanoparticles by Roychoudhury et al., (2014).

11.9. Genetic Engineering and GM Algae

Algal genome research is needed to get a new level of efficiency in biotechnological applications. Algal species are targets of genetic engineering to improve productivity, broaden environmental tolerance limits, or increase pest or pathogen resistance. DNA sequences are introduced into the algal cells with the goal of modifying biochemical pathways (Jyothi, 2018).

The most advanced microalgal genome projects are *Cyanidioschyzon merolae* (the red alga), *Thalassiosira pseudonana* (the diatom), and the three green algae *Chlamydomonas reinhardtii, Volvox carteri,* and *Ostreococcus tauri*. Cyanobacteria are preferable source materials in plant genetic engineering studies due to the genetic homology of chloroplasts in eukaryotic plants, prokaryotic genome organization, short generation time, and fast growth rate. Most studies have therefore been made commonly called 'green yeast', *Chlamydomonas reinhardtii.* Other cyanobacterial strains of *Synechocystis* sp., *Synechococcus* sp., and *Anabaena variabilis. Synechocystis* sp. and *Synechococcus* sp. are naturally transformable with exogenous DNA. The green alga *Haematococcus pluvialis* naturally accumulates large amounts of astaxanthin when exposed to unfavorable growth conditions. The gene coding for the enzyme that converts ß-carotene into astaxanthin has been identified from *H. pluvialis* and cloned into *Synechococcus*. After the transformation, both zeaxanthin and the attractive compound astaxanthin were produced (Sahu et al., 2012). Tremendous success has been achieved in establishing a functional association between plant seedling/calli/protoplast cultures and N_2 -fi xing cyanobacterial filaments through the application of basic tissue culture technique. *Nostoc, Anabaena,* and *Plectonema* are among the few N_2 -fixing cyanobacterial species successfully practiced in these studies. The use of transgenic cyanobacterium to control insects is a new technology. Various combinations of genes were introduced in Anabaena which showed toxicity against mosquito larvae. Further studies of genetically engineered algae and bioremediation have been performed on the green algae on *Chlamydomonas reinhardtii*. In the last few years, successful genetic transformations of ~25 algal species have been demonstrated; most of these were achieved by nuclear transformation. Ten species of green algae have been transformed; stable transformation has been shown for seven of them, one of which was the unicellular model organism *Chlamydomonas reinhardtii* (Debuchy et al., 1989) and transient transformation was demonstrated in the other three. All of these green algae are unicellular species except for *Volvox carteri*, for which stable transformation has been shown (Schiedlmeier et al. 1994). Many species of cyanobacteria, e.g., *Spirulina, Anabaena*, or *Synechocystis*, can be transformed by electroporation or conjugation (Koksharova and Wolk 2002). Algae have also demonstrated suitability for synthesizing vaccines. In this regard, the stable expression of the hepatitis B surface antigen gene has been shown in *Dunaliella salina* (Sayre et al., 2001; Sun et al., 2003). A further project aims at the application of antigen producing algae in the fish industry. It is intended to use an alga-produced antigen to vaccinate fish against the infectious hematopoietic necrosis virus (IHNV) which causes an infectious disease that kills 30% of the US trout population each year; vaccination is realized simply by feeding the fi sh with the algae (Banicki, 2004). F or further biotechnological exploitation of algae, several researchers are screening extracts from a multiplicity of algal species in order to find effective organic components like secondary metabolites (Kopecky et al., 2000; Lubián et al., 2000), antifungal or antibacterial biomolecules (Piccardi et al., 2000), algal toxins (Piccardi et al., 2000), or active pharmaceutical ingredients as drug candidates (Skulberg, 2000).

11.9.1. Production of Recombinant Proteins in Microalgae

There is significant research in the widespread production of recombinant proteins in microalgae (Potvin & Zhang, 2010). Microalgae combining both the high growth rate and ease of cultivation of microorganisms with the ability to perform posttranscriptional and post-translational modifications of plants. The development of economically viable microalgal expression systems is, however, hindered by low recombinant protein yields. Although there are still many obstacles to overcome before microalgae become standard expression systems, considerable progress has been made in recent years in regards to elucidating the causes for these low yields and in the development of strategies to improve them. Transgenes have successfully been expressed in both nuclear and chloroplast microalgal genomes, although at economically viable levels only in the latter. Genetically engineered microalgae systems are being used to make PHAs (Rasala et al., 2010; Specht et al., 2010). Through increased synthetic biology strategies, microalgal strains will be developed to produce specialty chemicals (Fu et al., 2015). *Chlamydomonas reinhardtii* is the most common strain that has been studied at the genetic level. The single-celled green alga, *Chlamydomonas reinhardtii*, has proven to be an outstanding model organism for studies of mechanisms of photosynthesis, cilia/flagella based motility and, more recently, biochemical pathway for the production of lipids and other materials of interest to the renewal biofuels community and specialty chemical companies. *C. reinhardtii* grows rapidly, has a fully sequenced and annotated genome, is easily genetically transformed, is amenable to classical genetic analyses and has associated with it both a wealth of biochemical, molecular, cellular information and numerous experimental tools to expedite research. Because it is a haploid organism, mutant phenotypes are easily created and identified through classical screening techniques or through new, high throughput DNA analysis approaches.

11.9.2. PHB Production through Genetic Manipulation Techniques

The genome of *C. reinhardtii* was fully sequenced and found to have similar phylogenetic properties as plants (Larkum et al., 2012; Merchant et al., 2007). *Cyanidioschyzon merolae*, *Nannochloropsis* sp., *Ostreococcus tauri*, *Phaeodactylum tricornutum,* and *Thalassiosira pseudonana* are some other strains that are suitable for genetic manipulation (Hlavova et al., 2015). The phbB gene was successfully incorporate into *C. reinhardtii* from *R. eutropha* by Chaogang et al., 2010. *Ralstonia eutropha* is a facultatively chemolithoautotrophic bacterium able to grow with organic substrates or H_2 and CO_2 under aerobic conditions. Under conditions of nutrient imbalance, *R.eutropha* produces copious amounts of poly (R)-3-hydroxybutyrate (PHB). Wang et al., (2004) incorporated part of a native bioplastic synthesis pathway into a non-native host by this study. Chaogang et al., (2010) demonstrated the successful incorporation of phbB and phbC genes from *R. eutropha* into *C. reinhardtii*, with the third enzyme phbA, already present in *C. reinhardtii*. This study demonstrated detectable quantities of PHB (6 mg/g dry cell weight) produced by *C. reinhardtii*. Hempel et al., (2011) incorporated the full PHB pathway from *R. eutropha* H16 into *P. tricornutum*. It was the first instance of a complete PHA pathway being incorporated into a microalgal strain. The results of this study were impressive as it was reported that approximately 10.6% PHB was achieved in this engineered strain after Seven days of culturing. Further optimization of the system for increased PHB production could be conducted in the future Codon optimization and extensive screening. The demonstration of PHA production in a microalgae strain allows for the possible biological manufacturing of other PHAs such as scl PHAs and mcl PHAs. Atmospheric CO_2 is essentially a free carbon source in the case of P. tricornutum and C. reinhardtii which makes PHB production economically viable in microalgae (Choi & Lee, 1999).Though the cost of the carbon substrate is significant, the other fact that needs to be noticed is the of purification of PHB from microalgae biomass. Generally, cells are lysed and PHB is extracted. This is a major expensive step in biologically produced PHBs and alternative methods need to be developed to make PHB production economically feasible (Jacquel et al., 2008).

Noor-Mohammadi et al. (2014) developed the advanced technologies for genetic manipulation engineering in the baker's yeast *Saccharomyces cerevisiae* and then the successful transfer of the assembled pathway into the nucleus of *C. reinhardtii*. Such techniques can be used to assemble genes in the chloroplast of *C. reinhardtii* (Noor et al., 2012). There could be potentiality to genetically engineered microalgae strains to produce PHAs in the near future, by using clustered regularly interspaced short palindromic repeats (CRISPR) and Cas9 (an RNA guided cleaving enzyme) cloning systems (Sternberg & Doudna, 2015). Studies using the CRISPR-Cas9 system have already been developed for use with the microalgae strain *C. reinhardtii* (Jinkerson & Jonikas, 2015; Jiang et al., 2014). The advent of CRISPR/Cas9 technology for facile gene editing in numerous organisms has heralded rapid progress in creating targeted gene knockouts and gene replacement by homologous recombination. Transcription activator-like effectors and zinc finger nucleases have also been successfully used in *C. reinhardtii* (Gao et al., 2014).

Conclusion

Some algae have been exploited for millennia. Currently, they have several applications from human and animal nutrition to cosmetics and the production of high-value molecules (e.g., fatty acids, pigments, stable isotope biochemicals). However, algae are still not a well-studied group from a biotechnological point of view. Indeed, among the 10,000 species that are believed to exist, only a few thousand strains are kept in collections, a few hundred are investigated for chemical content and just a handful are cultivated in industrial quantities. The development of algal biotechnology has been slowed by the limited growth performance of algae in industrial photobioreactors. Currently, the majority of microalgal production occurs in outdoor cultivation. However, closed-system commercialization has begun with Haematococcus in Japan and Israel and with Chlorella in Germany. Algal production systems need to be further improved in order to become more competitive and more economically feasible. Heterotrophic and mixotrophic cultivation could be a possible avenue of research. The genetic improvement of algal strains is also a present challenge. The use of transgenic microalgae for commercial applications has not yet been reported but holds significant promise. Despite the significant global progress which has further outlined the potential of algal biomass as a source of natural products, some challenges remain and further research needs can be recognized.

REFERENCES

Abomohra, A. E. F., Naggar, E.I., Hamed, A., Baeshen and Areej,Ali.(2017). Potential of macroalgae for biodiesel production: Screening and evaluation studies, *J. Biosci. Bioeng*. http://dx.doi.org/10.1016/j.jbiosc.2017.08.020.

Amin, S. (2009). Review on biofuel oil and gas production processes from microalgae. *Energy Convers Manag*. 50:1834–1840.

Ayoola, S. O. and Idowu, A. A. (2008). *Biotechnology and species development in aquaculture*. pp. 4722-4725.

Bansemir, A., Blume, M., Schröder, S.and Lindequist, U. (2006). Screening of cultivated seaweeds for antibacterial activity against fi sh pathogenic bacteria. *Aquaculture*, 252:79–84.

Banicki, J.J. (2004). An alga a day keeps the doctor away. Engineered algae as a new means to vaccinate fish. *Twine Line,* 26:1–5.

Becker, W. (2004). Microalgae in human and animal nutrition. In: Richmond A (ed.) *Handbook of microalgal culture.* Blackwell, Oxford, pp 312–351.

Bedoux, G., Hardouin, K. and Burlot, A.S. (2014). Bioactive components from seaweeds: cosmetic applications and future development. *Adv Bot Res,* 71:345–378.

Bhattacharya, P. and Pal, R. (2011). Response of cyanobacteria to arsenic toxicity. *J Appl Phycol,* 23:293–299.

Borowitzka, M.A. (1999). Commercial production of microalgae: ponds, tanks, tubes and fermenters. *J Biotechnol,* 70:313–321.

Brennan, L.and Owende, P. (2010). Biofuels from microalgae – a review of technologies for production, processing, and extractions of biofuels and co-products. *Renew Sust Energy Rev*. 14:557–577.

Bursali, E.A., Cavas, L., Seki, Y., Bozkurt, S.S and Yurdakoc, M. (2009). Sorption of boron by invasive marine seaweed: Caulerpa racemosa var. cylindracea. *Chem Eng J*, 150:385.

Cai, X.H., Logan, T., Gustafson, T., Traina, S.andSayre, R.T. (1995). Application of eukaryotic algae for removal of heavy metal from water. *Mol Mar Biol Biotechnol,* 4:338 C.

Certik, M.and Shimizu, S. (1999). Biosynthesis and regulation of microbial polyunsaturated fatty acid production. *J Biosci Bioeng,* 87:1–14.

Chakraborty, N., Banerjee, A. and Pal, R. (2011). Biomonitoring of lead, cadmium, and chromium in environmental water from Kolkata, North and South 24-Parganas using algae as bioreagent. *J Algal Biomass Utln,* 2(3):27–41.

Choi, S. and Lee, Y. (1999). Factors affecting the economics of polyhydroxyalkanoate production by bacterial fermentation. *Appl Microbiol Biotechnol,* 51: 13-21.

Chen, Y.C. (2003). Immobilized *Isochrysis galbana* (Haptophyta) for long-term storage and applications for feed and water quality control in clam (*Meretrix lusoria*) cultures. *J Appl Phyco*, 15:439–444.

Craigie, J.S. (2011) Seaweed extract stimuli in plant science and agriculture. *J Appl Phycol* 23:371–393.

Cynthia, L. F., de. A., Heloina, S. F.,, Gedson, R., Lima,M., Camila, A., Montenegro., Narlize, S., Lira., Petronio, F. D. A., Luis, C. R., Maria, D. F.V. D. S., Jose, M. B. and Leonia M. B. (2011).· Bioactivities from Marine Algae of the Genus *Gracilaria. Int J Mol Sci.;* 12(7): 4550–4573.

Daniel, M.C. and Astruc, D. (2004). Gold nanoparticles: assembly, supramolecular chemistry, quantum-size-related properties, and applications toward biology, catalysis, and nanotechnology. *Chem Rev,* 104:293–346.

Dawczynski, C., Schubert, R. and Jahreis, G. (2007) Amino acids, fatty acids, and dietary fibre in edible seaweed products. *Food Chem.* 103(3): 891–899.

Debuchy, R., Purton, S. and Rochaix, J.D. (1989). The argininosuccinate lyase gene of Chlamydomonas reinhardtii: an important tool for nuclear transformation and for correlating the genetic and molecular maps of the ARG7 locus. *EMBO J,* 8:2803–2809.

Deniz, F and Ersanli, E.T. (2018). An ecofriendly approach for bioremediation of contaminated water environment: Potential contribution of a coastal seaweed community to environmental improvement. *Int J Phytoremediation.* 23;20(3):256-263.

Desmorieux, H. and Decaen, N. (2005). Convective drying of Spirulina in thin layer. *J Food Eng* 66:497–503.

Duerr, E.O., Molnar, A. and Sato, V. (1998). Cultured microalgae as aquaculture feeds. *J Mar Biotechnol,* 75:65–70.

Freitas, A.C., Rodrigues, D. and Rocha-Santos, T.A. (2012). Marine biotechnology advances towards applications in new functional foods. *Biotechnol Adv,,* 30:1506–1515.

Gao, H., Wright, D.A., Li, T., Wang, Y., Horken, K. and Weeks, D.P. (2014). TALE activation of endogenous genes in *Chlamydomonas reinhardtii. Algal Res* 5:52-60.

Gao, C., Zhai, Y., Ding, Y. and Wu, Q. (2010). Application of sweet sorghum for biodiesel production by heterotrophic microalga *Chlorella protothecoides. Appl Energy,* 87:756–761.

Harel, M. and Clayton, D. (2004). Feed formulation for terrestrial and aquatic animals. US Patent 20070082008 (WO/2004/080196).

Hauck, T.S., Jennings, T.L., Yatsenko, T., Kumaradas, J.C. andChan, W.C.W. (2008). Enhancing the toxicity of cancer chemotherapeutics with gold nanorod hyperthermia. *Adv Mater,* 20:3832–3838.

Hlavova, M., Turoczy, Z.and Bisova, K. (2015). Improving microalgae for biotechnologydfrom genetics to synthetic biology. *Biotechnol Adv,* 33:1194-1203.

Huang, X, EL-sayed, I.H., Qian, W. and EL-sayed, M.A. (2006). Cancer cell imaging and photothermal therapy in the near-infrared region by using gold nanorods. *J Am Chem Soc,* 128:2115–2120.

Huang, G., Chen, F., Wei, D., Zhang, X. and Chen, G. (2010). Biodiesel production by microalgal biotechnology. *Appl Energy,* 87:38–46. Ioannou, E.and Roussis, V. (2009).

Natural products from seaweeds. In: Osbourn AE, Lanzotti V (eds.) *Plant-derived natural products: synthesis, function, and application.* Springer, New York, pp 51–81.

Iwamoto, H. (2004). Industrial production of microalgal cell-mass and secondary products— major industrial species— Chlorella. In: Richmond A (ed.) *Handbook of microalgal culture.* Blackwell, Oxford, pp 255–263.

Jackson, A. E., Ayer, S. W.and Laycock, M. V. (1992). The effect of salinity on growth and amino acid composition in the marine diatom Nitzschia pungens. *Canadian Journal of Botany,* 70, 2198 –2201.

Jacquel, N., Lo, C.W., Wei, Y.H., Wu, H.S.and Wang, S.S. (2008). Isolation and purification of bacterial poly (3- hydroxyalkanoates). *Biochem Eng, J* 39: 15-27.

Jinkerson, R.E. and Jonikas, M.C. (2015). Molecular techniques to interrogate and edit the Chlamydomonas nuclear genome. *Plant J*, 82:393-412.

Jyothi, K. (2018). Polyhydroxyalkanoate (PHA) Production by Genetically Engineered Microalgae: A Review. *J New Biol Report,* 7(2) 68 – 73.

Jyothi, K. (2018). Application of Microalgae in Aquaculture. *Phykos* 48 (1): 21-26.

Jyothi, K. (2017). Algae in Formulated Feeds. *J. Algal Biomass Utln.* 2017, 8(3): 23- 28.

Kadam, K.L. (2002). Environmental implications of power generation via coal microalgae cofiring. *Energy* 27:905–922.

Khan, W., Rayir, U.P. and Subramanian, S. (2009). Seaweed extracts as biostimulants of plant growth and development. *J Plant Growth Regul,* 28:386–399.

Khatoon, N., Chattopadhyay, P., Mukhopadhyay, A., Mukhopadhyay, M.andPal, R. (2009). Algal diet in prawn aquaculture. *Fishing Chimes*, 28:44–47.

Khatoon, N., Chaudhuri, A., Sen, R.S., Kundu, N., Mukherjee, S., Majumdar, D., Homechaudhuri, S. and Pal, R. (2010a). Algae as feed supplement in fi sh nutrition. *J Bot Soc*, 64(2):85–93.

Khatoon, N., Sengupta, P., Homechaudhuri, S. and Pal, R. (2010b). Evaluation of algae based feed in goldfi sh (Carassius auratus) nutrition. *Proc Zool Soc,* 63(2):109–114.

Koksharova, O.A and Wolk, C.P. (2002). Genetic tools for cyanobacteria. *Appl Microbiol Biotechnol,* 58:123–137.

Kumar, A., Mandal, S., Selvakannan, P.R., Parischa, R., Mandale, A.B. and Sastry, M. (2003). Investigation into the interaction between surface-bound alkylamines and gold nanoparticles. *Langmuir,* 19:6277–6282.

Larkum, A.W.D., Ross, I.L., Kruse, O.and Hankamer, B. (2012). Selection, breeding and engineering of microalgae for bioenergy and biofuel production. *Trends Biotechnol,* 30: 198-204.

Lazo, P, Dinis, M.T., Holt, J., Faulk, C.and Arnold, C. (2000). Co-feeding microparticulate diets with algae: toward eliminating the need of zooplankton at fi rst feeding in larval red drum (Sciaenops ocellatus). *Aquaculture,* 188:339–351.

Lee, Y.K. (2001). Microalgal mass culture systems and methods: their limitation and potential. *J Appl Phycol,* 13:307–315.

Liang, S., Xueming, L., Chen, F. and Chen, Z. (2004). Current microalgal health food R&D activities in China. *Hydrobiologia* 512:45–48.

Lin, Z., Wu, J., Xue, R. andYang, Y. (2005). Spectroscopic characterization of Au3þbiosorption by waste biomass of Saccharomyces cerevisiae. *Spectrochim Acta,* 61:761–765.

Liu, Y., Yang, S., Tan, S., Lin, Y. and Tay, J. (2002). Aerobic granules: a novel zinc biosorbent. *Lett Appl Microbiol,* 35:548–551.

Lubián, L.M., Montero, O., Moreno-Garrido, I., Huertas, I.E., Sobrino, C., Gonzalez, D. V. M. and Pares, G. (2000). Nannochloropsis (Eustigmatophyceae) as source of commercially valuable pigments. *J Appl Phycol*, 12:249–255.

Merchant, S.S., Prochnik, S.E., Vallon, O., Harris, E.H., Karpowicz, S.J. and Witman, G.B. (2007). The Chlamydomonas genome reveals the evolution of key animal and plant functions. *Science.* 318: 245-250.

Maqubela, M.P. Mnkeni, P.N.S., Malamissa, O., Pardo, M.T. and Acqui, L.P.D. (2008). Nostoc cyanobacterial inoculation in South African agricultural soils enhances soil structure, fertility and maize growth. *Plant Soil,* 315:79–92.

Marinho-Soriano, E., Azevedo, C.A.A., Trigueiro, T.G., Pereira, D.C., Carneiro M.A.A and, Camara, M.R. (2011). Bioremediation of aquaculture wastewater using macroalgae and Artemia, *International Biodeterioration & Biodegradation,* 65 (2011): 253-257.

Melis, A. (2002). Green alga hydrogen production: progress, challenges and prospects. *Int J Hydrogen Energy*, 27:1217–1228.

Michalak, I., Chojnacka, K. and Dobrzański, Z. (2011). Effect of Macroalgae Enriched with Microelements on Egg Quality Parameters and Mineral Content of Eggs, Eggshell, Blood, Feathers and Droppings. In *Journal of Animal Physiology and Nutrition,* 95(3): 374–387.

Misra, S. and Kaushik, B.D. (1989). Growth promoting substances of cyanobacteria II detection of amino acids, sugars and auxins. *Proc Indian Sci Acad*, B55:499–504.

Moreno-Garrido, I. (2008). Microalgae immobilization: current techniques and uses – review. *Biores Technol,* 99:3949–3964.

Munoz, R and Guieysse, B. (2008). Algal–bacterial processes for the treatment of hazardous contaminants: a review. *Water Res,* 40:2799–2815.

Naas, K.E., Naess, T. and Harboe, T. (1992). Enhanced 1st feeding of Halibut Larvae (Hippoglossus hippoglossus L) in green water. *Aquaculture,* 105:143–156.

Narro, M.L. (1987). Petrolium toxicity and oxidation of the aromatic hydrocarbons. In: Fay P, Van Baalen C (eds.) *The cyanobacteria.* Elsevier, Amsterdam, 491–511.

Nath, P.R., Khozin-Goldberg, I. and Cohen, Z. (2012). Dietary Supplementation with the Microalgae Parietochloris incisa Increases Survival and Stress Resistance in Guppy Poecilia reticulata Fry. In *Aquaculture Nutrition,* 18(2), pp. 167–180.

Nayak, D., Nag, M., Banerjee, S., Pal, R., Laskar, S. and Lahiri, S. (2006). Preconcentration of 198Au in a green alga, Rhizoclonium. *J Radioanal Nucl Ch,* 268:337–340.

Noor-Mohammadi, S., Pourmir, A. and Johannes, T.W. (2014). Method for assembling and expressing multiple genes in the nucleus of microalgae. *Biotechnol Lett,* 36: 561-566.

Opik, H. and Flynn, K.J. (1989). The digestive process of the dinoflagellate, Oxyrrhis marina Dujardin, feeding on the chlorophyte, Dunaliella primolecta Butcher a combined study of ultrastructure and free amino acids. *New Phytology.* doi: 10.1111(j.1469-8137.1989. tb04700.

Oswald, W.J. and Golueke, C. (1960). Biological transformation of solar energy. *Adv Appl Microbiol,* 2:223–262.

Parial, D. and, Pal, R. (2014) .Green synthesis of gold nanoparticles using cyanobacteria and their characterization. *Indian J Appl Res,* 4:69–72.

Planas, M. and Cunha, I. (1999). Larviculture of marine fi sh: problems and perspectives. *Aquaculture*, 177:171–190.

Piccardi, R., Frosini, A., Tredici, M.R. and Margheri, M.C. (2000). Bioactivity in free-living and symbiotic cyanobacteria of the genus Nostoc. *J Appl Phycol,* 12:543–547.

Potvin, G. and Zhang, Z. (2010). Strategies for high-level recombinant protein expression in transgenic microalgae: a review. *Biotechnol Adv*, 28: 910-918.

Pulz, O. and Gross, W. (2004). Valuable products from biotechnology of microalgae. *Appl Microbiol Biotechnol*, 65:635–648.

Pulz, O. and Scheibenbogen, K. (1998). Photobioreactors: design and performance with respect to light energy input. *Adv Biochem Eng Biotechnol,* 59:123–151.

Potvin, G. and Zhang, Z. (2010). Strategies for high-level recombinant protein expression in transgenic microalgae: a review. *Biotechnol Adv,* 28: 910-918.

Roychoudhury, P. and Pal, R. (2014) *Spirogyra submaxima* -a green alga for nanogold production. *J Algal Biomass Utln,* 5(1):15–19.

Saadatnia, H. and Riahi, H. (2009). Cyanobacteria from paddy- fields in Iran as a biofertilizer in rice plants. *Plant Soil Environ,* 55(5):207–212.

Sahu, D., Priyadarshani, I. andRath, B. (2012). Cyanobacteria – as potential biofertilizer CIBTech. *J Microbio,* 1:20–26.

Satıroglu, N., Yalcınkaya, Y., Denizli, A., Arica, M.Y., Bektas, S. and Genc, O. (2002). Application of NaOH treated *Polyporus versicolor* for removal of divalent ions of Group IIB elements from synthetic wastewater. *Process Biochem,* 38:65–72.

Sayre, R.T., Wagner, R.E., Sirporanadulsil, S. and Farias, C. (2001). Transgenic algae for delivery antigens to animals. Int. Patent Number W.O. 01/98335 A2.

Schiewer, S. and Volesky, B. (2000). In: Lovley DR (ed.) *Environmental microbe–metal interactions.* ASM Press, Washington, DC, pp 329–362.

Schlegel, I., Doan, N.T., Chazal, N. and Smith, G.D. (1999). Antibiotic activity of new cyanobacterial isolates from Australia and Asia against green algae and cyanobacteria. *J Appl Phycol*, 10:471–479.

Schneider, D. (2006). Grow your own: would the wide spread adoption of biomass-derived transportation fuels really help the environment. *Am Sci,* 94:408–409.

Schiedlmeier, B., Schmitt, R., Muller, W., Kirk, M.M., Gruber, H., Mages, W. and Kirk, D.L. (1994). Nuclear transformation of Volvox carteri. *Proc Natl Acad Sci USA* 91:5080–5084 Schiewer S, Volesky B (2000) In: Lovley DR (ed.).

Skulberg, O.M. (2000). Microalgae as a source of bioactive molecules – experience from cyanophyte research. *J Appl Phyco,* 12:341–348.

Soletto, D., Binaghi, L., Lodi, A., Carvalho, J.C.M. and Converti, A. (2005). Batch and fed-batch cultivations of Spirulina platensis using ammonium sulphate and urea as nitrogen sources. *Aquaculture,* 243:217–224.

Song, T., Martensson, L., Eriksson, T., Zheng, W. and Rasmussen, U. (2005). Biodiversity and seasonal variation of the cyanobacterial assemblage in a rice paddy fi eld in Fujian China. *Fed Eur Mat Soc Microbiol Ecol,* 54:131–140.

Specht, E., Miyake-Stoner, S. and Mayfield S. (2010). Micro-algae come of age as a platform for recombinant protein production. *Biotechnol Lett,* 32: 1373-1383.

Spolaore, P., Joannis-Cassan, C., Duran, E and Isambert, A. (2006). Commercial applications of microalgae. *J Biosci Bioeng,* 101:87–96.

Sternberg, S.H. and Doudna, J.A. (2015). Expanding the biologist's toolkit with CRISPR-Cas9. *Mol Cell,* 58: 568-574.

Stengel, D.B., Connan, S. and Popper, Z.A. (2011). Algal chemodiversity and bioactivity: sources of natural variability and implications for commercial application. *Biotechnol Adv,* 29(5):483–501.

Sun, M., Qian, K., Su, N., Chang, H., Liu, J. and Shen, G. (2003). Footand-mouth disease virus VP1 protein fused with cholera toxin B subunit expressed in Chlamydomonas reinhardtii chloroplast. *Biotechnol Lett*, 25:1087–1092.

Ting, Y.P., Teo, W.K.and Soh, C.Y. (1995). Gold uptake by *Chlorella vulgaris. J Appl Phycol*, 7:97–100.

Xu, X., Stevens, M. and, Cortie, M.B. (2004). In situ precipitation of gold nanoparticles onto glass for potential architectural applications. *J Mater Chem,* 16:2259–2266.

Viskari, P.J. and Colyer, C.L. (2003). Rapid extraction of phycobiliproteins from cultured cyanobacteria samples. *Anal Biochem,* 319:263–271.

Volkman, J.K., Jeffery, S.W., Nichols, P.D., Rogers, G.I. and Garland, C.D. (1989). Fatty acid and lipid composition of 10 species of microalgae used in mariculture. *J Exp Mar Biol Ecol,* 128:219–240.

Walker, T.L., Purton, S., Becker, D.K.and Collet, C. (2005). Microalgae as bioreactors. *Plant Cell Rep,* 24:629–641.

Wang, C., Hu, Z., Hu, W. and Lei, A. (2004). Expression and molecular analysis of phbB gene in *Chlamydomonas reinhardtii. Chin Sci Bull,* 49: 1713-1717.

Yuan, Y.V., Carrington, M.F. and Walsh, N.A. (2005). Extracts from dulse (*Palmaria palmata*) is effective antioxidants and inhibitors of cell proliferation in vitro. *Food Chem Toxicol* 43:1073–1081.

In: Applied Algal Biotechnology
Editors: M. Arumugam, S. Kathiresan et al.
ISBN: 978-1-53617-524-0

Chapter 12

NUTRACEUTICALS AND THERAPEUTIC APPLICATIONS OF ALGAE

Aswathy Udayan[1,2], B. Jeyakumar[3], Shanmugam Kathiresan[3,*] and Muthu Arumugam[1,2]

[1]Microbial Processes and Technology Division,
National Institute for Interdisciplinary Science and Technology (NIIST),
Council of Scientific and Industrial Research (CSIR), Trivandrum, Kerala, India
[2]Academy of Scientific and Innovative Research (AcSIR), CSIR, India
[3]Department of Molecular Biology, School of Biological Sciences,
Madurai Kamaraj University, Madurai, Tamilnadu, India

ABSTRACT

Among the pioneering entries in the food supplements, algae play a major role in nutraceutical industry with excellent market value. The products derived from algae are proteins, carotenoids, polyunsaturated fatty acids (PUFAs), polysaccharides, vitamins, sterols, and biologically active molecules. These pharmacologically active metabolites and biomolecules possess cardiovascular protection, neuroprotective role, anticancer and anti-inflammatory activities, etc. The metabolites derived from algae have wide array of therapeutic potential in human. They act as inhibitors, activators, transporters and govern transcription factors that can modulate several metabolic pathways in disease progression. This chapter describes the available algal derived nutraceuticals and recent research accomplished towards the use of bioactive metabolites in various therapeutic studies *in vitro* and *in vivo*.

Keywords: algae, nutraceuticals, cosmetics, PUFAs, anti- inflammatory, neuroprotective

[*] Corresponding Author's Email: kskathir@gmail.com.

12.1. INTRODUCTION

In the current global scenario, achieving an adequate nutrition is a major concern with increasing population and urbanization. It is indeed necessary to find cost effective sources of nutrients that can rapidly produce significant amounts of high value nutritional products. Algae are rich source of high value metabolites that can act as essential nutrients to support human health which includes a diverse group of photosynthetic organisms both unicellular (microalgae) and multicellular (macroalgae). These organisms have the ability to grow rapidly using light energy and atmospheric carbon dioxide and produce more biomass per hectare than vascular plants. Since algal biomass has been used for centuries as food and medicine and the health promoting abilities of algae were discovered in early 1500 BC. Recently algal biomass gained more interest as a source of neutraceutically and pharmaceutically important products. Algae contains a wide range of high value metabolites including proteins, carbohydrates, lipids, omega 3 fatty acids, pigments, amino acids, vitamins, antioxidants and trace elements(Figure 12.1) (Udayan et al., 2017). These high value metabolites serve as natural nutritional supplements in human and animal feed with a number of health promoting effects and nutraceutical activities.

The term nutraceutical comes from the integration of "nutrition" and pharmaceutical and refers to substances related to nutrition consists of physiological benefits which are capable of prevention and protection against chronic diseases (Kalra, 2003). Nutraceuticals include mixtures and pure compounds extracted from herbs, beverages, dietary derivatives, etc. with beneficial effects on human health, prevention of diseases, increase in life expectancy, delay aging process and provides structural and functional support to the body. A functional is defined as food that can deliver additional or enhanced benefits over and above its basic nutritional value. This additional value is generally related to disease prevention and health promotion by adding new ingredients or more of the already existing ones. Functional foods contribute to the physiological benefits and reduce the risk of chronic diseases beyond basic nutritional functions, which have similar appearance to conventional food and can be consumed as a part of routine diet as defined by the US Department of Agriculture, Agriculture Research Service (Functional Foods Research in ARS, 2019). In recent years, market has been focused more on functional foods and nutraceuticals with increasing request and consumption of nutraceuticals sold in pharmacies and drug stores.

Microalgae have been considered as new model organisms for different biotechnological applications such as biodiesel production, bioremediation, heavy metal removal and dietary supplements for human and animal nutrition. Recent economic studies has proven that due to less biomass productivity and relatively high costs of commercial algal production, biofuel production from algae is not cost effective unless the process is coupled with the commercialization of high value co metabolites(Richardson et al., 2014). The use of microalgal biomass for the development of products with health benefits such as functional foods and nutraceuticals is considered as a rapidly growing market. The worldwide carotenoid market is expected to increase from \$1.24 billion USD in 2016 to \$1.53 billion USD by 2021 (Markets and Markets, 2017).

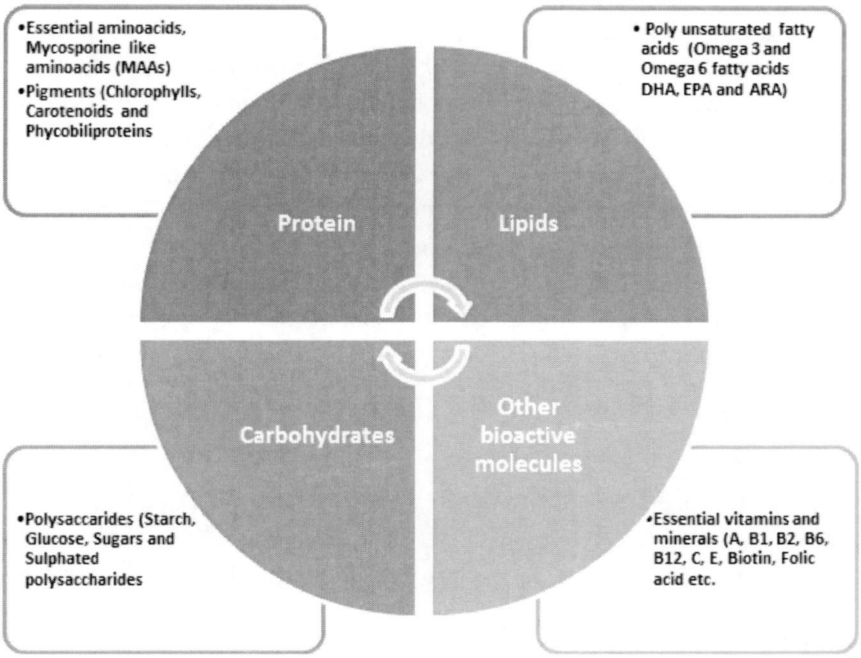

Figure 12.1. High value metabolites from microalgae.

More funding and research works has been focused over the last few decades to screen microalgal bioactive compounds. Different bioactive compounds were identified and extracted from microalgae and macroalgae such as sulphated polysaccharides, carotenoids, beta carotene, omega 3 fatty acids, polyphenols, vitamins and proteins(Pouvreau et al., 2008; Peng et al., 2011; Goiris et al., 2012; Haimeur et al., 2012; Raposo et al., 2013; Gammone et al., 2015; Udayan and Arumugam, 2017; Udayan et al., 2018). These high value bioactive metabolites have great potentials as supplement in the human diet for the prevention and treatment of several diseases. Thus microalgae and macroalgae derived compounds have showed different biological activities such as, anti inflammatory, anti diabetes, antioxidant, anticancer and antiviral properties (Lee et al., 2006; Xia et al., 2014; Lauritano et al., 2016).

Algae serve as a viable source for defending several diseases and nutrient deficiencies experienced by people around the world. The four most prevalent global diseases such as malnutrition, nutritional anaemia, xerophthalmia and endemic goiter were reported very recently (Berquin et al., 2011). These deficiencies can be cured by the consumption of single tablespoon of appropriate edible algae per day. Consumption of algae promote regular healthy functioning of major body processes including the cardiovascular system, immune system, respiratory system, humoral system and central nervous system(Chew and Park, 2004; Kau et al., 2011). Algae have also been used in cosmetics as a moisturizing agent, sun protectant, and as a treatment for wounds, burns, and bruises, due to its anti-inflammatory properties. In addition to these, algae offer variety of possibilities for enhancing health and wellbeing of human and animals. Recent research has even suggested algae as a therapeutic solution for a number of serious conditions, including diabetes, heart disease, autoimmune diseases, cancer and cognitive decline, in the form of dementia and Alzheimer's disease (Riediger et al., 2009; Lordan et al., 2011). More importance has been given for the development of value added functional food products to treat different diseases like diabetes, neurodegenerative diseases, anaemia, obesity and other chronic diseases. Besides obesity and diabetes, development of cardiovascular diseases and obesity are strictly dependent on dietary and lifestyle factors. Cancer is the second leading cause of death globally and it resulted in 9.6 million deaths in 2018 (WHO) (Ferlay et al., 2015). It is showed that almost 30% tumours can be prevented by lifestyle changes like poor dietary habits, use of tobacco, ultraviolet exposure, infectious agents, and lack of exercise, overweight and obesity (Stewart et al., 2015). Promotion of healthy dietary habits and exercise are useful strategies to reduce the occurrence of the above mentioned diseases. Considering the above said fact that alga offers an attractive health benefits in a commercially viable manner. The present chapter discuss the health benefits of nutraceuticals, algae as functional foods, feeds, and cosmetics in comprehensive manner with appropriate illustration.

12.2. Functional Foods from Algae

From ancient times, microalgae have been considered as nutritional supplements and food source. Mexicans used *Spirulina* from Lake Texcoco from AD 1300 as food ingredient. The population in North-Central Africa has been using *Spirulina* from Lake Kossorom as a daily diet. It is also reported that local fishermen from Spain has been collecting blue green biomass from lakes for the preparation of a dry cake known as 'tecuitlatl'. Filamentous cyanobacteria *Nostoc* has also consumed by the ancient people in Mongolia, China and South America. *Spirogyra* has also used as a food component in Thailand, Vietnam, Burma and India (García et al., 2017). 'Suizenji-nori' native food in Japan also contains edible cyanobacterium *Aphanotheca sacrum*. It is very well known that the modern era of microalgal and cyanobacterial uses in biotechnological developments has starter in early 1940s and gained special interest with the first Algal Mass Culture Symposium conducted at Stanford University in United States during 1952. This symposium has opened a new gateway for the mass cultivation of microalgae for their commercial exploitation from being harvested from their natural habitats. Microalgae can be used as alternatives for high food quality and more environmental friendly approach because microalgae can be cultivated in non arable lands even in waste waters with high productivity. Microalgae are used in human nutrition nowadays and commercialized products from microalgae have listed in Table 12.1.

Table 12.1. List of commercialized microalgae for human nutrition

Algae	Major producers	Products	World production (ton/year)	References
Spirulina	Hainan Simai Pharmacy Co. (China), Cyanotech Corp. (USA) Earthrise nutritionals (USA)	Powders, extracts, tablets, beverages	3000	(Pulz and Gross, 2004; Spolaore et al., 2006)
Chlorella	Taiwan Chlorella Manufacturing Co. (Taiwan)	Tablets, powders, noodles	2000	(Pulz and Gross, 2004; Spolaore et al., 2006)
Dunaliella	Cognis Nutrition and Health (Australia)	Powders, β-carotene	1200	(Spolaore et al., 2006)

12.2.1. Algae as Source of Proteins

Proteins are essential nutrients in humans especially in the diet of athletes and people who are doing high exercises for the building and repair of muscle tissues. Seaweed and microalgae have been considered as bulk source of proteins from ancient years. Microalgal derived proteins have several advantages over the conventional protein sources and it significantly contribute to meet the populations need for protein. Algal derived proteins are suitable for vegetarians for whom egg and dairy whey proteins are not suitable. From the ancient times people are depending sea weed and microalgae because of its significant protein content which is greater than other plant sources (E. W. Becker, 2007). *Spirulina*, *Chlorella*, *Dunaliella*, *Haematococcus* and *Schizochtrium* have been considered as food sources by the US Food and Drug Administration which is falling into the GRAS category of foods.

Due to high protein content and additional nutritional benefits, *Spirulina* is the highly consumed microalgae in the world. It has health benefits including renal protection, anti hyperglycaemic, anti hypertension, anti-hyperlipidaemia, etc. (Spolaore et al., 2006). Along with high content of proteins, it also contains free radical scavenging phycobiliproteins, vitamins and gamma linolenic acid [GLA] (Sajilata et al., 2008). Because of the high value energy rich compounds in *Spirulina*, World Health Organization (WHO) has been labelled it as a 'super food' and has been sent to space by National Aeronautics and Space Administration (NASA). It is an interesting fact that *Spirulina* has 5100% more iron than spinach, 3100% more carotene than carrots, 670% more protein than tofu, 180% more calcium than milk and 180% more calcium than milk (Khan et al., 2005). When look into the protein content, it contains up to 70% protein of dry weight which is 10 times more when compared to soybean and three times to that of beef protein (Capelli and Cysewski, 2010). It also provides additional compliment of nine essential amino acids. When consumed, *Spirulina* are readily absorbed by the body and brings a normal level of nutrient status especially for minerals and vitamins (Dillon et al., 1995). *Spirulina* improves the iron content during pregnancy and lactation in rat models (Kapoor and Mehta, 1998). Due to its effective bioavailability of nutrients, it can be considered as a good food for malnourished children and women during pregnancy and lactation. Hainan Simai Enterprising Ltd, located in China is the world's largest commercial producer of *Spirulina* (Spolaore et al., 2006). This company produces 200 tonnes of dried biomass annually which accounts for the 25% of the national output and 10% of the global production. Earthrise Company in California, USA has the largest *Spirulina* production plant which covers an area of 440000m^2.

Chlorella is another widely consumed microalga with a global production market with sales of more than US $38 billion (Yamaguchi, 1996). The largest producer of *Chlorella* is in Chlorella Manufacturing and Co. which has an annual production of 400 ton dried biomass (Yamaguchi, 1996). *Chlorella* contains 60% proteins with most of the essential amino acids and also contains poly unsaturated fatty acids and phosphorous in a significant amount. The protein content in *Chlorella* is three times more compared to beef which is considered as the major concentrated forms of protein available. Peptides from *Chlorella vulgaris* are effective to prevent cellular damage. Extracts from *Chlorella* shows potential activities like anti oxidant, anti inflammatory, anti cancer and anti microbial (Guzman et al., 2003; Lordan et al., 2011). It has been shown that *Chlorella* has the ability to lower cholesterol levels and blood pressure, improves wound healing and stimulates the immune system (Holdt and Kraan, 2011). Microalgae are mainly consumed as a dietary supplement in the form of powder, pills or tablets. It has also been incorporated in different functional foods including bread, noodles, biscuits, sweets, drinks and beer. AlgaVia is the major manufacturer of microalgae based products which produces protein and lipid rich algal flour from *Chlorella protothecoides*.

Several species of macroalgae like red sea weeds are also reported to have high content of proteins which are largely consumed by the humans. *Pyropia tenera* (nori) which belongs to the red algal species is used in sushi wraps in several Asian countries. Monosodium glutamate, the taste enhancer was first extracted and discovered from the brown seaweed *Laminaria japonica* (MacArtain et al., 2007). Sea weed is considered as a rich source of protein and essential amino acids, which makes it as a potential source for future uses. But in Western countries the consumption of sea weed in humans is under developed in the current scenario. In laboratory scale, sea weed has been successfully incorporated as an ingredient in several food products. Bread containing 4% brown algae *Ascophyllum nodosum* significantly lowers energy intake in individuals with overweight (Hall et al., 2012). Similarly, pasta made of *Undaria pinnatifida* has greater antioxidant activity.

12.2.2. Algae as Source of Polyunsaturated Fatty Acids (PUFAs)

Microalgal lipids are considered as high value energy rich compounds with many health benefits. Among the high energy bioactive compounds from microalgae production of PUFA or long chain fatty acids have potential health benefits on human health. Eicosapentaenoic acid (EPA) and Docosahexaenoic acid (DHA) are two most important omega 3 long chain PUFAs. Gamma linolenic acid (GLA) and Arachidonic acid (ARA) are important omega 6 fatty acids which act as precursors for the synthesis of prostaglandins. Currently fish and fish derived oils are the major sources of omega 3 fatty acids. But fish oils contain many toxic contaminants which is not suitable for infants and pregnant women. Moreover fish obtain these LC-PUFAs through the consumption of microalgae. Microalgae are considered as the primary producers of omega 3 fatty acids in the aquatic food chain. Since microalgae can attain higher productivities compared to other possible sources, PUFAs from microalgal origin have a very important biotechnological market for food and feed particularly suitable for infant formulae. DHA is a vital component involved in foetal brain development and retinal development in infants.

DHA rich algal oil is derived from the heterotrophic fermentation of the marine alga, *Schizochytrium* sp. It neither produces any toxic chemicals nor is it pathogenic, thus it is commercially exploited for the production of omega 3 fatty acids. Among the other marine microalgae *Cryptothecodinium cohnii* is the prolific producer of DHA. Since it produces only DHA, and no other PUFAs in its lipid content makes easy separation of DHA. Successful cultivation of *C. cohnii* for commercial edible oil containing DHA has been achieved by Martek Corporation in Maryland, USA. Martek has developed a GMP process utilizing a wild type *C. cohnii* growing on glucose as the main carbon source to produce single cell oil (DHASCO) with a DHA enrichment of above 40% (Kyle, 1996). Single cell edible oil containing DHA can be used in infant formulas and baby foods, pharmaceutical products and dietary supplements. As *C. cohnii* is non-photosynthetic organism has the potential as model organism for the characterization of the enzymes involved in the production of DHA. *C. cohnii* possesses all the enzymes necessary for *de-novo* synthesis of DHA thus becomes an ideal candidate for engineering of the fatty acid elongation and saturation metabolic pathways. *Nannochloropsis* is considered as a promising alga for industrial applications because of its ability to accumulate high levels of PUFAs. It is used as both food for humans and animals. EPA content of *Nannochloropsis* is highly influenced by the culturing conditions (Pulz and Gross, 2004; Gouveia et al., 2008). A *diatom*, *Phaeodactylum tricornutum,* naturally accumulates high levels of EPA (but trace amount of DHA) and is considered as a good source for large scale commercial production. Omega 3 rich microalgae *Isochrysis galbana* has reported to use as ingredient in biscuits (Gouveia et al., 2008). These biscuits contain high amounts of EPA and DHA upon microalgal incorporation which reflect an important source of omega 3 PUFAs with moderate biscuit consumption. Feeding hens with omega 3 fatty acid enriched microalgae significantly increases the omega 3 fatty acid content in eggs which is commercially profitable (Pulz and Gross, 2004).

12.2.3. Algae as a Source of Natural Food Colorants

It is very important to keep the food good looking and prevent from discoloration for proper maintenance of food items. Coal tar derivatives are mainly used as synthetic colorants in food industry, which is banned in several countries due to its negative impact on health. Main reason behind selection of synthetic colorants is low yield of natural colours from conventional plant sources. In order to substitute the synthetic colorants, extraction of pigments from plant sources require higher amount of biomass and high operational cost. Microalgae are the major photosynthetic organisms on earth which contains chlorophylls, carotenoids and phycobiliproteins. Hence microalgae can be considered as the potential alternative source for natural colorants. More over microalgae can be cultivated in large scales in non arable lands for higher biomass and pigment production. Microalgal pigments are widely used in different industries to substitute synthetic colorants including food, pharmaceutical, nutraceutical, cosmetic and aquaculture industry. It has also been used in clinical research laboratories which can be used to label antibodies and other receptors. Phycobiliproteins from microalgae exhibited anti-inflammatory, antioxidant, hepatoprotective and neuroprotective effects. There is an increasing interest in microalgae to use in aquaculture/feed formulation, production of nutrient rich organisms, and enhancement of animal colour. As an ecofriendly and renewable source, microalgae have exploited for centuries as potential source for food and health care.

In Brazil, chlorophyll obtained from *Spirulina* biomass is used as a natural food colorant. *Spirulina sp.* contains 1.15mg/g chlorophyll when compared with spinach with the content of approximately 0.06mg/g (Riediger et al., 2009). Incorporation of whole *Spirulina* biomass in food will impart green colour together with enrichment of other high value metabolites which will increase the nutritional value of food (Santiago-Santos et al., 2004). Algae are rich source of carotenoids which can be used to substitute synthetic food colorants, as an additive for feed, vitamin supplements and other health care products. *Chlorella vulgaris* used in poultry feed increases the pigmentation of egg yolk which is comparable with the synthetic pigments used (Gouveia et al., 1996). β-carotene is widely used in food products due to its antioxidant properties. Natural form of β-carotene is effectively absorbed by the body in comparison with synthetic pigments. *Dunaliella salina* is a potent producer of β-carotene with an accumulation up to 16% dry weight (Del Campo et al., 2007). *Dunaliella* powder is widely used as food and feed additives. Astaxanthin is a reddish pigment belongs to a group of chemicals called carotenoids. It is widely used in aquaculture industry as a pigmentation source. Astaxanthin is synthesized naturally by microalgae or phytoplankton. Consumption of astaxanthin rich microalgae by fish imparts pinkish-red colour in the flesh of these fishes which has been considered as super vitamin E due to its natural antioxidant activities. The major source of astaxanthin is the green alga *Haematococcus pluvialis*, which is also referred as *Haematococcus lacustris* or *Sphaerella lacustris* belongs to Haematococcaceae family. It is proved that during unfavourable environmental conditions *Haematococcus* will be exposed to harsh ultraviolet rays, which will produce astaxanthin as a protective barrier. The astaxanthin level in *Haematococcus* is about 1.5-3% of its dry weight (Sheikhzadeh et al., 2012). Astaxanthin is widely used in nutraceuticals, cosmetics, food and feed industries due to its potent antioxidant activity which is more than 10 times stronger than β-carotene and close to 1000 times more effective than vitamin E. Administration of *Haematococcus pluvialis* especially in dose of 3 g per Kg, effectively enhance the antioxidant system and some biochemical parameters in rainbow trout. Although the natural form cannot compete commercially with synthetic ones, many companies have been focused on the commercial production of astaxanthin from microalgae to make it as a marketable product.

Phycobiliproteins are a group of accessory light harvesting pigment molecules and are important water soluble fluorescent pigment-protein complexes organised in supramoleularphycobilisome complexes. Phycocyanin is a major phycobiliprotein with a wide range of application in food industry as a colouring agent. Cyanobacteria rich in c-phycocyanin have an important role in improving the immune system of humans. *Spirulina* rich in phycocyanin is a potential free radical scavenger and it also inhibits lipid peroxidation in microsomes. Phycocyanin helps to protect renal failure caused by certain drugs. It also has ant cancerous activity. C-phycocyanin also inhibits the growth of cancer cell lines. Phycocyanin is one of the major commercial substances produced from *Sprulina sp.* which is used as a natural food colorant in Japan under the name 'Lina-blue'. Japan and China uses *Spirulina* for the production of candies, chewing gums, jellies, soft drinks, dairy products, etc. (Gantar and Svirčev, 2008). Production of phycobiliproteins from cyanobacterial species isolated from Serbia represents an excellent source for different phycobiliproteins.

12.2.4. Algae as a Source of Vitamins and Minerals

A vitamin is an organic compound required as a nutrient in tiny amounts by an organism. A compound is called a vitamin when it cannot be synthesized in sufficient quantities by an organism, and must be obtained from the diet. Microalgae are considered as rich sources of vitamins and minerals, which make them suitable for nutritional supplements. *Spirulina* and *chlorella* are rich in vitamin B12. A number of vitamins are presented in higher concentrations in the microalgae than in conventional food, traditionally considered rich in vitamins such as vitamin E – carotene, pro-vitamin β (tocopherols), C (ascorbic acid), A) and B12. *Dunaliella* is rich in lipid soluble vitamins. *D. tertiolecta* is very rich in vitamin B2, B12, folic acid, vitamin C, nicotinic acid and E, which accounted 31.2, 0.7, 4.8, 163.2, 79.3, and 116.3 (mg/Kg d.w), respectively (Udayan et al., 2017). *Spirulina* contains more vitamin B12 compared to other plant and animal sources. Compared to spinach, liver and other green algae, *Spirulina* is considered as the richest source of vitamin E, thiamine, cobalamine, biotin and inositol. Microalgae have the ability to produce α-tocopherol, the most biologically active form of vitamin E in high concentrations. Heterotrophically grown microalgae *Euglena gracilis* resulted in a production of 3.7mg/g of α-tocopherol in 120 hours which is comparatively higher productivity than the common natural sources of vitamin E like sunflower, olive, soybean and corn. It is reported that seaweeds contains higher amounts of folate than the conventional sources (Rupérez, 2002).

Algae can also be used as potential sources of minerals in food and feed industry. Many edible microalgae contains higher amounts of macro minerals like sodium, potassium, calcium, magnesium, etc. and trace elements like iron, zinc, manganese and copper than the edible land plants. Hence edible brown and red seaweeds can be used as a food supplement to meet the recommended daily intake of macro minerals and trace elements. But algal biomass offers only as a supplementary source rather than a complete replacement for the synthetic minerals or vitamins in food and feed industries.

12.2.5. Algae as a Source of Antioxidants

Antioxidants belong to the major compound of interest in algae which is considered as a multi component antioxidant system (Plaza et al., 2008). Microalgal antioxidants are considered to be more effective due to its interactions between other antioxidant components. The major important water soluble antioxidants from algae are polyphenols, phycobilproteins and vitamins (Pulz and Gross, 2004). Cultivation of algae in closed photobioreactors with high light and oxygen concentrations promote the accumulation of antioxidant molecules to protect the cells from damage. During oxygen stress antioxidative potential of *Spirulina platensis* has resulted in an increase of 2.3 fold (Pulz and Gross, 2004). Vitamin C which are abundant in photosynthetic organism have the ability to reduce reactive oxygen species (ROS). In chloroplast vitamin C has a major role in the regeneration of membrane bound carotenoids and tocopherols to protect the photosynthetic apparatus. High levels of vitamin C was detected in *Chlorella sp.*, *Dunaliella sp.*, *Chaetoceros calcirans* and *Skeletonema sp.* (Brown et al., 1999; Running et al., 2002). The free radical scavenging of microalgal products are gaining more interest for functional foods and nutraceuticals especially in the beverages industry.

12.2.6. Algae in Animal Nutrition

To meet the food demands of the increasing population, aquaculture and animal husbandry has increased exceedingly. Use of small fishes and agricultural products for feed production will eventually lead to the over exploitation of the resources and competition with food. In this context, microalgae can be considered as the next alternative source for animal and aquaculture feed production in an eco-friendly and cost effective manner. When compared with other conventional feeds microalgae are richer in high value metabolites. Researchers have been proved the high nutritional value of microalgae in the diet of ruminants, chicken and aquatic organisms (Sousa et al., 2008). Even when used in small amounts algae have the ability to improve the immune system, gut function, lipid metabolism, stress resistance along with increase in weight, appetite number of eggs, reducing cholesterol levels and reproductive performance of animals (Sousa et al., 2008). Different nutritional and toxicological studies proved the suitability of algal biomass as a substitute for conventional feed supplements. Several microalgae such as *Spirulina*, *Chlorella*, *Tetraselmis*, *Chaetoceros*, *Scenedesmus*, *Nannochloropsis*, *Haematococcus*, *Dunaliella*, *Nitzchia*, *Schizochytrium*, *Cryptothecodinium,* etc. and macroalgae like *Laminaria*, *Gracilaria*, *Padina*, *Ulva*, *Pavonica*, etc. can be used as a feed for terrestrial and aquatic animals (Baker and Günther, 2004).

12.2.7. Aquaculture

Microalgae are considered as potential sources of high value metabolites such as proteins, lipids, carbohydrates, vitamins, amino acids; pigments, etc. microalgae have been mainly used as fish meal, oil replacement, food additives, colouring of salmonoids, enrichment of nutritional value of zooplankton fed to fish larvae and inducers of biological activities. Nutrient composition of different microalgal species has been shown in Table 12.2. Hence microalgae as aquaculture feed has gained more interest due to the stimulatory effects on weight gain, enhanced protein and triglyceride deposition in muscle, disease resistance, decreased nitrogen output into the environment, improved taste and consistency of flesh, increased omega 3 fatty acid content, starvation tolerance, physiological activity, carcass quality, and increase in the rate of growth of aquatic species to better digestibility (Segner et al., 1989).

The most commonly used species in aquaculture are *Chlorella*, *Tetraselmis*, *Pavlova*, *Phaeodactylum*, *Nannochloropsis*, *Chaetoceros*, *Skleletonema* and *Thalassiosira* (Lemahieu et al., 2016). A better balanced nutrition is achieved by the use of mixed microalgal cultures which gives good protein profile, high PUFA content and adequate vitamin content. Aquatic organisms like salmonoids, lobster, gold fish and shrimp under intensive conditions require carotenoid supplementation in their food to achieve characteristic muscle colour. In addition to this astaxanthin and canthaxanthin have positive effects on animal welfare and health which promotes larval development and exerts stimulatory effects on farmed fish. Carotenoid supplementation also has reported growth benefits in carp and tilapia fishes (Khan et al., 2005). Astaxanthin supplementation also promoted skin pigmentation of ornamental gold fish and shrimps. *Spirulina* rich in β-carotene is mainly used in aquaculture feeds up to 5-20% as fish fees which enhance the red and yellow patterns in the fish leaving a brilliant white colour (Khan et al., 2005). Application of diatom *Haslea* imparts a blue green colour on the gilla of oysters which increases the market value by 40%. *Spirulina* have been used as food for koi culture to provide bright colours for which these fish are famous. The largest use of aquaculture of *Spirulina* is mainly for koi feeding. More over to the pigment content *Spirulina* contains other high value metabolites such as GLA and vitamins. Several fish feeding studies in Japan proved that incorporating even small amounts of *Spirulina* in fish feed increased growth, flavour and survival rates of fishes.

The major advantages of using microalgae as aquaculture feeds includes increased health benefits and improved taste which makes them similar to the taste of fish and shrimp caught in the wild which increases their value in food markets. In addition to this microalgae have great potential for use as aquaculture feeds not only because of the high protein and lipid content but also their phototrophic ability because they can produce high value metabolites directly from sunlight. The major reason for considering microalgae as valuable components in fish diets is mainly their high protein content and amino acid profiles to replace plant protein concentrates. Scientists have demonstrated that tilapia diets are capable of substituting soybean meal for 6% of fish meal or 10% meat and blood meal without compromising the growth of fish.

12.2.8. Poultry

Poultry has a major role in diet which helps in increasing income and gave good quality of proteins in the diets of rural people in developing countries. Compared to other meat products, chicken is healthier with desirable amount of monounsaturated fatty acids. The major problem faced by farmers is the price for feeds and chemical contents from the supplements fed to chicken to fasten its growth which can cause several health risks in consumers. Here comes the importance of an alternative source like microalgae over the traditional feed sources. Algae are potential sources of proteins, lipids, PUFAs, polysaccharides, vitamins, minerals, pigments and enzymes. The replacement of traditional feed sources in poultry was done by several feeding trials using different microalgae like *Scenedesmus*, *Chlorella*, *Euglena*, *Oocystis* and *Spirulina* in different concentrations. No difference was observed in egg quality and egg production rate between control and *Chlorella* fed birds (Lemahieu et al., 2016). Hens fed with algae have eggs with a distinct dark orange colour yolk (Lemahieu et al., 2016). Carotenoid serves as the reason for pigmentation in broilers and egg yolks. This purpose was fulfilled by using dehydrated alfalfa meal and yellow corn in traditional practices. *Dunaliella sp.* has used as a yolk enhancing agent and a source of vitamin A in laying hens. *Chlorella vulgaris* biomass enhances the pigmentation of egg yolk compared with the synthetic pigments used (Baker and Günther, 2004). Supplementation of *Porphyridium* biomass showed a modified fatty acid composition and reduced blood cholesterol level in egg yolk without any differences in egg number, weight and body weight (Gouveia et al., 1996). Egg yolk of chickens fed with algae showed reduced cholesterol levels and increased linolenic acid and arachidonic acid contents together with a dark coloured egg yolk with high carotenoid content. Supplementation of *Nannochloropsis* in the diet of hens has increased the omega-3 fatty acid content of egg yolks (Lemahieu et al., 2016). Selenium is considered as an important trace mineral for poultry that increases thymus weight and increases immunity in broilers (Hussain et al., 2004). *Chlorella* is a rich source of selenium and supplementation of *Chlorella* to hens increase the immune system due to high selenium content (Becker, 2004). Hence algae are officially approved in several countries as poultry feed which do not require new testing or approval.

12.2.9. Algae in Cosmetics

Microalgae have the ability to regenerate and protect themselves under adverse environmental conditions by preventing the accumulation of free radicals and reactive oxygen species, thereby counteracting cell damaging activities. Therefore much focus has been given to use microalgae as a natural substitute for synthetic components in cosmetics. Many products were commercialized using microalgae as cosmetics and cosmeceuticals. Microalgae can be used in two ways, based on the activity of microalgae itself or based on the activity of the ingredients. Different microalgae have been commercialized cosmetics products include anti aging creams, regenerating care products, anti irritant peelers, hair care products and sunscreen. These products are potential and innovative substitutes to existing cosmetics and paves new path for the development of cosmetic products. Companies from US and Europe have used microalgal extracts from *Chlorella*, *Spirulina*, *Nannochloropsis*, *Dunaliella*, *Anacystis* and *Arthrospira* to launch their cosmetic products. These algae have the ability to minimize vascular imperfections, prevent wrinkle formation and boost collagen synthesis. Skinicer produced by Ocean Pharma in Germany used extracts from *Arthrospira* to maintain the youthful characteristics of the skin. Pentapharma in Switzerland produced Pepha-Tight using *Nannochloropsis* extract which have anti aging properties. Phytomer, a company in France used *Chlorella vulgaris* extract to improve skins natural protection (Table 12.3). These cosmetics contain bioactive molecules from microalgae like PUFAs, phycobiliproteins, carotenoids, pigments and polysaccharides which are incorporated in creams, lotions and ointments (Kim et al., 2015). Mycosporine like amino acids (MAAs) from *Dunalilella*, *Chlorella and Spirulina* have the ability to act as sunscreens to reduce ultraviolet light induced damage (Udayan et al., 2017). MAAs is small molecules produced by marine microorganisms that live in high sunlight environments. They have radical scavenging activity against free radicals and ROS. PUFAs from *Odontella aurita* have showed antioxidant activity and help to maintain youthful skin (Ryu et al., 2015).

Table 12.2. Nutrient composition of important microalgae used in aquaculture

Algae	Protein %	Carbohydrate %	Lipid %	References
Spirulina maxima	60-71	13-16	6-7	(Becker, 2004; E. W. Becker, 2007)
Spirulina platensis	50-65	8-14	4-9	(Becker, 2004; E. W. Becker, 2007)
Chlorella sp.	51-58	12-17	14-22	(E W Becker, 2007)
Scenedesmus sp.	50-56	10-52	12-14	(E W Becker, 2007)
Dunaliella sp.	49-57	4-32	6-8	(E W Becker, 2007)
Euglena	39-61	14-18	14-20	(Becker, 2004; E. W. Becker, 2007)

Table 12.3. Application of microalgae in cosmetics products and major companies involved in cosmetics production using microalgae

Microalgae	Major producers	Products	Activity
Spirulina	Optimum Derma Acidate, (ODA, Britain), Ferenes cosmetics	Skin whitening facial mask, Algae mask	Improves moisture balance of skin, immunity, complexion and reduce wrinkles
Spirulina maxima	Ocean Pharma (Germany)	Skinicer	Skin regeneration and natural protection
Chlorella vulgaris	Phytomer (France)	Phytomer	Strengthens skins natural protection and neutralizes inflammation
Nannochloropsis oculata	Pentapharm Ltd. (Switzerland)	Pepha-Tight	Improves skin tightening properties
Dunaliella salina	Pentapharm Ltd. (Switzerland)	Pepta-Ctive	Stimulates skin proliferation and energy metabolism of skin
Anacystis nidulans	Estee Lauder(USA) Solazyme (France)	Photosomes Algenist	Improves skin immunity and sun protection
Spirulina platensis	Exsymol S.A.M (Monaco) Sanatur GmbH (Germany)	Protulines	Prevents early skin ageing and wrinkle formation

12.3. Microalgae - Its Therapeutic Applications

Microalgae are currently being used as functional foods due to their high polyunsaturated fatty acids essential amino acids, pigments, and vitamins. In recent times, microalgal biomass has been used an additive in pasta and baby soups as nutritional content. The term "nutraceutical" arises from the integration of "nutrition" and "pharmaceutical" and refers to substances related to nutrition, endowed with physiological benefits associated with prevention and/or protection against chronic diseases. Apart from food, the trends towards raising worldwide dietary demand for algal products from a higher focus on health and nutraceutics. In relation to their dietary importance, algae are increasingly being sold as functional foods or nutraceuticals. These products have not explored in therapeutic applications. The bioactive compounds or phytochemicals have several important roles in human health and disease. Several studies report the nutritional and bioactive contents of algae, but only few studies showed its therapeutic applications. Here, an overview of known bioactive molecules discussing the developments of microalgal biotechnology in terms of nutraceuticals and functional ingredients for health benefit and disease prevention.

12.3.1. Microalgal Derived Products in Cardiovascular Protection

Cardiovascular disorders (CVDs) are a group of disorders that affects the blood vessels and important parts of heart. They are mainly are coronary heart disease, Atherosclerosis, rheumatic heart disease, congenital heart disease and thrombosis and pulmonary embolism (Siti et al. 2015). CVDs are the world's leading cause of death, with an estimated 17.5 million in 2012. About 6.7 and 7.4 million fatalities, reported due to coronary heart disease and stroke. They are regarded as leading causes of death in the coming years and, according to the World Health Organization (WHO), in 2030 around 23.6 million people will die from CVDs (http://www.who.int accessed on 21st September 2019). Recent studies and clinical evidence suggests that obesity contributes directly contributing to the onset and growth of the disease. Other risk factors, including dyslipidemia, hypertension, atheroclerosis and thrombosis are closely related to CVDs (Zalesin et al. 2008). Macroalgae and microalgae represent promising candidates enriched with key nutrients having several health-promoting compounds. Polyunsaturated fatty acids, pigments, dietary fibers, and minerals are directly linked to cardiovascular-health promotion. Recent research accomplished towards CVDs from algae in targeting reducing oxidative stress, vascular inflammatory condition and improving the lipid profile.

Several studies reported the role of algal derived PUFAs or whole biomass in terms of cardiovascular protection. A Comparative study on fish oil and microalgal omega-3 dietary supplementation on cardiovascular risk factors in high fat fed rats was studied. Interestingly microalga *O. aurita* fed rats showed significantly lower bodyweight than other groups. Microalgafed rats displayed significant reduction in insulinemia, platelet aggregation, oxidative status and hepatic triacyglycerol levels in high-fat diet induced rats. Microalgal omega-3 supplementation showed greater biological effects than the fish oil (Haimeur et al. 2016). Similarly in another study Haimeur et al. (2012) reported that possible health benefits of EPA rich marine diatom *Odontella aurita* in high fat induced metabolic syndrome in

Wistar rats. Results revealed that *Odontella aurita* supplementation resulted in lower glycemia, oxidative stress and lipid levels in the plasma and liver of high fat fed rats. Further, n-3 fatty acid levels in the tissues were higher than other groups. The synergistic effect of EPA other bioactive molecules in *O. aurita* reduced the risk factors for high-fat induced metabolic syndrome. The intake of long-chain n-3 PUFAs is associated with reduction of cardiovascular risk factors. *Schizochytrium* sp. oil containing docosahexaenoic acid (22: 6n-3) and docosapentaenoic acid (22: 5n-6) on in healthy men and women was evaluated on cardiovascular risk factors. Subjects received 4 g oil/d for 4 weeks; it contained 1·5 g DHA and 0·6 g DPA. Results showed that plasma concentrations of arachidonic acid, adrenic acid, DPA, and DHA levels were increased by 21, 11, 11 and 88 mg/l respectively. In erythrocytes DPA and DHA phospholipids were observed by 78 and 27% respectively. Further results revealed that the oil did not adversely affect cardiovascular risk factors in human subjects (Sanders et al. 2006). A meta-analysis in human subjects showed that docosahexaenoic acid from algal oil increases HDL cholesterol and reduces serum triglycerides and LDL-cholesterol in persons without coronary heart disease (Bernstein et al. 2012).

Pigments from algae have wide array of biological activities from reducing the oxidative stress in pathological condition in CVDs. In a study beneficial effect of 9-*cis* carotene-rich powder of the alga *Dunaliella bardawil* was reported in human trials. The combined treatment of fibrate plus 9-*cis* carotene-rich powder of microalga efficiently increased serum HDL levels. Results showed that 9-*cis* carotene-rich powder efficiently amplifies the fibrate drug and improves HDL cholesterol levels (Shaish et al. 2006). Pigment such as phycobiliprotein (c-phycocyanin) from *spirulina platensis* known to reduce oxidative stress and its therapeutic potential was studied in an atherogenic model of hamsters. Hamsters treated with selenium rich phycocyanin showed reduction in plasma cholesterol and superoxide anion. Plasma antioxidant capacity, liver glutathione peroxidase and superoxide dismutase activity was significantly increased and prevents the development of atherosclerosis (Riss et al. 2007). Astaxanthin was known to improve dyslipidemia and metabolic syndrome in animals, but this was not studied in human trials. Placebo-controlled astaxanthin supplementation at doses of 0, 6, 12, 18 mg/day for 12 weeks was assigned to 61 non-obese human subjects with fasting serum triglyceride levels of 120–200 mg/dL and without diabetes and hypertension, age group of 25–60 years. Results revealed that 12 and 18 mg/day doses significantly reduced triglyceride, and 6 and 12mg doses significantly increased HDL-cholesterol and serum adiponectin levels (Yoshida et al. 2010). Several studies reported the possible health benefits of *Spirulina* sp. in cardiovascular protection by improving the lipid profile and act as hypocholesterolemic drug. Effect of microalga *Spirulina platensis* was studied on hippocampus lipoperoxidation and hypercholesterolemia induced rats. Results showed that no significant difference in hippocampus lipoperoxidation was observed when *Spirulinaplatensis*was added to hypercholesterolemic diet. The lipid profile showed that the supplementation of the microalga significantly reduced serum lipids in a protective way from hypercholesterolemia (Bertolin et al. 2009). In human trial possible health benefits of *Spirulina* was studied in patients with hyperlipidemic nephrotic syndrome. *Spirulina* is rich in essential fatty acids like gamm linolenic acid (GLA), it can able to reduce cholesterol accumulation in the body. In this study 23 patients with nephrotic syndrome supplemented either medication (group I) or medication plus (group II) received 1 g/day of *Spirulina* for about 2 months Height, weight, and serum levels of fasting blood sugar, triglycerides, total cholesterol. Total cholesterol, LDL cholesterol and triglycerides levels were significantly reduced to 116.33 mg/dl, 94.14 mg/dl, and 67.72 mg/dl respectively when compared to control (Samuels et al. 2002). Ponce et al. (2010) reported that dietary supplementation of *Spirulina maxima* reduced the hyperlipidemia and oxidative-stress. The results showed that microalga fed rats exhibited improved antioxidant status in liver and kidney in lead acetate induced toxicity.

Eicosapentaenoic acid rich *P. tricornutum* was used to treat metabolic syndrome and obesity. Wistar rats (n-6) were fed with high-fat diet (HF) with 10% fructose in drinking water, supplemented or not with 12% of *P. tricornutum* (HF-Phaeo). Supplementation results in LC-PUFA enrichment in liver, plasma and erythrocytes. Body weight, fat mass, inflammatory markers and insulinemia decreased in HF-Phaeo rats when compared to HF group. Plasma total cholesterol, triacylglycerols and leptine reduced in HF-Phaeo rats, while HDL-cholesterol increased (Mayer et al. 2019). The possible health benefits of PUFA rich microalga *C. onubensis* were studied in rats. Results revealed that no histological changes in liver and kidney tissue. Furthermore, rats fed with microalgae-enriched diet showed statistically significant decrease in blood cholesterol and triglyceride levels decreased by about 50% in rats (Navarro et al. 2016).Indeed, there is proof that dietary supplementation with algal derived products can improve cardiovascular protection. Still, there is lot of challenges in using these ingredients in foods and possible health benefits of the new products should be tested.

12.3.2. Use of Microalgal Derived Omega-3 PUFAs in Neurological Applications

Microalgae represent a potential source of renewable nutrition and there is growing interest in algae-based dietary supplements in the form of whole biomass. Microalgae are a source of long-chain PUFAs especially of the omega-6 family, such as linoleic acid (GLA, 18:3), arachidonic acid (AA, 20:4), omega-3 family as EPA and DHA. Several studies reported that the use of fish oil derived PUFAs in rodents as dietary supplements reduced maternal dyslipidemia and inflammation (Breetha et al. 2016& 2018). Only few studies reported the use of whole biomass/algal oil in animal models as a source of omega-3 fatty acids (Kavitha et al. 2014 & Vidyasankar et al. 2015). The use of microalgal derived oils/ whole biomass as an omega-3 supplements in animal models were reported very less.

Increasing evidence suggested that eukaryotic microalga *Chlorella sorokiniana*, can be used as PUFA supplement in terms of omega-3 and omega-6 PUFAs. The oral administration of a lipid extract of *Chlorella*, improved short-term memory compared to controls. Neurochemical investigations showed elevated levels of noradrenaline (NA) and serotonin (5-HT) in hippocampus alone (Morgese et al. 2016). The therapeutic potential of *Schizochytrium* microalga oil was studied to alleviate possible negative effects of high-fat-high-energy diets in Wistar rats. Rats were supplemented with beef tallow + evaporated sweetened milk (diet T) for 7 weeks results in mild obesity and low-grade inflammation. Subsequently, the animals were divided into four groups ($n = 10$) and fed either the T-diet (control) or 6% of safflower oil diet (S), 6% of fish oil diet (F) and 6% of *Schizochytrium* microalga oil diet (A), respectively, for 7 weeks of experimental study. Results showed that the A-diet decreased ($p < 0.05$) liver weight to 86% and glycaemia to 85% of control, respectively. In comparison with control, higher ($p < 0.05$) deposition of eicosapentaenoic acid (EPA) + docosahexaenoic acid (DHA) in the epididymal adipose tissue (EAT) of the microalgal oil treated rats. Furthermore, higher EPA and DHA levels were observed in the liver and correlated with higher expression PPAR gamma levels, anti-inflammatory cytokine TGFb1 and lower expression of adiponectin rectors (Komprda et al. 2015). The use of marine microalga *Diacronema vlkianum* whole biomass was studied in single dose CD1 mice and followed by Wistar rats. Microalgae-supplementation altered EPA and docosapentaenoic acid levels in erythrocytes and liver. Furthermore, it improves EPA content of heart and adipose tissues of rats. Oral supplementation significantly improved serum lipid profile, omega-3 index and erythrocyte plasticity in microalga treated rats. It indicates the use of *D. vlkianum*-biomass as a "heart-healthy" food supplement for a safe and renewable source of omega-3 fatty acids (de Mello-Sampayo et al. 2017). In a study the bioequivalence and therapeutic potential of microalgal diet and fish oil diets were studied in rats. The mRNA expression levels of serotonin receptors, HT1A and HT2A along with CREB, the transmission factor for learning and memory, were higher in the hippocampus of rats fed n-3 LCPUFA diets comparative to other diets showed the therapeutic potential of microalgal diet (Lopes et al. 2017). *Crypthecodinium cohnii* is rich in DHA among all the microalgae and the fermentation process produce elevated levels of DHA. The commercial infant supplements are enriched with *C. cohnii* DHA for cognitive development. Docosahexaenoic acid (DHA) derived from microalgae (Commercially available-DHASCO) and highly purified ethyl ester from other commercial source (MATK-90) with equivalent doses were supplemented to Sprague–Dawley rats. Effects of MATK-90 were compared with DHASCO and control (corn oil). Doses of MATK-90 (1.3, 2.5 and 5.0 g/kg/day) and DHASCO (5.0 g/kg/day = 2 g of DHA) were administered by oral gavage at a volume of 10 mL/kg body weight. Results showed that no treatment-related adverse effects in any of the parameters measured at doses of MATK-90 or of DHASCO. But, in the mesenteric lymph node, distinct macrophage infiltration was observed in 3 of 19 animals in the MATK-90 treatment group and was observed. The no-observable-adverse-effect level (NOAEL) for MATK-90 under the conditions of this study was 2.5 g/kg/day (Hadley et al. 2010). In another study a concentrated ethyl ester of DHA from microalgae (MATK-90, 900 mg/g) was used to treat hypertriglyceridemia rats. Effects of MATK-90 were compared with DHASCO, algal-DHA, and control (corn oil). Doses of MATK-90 (0.6, 1.3, 2.5, 5.0 g kg/Day), algal-DHA (2 g DHA kg/Day), and DHASCO (5.0 g kg/Day) were administered through oral gavage for 28 days. A significant reduction in TAG and cholesterol levels was observed in MATK- 90 groups with control rats. The high-dose group of MATK-90 and DHASCO produced reductions in TAG levels (Ryan et al. 2009). The safety of DHA-rich oil from *Schizochytrium* sp. and ARA-rich oil from *Mortierella alpina* was evaluated in Wistar rats with dietary exposure to 1000, 2500, and 5000 mg per kg bw of the DHA-rich and ARA-rich oils and two control diets: water and corn oil (vehicle control). The results showed that no treatment related effects of both DHA-rich and ARA-rich oils on physical parameters, behavior, hematology, clinical biochemistry and necropsy. No observable adverse effect level (NOAEL) for both the DHA-rich and ARA-rich oils of highest dose tested. The results confirm that use of oils from *Schizochytrium* sp. and ARA-rich oil from *Mortierella alpina* as an omega-3 supplement (Lewis et al. 2014). Diabetes mellitus is considered a chronic non communicable disease in which inflammation plays a main role in the progression of the disease and it is known that *n*-3 fatty acids have anti-inflammatory properties. In this study *n*-3 fatty acids extracted from microalgae supplementation on the inflammatory markers from two different strains of mice was studied.

Mice of two strains, db/db and CD1, were fed with *n*-3 fatty acids extracted from microalgae in lyophilized form and dietary form for 8 weeks. Results showed that microalgae supplementation decreased the percentage of TCD4+ cells producing IFN-γ and TNF-α and increased the IL-17A and IL-12 in both strains. Furthermore, microalgae supplementation decreased percentage of TCD4+ cells producing IL-4 and increased the TGF-β showed its therapeutic potential of reducing the inflammatory status (Guiterrez et al. 2018).

A monogalactosylmonoacylglycerols derived from freshwater microalgae *Chlorella sorokiniana* and *Tetraselmis chui* haspotentnitric oxide inhibitory activity in macrophage cells (Banskotta et al. 2013 & 2012). Two anti-inflammatory omega-3 polyunsaturated fatty acids (PUFAs) of stearidonic acid (SA) and eicosapentaenoic acid (EPA) and one pro-inflammatory omega- 6PUFA of arachidonic acid (AA) were isolated from the edible brown seaweed *Undaria pinnatifida* was active against mouse ear inflammation induced by phorbol myristate acetate (Khan et al. 2007). The anti-inflammatory activity of two DHA-rich algal oils, were examined the anti-inflammatory effects in rats. DHA/DPAn-6-rich DHA-STM (DHA-S) algal oil was more effective at reducing edema in rats than DHA-rich DHA-TTM (DHA-T), suggesting that DPAn-6 has anti-inflammatory properties. Further *in vivo* analyses demonstrated that feeding DPA n-6 alone and combination with DHA enhanced the anti-inflammatory activity. DPA n-6 has anti-inflammatory activity and enhances the effect of DHA *in vitro* and *in vivo.* It can be used a potential for anti-inflammatory applications (Nauroth et al. 2010).

12.3.3. Anticancer Agents Derived from Algae and Its Use in Cancer Therapeutics

Cancer is one of the most deadly diseases in humans and has a variety of etiologies, including lifestyle, culture and environment. Since, there is lot of synthetic anticancer drugs on the market; there is still a need for powerful naturally occurring drugs from algae with less toxicity and side effects. Anticancer drugs from algal origin have been widely reported in last two decades. They are mainly fucoidans, crude extracts, terpenoids, and secondary metabolites.

A sulfated polysaccharide fucoidan isolated from brown alga *Fucus evanescens* was studied in C57Bl/6 mice for antitumor and antimetastatic activities in Lewis lung adenocarcinoma. Single and repeated administration of fucoidan in a dose of 10 mg/kg alone or in combination with CP produced moderate antitumor and antimetastatic effects in tumor tissue (Alekseyenko et al. 2007). A fucose rich sulphated polysaccharide from *Ecklonia cava* has *in vitro* antiproliferative effects on murine colon carcinoma (CT-26), human leukaemic monocyte lymphoma (U-937), human promyelocytic leukaemia (HL-60), and mouse melanoma (B-16) cell lines (Athukorala et al. 2009). Fucoidans isolated from *U. pinnatifida* showed antitumour activity of in PC-3, HeLa, A549 and HepG2 cancer cells and it biological activity in tumour cells was very similar to commercial fucoidans (Synytsya et al. 2010). Similarly in other study fucoidans isolated from *S. japonica* and *U. pinnatifida* showed antiproliferative effects and anti-colony forming capability in breast cancer and melanoma cell lines in a dose-dependent manner (Stepanenkoet al. 2011). Tripathi et al. (2012) reported that extracts of *Nostoc muscorum* and *Oscillatoria* sp. possess anticancer activity against Ehrlich's Ascites Carcinoma Cell (EACC) and human hepatocellular cancer cell line

(HepG2). Anticancer compounds such as anatoxin-A, microcystins isolated from cyanobacteria have the potential to treat bone cancer (Mahdi and Fariba, 2012). Lin et al. (2012) reported that enzyme extraction of *Caulerpa microphysa* extracts has *in vitro* anticancer activity against human promyelocytic leukemia cells and *in vivo* BALB/c mice with transplanted leukemia were effectively controlled. The extract possesses DNA damaging activity in cancer cell lines as compared to control. The crude extract from *Ulva fasciata* shows *in vivo* anticancer activity against implanted lymphoma cells in Swiss albino mice. The extracts considerably reduce the number of cancer cells, and enhanced the lifespan of the treated mice (Abirami & Kowsalya 2012).

Red algae are rich in secondary metabolites and halogenated derivatives have wide array of biological activity. Campos et al. (2012) studied the antitumour activity of sesquiterpene elatol isolated from red alga *Laurencia microcladia* in cancer cell lines. Results revealed that elatol significantly inhibiting the cell cycle and arrest the cells in G1 and sub G1 phases and results in apoptosis. In animal model studies, elatol significantly reduce the tumour growth in C57Bl6 mice. In another study, water crude extract of red alga *Gracilaria corticata* possess anticancer and antitumour activity on Jurkat and molt-4 human lymphoblastic leukaemic cell lines (Zandi et al. 2010). Aqueous extract of Canadian marine microalgal preparation possess antimetastatic activity in cancer cell lines through preferential killing of suspended cancer cells. *In vivo* antimetastatic activity was significantly reduced in NOD-SCID mice carrying PC3 prostate cancer cells (Somasekharan et al. 2016). Akao et al. (2009) reported that oral supplementation of *Spirulina* extract activates tumouricidal natural killer cells and exerts a synergistic antitumour activity and used as immunotherapy in melanoma. A novel fatty alcohol ester isolated from cultured marine diatom, *Phaeodactylum tricornutum* has apoptotic anticancer activity in leukemia, lung carcinoma and mouse melanoma. The compound has strong suppression on cancer cells and DNA damage, nuclear condensation was observed in dose dependent manner (Samarakoon et al. 2014). Li et al. (2010) reported that the pro-apoptotic mechanism of C-phycocyanin (C-PC) was studied in a murine tumor model and cultured MCF-7 cells. *In vivo* results showed C-PC-treated mice, the expression of pro-apoptotic Fas protein were increased and anti-apoptotic proteins expressions (NF-kB and P53) were significantly decreased. *In vitro*, C-PC significantly inhibited MCF-7 cell proliferation through ultra-structural changes and chromatin condensation and leading to the apoptosis of MCF-7 cells of the anti-tumor activity. In a study anticancer activity of *Chlorella sorokiniana* extracts was studied in human non-small cell lung cancer cells and *in vivo*. Our results showed that CS extracts resulted in a concentration-dependent reduction in cell viability of cell lines. *In vivo* results demonstrated that tumor growth of subcutaneous xenograft was markedly inhibited in CS treatment (Lin et al. 2017). Cytotoxic ethanol extract of microalga, *chaetoceros calcitrans*, and its mechanisms in inducing apoptosis in human breast cancer cell line was studied. It was found that cell growth was inhibited through the induction of apoptosis without cell cycle arrest in cancer lines (Ebrahimi Nigjeh et al. 2013). The chemo preventive effects of high fat microalgal oil diet on azoxymethane (AOM)-induced colonic aberrant crypt foci (ACF) were studied in male Fischer 344 rats for the experimental period of 8 weeks of dietary intervention. High fat corn oil diet, fish oil was used a positive control and alternative to study the colorectal cancer. Experimental results showed that the microalgal oil and fish oil, both these oils showed 50% reduction of AOM-induced ACF and lipid peroxidation when compared to control rats. Furthermore, n-3 PUFA-rich oil from microalgae is as good as fish oil regarding chemoprevention (Van beelan et al. 2009).

12.3.4. Anti-Inflammatory Compounds Derived from Algae and Its Use in Therapeutic Applications

Inflammation underlies a variety of human illnesses, including cancer, Alzheimer's disease, arthritis, allergies, asthma, atherosclerosis, and myocardial ischemia, there is a huge demand for new and powerful anti-inflammatory drugs. Inflammation is the biological response of tissues to harmful stimulants such as tissue injury, pathogens or irritants. Inflammation plays a major role in defence mechanism in higher animals; unregulated inflammatory conditions are detrimental to large number of diseases. The process of inflammation is governed by a group of chemicals called mediators (Ward, 1994). The chemical mediators are vasoactive amines, cytokines, complement, eicosanoids, platelet activating factor (PAF), nitric oxide (NO) and neuropeptides. Eicosanoids from arachidonic acid (AA) formed by distinct enzymatic cascades and produce prostanoids, and thromboxanes (TXs), derive from the conversion of AA by COX (cyclooxygenases), whereas leukotrienes (LTs) are formed via the 5-lipooxygenase (5-LOX) pathway involved in pro-inflammatory conditions. Inhibition of these pro-inflammatory markers in the treatment of diseases is important to reduce the disease pathology. Algae are promising organisms with novel bioactive substances having wide array of anti-inflammatory activities. Several studies reported the role of anti-inflammatory activity of algae and its biological activities in animal models and cell lines. The recent research accomplished towards anti-inflammatory activity was discussed.

A study reported the anti-inflammatory effect of ethanolic extract of *Myagropsis myagroides* in lipopolysaccharide-stimulated BV-2 microglial cells was studied. Results revealed that prostaglandin E2 (PGE2) and pro-inflammatory cytokine levels were significantly reduced showed its anti-inflammatory drug to treat against neuroinflammatory diseases (Kim et al. 2013). Similarly, in other study anti-inflammatory effects of sargachromenol-rich ethanolic extract of *Myagropsis myagroides* on lipopolysaccharide-stimulated BV-2 cells was observed. Ethanolic extract significantly inhibited the mRNA expression of pro-inflammatory cytokines iNOS, COX, NO and PGE2 production and showed its anti- inflammatory potential (Kim et al. 2014). Similarly in a study, ethanol extract of *Nannochloropsis oceanica* on lipopolysaccharide-induced neuroinflammation was studied. In-vivo experiments resulted in suppression of NFκβ and Aβ1 through down regulation of APP and BACE1 expression and ameliorated anti-amyloidogenic mechanisms (Choi et al. 2017). Phlorotannin dieckol isolated from marine brown alga possess anti-inflammatory activity in microglial cells. The results showed that dieckol significantly reduced LPS-induced production of nitric oxide (NO), (iNOS), cyclooxygenase-2 (COX-2) and prostaglandin E2 (PGE2) in a dose-dependent manner and reduced proinflammatory cytokines, such as interleukin (IL)-1β and tumor necrosis factor (TNF-α) without causing cytotoxicity (Jung et al. 2009). In a study neuroprotective role of fucoidan against lipopolysaccharide induced neuronal damage was studied in experimental rats. Results showed that fucoidan significantly reduced the loss of dopaminergic neurons in the substantia nigra pars compacta of LPS induced rats. Furthermore, *in vitro* results displayed the significant reduction in TNF-α and ROS in LPS-induced primary microglia (cui et al. 2012).

Lee et al. (2018) reported that phlorotannins from *Ecklonia cava* has anti-neuroinflammatory property of on a$\beta_{25\text{-}35}$ induced damage in PC12 cells in Alzheimer disease (AD) model. Dieckol showed the intracellular oxidative stress and mitochondrial dysfunction was suppressed. Further, of pro-inflammatory enzymesTNF-α, IL-1β, iNOS and COX-2 and PGE2 production was significantly reduced and the results of this study proved the therapeutic potential of dieckol against AD. Anti-inflammatory effect of fucoxanthin isolated from brown algae was studied in lipopolysaccharide-stimulated RAW 264.7 macrophages. Results showed that fucoxanthin significantly inhibited the nitric oxide, cyclooxygenase 2 (COX-2) and prostaglandin E2 production. The mRNA expression levels of pro-inflammatory cytokines were altered by dose-dependent manner of fucoxanthin and it offers anti-inflammatory against various inflammatory diseases (Lee et al. 2013). Wong et al. (2018) studied the protective effect of fucosterol amyloid-induced microglial cells. Fucosterol displayed dose-dependent manner in reducing pro-inflammatory cytokines (IL-6, IL-1β), tumor necrosis factor-α (TNF-α), NO, and PGE2 in LPS- or Aβ-induced microglial cells and reduced the Aβ-mediated neuroinflammation. Kim et al. (2013) reported that floridoside derived from marine red algae possess anti-inflammatory activity in activated microglia. The results show that floridoside has the ability to suppress pro-inflammatory responses in microglia by markedly inhibiting the production of nitric oxide (NO) and reactive oxygen species (ROS). Furthermore, floridoside down-regulated the iNOS and COX-2 mRNA and protein levels indicated the use of floridoside for neuroinflammation. In a study, possible health benefits of astaxanthin inhibit glutamate release in cerebral cortex of rats was studied. Astaxanthin significantly blocks the Ca^{2+} entry in nerve terminals through mitogen-activated protein kinase signaling pathway (Lin et al. 2010). Similarly, in other study the neuroprotective role of astaxanthin was studied in experimental subarachnoid hemorrhage model. Results indicated that astaxanthin with high dose significantly downregulated the nuclear factor kappa β activity and pro-inflammatory cytokines and suppress cerebral inflammation (Zhang et al. 2014). Presently, the drugs available for treating neurological disorders are limited and side effects. Algae are considered as a potential natural source with excellent neuroprotective role and used in various and used in the treatment of various neurodegenerative disorders in animal studies.

CONCLUSION

In general, algal species posses many nutraceuticals metabolites which are useful for one or in another way to human. They have direct roles in intracellular processes i.e., reduces cardiovascular risk factors such as triglyceriemia, and improve cerebral blood flow, attenuate inflammation and might directly limit the different metabolic related disease pathology. Medical evidence suggests that the low consumption micro algal food leads to several metabolic disorders such as increased risk of cardiovascular disorders, neurocognitive disorders, cancer, stroke, diabetes and many other disorders. To overcome these therapeutic disorders, we need to search for alternate sources and one such source is the algal species. The increased demand for these nutraceuticals in recent years showed the essential of increasing the production of these compounds from available biological sources. Currently, the principal source of these compounds is from other plant based or microbial based which might not be sufficient for the enormous population growth and hence, there is a need to produce many such nutraceuticals compounds that will be reduce the demand in medical and health sector.

REFERENCES

Abirami, R. G. & Kowsalya, S. (2012). Anticancer activity of methanolic and aqueous extract of *Ulva fasciata* in albino mice. *Int J Pharma Pharmaceut Sci.*, *4*, 681-84.

Akao, Y., Ebihara, T., Masuda, H., Saeki, Y., Akazawa, T., Hazeki, K., Hazeki, O., Matsumoto, M. & Seya, T. (2009). Enhancement of antitumor natural killer cell activation by orally administered *Spirulina* extract in mice. *Cancer science.*, *100*(8), 1494-501.

Alekseyenko, T. V., Zhanayeva, S. Y., Venediktova, A. A., Zvyagintseva, T. N., Kuznetsova, T. A., Besednova, N. N. & Korolenko, T. A. (2007). Antitumor and antimetastatic activity of fucoidan, a sulfated polysaccharide isolated from the Okhotsk Sea Fucus evanescens brown alga. *Bulletin of experimental biology and medicine.*, *143*(6), 730-2.

Athukorala, Y., Ahn, G. N., Jee, Y. H., Kim, G. Y., Kim, S. H., Ha, J. H., Kang, J. S., Lee, K. W. & Jeon, Y. J. (2009). Antiproliferative activity of sulfated polysaccharide isolated from an enzymatic digest of Ecklonia cava on the U-937 cell line. *Journal of applied phycology.*, *21*(3), 307-14.

Baker, R. & Günther, C. (2004), The role of carotenoids in consumer choice and the likely benefits from their inclusion into products for human consumption. *Trends Food Sci. Technol.*, *15*(10), 484–488.

Banskota, A. H., Gallant, P., Stefanova, R., Melanson, R. & O'Leary, S. J. (2013). Monogalactosyldiacylglycerols, potent nitric oxide inhibitors from the marine microalga Tetraselmis chui. *Natural product research.*, 2013, Jun 1, 27(12), 1084-90.

Banskota, A. H., Stefanova, R., Gallant, P., Osborne, J. A., Melanson, R. & O'Leary, S. J. (2013). Nitric oxide inhibitory activity of monogalactosylmonoacylglycerols from a freshwater microalgae *Chlorella sorokiniana. Natural product research.*, *27*(11), 1028-31.

Becker, E. W., (2007), Micro-algae as a source of protein. *Biotechnol. Adv.*, *25*(2), 207–210.

Becker, W. (2004), 18 Microalgae in human and animal nutrition, in *Handbook of microalgal culture: biotechnology and applied phycology*, Wiley Online Library.

Bernstein, A. M., Ding, E. L., Willett, W. C. & Rimm, E. B. (2011). A meta-analysis shows that docosahexaenoic acid from algal oil reduces serum triglycerides and increases HDL-cholesterol and LDL-cholesterol in persons without coronary heart disease. *The Journal of Nutrition.*, *142*(1), 99-104.

Berquin, I. M., Edwards, I. J., Kridel, S. J. & Chen, Y. Q. (2011). Polyunsaturated fatty acid metabolism in prostate cancer. *Cancer Metastasis Rev.*, *30*(3–4), 295–309.

Bertolin, T. E., Pilatti, D., Giacomini, A. C., Bavaresco, C. S., Colla, L. M. & Costa, J. A. (2009). Effect of microalga *Spirulina platensis* (*Arthrospira platensis*) on hippocampus lipoperoxidation and lipid profile in rats with induced hypercholesterolemia. *Brazilian Archives of Biology and Technology.*, *52*(5), 1253-9.

Breetha, R. & Ramaprasad, T. R. (2018). Dietary n-3 but not n-6 fatty acids down-regulate maternal dyslipidemia induced inflammation: A three-generation study in rats. *Prostaglandins, Leukotrienes and Essential Fatty Acids.*, *1*, 135, 83-91.

Brown, M. R., Mular, M., Miller, I., Farmer, C. & Trenerry, C. (1999). The vitamin content of microalgae used in aquaculture. *J. Appl. Phycol.*, *11*(3), 247–255.

Campos, A., Souza, C. B., Lhullier, C., Falkenberg, M., Schenkel, E. P., Ribeiro-do-Valle, R. M. & Siqueira, J. M. (2012). Anti-tumour effects of elatol, a marine derivative compound obtained from red algae Laurencia microcladia. *Journal of Pharmacy and Pharmacology.*, *64*(8), 1146-54.

Capelli, B. & Cysewski, G. R. (2010). Potential health benefits of spirulina microalgae*. *Nutrafoods*, *9*(2), 19–26.

Chew, B. P. & Park, J. S. (2004). Carotenoid action on the immune response. *J. Nutr.*, *134*(1), 257S–261S.

Choi, J. Y., Hwang, C. J., Lee, H. P., Kim, H. S., Han, S. B. & Hong, J. T. (2017). Inhibitory effect of ethanol extract of Nannochloropsis oceanica on lipopolysaccharide-induced neuroinflammation, oxidative stress, amyloidogenesis and memory impairment. *Oncotarget.*, *11*, 8(28), 45517.

Cui, Y. Q., Jia, Y. J., Zhang, T., Zhang, Q. B. & Wang, X. M. (2012). Fucoidan protects against lipopolysaccharide-induced rat neuronal damage and inhibits the production of proinflammatory mediators in primary microglia. *CNS neuroscience & therapeutics.*, *18*(10), 827-33.

De Mello-Sampayo, C., Paterna, A., Polizzi, A., Duarte, D., Batista, I., Pinto, R., Gonçalves, P., Raymundo, A., Batista, A., Gouveia, L. & Silva-Lima, B. (2017). Evaluation of marine microalga *Diacronema vlkianum* biomass fatty acid assimilation in Wistar rats. *Molecules.*, *1*, 22(7), 1097.

Del Campo, J. A., García-González, M. & Guerrero, M. G. (2007). Outdoor cultivation of microalgae for carotenoid production: current state and perspectives. *Appl. Microbiol. Biotechnol.*, *74*(6), 1163–1174.

Dillon, J. C., Phuc, A. P. & Dubacq, J. P. (1995). Nutritional value of the alga Spirulina. *World Rev. Nutr. Diet.*, *77*, 32–46.

Dvir, I., Stark, A. H., Chayoth, R., Madar, Z. & Arad, S. M. (2009). Hypocholesterolemic effects of nutraceuticals produced from the red microalga Porphyridium sp. in rats. *Nutrients.*, *1*(2), 156-67.

Ebrahimi Nigjeh, S., Yusoff, F. M., Alitheen, M., Banu, N., Rasoli, M., Keong, Y. S. & Omar, A. R. (2013). Cytotoxic effect of ethanol extract of microalga, Chaetoceros calcitrans, and its mechanisms in inducing apoptosis in human breast cancer cell line. BioMed research international.

Ferlay, J., Soerjomataram, I., Dikshit, R., Eser, S., Mathers, C., Rebelo, M., et al., (2015). Cancer incidence and mortality worldwide: sources, methods and major patterns in GLOBOCAN 2012. *Int. J. cancer*, *136*(5), E359–E386.

Gammone, M. A., Riccioni, G. & D'Orazio, N. (2015). Marine carotenoids against oxidative stress: effects on human health. *Mar. Drugs*, *13*(10), 6226–6246.

Gantar, M. & Svirčev, Z. (2008). MICROALGAE AND CYANOBACTERIA: FOOD FOR THOUGHT(1). *J. Phycol.*, *44*(2), 260–8.

García, J. L., de Vicente, M. & Galán, B. (2017). Microalgae, old sustainable food and fashion nutraceuticals. *Microb. Biotechnol.*, *10*(5), 1017–1024.

Goiris, K., Muylaert, K., Fraeye, I., Foubert, I., De Brabanter, J. & De Cooman, L. (2012). Antioxidant potential of microalgae in relation to their phenolic and carotenoid content. *J. Appl. Phycol.*, *24*(6), 1477–1486.

Gouveia, L., Coutinho, C., Mendonça, E., Batista, A. P., Sousa, I., Bandarra, N. M. & Raymundo, A. (2008). Functional biscuits with PUFA-ω3 from Isochrysis galbana. *J. Sci. Food Agric.*, *88*(5), 891–896.

Gouveia, L., Veloso, V., Reis, A., Fernandes, H., Novais, J. & Empis, J. (1996). Chlorella vulgarisused to Colour Egg Yolk. *J. Sci. Food Agric.*, *70*(2), 167–172.

Gutierrez-Pliego, L. E., Martinez-Carrillo, B. E., Reséndiz-Albor, A. A., Arciniega-Martínez, I. M., Escoto-Herrera, J. A., Rosales-Gómez, C. A. & Valdés-Ramos, R. (2018). Effect of supplementation with n-3 fatty acids extracted from microalgae on inflammation biomarkers from two different strains of mice. *Journal of lipids.*, 2018.

Guzman, S., Gato, A., Lamela, M., Freire-Garabal, M. & Calleja, J. M. (2003). Anti-inflammatory and immunomodulatory activities of polysaccharide from Chlorella stigmatophora and Phaeodactylum tricornutum. *Phyther. Res.*, *17*(6), 665–670.

Hadley, K. B., Ryan, A. S., Nelson, E. B. & Salem, Jr. N. (2010). Preclinical safety evaluation in rats using a highly purified ethyl ester of algal-docosahexaenoic acid. *Food and Chemical Toxicology.*, *48*(10), 2778-84.

Haimeur, A., Mimouni, V., Ulmann, L., Martineau, A. S., Messaouri, H., Pineau-Vincent, F., Tremblin, G. & Meskini, N. (2016). Fish oil and microalga omega-3 as dietary supplements: a comparative study on cardiovascular risk factors in high-fat fed rats. *Lipids.*, *51*(9), 1037-49.

Haimeur, A., Ulmann, L., Mimouni, V., Guéno, F., Pineau-Vincent, F., Meskini, N. & Tremblin, G. (2012). The role of Odontella aurita, a marine diatom rich in EPA, as a dietary supplement in dyslipidemia, platelet function and oxidative stress in high-fat fed rats. *Lipids in health and disease.*, *11*(1), 147.

Haimeur, A., Ulmann, L., Mimouni, V., Guéno, F., Pineau-Vincent, F., Meskini, N. & Tremblin, G. (2012). The role of Odontella aurita, a marine diatom rich in EPA, as a dietary supplement in dyslipidemia, platelet function and oxidative stress in high-fat fed rats. *Lipids Health Dis.*, *11*(1), 147.

Hall, A. C., Fairclough, A. C., Mahadevan, K. & Paxman, J. R. (2012). Ascophyllum nodosum enriched bread reduces subsequent energy intake with no effect on post-prandial glucose and cholesterol in healthy, overweight males. A pilot study. *Appetite*, *58*(1), 379–86.

Holdt, S. L. & Kraan, S. (2011). Bioactive compounds in seaweed: Functional food applications and legislation. *J. Appl. Phycol.*, *23*(3), 543–597.

Hussain, M. I., Khan, S. A., Chaudhary, Z. I., Aslam, A., Ashraf, K. & Rai, M. F. (2004). Effect of organic and inorganic selenium with and without vitamine E on immune system of broilers. *Pak. Vet. J.*, *24*(1), 1–4.

Jung, W. K., Heo, S. J., Jeon, Y. J., Lee, C. M., Park, Y. M., Byun, H. G., Choi, Y. H., Park, S. G. & Choi, I. W. (2009). Inhibitory effects and molecular mechanism of dieckol isolated from marine brown alga on COX-2 and iNOS in microglial cells. *Journal of agricultural and food chemistry.*, *57*(10), 4439-46.

Kalra, E. K. (2003). Nutraceutical-definition and introduction. *AAPS PharmSci*, *5*(3), 27–28.

Kapoor, R. & Mehta, U. (1998). Supplementary effect of spirulina on hematological status of rats during pregnancy and lactation. *Plant Foods Hum. Nutr.*, *52*(4), 315–324.

Kau, A. L., Ahern, P. P., Griffin, N. W., Goodman, A. L. & Gordon, J. I. (2011). Human nutrition, the gut microbiome and the immune system. *Nature*, *474*(7351), 327.

Kavitha, M. D., Shree, M. S., Vidyashankar, S. & Sarada, R. (2016). Acute and subchronic safety assessment of *Porphyridium purpureum* biomass in the rat model. *Journal of applied phycology.*, *28*(2), 1071-83.

Khan, M. N., Cho, J. Y., Lee, M. C., Kang, J. Y., Park, N. G., Fujii, H. & Hong, Y. K. (2007). Isolation of two anti-inflammatory and one pro-inflammatory polyunsaturated fatty acids from the brown seaweed *Undaria pinnatifida. Journal of agricultural and food chemistry.*, *55*(17), 6984-8.

Khan, Z., Bhadouria, P. & Bisen, P. S. (2005). Nutritional and Therapeutic Potential of Spirulina.

Kim, J. K., Kottuparambil, S., Moh, S. H., Lee, T. K., Kim, Y. J., Rhee, J. S., et al., (2015). Potential applications of nuisance microalgae blooms. *J. Appl. Phycol.*, *27*(3), 1223–1234.

Kim, M., Li, Y. X., Dewapriya, P., Ryu, B. & Kim, S. K. (2013). Floridoside suppresses pro-inflammatory responses by blocking MAPK signaling in activated microglia. *BMB reports.*, *46*(8), 398.

Kim, S., Kim, J. I., Choi, J. W., Kim, M., Yoon, N. Y., Choi, C. G., Choi, J. S. & Kim, H. R. (2013). Anti-inflammatory effect of hexane fraction from Myagropsis myagroides ethanolic extract in lipopolysaccharide-stimulated BV-2 microglial cells. *Journal of Pharmacy and Pharmacology.*, *65*(6), 895-906.

Kim, S., Lee, M. S., Lee, B., Gwon, W. G., Joung, E. J., Yoon, N. Y. & Kim, H. R. (2014). Anti-inflammatory effects of sargachromenol-rich ethanolic extract of Myagropsis myagroides on lipopolysaccharide-stimulated BV-2 cells. *BMC complementary and alternative medicine.*, *14*(1), 231.

Komprda, T., Sládek, Z., Škultéty, O., Křížková, S., Rozíková, V., Němcová, B., Šustrová, T. & Valová, M. (2016). Effect of dietary Schizochytrium microalga oil on selected markers of low-grade inflammation in rats. *Journal of animal physiology and animal nutrition.*, *100*(6), 1169-78.

Kyle, D. J. (1996). Production and use of a single-cell oil which is highly enriched in docosahexaenoic acid. *Lipid Technol.*, *8*, 107–110.

Lauritano, C., Andersen, J. H., Hansen, E., Albrigtsen, M., Escalera, L., Esposito, F., et al., (2016). Bioactivity screening of microalgae for antioxidant, anti-inflammatory, anticancer, anti-diabetes, and antibacterial activities. *Front. Mar. Sci.*, *3*, 68.

Lee, J. B., Hayashi, K., Hirata, M., Kuroda, E., Suzuki, E., Kubo, Y. & Hayashi, T. (2006). Antiviral sulfated polysaccharide from Navicula directa, a diatom collected from deep-sea water in Toyama Bay. *Biol. Pharm. Bull.*, *29*(10), 2135–2139.

Lee, J. Y., Lee, M. S., Choi, H. J., Choi, J. W., Shin, T., Woo, H. C., Kim, J. I. & Kim, H. R. (2013). Hexane fraction from Laminaria japonica exerts anti-inflammatory effects on lipopolysaccharide-stimulated RAW 264.7 macrophages via inhibiting NF-kappaB pathway. *European journal of nutrition.*, *1*, 52(1), 409-21.

Lee, S., Youn, K., Kim, D., Ahn, M. R., Yoon, E., Kim, O. Y. & Jun, M. (2019). Anti-Neuroinflammatory Property of Phlorotannins from Ecklonia cava on Aβ25-35-Induced Damage in PC12 Cells. *Marine drugs.*, *17*(1), 7.

Lemahieu, C., Bruneel, C., Dejonghe, C., Buyse, J. & Foubert, I. (2016). The cell wall of autotrophic microalgae influences the enrichment of long chain omega-3 fatty acids in the egg. *Algal Res.*, *16*, 209–215.

Lewis, K. D., Huang, W., Zheng, X., Jiang, Y., Feldman, R. S. & Falk, M. C. (2016). Toxicological evaluation of arachidonic acid (ARA)-rich oil and docosahexaenoic acid (DHA)-rich oil. *Food and Chemical Toxicology.*, *96*, 133-44.

Li, B., Chu, X., Gao, M. & Li, W. (2009). Apoptotic mechanism of MCF-7 breast cells *in vivo* and *in vitro* induced by photodynamic therapy with C-phycocyanin. *Acta Biochim Biophys Sin.*, *42*(1), 80-9.

Lin, H. C., Chou, S. T., Chuang, M. Y., Liao, T. Y., Tsai, W. S. & Chiu, T. H. (2012). The effects of *Caulerpa microphysa* enzyme-digested extracts on ACE-inhibitory activity and *in vitro* anti-tumour properties. *Food chemistry.*, *134*(4), 2235-41.

Lin, P. Y., Tsai, C. T., Chuang, W. L., Chao, Y. H., Pan, I. H., Chen, Y. K., Lin, C. C. & Wang, B. Y. (2017). *Chlorella sorokiniana* induces mitochondrial-mediated apoptosis in human non-small cell lung cancer cells and inhibits xenograft tumor growth *in vivo. BMC complementary and alternative medicine.*, *17*(1), 88.

Lin, T. Y., Lu, C. W. & Wang, S. J. (2010). Astaxanthin inhibits glutamate release in rat cerebral cortex nerve terminals via suppression of voltage-dependent Ca2+ entry and mitogen-activated protein kinase signaling pathway. *Journal of agricultural and food chemistry.*, *58*(14), 8271-8.

Lin, Y. H., Shah, S. & Salem, Jr. N. (2011). Altered essential fatty acid metabolism and composition in rat liver, plasma, heart and brain after microalgal DHA addition to the diet. *The Journal of nutritional biochemistry.*, *22*(8), 758-65.

Lopes, P. A., Bandarra, N. M., Martins, S. V., Martinho, J., Alfaia, C. M., Madeira, M. S., Cardoso, C., Afonso, C., Paulo, M. C., Pinto, R. M. & Guil-Guerrero, J. L. (2017). Markers of neuroprotection of combined EPA and DHA provided by fish oil are higher than those of EPA (Nannochloropsis) and DHA (Schizochytrium) from microalgae oils in Wistar rats. *Nutrition & metabolism.*, *14*(1), 62.

Lordan, S., Ross, R. P. & Stanton, C. (2011). Marine bioactives as functional food ingredients: potential to reduce the incidence of chronic diseases. *Mar. Drugs*, *9*(6), 1056–1100.

MacArtain, P., Gill, C. I. R., Brooks, M., Campbell, R. & Rowland, I. R. (2007). Nutritional value of edible seaweeds. *Nutr. Rev.*, *65*(12 Pt 1), 535–43.

Mahdi, E. & Fariba, K. (2012). Cancer treatment with using cyanobacteria and suitable drug delivery system. *Annals of Biological research.*, *3*(1), 622-7.

Mayer, C., Côme, M., Ulmann, L., Chini Zittelli, G., Faraloni, C., Nazih, H., Ouguerram, K., Chénais, B. & Mimouni, V. (2019). Preventive Effects of the Marine Microalga Phaeodactylum tricornutum, Used as a Food Supplement, on Risk Factors Associated with Metabolic Syndrome in Wistar Rats. *Nutrients.*, *11*(5), 1069.

Morgese, M., Mhillaj, E., Francavilla, M., Bove, M., Morgano, L., Tucci, P., Trabace, L. & Schiavone, S. (2016). *Chlorella sorokiniana* extract improves short-term memory in rats. *Molecules.*, *21*(10), 1311.

Nauroth, J. M., Liu, Y. C., Van Elswyk, M., Bell, R., Hall, E. B., Chung, G. & Arterburn, L. M. (2010). Docosahexaenoic acid (DHA) and docosapentaenoic acid (DPAn-6) algal oils reduce inflammatory mediators in human peripheral mononuclear cells *in vitro* and paw edema *in vivo. Lipids.*, *45*(5), 375-84.

Navarro, F., Forján, E., Vázquez, M., Montero, Z., Bermejo, E., Castaño, M. Á., Toimil, A., Chagüaceda, E., García-Sevillano, M. Á., Sánchez, M. & Domínguez, M. J. (2016). Microalgae as a safe food source for animals: nutritional characteristics of the acidophilic microalga *Coccomyxa onubensis*. *Food & nutrition research.*, *60*(1), 30472.

Peng, J., Yuan, J. P., Wu, C. F. & Wang, J. H. (2011). Fucoxanthin, a marine carotenoid present in brown seaweeds and diatoms: metabolism and bioactivities relevant to human health. *Mar. Drugs*, *9*(10), 1806–1828.

Plaza, M., Cifuentes, A. & Ibáñez, E. (2008), In the search of new functional food ingredients from algae. *Trends Food Sci. Technol.*, *19*(1), 31–39.

Ponce-Canchihuamán, J. C., Pérez-Méndez, O., Hernández-Muñoz, R., Torres-Durán, P. V. & Juárez-Oropeza, M. A. (2010). Protective effects of Spirulina maxima on hyperlipidemia and oxidative-stress induced by lead acetate in the liver and kidney. *Lipids in health and disease.*, *9*(1), 35.

Pouvreau, J. B., Taran, F., Rosa, P., Pin, S., Fleurence, J. & Pondaven, P. (2008). Antioxidant and free radical scavenging properties of marennine, a blue-green polyphenolic pigment from the diatom Haslea ostrearia (Gaillon/Bory) Simonsen responsible for the natural greening of cultured oysters. *J. Agric. Food Chem.*, *56*(15), 6278–6286.

Pulz, O. & Gross, W. (2004), Valuable products from biotechnology of microalgae. *Appl. Microbiol. Biotechnol.*, *65*(6), 635–48.

Ramaiyan, B., Bettadahalli, S. & Talahalli, R. R. (2016). Dietary omega-3 but not omega-6 fatty acids down-regulate maternal dyslipidemia induced oxidative stress: a three generation study in rats. *Biochemical and biophysical research communications.*, *2*, 477(4), 887-94.

Raposo M. F. de J., De Morais, R. M. S. C., de Morais, B. & Miranda, A. M. (2013). Bioactivity and applications of sulphated polysaccharides from marine microalgae. *Mar. Drugs*, *11*(1), 233–252.

Richardson, J. W., Johnson, M. D., Zhang, X., Zemke, P., Chen, W. & Hu, Q. (2014). A financial assessment of two alternative cultivation systems and their contributions to algae biofuel economic viability. *Algal Res.*, *4*, 96–104.

Riediger, N. D., Othman, R. A., Suh, M. & Moghadasian, M. H. (2009). A Systemic Review of the Roles of n-3 Fatty Acids in Health and Disease. *J. Am. Diet. Assoc.*, *109*(4), 668–679.

Riss, J., Décordé, K., Sutra, T., Delage, M., Baccou, J. C., Jouy, N., Brune, J. P., Oréal, H., Cristol, J. P. & Rouanet, J. M. (2007). Phycobiliprotein C-phycocyanin from *Spirulina platensis* is powerfully responsible for reducing oxidative stress and NADPH oxidase expression induced by an atherogenic diet in hamsters. *J. of Agricultural and Food Chemistry.*, *16*, 55(19), 7962-7.

Running, J. A., Severson, D. K. & Schneider, K. J. (2002). Extracellular production of L-ascorbic acid by Chlorella protothecoides, Prototheca species, and mutants of P. moriformis during aerobic culturing at low pH. *J. Ind. Microbiol. Biotechnol.*, *29*(2), 93–8.

Rupérez, P. (2002). Mineral content of edible marine seaweeds. *Food Chem.*, *79*(1), 23–26.

Ryan, A. S., Bailey-Hall, E., Nelson, E. B. & Salem, N. (2009). The hypolipidemic effect of an ethyl ester of algal-docosahexaenoic acid in rats fed a high-fructose diet. *Lipids.*, *1*, 44(9), 817-26.

Ryan, L. & Symington, A. M. (2015). Algal-oil supplements are a viable alternative to fish-oil supplements in terms of docosahexaenoic acid (22: 6n-3; DHA). *Journal of functional foods.*, *1*, 19, 852-8.

Ryu, B. M., Himaya, S. W. A. & Kim, S. K. (2015). Applications of Microalgae-Derived Active Ingredients as Cosmeceuticals, in *Handbook of Marine Microalgae: Biotechnology Advances*, Elsevier Inc., pp. 309–316.

Sajilata, M. G., Singhal, R. S. & Kamat, M. Y. (2008). Fractionation of lipids and purification of γ-linolenic acid (GLA) from Spirulinaplatensis. *Food Chem.*, *109*(3), 580–586.

Samarakoon, K. W., Ko, J. Y., Lee, J. H., Kwon, O. N., Kim, S. W. & Jeon, Y. J. (2014). Apoptotic anticancer activity of a novel fatty alcohol ester isolated from cultured marine diatom, *Phaeodactylum tricornutum. Journal of functional foods.*, *1*, 6, 231-40.

Samarakoon, K. W., Ko, J. Y., Shah, M. D., Mahfuzur, R., Lee, J. H., Kang, M. C., Kwon, O. N., Lee, J. B. & Jeon, Y. J. (2013). *In vitro* studies of anti-inflammatory and anticancer activities of organic solvent extracts from cultured marine microalgae. *Algae.*, *28*(1), 111-9.

Sanders, T. A., Gleason, K., Griffin, B. & Miller, G. J. (2006). Influence of an algal triacylglycerol containing docosahexaenoic acid (22: 6n-3) and docosapentaenoic acid (22: 5n-6) on cardiovascular risk factors in healthy men and women. *British journal of nutrition.*, *95*(3), 525-31.

Sansone, C., Braca, A., Ercolesi, E., Romano, G., Palumbo, A., Casotti, R., Francone, M. & Ianora, A. (2014). Diatom-derived polyunsaturated aldehydes activate cell death in human cancer cell lines but not normal cells. *PLoS One.*, *3*, 9(7), 1012-20.

Santiago-Santos, M. C., Ponce-Noyola, T., Olvera-Ramírez, R., Ortega-López, J. & Cañizares-Villanueva, R. O. (2004). Extraction and purification of phycocyanin from Calothrix sp. *Process Biochem.*, *39*(12), 2047–2052.

Segner, H., Arend, P., Von Poeppinghausen, K. & Schmidt, H. (1989). The effect of feeding astaxanthin to Oreochromis niloticus and Colisa labiosa on the histology of the liver. *Aquaculture*, *79*(1–4), 381–390.

Shaish, A., Harari, A., Hananshvili, L., Cohen, H., Bitzur, R., Luvish, T., Ulman, E., Golan, M., Ben-Amotz, A., Gavish, D. & Rotstein, Z. (2006). 9-cis β-carotene-rich powder of the alga Dunaliella bardawil increases plasma HDL-cholesterol in fibrate-treated patients. Atherosclerosis., *1*, 189(1), 215-21.

Sheikhzadeh, N., Tayefi-Nasrabadi, H., Oushani, A. K. & Enferadi, M. H. N., (2012). Effects of Haematococcus pluvialis supplementation on antioxidant system and metabolism in rainbow trout (Oncorhynchus mykiss). *Fish Physiol. Biochem.*, *38*(2), 413–9.

Siti, H. N., Kamisah, Y. & Kamsiah, J. (2015). The role of oxidative stress, antioxidants and vascular inflammation in cardiovascular disease (a review). *Vascular pharmacology.*, *1*, 71, 40-56.

Somasekharan, S. P., El-Naggar, A., Sorensen, P. H., Wang, Y. & Cheng, H. (2016). An aqueous extract of marine microalgae exhibits antimetastatic activity through preferential killing of suspended cancer cells and anticolony forming activity. Evidence-Based Complementary and Alternative Medicine.

Sousa, I., Gouveia, L., Batista, A. P., Raymundo, A. & Bandarra, N. M. (2008). Microalgae in novel food products. *Food Chem. Res. Dev.*, 75–112.

Spolaore, P., Joannis-Cassan, C., Duran, E. & Isambert, A. (2006). Commercial applications of microalgae. *J. Biosci. Bioeng.*, *101*(2), 87–96.

Stepanenko, O. V., Stepanenko, O. V., Shcherbakova, D. M., Kuznetsova, I. M., Turoverov, K. K. & Verkhusha, V. V. (2011). Modern fluorescent proteins: from chromophore formation to novel intracellular applications. *Biotechniques.*, *51*(5), 313-27.

Stewart, B. W., Bray, F., Forman, D., Ohgaki, H., Straif, K., Ullrich, A. & Wild, C. P. (2015). Cancer prevention as part of precision medicine: 'plenty to be done.' *Carcinogenesis*, *37*(1), 2–9.

Synytsya, A., Kim, W. J., Kim, S. M., Pohl, R., Synytsya, A., Kvasnička, F., Čopíková, J. & Park, Y. I. (2010). Structure and antitumour activity of fucoidan isolated from sporophyll of Korean brown seaweed Undaria pinnatifida. *Carbohydrate Polymers.*, *81*(1), 41-8.

Tang, X., Li, Z. J., Xu, J., Xue, Y., Li, J. Z., Wang, J. F., Yanagita, T., Xue, C. H. & Wang, Y. M. (2012). Short term effects of different omega-3 fatty acid formulation on lipid metabolism in mice fed high or low fat diet. *Lipids in health and disease.*, *11*(1), 70.

Tripathi, A., Fang, W., Leong, D. T. & Tan, L. T. (2012). Biochemical studies of the lagunamides, potent cytotoxic cyclic depsipeptides from the marine cyanobacterium Lyngbya majuscula. *Marine drugs.*, *10*(5), 1126-37.

Udayan, A. & Arumugam, M. (2017). Selective enrichment of Eicosapentaenoic acid (20:5n-3) in N. oceanica CASA CC201 by natural auxin supplementation. *Bioresour. Technol.*, *242*, 329–333.

Udayan, A., Arumugam, M. & Pandey, A. (2017). Nutraceuticals From Algae and Cyanobacteria, in *Algal Green Chemistry: Recent Progress in Biotechnology*, Elsevier, pp. 65–89.

Udayan, A., Kathiresan, S. & Arumugam, M. (2018). Kinetin and Gibberellic acid (GA3) act synergistically to produce high value polyunsaturated fatty acids in Nannochloropsis oceanica CASA CC201. *Algal Res.*, *32*, 182–192.

Van Beelen, V. A., Spenkelink, B., Mooibroek, H., Sijtsma, L., Bosch, D., Rietjens, I. M. & Alink, G. M. (2009). An n-3 PUFA-rich microalgal oil diet protects to a similar extent as a fish oil-rich diet against AOM-induced colonic aberrant crypt foci in F344 rats. *Food and Chemical Toxicology.*, *1*, 47(2), 316-20.

Vidyashankar, S., VenuGopal, K. S., Chauhan, V. S., Muthukumar, S. P. & Sarada, R. (2015). Characterisation of defatted Scenedesmus dimorphus algal biomass as animal feed. *Journal of applied phycology.*, *1*, 27(5), 1871-9.

Wong, C. H., Gan, S. Y., Tan, S. C., Gany, S. A., Ying, T., Gray, A. I., Igoli, J., Chan, E. W. & Phang, S. M. (2018). Fucosterol inhibits the cholinesterase activities and reduces the release of pro-inflammatory mediators in lipopolysaccharide and amyloid-induced microglial cells. *Journal of applied phycology.*, *1*, 30(6), 3261-70.

Xia, S., Gao, B., Li, A., Xiong, J., Ao, Z. & Zhang, C. (2014). Preliminary characterization, antioxidant properties and production of chrysolaminarin from marine diatom Odontella aurita. *Mar. Drugs*, *12*(9), 4883–4897.

Yamaguchi, K. (1996). Recent advances in microalgal bioscience in Japan, with special reference to utilization of biomass and metabolites: A review. *J. Appl. Phycol.*, *8*(6), 487–502.

Yoshida, H., Yanai, H., Ito, K., Tomono, Y., Koikeda, T., Tsukahara, H. & Tada, N. (2010). Administration of natural astaxanthin increases serum HDL-cholesterol and adiponectin in subjects with mild hyperlipidemia. *Atherosclerosis.*, *1*, 209(2), 520-3.

Zalesin, K. C., Franklin, B. A., Miller, W. M., Peterson, E. D. & McCullough, P. A. (2008). Impact of obesity on cardiovascular disease. *Endocrinology and metabolism clinics of North America.*, *1*, 37(3), 663-84.

Zandi, K., Tajbakhsh, S., Nabipour, I., Rastian, Z., Yousefi, F., Sharafian, S. & Sartavi, K. (2010). *In vitro* antitumor activity of Gracilaria corticata (a red alga) against Jurkat and molt-4 human cancer cell lines. *African Journal of Biotechnology.*, *9* (40), 6787-90.

Zhang, X. S., Zhang, X., Wu, Q., Li, W., Wang, C. X., Xie, G. B., Zhou, X. M., Shi, J. X. & Zhou, M. L. (2014) Nov Astaxanthin offers neuroprotection and reduces neuroinflammation in experimental subarachnoid hemorrhage. *Journal of Surgical Research.*, *1*, 192(1), 206-13.

In: Applied Algal Biotechnology
Editors: M. Arumugam, S. Kathiresan et al.
ISBN: 978-1-53617-524-0

Chapter 13

MICROALGAE AND CYANOBACTERIA: ROLE AND APPLICATIONS IN AGRICULTURE

Geetha Thanuja Kalyanasundaram[1], Anupriya Ramasamy[1], Suchitra Rakesh[2] and Karthikeyan Subburamu[1,3,*]

[1]Department of Agricultural Microbiology, Tamil Nadu Agricultural University, Coimbatore, Tamil Nadu, India
[2]Department of Microbiology, School of Life Sciences, Central University of Tamil Nadu, Thiruvarur, Tamil Nadu, India
[3]Department of Renewable Energy Engineering, Agricultural Engineering College and Research Institute, Tamil Nadu Agricultural University, Coimbatore, Tamil Nadu, India

ABSTRACT

Microalgae and Cyanobacteria are ubiquitous in nature, as a primary microbial photosynthetic agent. With the most productive agricultural areas already under cultivation, crop cultivation becomes more critical so as to meet the expected demands of food without harming the environment. Agricultural bioinoculant microorganisms are natural soil nutritionists and stimulants that, improve the soil chemical and biological properties, stimulate plant growth and restore soil fertility. The mainstream algae biofertilizer research has been intensive on seaweeds and on plant growth promoting microorganisms like cyanobacteria predominantly in rice fields, due to their abundance in soil habitats. Equally a number of microalgae species have also been investigated for their ability to produce bioactive compounds. Several studies have established an association between greater nutrient uptake, higher biomass accumulation and greater crop yields to the incorporation of microalgae biofertilizers. The beneficial effects of microalgae in agriculture involves production of organic acids for phosphorus uptake, biological nitrogen fixation, increases soil organic matter, concentrate heavy metal ions, production and release of bioactive compounds like plant growth regulators, vitamins, amino acids, antibacterial or antifungal substance etc, that exert phytopathogenic biocontrol and improve soil structure by exopolysaccharides production. Cyanobacteria

[*] Corresponding Author's Email: skarthy@tnau.ac.in.

belong to the photosynthetic prokaryotic group with wide morphological diversity, ranging from unicellular to branched filamentous organization and differentiated by akinetes for survival, hormogonia for taxis and heterocyst for nitrogen fixation. The role of cyanobacteria as nitrogen supplements for paddy and other crops is well established. Cyanobacteria consortia with agriculturally important plant growth promoting rhizobacteria for biofilm production exhibits an eco-friendly approach for integrated nutrient management practices for selected crops. This book chapter focuses on the research achievements on microalgae and cyanobacteria based plant biofertilizers and biostimulants in the agricultural applications. The challenges to commercializing these kinds of inoculants in biofertilizers are also discussed. The applications of algal bioinoculants will meet the needs of sustainable agriculture with the key goals of enhanced soil health and providing an eco-technological solution.

Keywords: agriculture, exopolysaccharide, biofilm, nitrogen fixation, phytopathogenic

13.1. Introduction

Cyanobacteria are gram negative oxygenic photosynthetic prokaryotes with morphologically diverse, occurring as unicellular, filamentous, planktonic or benthic and colonial forms and found in hot, cold, alkaline, acidic, freshwater, marine, saline and terrestrial environments (Gupta et al., 2013). As per endosymbiotic theory, cyanobacteria are the first oxygen evolving photoautotrophic prokaryotic organisms, which comprises incipient type of nucleus, gas vesicles, loosely arranged thylakoids, hexagonal carboxysomes, phycobilisomes, ribosomes and various storage granules. Cyanobacteria mainly possess chlorophyll a, along with accessory pigments and with active PSII reaction centre (Ahmed et al., 2010). Cyanobacteria utilize inorganic nitrogen which is facilitated by permeases that aids in uptake of nitrite, nitrate and ammonium (Bryant, 2006). Most of the cyanobacteria comprises heterocysts (differentiated thick walled cells) containing nitrogenase enzyme for nitrogen fixation and only photosystem I for ATP generation. Essentially photosystem II is degraded in heterocysts to prevent oxygen production. The diversity of cyanobacteria ranges from coccoid to branched filaments, unicellular to multicellular, nearly colorless to intensely pigmented, psychrophilic to thermophilic, autotrophic to heterotrophic, acidophilic to alkaliphilic, planktonic to barophilic, freshwater to marine (Thajuddin and Subramanian, 2005).

13.2. Diversity of Microalgae and Cyanobacteria in Crop Fields

Rice fields provide a suitable environment for the growth of diazotrophic oxygenic cyanobacteria and the common genera involved are *Anabaena, Tolypothrix, Aulosira, Nostoc, Westiellopsis, Cylindrospermum, Scytonema* etc in Indian rice field soils, contributing significantly to the maintenance of fertility (Nayak et al., 2004). Singh et al., (2014) has reported both heterocystous and non-heterocystous cyanobacteria from different rice fields of Chhattisgarh state. Irrisari et al., (2001) have reported that *Nostoc, Synechocystis* and *Phormidium* as common cyanobacteria found in rice fields of India, Spain, Bangladesh and

Uruguay. Srivastava et al., (2009) reported many species of cyanobacteria that fixes atmospheric nitrogen in rice fields of eastern Uttar Pradesh like *Nostoc* and *Anabaena*. Such diversity studies provide useful information for selection of region specific inocula for maximizing the promise of cyanobacterial inoculants.

The Cyanobacterial diversity of wheat plant is less explored compared to paddy field. Jaiswal et al., (2008) reported four heterocystous strains of *Nostoc* and *Anabaena* from soil samples of wheat field. The dominant cyanobacteria genera found in wheat field are *Nostoc, Calothrix, Anabaena, Oscillatoria* and *Phormidium* (Karthikeyan et al., 2009). Wheat is moderately tolerant to salinity; increasing salinity causes reduction in yield and declines nitrogenase activity as well in cyanobacteria. Cuddy et al., (2012) has isolated nine cyanobacteria isolates from wheat field and found that out of the 9 isolates, only *Cylindrospermum* has highest nitrogenase activity.

13.3. Functionality Framework of Algae and Cyanobacteria

13.3.1. Plant Related Frameworks

Algae impart benefits to the plant by various growth substances and bio-stimulants. Their active ingredients like protein hydrolysates, aminoacids, polysaccharides are described to involve in growth and development (Calvo et al., 2014), resistance against abiotic stresses (Kowalczyk et al., 2008) and regulation of cell expansion genes respectively. On the hand, they are meant for slow release of fertilizers like N, P and K. Further Mercier and co-workers (2001) have demonstrated the role of algal polysaccharides carrageenans and laminarin in defense response and signaling pathway. Ability of λ carrageenans to elicit range of defense related genes evidences the plant protection characters of algae. In harmony, studies of Agarwal et al., (2016) also unveiled role of *Kappaphycus alvarezii* sap on regulation of defense gene expression rendering resistance to tomato plants. Yang and coworkers, 2017 have succeeded in improvement for mitigation of CO_2 with transgene expression with increased aldolase activity. Further many transgenic approaches are being attempted for enhancing the photosynthetic efficiency.

13.3.2. Soil Related Frameworks

Owing to minimization of chemical fertilizers and increased yield, application of algae and cyanobacteria to different crops has gained importance as biostimulant and biofertilizer. They plays pivotal role in refining soil structure, cutting over intensive chemical application, checking soil erosion, retrieval of degraded habitats, sequestration of carbon and mitigation of greenhouse gases (Renuka et al., 2018). They were found to positively influence soil characteristics like water holding capacity by increasing pore volume, aggregate stability, biomass and activity of soil microorganisms, thereby proving their role as soil conditioner. They assist growth of beneficial microbes, support microbial interactions, enhance biomass and soil carbon hence they contribute to overall soil fertility for sustainable agriculture.

Gougoulias et al., (2018) have documented the role of *Chlorella vulgaris* with sheep manure on various soil-physico chemical properties and stimulated soil organic carbon biodegradation.

Table 13.1. Reported role of algae and cyanobacteria in different plants

Algae/cyanobacteria	Plants	Contribution	Reference
Cyanobacteria	Rice	Biological nitrogen fixation & increased nitrogen efficiency	Pereira et al., 2009
Spirulina platensis, Ascophyllum nodosum & Baltic green macroalgae	Wheat	Increase of 100 grain weight and dry mass of shoot	Michalak et al., 2016
Jania rubens, Laurencia obtuse, Corallina elongate	Maize	Improvement in root and polyphenolic and anti-oxidant contents	Al saman et al., 2015
Ulva lactuca, Caulerpa sertularioides, Padina gymnospora, and *Sargassum liebmannii*	Tomato	Biostimulants-germination & growth	Hernandez – Herrera et al., 2014
Spirulina platensis, Amphora cofeaeformis	Cucumber	Significant increase in yield and fruit quality Nematode control	El Eslamboly et al., 2019
Chlorella vulgaris, Scenedesmus quadricauda	Sugar beet	Increased root length, upregulation of genes involved in nutrient acquisition	Barone et al., 2018
Ascophyllum nodosum	Wine grape	Increased anthocyanin content	Frioni et al., 2018
Chlorella vulgaris	Banana	Improved yield, bunch and hand weight	El Moniem et al., 2008

13.3.2.1. Soil Fertility Improvement

Algae help to reduce atmospheric nitrogen *via* nitrogen fixation. Heterocystous cells of cyanobacteria facilitate the oxygen-sensitive process of nitrogen fixation to continue. Cyanobacteria have great potential as biofertilizers and will reduce the demand for chemical fertilizer production. Agricultural potential of heterocystous cyanobacteria, either symbiotic with *Azolla* or free-living, has long been recognized (El-Zeky et al., 2005). Hence, the use of cyanobacterial strains as natural biofertilizer for paddy cultivation has been reported in Chine, Philippines, Egypt and India (Ladha and Reddy, 2003).

13.3.2.2. Uptake of N and P

Some of the cyanobacterial species are phosphate solubilizing. Phosphorus and nitrogen are the primary limiting nutrients in most of the aquatic systems. A competitor should be able maintain net population growth at resource levels less than those required by other species (Silke et al., 2007). Algae usually withdraw nutrients from their environments via structural changes, storage or increased resource utilization efficiency (Singh and Dhar, 2007). Some algae excrete substances to enhance nutrient availability, along with biochemical and physiological adaptations. Algae changes the pH of their surroundings, excrete extracellular phosphatases and other compounds for P absorption (Healy, 1973; Grobbelaar, 1983). In addition, algae also store resources in excess like P for their immediate nutrient requirements. It was evidenced by Schreiber et al., (2018) with *Chlorella vulgaris* and their ability to release N and P from their biomass. Studies have also witnessed phosphate solubilizing activity of *Anabaena variabilis* and *Westiellopsis prolifica* (Yandigeri et al., 2010). Cyanobacterial biofertilizers have greater significance over chemical fertilizers for improving nutrients uptake by cereal crops due to high nutrient turn-over in cereal production system, reduces the cost of fertilizers and protects environment.

13.3.2.3. Nitrogen Fixation

Cyanobacteria with heterocysts are aerobic photo-diazotrophic in nature. The different forms of cyanobacteria involved in nitrogen fixation are mentioned in Table 13.2. Algae are the major nitrogen-fixing agents in agricultural soils. Algae fix atmospheric nitrogen and then re-assimilated by the higher plants (Stewart, 1970). Most of the cyanobacterial and algal species are nitrogen fixing and improves the soil fertility in paddy cultivation (Saikai and Bordoloi, 1994). Cyanobacterial biofertilizer for paddy cultivation is highly promising but limited due to changes in quantity and quality of inoculum and physiological conditions in different agro-ecological regions. Cyanobacteria is widely used biofertilizer for paddy cultivation throughout Asia and improves soil fertility by biological nitrogen fixation (Halperin et al., 1981)

Table 13.2. Nitrogen fixing Cyanobacteria with various morphological forms

Form	Cyanobacteria example	Reference
Filamentous heterocystous form	*Anabaena, Aulosira, Calothrix, Cylindrospermum, Fischerella, Hapalosiphon, Mastigocladus, Nostoc, Scytonema, Stigonema, Tolypothrix, Westiella and Westiellopsis*	Prasanna and Kaushik, 1995
Unicellular and colonical form	*Aphanothece, Chrococcidiopsis, Cyanothece, Dermocarpa, Gloeothece, Myxosarcina, Synechococcus, Synechocystis and Xenococcus*	Prasanna and Kaushik, 1994
Non-heterocystous filamentous form	*Oscillatoria, Trichodesmium, Microcoleus, Pseudanabaena, Plectonema and Lyngbya*	Stal and Krumbein, 1985

13.3.2.4. Contribution to Organic Matter

Soils rich in organic carbon are considered as more productive with high nutrient utilization efficiency and also check the nutrient loss via leaching and volatilization. Cyanobacteria play an important role in contributing high organic carbon into the soil. Algae are rich source of organic matter in soil (Ibraheem, 2007). Algal residues upon death and decay are mixed in soil wherein the mucilage acts as binding agent for improving soil texture, humus content and make the soil more habitable for other plants. The moisture retention capacity of soil will also improves as humus accumulation increases (Bolyshev and Novichkova-Ivanova, 1978). The most important contributions of algae added to soil are incorporation of organic carbon via photosynthesis and of organic nitrogen *via* nitrogen fixation. Nekrasova and Aleksandrova (1982) confirmed that algal biomass contributed significantly to humus formation in soils. Under favourable environmental conditions, algal bloom in paddy field yields around 6-8 tonnes of fresh biomass.

13.3.2.5. Soil Reclamation

Algae play an important role in soil reclamation by improving the plant growth and development, and also increase the soil fertility under adverse environmental conditions (Nisha et al., 2007; Prabu and Udayasoorian, 2007). Many studies have reported the effect of salinity on metabolism, growth and yield of algae and plant (Tang et al., 2007). Some of the growth regulators like gibberellic acid released by algae will improves the salt tolerance in plants. Inoculation of cyanobacterial consortia (*Nostoc kihlmani* and *Anabaena cylindrica*) were found to enhance soil structure, soil nutrients, germination of wheat with improved root and shoot biomass. Usage of algae in soil reclamation is not only economical but will also improve the soil fertility status as well as plant growth conditions under adverse environmental stresses (Pandey et al., 2005).

13.4. Importance of Microalgal Extracellular Substances

Algae releases variety of substances which will improve the growth and development of the plants (Rodriguez et al., 2006); growth promoting substances like vitamins, gibberellin, amino acids, polypeptides; antifungal and antibacterial substances exerting phytopathogenic or biocontrol and also release exopolysaccharides for exoenzyme activity and improving soil structure.

13.4.1. Plant Growth Promoting Activity

Many studies have reported the influence of algae for accelerating the growth of plants. Gupta and Lata (1964) observed the effect of algae in improving seed germination and seedling growth promotion. In addition, they also improve the crop yield and quality. Biofertigation *via* algae increases the rate of seed germination and growth parameters of many plants (Strick et al., 1997). Algae act as potential biofertilizer and soil conditioner. Most of the algae have capacity to release plant growth hormones like auxins like substances, gibberellins, cytokinin like substances, antibiotics, algicide, toxins, organic acids, siderophore, etc. (Strick et al., 1997). Mahmoud and Amara (2000) have reported that in tomato plants biofertilizers improves the growth parameters and carbon dioxide evolution. Amin and co-workers (2009) have reported high volume of indole 3 acetic acid, gibberllic acid, benzyladenine, abscisic acid, jasmonic acid from aqueous extract of *Spirulina*.

13.4.2. Role of Extra Cellular Polysaccharides Produced

Most of the cyanobacteria release extracellular polymers with varied composition. These polymers are usually exopolysaccharides that enhance microbial activity and improves soil structure and exoenzyme activity (Ibraheem, 2007). Rogers and Burns (1994) has reported the role of cyanobacteria in maintaining the soil structure and nutrient status of the soil, as most of the cyanobacteria releases the slime or mucilage that helps in soil aggregate formation and preserves the nutrients from runoff. Algal biofertilizers facilitates reclamation of marginal saline-alkaline and calcareous soils (Hedge et al., 1999). Algal biomass and exopolysaccharides released in soil, facilitates plant growth promotion activities (De Caire et al., 2000). They ensure uptake of nutrients and nutrients facilitating moisture and water stable aggregates thereby stimulates soil community. In addition, they assist nitrogen fixation by preventing the entry of oxygen in heterocysts and confer protection against various abiotic stresses. They take part against oxidative stress by enhancing activity of antioxidant enzymes, phenolics and metabolites (Arroussi et al., 2018)

13.5. Role of Microalgae for Phytopathogen Biocontrol

Most of the chemical pesticides are highly phytotoxic and are not desired. The application of these chemical pesticides is unsafe for soil and plants, as it leads to accumulation of various undesirable chemicals in the crops. Cyanobacteria suppress the growth of phytopathogens by the production of several hydrolytic enzymes like CMCase, chitosanase, β-1,3-endoglucanase. Prasanna et al., (2013) has reported that *A. laxa* and *A. variabilis* produce hydrolytic enzymes which inhibit the growth of tomato wilt caused by *Fusarium sp.* Cyanobacteria amended composts supresses disease in tomato (Dukare et al., 2011), reduces fungal load in soil and improves plant growth. *Anabaena sp.* and *Calothrix elenkinii* exhibits fungicidal and biocidal activity against many of the plant pathogens (Manjunath et al., 2010; Chaudhary et al., 2012).

13.6. Role of Cyanobacteria as Biofertilizer and Biocontrol Agents

Use of chemical fertilizers and pesticides cause groundwater contamination and serious environmental and health issue. Whereas, use of cyanobacteria as biofertilizer helps in biological nitrogen fixation and increases the soil organic matter content. In the plant, they enhance root and shoot length, fresh and dry weight, micronutrient uptake, grain weight and yield etc. (Prasanna et al., 2015; Bidyarani et al., 2016; Rana et al., 2012). The use of cyanobacteria in biocontrol of plant diseases is recently gaining importance in agriculture (Kumar et al., 2016). Kulik (1995) has reported that cyanobacterial extract can be used for control of many of the plant diseases like *Erysiphe polygoni* (Powdery mildew) on turnips, damping-off in tomato, *Botrytis cinerea* on strawberries etc. The major biological agents for control of plant pathogens involve bacteria and fungi. Recently cyanobacteria have also received enough attention as biocontrol agent for many of the plant diseases (Hassan, 2007). The nitrogen fixing cyanobacteria *N. muscorum* at varied concentrations can be used for damping-off disease control in plants (De Caire et al., 1997). It was also reported that many of the plant pathogens like *Sclerotinia sclerotiorum* and *Rhizoctonia solani* can be controlled by the cyanobacterial extracellular material (De Mule et al., 1991).

13.7. Agricultural Wastewater Treatment by Microalgae

Most of microalgal species are capable to remove nitrogen, phosphorus, pesticides, heavy metals, organic and inorganic toxins, and pathogens from the wastewater. Recently many studies have reported the use of microalgae like *Chlorella, Phormidium, Botryococcus, Chlamydomonas, Arthrospira* and *Scenedesmus* for wastewater treatment as one the promising technology (Chinnasamy et al., 2010; Pittman et al., 2011). Compared with expensive chemical wastewater treatment processes, microalgae based wastewater treatment plants are more viable and economical with added benefits of biomass recovery for biofuel production (Graham et al., 2009). Ruiz-Martin et al., (2010) has reported that most of the unicellular chlorophytic microalgae are tolerant to wastewater conditions and are very efficient in nutrient accumulation from wastewater. *Chlorella* and *Scenedesmus* are the predominant algae in oxidation ponds with high efficiency of accumulating nitrogen and phosphorus from wastewater (Masseret et al., 2000). Bhatnagar et al., (2010) have identified and characterized *Chlorella minutissima* from wastewater treatment oxidation ponds in India. This microalga has high efficiency of utilizing available nutrient from wastewater and shows high biomass productivity.

13.7.1. Remediation of Heavy Metal Contamination

Most of the cyanobacteria have a capacity to concentrate metal ions from the environment (Samhan, 2008). Resistance to heavy metals via cyanobacteria and microalgae involves environmental conditions, specific and non-specific protective mechanisms of the cell with toxic metal ions. They were meant to exploit various mechanisms such as adsorption-biosorption, exclusion, complex metal ion formation, synthesis of heavy metal binding peptides. Many of the diazotrophic cyanobacteria can be used in remediation of heavy metal contaminated soils like *M. tenera* can be used as bioinoculant for remediating lead contaminated soils and waters (Zaccaro, 2000). The efficiency of heavy metal removal by microalgae depends on the algal species. For example, *Oscillatoria* can be used for Chromium removal; *Chlorella vulgaris* for cadmium, copper and zinc removal; *Chlamydomonas* spp. for lead removal and *Scenedesmus chlorelloides* for molybdenum removal (Sakaguchi et al., 1981; Ting et al., 1991). Algal genera tolerant to organic pollutants in wastewater include *Oscillatoria, Chlamydomonas, Scenedesmus, Euglena, Chlorella, Navicula* and *Stigeoclonium* (Palmer, 1969).

13.7.2. Soil Aggregate Stabilization

Inoculation of cyanobacteria to rice fields helps to regenerate and improve the soil structure. Most of the cyanobacteria excrete extracellular compounds like peptides, polysaccharides, lipids etc. during their growth in soil. These compounds diffuse around soil particles, glue and hold them together in the form of microaggregates. Polysaccharides are made of fibres, which entangle clay particles and form cluster. The interwoven nature of growing algal filaments also helps in binding the soil particles along with the organic carbon added through algal biomass which ultimately helps in improvement of water holding capacity. Algal proteoglycans adhere cells to solid surface and serves add-on to soil aggregate formation. The role of algae in soil C dynamics, enhanced soil organic carbon storage through soil aggregate formation have been studied (Lewis et al., 2017)

13.8. Recent Trends in Microalgae and Cyanobacterial Bio-Inoculants

Use of microalgae and cyanobacteria as bio-inoculant in rice fields can results in an addition of 30 Kg N/ha. Also the biomass from cyanobacteria enriches soil with organic matter content. Cyanobacteria provides 30-40 kg K/ha per season and sustain higher dehydrogenase activity, more chlorophyll, better nutrient availability and phosphorus content in soil of various crops like maize, chickpea, rice, wheat, soybean, cotton etc (Karthikeyan et al., 2007; Rana et al., 2012; Prasanna et al., 2015 and Verma et al., 2016). In plants, cyanobacteria enhance the root and shoot length, micronutrient uptake, fresh and dry weight, grain weight and yield etc. (Prasanna et al., 2016; Bidyarani et al., 2016).

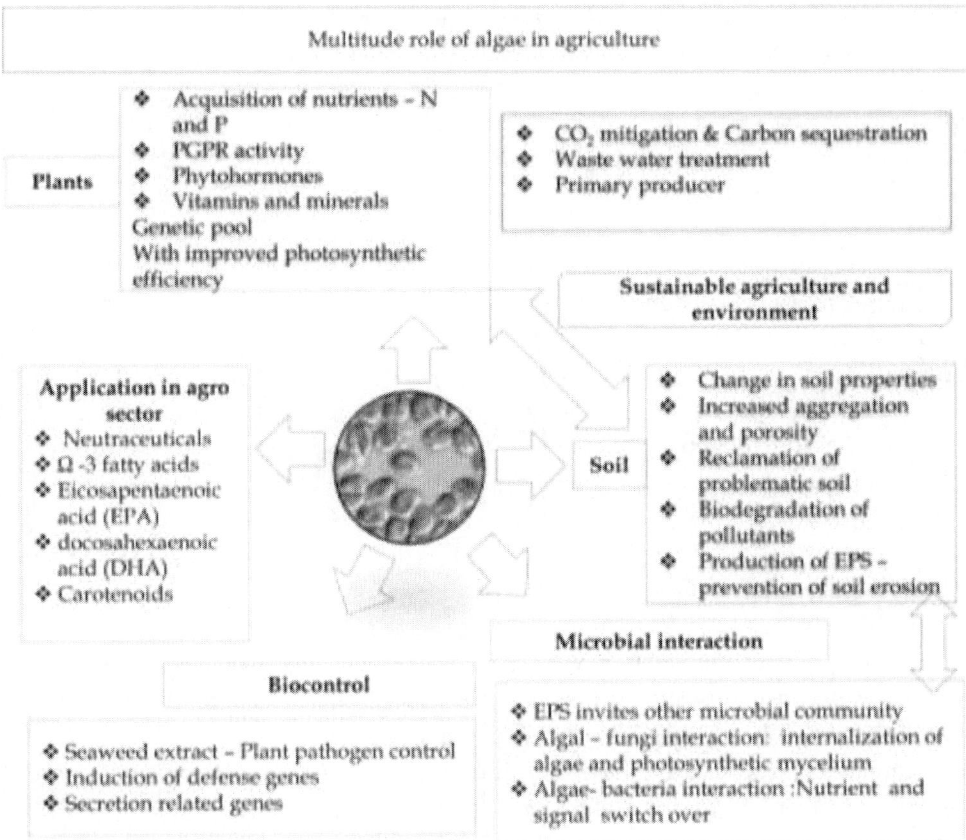

Figure 13.1 Multitude utilities of algae in different components of agriculture contributing towards sustainability in environment.

13.8.1. Microalgae/Cyanobacteria and Irrigation Water

Thind and Rowell, 1999 have analyzed the fate of algal nitrogen under flooded system with labelled ^{15}N. However assimilation of N by algae along with storage and slow release is crucial in N dynamics, various soil parameters plays pivotal role. Reduction in ammoniacal nitrogen under flooded paddy system is contributed by algal immobilization to certain extent. Apart from supplying nitrogen to plants, Azolla were reported to prevent the increase in pH, maintain water temperature and limits volatilization of NH_3 (Mandal et al., 1999). The role of algae in agriculture and its allied enterprises is copious and Figure 1 represents the gross functions.

13.9. Algae as Bioinoculants

Cyanobacteria in rice field helps in maintaining soil fertility and improves rice growth and yield (Song et al., 2005). These algae increase the soil pores with its filamentous structure and produces adhesive substances. Algae improve the water holding capacity of soil via exopolysaccharide compounds released in soil. Rodriguez et al., (2006) reported that cyanobacteria excrete growth promoting substances like hormones, vitamins and amino acids. Cyanobacteria being rich in organic carbon, increase soil biomass after death and decomposition (Saadatnia and Riahi, 2009). Cyanobacteria also decrease soil salinity, prevents weed growth and increase soil phosphate by excretion of organic acids (Saadatnia and Riahi, 2009). Ismail and Ismail, 2011 have elucidated the benefit of inoculation of *Nostoc entophytum* and *Oscillatoria angustissima* which exhibited increased seed germination, metabolic process of pea.

13.9.1. Large Scale Algal Growth

Compared to chemical fertilizers, algal production is highly economical and easy in adaptability by Indian farmers. Algal biofertilizer can provide an additional income for the farmers. There are four methods are followed for the algal production i.e., tank method, pit method, field method and nursey cum algal production method.

13.9.2. Formulations of Algal Inoculants

In markets, ready to use algal biofertilizers packets are available with clay as carrier material. One packet usually weighs 500 g that has to be mixed with 4 kg of sieved farm soil. The mixture has to be broadcasted on standing water 3-6 days after transplanting. The field has to be maintained in waterlogged situation for around 10-12 days after inoculation to allow good growth of algae. If chemical fertilizer or pesticides are applied in the field, algal application should be followed after a gap of 3-4 days.

13.9.3. Challenges in Commercialization

Hitherto, there is no complete and commercially viable technology available for production of algal biofertilizers. Algae/cyanobacteria however find numerous applications; their cultivation in simple production system incurs high cost. Their utilization should surmount various issues for acquiring economic feasibility and sustainability. The production towards commercialization should focus on the strain selection; ensure temperature, light and water, proper management without contamination. Cyanobacterial biofertilizer have been widely used for paddy field whose habitat supports their growth. In such a way, various factors like soil, climatic conditions, growth requirement for individual algae have to be optimized to be applied for selective crop. In order to improve the cost effectiveness various optimization strategies are followed which also includes effective downstream processes like extraction, purification, selection of suitable carrier material and packaging.

Conclusion and Future Perspectives

Algae/cyanobacteria can be considered as natural biofertilizer, which provides benefit not only to the crops but also to the soil as well. Algae as biofertilizer provide growth promoting compounds to plants, controls soil borne pathogens and improves soil health. Algae is commonly used as biofertilizer in paddy cultivation, where it facilitates nitrogen fixation, soil stabilization, add organic matter to soil, release plant growth promoting substances and improves the physico-chemical properties of soil.

References

Agarwal, P., K. Patel, A. K. Das, A. Ghosh and P. K. Agarwal. 2016. Insights into the role of seaweed *Kappaphycus alvarezii* sap towards phytohormone signalling and regulating defence responsive genes in *Lycopersicon esculentum*. *J Appl Phycol. 28*: 2529-2537.

Ahmed, M., Stal, L. J., & Hasnain, S. (2010). Association of non-heterocystous cyanobacteria with crop plants. *Plant and Soil. 336*(1), 363-375.

Al-Saman, M. A., Farfour, S. A., & Hamouda, R. A. (2015). Effects of some red Algae on antioxidant and phytochemical contents of Maize (*Zea mays* L.) plants. *Int J Agric Sci. 5*, 393-398.

Amin, G. H., Al-Gendy, A. A., El-Ayouty, Y., & Abdel-Motteleb, A. (2009). Effect of *Spirulina platensis* extract on growth, phenolic compounds and antioxidant activities of *Sisymbrium irio* callus and cell suspension cultures. *Aust J Basic Appl Sci. 3*, 2097-2110.

Arroussi, H. E., Benhima, R., Elbaouchi, A., Sijilmassi, B., Mernissi, N. E., Aafsar, A., Smouni, A. (2018). *Dunaliella salina* exopolysaccharides: a promising biostimulant for salt stress tolerance in tomato (*Solanum lycopersicum*). *J Appl Phycol. 30*(5), 2929-2941.

Barone, V., Baglieri, A., Stevanato, P., Broccanello, C., Bertoldo, G., Bertaggia, M., Mandolino, G. (2018). Root morphological and molecular responses induced by microalgae extracts in sugar beet (*Beta vulgaris* L.). *J Appl Phycol. 30*(2), 1061-1071.

Bhatnagar, A., Bhatnagar, M., Chinnasamy, S., & Das, K. C. (2010). *Chlorella minutissima*—A Promising Fuel Alga for Cultivation in Municipal Wastewaters. *Appl Biochem Biotechnol. 161*(1), 523-536.

Bidyarani, N., Prasanna, R., Babu, S., Hossain, F., & Saxena, A. K. (2016). Enhancement of plant growth and yields in Chickpea (*Cicer arietinum* L.) through novel cyanobacterial and biofilmed inoculants. *Microbiol Res.188-189*, 97-105.

Bolyshev, N., & Novichkova-Ivanova, L. (1978). Algal flora of initial stages of soil formation on covering loams under the conditions of lysimeter. *Botanicheskii zhurnal*.

Bryant, D. A. (2006). *The molecular biology of cyanobacteria* (Vol. 1): Springer Science & Business Media.

Calvo, P., Nelson, L., & Kloepper, J. W. (2014). Agricultural uses of plant biostimulants. *Plant Soil. 383*(1-2), 3-41.

Chaudhary, V., Prasanna, R., & Bhatnagar, A. (2012). Modulation of fungicidal potential of *Anabaena* strains by light and temperature. *Folia microbiol. 57*(3), 199-208.

Chinnasamy, S., Bhatnagar, A., Hunt, R. W., & Das, K. (2010). Microalgae cultivation in a wastewater dominated by carpet mill effluents for biofuel applications. *Bioresour Technol. 101*(9), 3097-3105.

Cuddy, W. S., Neilan, B. A., & Gehringer, M. M. (2012). Comparative analysis of cyanobacteria in the rhizosphere and as endosymbionts of cycads in drought-affected soils. *FEMS Microbiol Ecol. 80*(1), 204-215. doi: 10.1111/j.1574-6941.2011.01288.x.

De Caire, G. Z., De Cano, M. S., De Mule, M. Z., Palma, R., & Colombo, K. (1997). Exopolysaccharide of *Nostoc muscorum* (Cyanobacteria) in the aggregation of soil particles. *J Appl Phyco. 9*(3), 249-253.

De Caire, G. Z., de Cano, M. S., Palma, R. M., & de Mulé, C. Z. (2000). Changes in soil enzyme activities following additions of cyanobacterial biomass and exopolysaccharide. *Soil Biol Biochem. 32*(13), 1985-1987.

De Mule, M., De Caire, G., De Cano, M., & De Halperin, D. (1991). Bioactive compounds from *Nostoc muscorum* (cyanobacteria). *Cytobios*.

Dukare, A. S., Prasanna, R., Chandra Dubey, S., Nain, L., Chaudhary, V., Singh, R., & Saxena, A. K. (2011). Evaluating novel microbe amended composts as biocontrol agents in tomato. *Crop prot. 30*(4), 436-442.

El-Eslamboly, A. A. S. A., Abd El-Wanis, M. M., & Amin, A. W. (2019). Algal application as a biological control method of root-knot nematode *Meloidogyne incognita* on cucumber under protected culture conditions and its impact on yield and fruit quality. *Egypt J of Biol Pest Control. 29*(1), 18. doi: 10.1186/s41938-019-0122-z.

El-Zeky, M., El-Shahat, R., Metwaly, G. S., & Elham, M. A. (2005). Using Cyanobacteria or Azolla as alternative nitrogen sources for rice production. *J. Agric. Mansoura Univ. 30*(9), 5567-5577.

El Moniem, E. A. A., Abd-Allah, A., & Ahmed, M. (2008). The combined effect of some organic manures, mineral N fertilizers and algal cells extract on yield and fruit quality of Williams banana plants. *J. Agric. Environ. Sci. 4*, 417-426.

Frioni, T., Sabbatini, P., Tombesi, S., Norrie, J., Poni, S., Gatti, M., & Palliotti, A. (2018). Effects of a biostimulant derived from the brown seaweed *Ascophyllum nodosum* on ripening dynamics and fruit quality of grapevines. *Sci Hortic. 232*, 97-106.

Gougoulias, N., Papapolymerou, G., Karayannis, V., Spiliotis, X., & Chouliaras, N. (2018). Effects of manure enriched with algae *Chlorella vulgaris* on soil chemical properties. *Soil and Water Res. 13*(1), 51-59.

Graham, L., GRAHAM, J., & WILcox, L. (2009). Algae. 2" Edition. 616 pp: Benjamin Cummings, San Francisco.

Grobbelaar, J. U. (1983). Availability to algae of N and P absorbed on suspended solids in turbid waters of the Amazon River. *Arch Hydrobiol. 96*, 302-316.

Gupta, A., & Lata, K. (1964). Effect of algal growth hormones on the germination of paddy seeds. *Hydrobiologia 24*(1-2), 430-434.

Gupta, V., Ratha, S. K., Sood, A., Chaudhary, V., & Prasanna, R. (2013). New insights into the biodiversity and applications of cyanobacteria (blue-green algae)—Prospects and challenges. *Algal Res. 2*(2), 79-97.

Halperin, D., De Cano, M., De Muele, M., & De Caire, G. (1981). Algas azueles fijadoras de nitrogenio atmosferico. *Cent. Invest. Biol. Mar. Contr. Tech. 3*(6).

Hassan, N. (2007). *Studies on the algal flora distributed at Wadi-Sannur of the Eastern-desert of Egypt.* M. Sc. Thesis, Faculty of Science, Beni-Suef University, Egypt.

Healey, F. P. (1973). Characteristics of phosphorus deficiency in *Anabaena*1. *J Appl Phycol. 9*(4), 383-394.

Hegde, D., Dwivedi, B., & Sudhakara Babu, S. (1999). Biofertilizers for cereal production in India: A review. *Ind J Agric Sci. 69*(2), 73-83.

Hernández-Herrera, R. M., Santacruz-Ruvalcaba, F., Ruiz-López, M. A., Norrie, J., & Hernández-Carmona, G. (2014). Effect of liquid seaweed extracts on growth of tomato seedlings (*Solanum lycopersicum* L.). *J Appl Phycol. 26*(1), 619-628.

Ibraheem, I. B. (2007). Cyanobacteria as alternative biological conditioners for bioremediation of barren soil. *Egyptian J. of Phycol. Vol. 8*(100).

Irisarri, P., Gonnet, S., & Monza, J. (2001). Cyanobacteria in Uruguayan rice fields: diversity, nitrogen fixing ability and tolerance to herbicides and combined nitrogen. *J Biotechnol. 91*(2-3), 95-103.

Ismail, A. E.-w. A., & Ismail, M. M. (2011). Antagonistic activity of some fungi and Cyanobacteria species against Rhizoctonia solani. *Int J Plant Pathol. 2*(3), 101-114.

Jaiswal, P., Prasanna, R., Nayak, S., Sood, A., & Suseela, M. (2008). Characterization of rhizo-cyanobacteria and their associations with wheat seedlings. *Egypt J Biol. 10*(1).

Karthikeyan, N., Prasanna, R., Nain, L., & Kaushik, B. D. (2007). Evaluating the potential of plant growth promoting cyanobacteria as inoculants for wheat. *Eur J Soil Biol. 43*(1), 23-30.

Karthikeyan, N., Prasanna, R., Sood, A., Jaiswal, P., Nayak, S., & Kaushik, B. D. (2009). Physiological characterization and electron microscopic investigation of cyanobacteria associated with wheat rhizosphere. *Folia Microbiol. 54*(1), 43-51.

Kowalczyk, K., Zielony, T., & Gajewski, M. (2008). Effect of Aminoplant and Asahi on yield and quality of lettuce grown on rockwool. *Biostimulators in Modern Agriculture. Vegetable Crops*. 35-43.

Kulik, M. M. (1995). The potential for using cyanobacteria (blue-green algae) and algae in the biological control of plant pathogenic bacteria and fungi. *Eur J Plant Pathol. 101*(6), 585-599.

Kumar, M., Karthikeyan, N., & Prasanna, R. (2016). Priming of Plant Defense and Plant Growth in Disease-Challenged Crops Using Microbial Consortia *Microbial-mediated Induced Systemic Resistance in Plants* (pp. 39-56): Springer.

Ladha, J. K., & Reddy, P. M. (2003). Nitrogen fixation in rice systems: state of knowledge and future prospects. *Plant Soil. 252*(1), 151-167.

Lewis, K., Foster, J., Hons, F., & Boutton, T. (2017). Initial aggregate formation and soil carbon storage from lipid-extracted algae amendment. *AIMS ENVIRONMENTAL SCIENCE. 4*(6), 743-762.

Mahmoud, H., & Amara, M. (2000). Response of tomato to biological and mineral fertilizers under calcareous soil conditions. *Bulletin of Faculty of Agriculture, University of Cairo. 51*(2), 151-174.

Mandal, B., Vlek, P. L. G., & Mandal, L. N. (1999). Beneficial effects of blue-green algae and Azolla, excluding supplying nitrogen, on wetland rice fields: a review. *Biol Fertil Soils, 28*(4), 329-342. doi: 10.1007/s003740050501.

Manjunath, M., Prasanna, R., Nain, L., Dureja, P., Singh, R., Kumar, A., Kaushik, B. D. (2010). Biocontrol potential of cyanobacterial metabolites against damping off disease caused by *Pythium aphanidermatum* in solanaceous vegetables. *Archives Phytopathol Plant Prot. 43*(7), 666-677. doi: 10.1080/03235400802075815.

Masseret, E., Amblard, C., Bourdier, G., & Sargos, D. (2000). Effects of a waste stabilization lagoon discharge on bacterial and phytoplanktonic communities of a stream. *Water Environ Res. 72*(3), 285-294.

Mercier, L., Lafitte, C., Borderies, G., Briand, X., Esquerré-Tugayé, M.-T., & Fournier, J. (2001). The algal polysaccharide carrageenans can act as an elicitor of plant defence. *New Phytologist. 149*(1), 43-51.

Michalak, I., Chojnacka, K., Dmytryk, A., Wilk, R., Gramza, M., & Rój, E. (2016). Evaluation of Supercritical Extracts of Algae as Biostimulants of Plant Growth in Field Trials. *Front Plant Sci. 7*(1591).

Nayak, S., Prasanna, R., Pabby, A., Dominic, T., & Singh, P. (2004). Effect of BGA-Azolla biofertilizers on nitrogen fixation and chlorophyll accumulation at different depths in soil cores. *Biol Fertil Soils. 40*, 67-72.

Nekrasova, K., & Aleksandrova, I. (1982). Participation of collembolas and earthworms in the transformation of algal organic matter. *Soviet Soil Science (USA).*

Nisha, R., Kaushik, A., & Kaushik, C. P. (2007). Effect of indigenous cyanobacterial application on structural stability and productivity of an organically poor semi-arid soil. *Geoderma. 138*(1), 49-56.

Palmer, C. M. (1969). A composite rating of algae tolerating organic pollution2. *J Phycol. 5*(1), 78-82.

Pandey, K. D., Shukla, P. N., Giri, D. D., & Kashyap, A. K. (2005). Cyanobacteria in alkaline soil and the effect of cyanobacteria inoculation with pyrite amendments on their reclamation. *Biol Fertil Soils. 41*(6), 451-457.

Pereira, I., Ortega, R., Barrientos, L., Moya, M., Reyes, G., & Kramm, V. (2009). Development of a biofertilizer based on filamentous nitrogen-fixing cyanobacteria for rice crops in Chile. *J Appl Phycol. 21*(1), 135-144.

Pittman, J. K., Dean, A. P., & Osundeko, O. (2011). The potential of sustainable algal biofuel production using wastewater resources. *Bioresour Technol. 102*(1), 17-25.

Prabu, P., & Udayasoorian, C. (2007). Native cyanobacteria *Westiellopsis* (TL-2) sp for reclaiming paper mill effluent polluted saline sodic soil habitat of India. *EJEAFChe, 6*(2), 1775-1786.

Prasanna, R., Adak, A., Verma, S., Bidyarani, N., Babu, S., Pal, M., Nain, L. (2015). Cyanobacterial inoculation in rice grown under flooded and SRI modes of cultivation elicits differential effects on plant growth and nutrient dynamics. *Ecol Eng. 84*, 532-541.

Prasanna, R., Chaudhary, V., Gupta, V., Babu, S., Kumar, A., Singh, R., Nain, L. (2013). Cyanobacteria mediated plant growth promotion and bioprotection against *Fusarium* wilt in tomato. *Eur J Plant Pathol. 136*(2). 337-353.

Prasanna, R., Kanchan, A., Ramakrishnan, B., Ranjan, K., Venkatachalam, S., Hossain, F., Nain, L. (2016). Cyanobacteria-based bioinoculants influence growth and yields by modulating the microbial communities favourably in the rhizospheres of maize hybrids. *Eur J of Soil Biol. 75*, 15-23.

Prasanna, R., & Kaushik, B. (1994). Physiological and molecular genetic aspects of nitrogen fixation in non-heterocystous cyanobacteria. *Ind J Exp Biol. 32*(4), 248-251.

Prasanna, R., & Kaushik, B. D. (1995). Nitrogen fixation andnif gene organization in branched heterocystous cyanobacteria: Variation in the presence of xisA. *Folia Microbiol. 40*(2), 176-180. doi: 10.1007/bf02815418.

Rana, A., Joshi, M., Prasanna, R., Shivay, Y. S., & Nain, L. (2012). Biofortification of wheat through inoculation of plant growth promoting rhizobacteria and cyanobacteria. *Eur J Soil Biol. 50*, 118-126.

Renuka, N., Guldhe, A., Prasanna, R., Singh, P., & Bux, F. (2018). Microalgae as multi-functional options in modern agriculture: current trends, prospects and challenges. *Biotechnol Adv.* 36(4), 1255-1273.

Rodríguez, A., Stella, A., Storni, M., Zulpa, G., & Zaccaro, M. (2006). Effects of cyanobacterial extracellular products and gibberellic acid on salinity tolerance in *Oryza sativa* L. *Saline Systems. 2*(1), 7.

Rogers, S. L., & Burns, R. G. (1994). Changes in aggregate stability, nutrient status, indigenous microbial populations, and seedling emergence, following inoculation of soil with *Nostoc muscorum*. *Biol Fertil Soils. 18*(3). 209-215. doi: 10.1007/bf00647668.

Ruiz-Marin, A., Mendoza-Espinosa, L. G., & Stephenson, T. (2010). Growth and nutrient removal in free and immobilized green algae in batch and semi-continuous cultures treating real wastewater. *Bioresour Technol. 101*(1), 58-64.

Ruppel, S., Torsvik, V., Daae, F. L., Øvreås, L., & Rühlmann, J. (2007). Nitrogen availability decreases prokaryotic diversity in sandy soils. *Biol Fertil Soils. 43*(4), 449-459.

Saadatnia, H., & Riahi, H. (2009). Cyanobacteria from paddy fields in Iran as a biofertilizer in rice plants. *Plant Soil Environ. 55*(5), 207-212.

Saikia, P., & Bordoloi, R. (1994). Bluegreen algal flora from the rice fields of Assam. *Phykos. 33*(1&2), 53-57.

Sakaguchi, T., Nakajima, A., & Horikoshi, T. (1981). Studies on the accumulation of heavy metal elements in biological systems. *Eur J Appl Microbiol and Biotechnol, 12*(2), 84-89.

Samhan, A. (2008). *Assessment of the ability of microalgae in removal of some industrial wastewater pollutants.* M. Sc. Thesis, Botany Dept., Fac. of Sci., Beni-Suef University, Egypt.

Schreiber, C., Schiedung, H., Harrison, L., Briese, C., Ackermann, B., Kant, J., Nedbal, L. (2018). Evaluating potential of green alga *Chlorella vulgaris* to accumulate phosphorus and to fertilize nutrient-poor soil substrates for crop plants. *J Appl Phycol. 30*(5), 2827-2836.

Singh, N. K., & Dhar, D. W. (2007). Nitrogen and phosphorous scavenging potential in microalgae.

Singh, S. S., Kunui, K., Minj, R. A., & Singh, P. (2014). Diversity and distribution pattern analysis of cyanobacteria isolated from paddy fields of Chhattisgarh, India. *J Asia-Pac Biodivers, 7*(4), 462-470.

Song, T., Mårtensson, L., Eriksson, T., Zheng, W., & Rasmussen, U. (2005). Biodiversity and seasonal variation of the cyanobacterial assemblage in a rice paddy field in Fujian, China. *FEMS Microbiol Ecol, 54*(1), 131-140.

Srivastava, A. K., Bhargava, P., Kumar, A., Rai, L. C., & Neilan, B. A. (2009). Molecular characterization and the effect of salinity on cyanobacterial diversity in the rice fields of Eastern Uttar Pradesh, India. *Saline Systems, 5*(1), 4.

Stal, L., & Krumbein, W. (1985). Isolation and characterization of cyanobacteria from a marine microbial mat. *Bot Mar. 28*(8), 351-366.

Stewart, W. D. P. (1970). Algal fixation of atmospheric nitrogen. *Plant Soil. 32*(1), 555-588.

Stirk, W. A., & van Staden, J. (1997). Screening of some South African seaweeds for cytokinin-like activity. *S Afr J Bot.* 63(3), 161-164.

Tang, D., Shi, S., Li, D., Hu, C., & Liu, Y. (2007). Physiological and biochemical responses of *Scytonema javanicum* (cyanobacterium) to salt stress. *J Arid Environ. 71*(3), 312-320.

Thajuddin, N., & Subramanian, G. (2005). Cyanobacterial biodiversity and potential applications in biotechnology. *Curr Sci. 89*(1), 47-57.

Thind, H., & Rowell, D. (1999). The fate of algal nitrogen in a flooded soil system. *Nutr Cycl Agroecosyst. 55*(1), 89-94.

Ting, Y. P., Prince, I. G., & Lawson, F. (1991). Uptake of cadmium and zinc by the alga *Chlorella vulgaris*: II. Multi-ion situation. *Biotechnol Bioeng. 37*(5), 445-455. doi: 10.1002/bit.260370506.

Vargas, R., & Novelo, E. (2007). Seasonal changes in periphyton nitrogen fixation in a protected tropical wetland. *Biol Fertil Soils. 43*(3), 367-372.

Verma, S., Adak, A., Prasanna, R., Dhar, S., Choudhary, H., Nain, L., & Shivay, Y. S. (2016). Microbial priming elicits improved plant growth promotion and nutrient uptake in pea. *Isr J Plant Sci. 63*(3), 191-207.

Yandigeri, M. S., Yadav, A. K., Meena, K. K., & Pabbi, S. (2010). Effect of mineral phosphates on growth and nitrogen fixation of diazotrophic cyanobacteria *Anabaena variabilis* and *Westiellopsis prolifica*. *Antonie Leeuwenhoek. 97*(3), 297-306.

Yang, B., Liu, J., Ma, X., Guo, B., Liu, B., Wu, T., Chen, F. (2017a). Genetic engineering of the Calvin cycle toward enhanced photosynthetic CO_2 fixation in microalgae. *Biotechnol Biofuels. 10*(1), 229. doi: 10.1186/s13068-017-0916-8.

Yang, B., Liu, J., Ma, X., Guo, B., Liu, B., Wu, T., Chen, F. (2017b). Genetic engineering of the Calvin cycle toward enhanced photosynthetic CO_2 fixation in microalgae. *Biotechnol Biofuels. 10*(1), 229.

Zaccaro, M. (2000). *Plant growth-promoting cyanobacteria.* Paper presented at the Proceedings of the 5th International PGPR Workshop, Córdoba, Argentina. http://www.ag.auburn.edu/argentina/pdfmanuscripts/zaccaro. pdf (viewed 8-6-2006).

In: Applied Algal Biotechnology
Editors: M. Arumugam, S. Kathiresan et al.
ISBN: 978-1-53617-524-0

Chapter 14

BIOFUELS AND HYDROCARBONS FROM ALGAE

Elamathi Vimali[1], Mohan Kumar Verma[1], Balasubramaniem Ashokkumar[2] and Perumal Varalakshmi[1,*]

[1]Department of Molecular Microbiology, School of Biotechnology, Madurai Kamaraj University, Madurai, Tamil Nadu, India
[2]Department of Genetic Engineering, School of Biotechnology, Madurai Kamaraj University, Madurai, Tamil Nadu, India

14.1. INTRODUCTION

The most plentiful and sustainable source of energy on Earth is the sun. It provides more than 3800 zettajoules (1 zettajoule = 1 × 1021 joules) of solar energy that is absorbed by Earth's atmosphere annually. Approximately only 0.05% of this energy is utilized by plant bodies for biomass production through photosynthesis (Hill et al., 2006, Zhu et al., 2012).

In a balanced ecosystem, carbon utilization by plant bodies, carbon deposition in soil oceans, and carbon discharge from biological and geological sources are in equilibrium. After the global industrialization, this equilibrium has been disturbed by massive carbon emissions by the combustion of fossil fuel and biomass, and reduced carbon uptake due to worldwide deforestation and the loss of arable land. The world consumes more than 80% of the energy from the combustion of fossil fuels, causing more than 24 gigatons generation of carbon dioxide (CO_2) annually (Lewis and Nocera, 2006).

As a consequence, atmospheric CO_2 concentrations have increased from 295 parts per million (ppm) to 380 ppm in the last 100 years. They have contributed significantly to climate change, global warming, drought (Lewis and Nocera 2006, Battisti and Naylor, 2009).

Fossil fuels such as oil, coal, and neutral gas cover 88% of total energy consumption in the world (Tabatabaeiet al., 2011). Informations from the United States Energy Information Administration (U.S. EIA), total energy production will reach 119.4 million barrels per day in the world by 2040 (EIA, 2013). As a consequence of this, the escalating of fossil fuel prices,

* Corresponding Author's Email: pvlakshmi.biotech@mkuniversity.org.

depletion of fossil fuels, and energy demands of countries led to a search for new energy sources (Christiansen, 2011 and Halim, 2012).

Biofuel is a renewable fuel, and there are various renewable energy sources such as bioethanol, methane, and biodiesel. The biofuel separate into three groups, such as solid, liquid, and gaseous. Solids biofuel made up of straw, trees, plants, and liquids biofuel are bioethanol, biodiesel. The gaseous biofuel is methane and known as biogas. Biofuels, mostly in the form of bioethanol, biodiesel, biogas, and biohydrogen, has created increasing attention (Antoniet al., 2007; Johnson and Wen 2010). Biodiesel is liquid and one of the essential biofuels (Fracet al., 2010).

Microalgae are harvesting cell unit of sunlight that exchanges carbon dioxide to potential biofuels, foods, chlorophyll, carotene, allophycocyanin, lutein, phycocyanin, astaxanthin. Additionally, they also produce phycoerythrin, polyphenols, polyunsaturated fatty acids (PUFAs), antioxidant molecules ergothioneine, ascorbate, and glutathione, antioxidant enzymes superoxide dismutase, peroxidase, biogas, feeds and high-value bioactive compounds (Saha and Murray, 2018).

Cyanobacteria are a species of bacteria that obtain their energy through photosynthesis. They are Gram-negative prokaryotic organisms with complex structures and play a significant role in nourishing the prosperity of this environment and very much useful for bioenergy production (Prasanna and Nayak, 2007).

Most important renewable biofuel methane produced by anaerobic digestion of microalgae (Spolaoreet al., 2006) biodiesel obtained from microalgal oil via transesterification and biohydrogen produced by the photo-biological method. Constant use of petroleum sourced fuel is now broadly predicted as unsustainable due to draining supplies and the involvement of these fuels to the assembly of CO_2 in the environment (Fedorovet al., 2005; Kapdan and Kargi, 2006).

14.2. Energy Crises

In the world energy crisis, global warming, and climate change, microalgal biodiesel productions have received much attention for sustainable development. Using microalgae as a biofuel feedstock carries several advantages concerning the environment, food security, and land use. The world's population has reached a level of 7 billion in the 21st century from 2 billion from the last few decades (Avni and Blazquez, 2011). There is no hesitation in the fact that energy is one of the most fundamental essential elements for the development and even the survival of human society.

Energy is the fundamental requirement for economic and social development of any country (Zhang et al., 2011). In countries like BRICS (Brazil, Russia, India, China, and South Africa), the net growth have found with emerging economies as China alone has accounted for 71% of the global energy consumption. In contrast, energy consumption by OECD (Organization for Economic Cooperation and Development) has reduced (BP, 2012).

In the present scenario, the primary energy consumption structure, fossil fuels account for 88.1% with a leading role. In contrast, crude oil consists of 34.8%, coal 29.2%, and natural gas 24.1%, while the percentage of hydroelectricity and nuclear energy are minute, only 6.4% and 5.5%, respectively (Ong et al., 2011).

Globally there was an increasing demand for diesel, which is being fulfilled by crude oil. The cars powered by gasoline have the dominant market share in the United States, whereas diesel car sales reduced to 2.68% from 2010 onwards (Fosten, 2012). While in European countries like Austria, Spain, France, and Italy, diesel-based cars has exceeded 50% since 2006 (Eichlseder and Wimmer 2003). Because of this reason, there will be an increasing demand for fossil fuel diesel; thus, there is an urgent need to develop alternative energy sources and put into practice (Schipper and Fulton 2013). International Energy Agency (IEA) reported that global energy consumption would witness a 53% increase by 2030 (Ong et al., 2011).

In response to energy crises in the world, renewable energies have got a lot of interest and will confidently become one of the leading energy sources for the world. According to one estimation, renewable energy in 2010 fulfilled only 13% of the global primary energy demand (IEA 2012b).

14.2.1. Global Warming and Climate Change

Excess CO_2 release and the running down of fossil fuels are the two leading causes that threaten the world. Increased population, prominent pollution, abrupt industrialization, and globalization are some crucial factors involved in the tremendous reduction of energy and cause global warming (Spolaoreet al., 2006).

Since the industrial revolution started, transportation has been a major key factor for economic growth worldwide. Demand for transportation fuel has moderately increased in developed but risen drastically in developing countries (European Expert Group on Future Transport Fuel, 2011).

The current transportation system mostly depends on (60%) petroleum products (IEA, 2009). The transport system accounts for 19% of global energy use and resulting in 23% carbon dioxide (CO_2) emissions and a 50% increase in CO_2 emissions by 2030 and more than 80% by 2050 (IEA, 2009). Carbon dioxide is the most significant anthropogenic greenhouse gas (IPCC, 2007). According to the United Nations Climate Change Conference in Copenhagen, there is an urgent need for "deep cuts in global emissions" to "stabilize greenhouse gas concentration in the atmosphere at a normal level that would prevent dangerous climate system" (UNFCCC, 2010).

14.3. Bioenergy

14.3.1. Fuel from Divergent Biomass

Biofuels are biodegradable, non-toxic, and have low GHG emissions when burned in combustion, then fossil fuels (Lam and Lee, 2012). According to the Renewable Energy Directive; Biofuel is the liquid or gaseous fuel for transport produced from biomass" (EC, 2009). Biofuels are bioethanol, biogas, jet-fuel, and biodiesel. Bioenergy found to be with the highest potential to satisfy the energy needs of modern society for both developed and developing countries (Ong et al., 2011). At present, bioenergy shares around 10–15% of the world's energy demand (Demirbaset al., 2009). Biofuels, mostly in the form of bioethanol, biodiesel, biogas, and biohydrogen, has created increasing attention (Antoni et al., 2007; Johnson and Wen 2010).

The bioethanol is an alcohol-based fuel produced from fermentation and distilling starch-, sugar- and lignocellulose-based materials. Biodiesel is a mixture of alkyl monoesters,fatty acid extracted through the chemical transesterification of triglycerides from vegetable oils and fats (Cheng &Timilsina, 2011). Biofuels are generally grouped into three categories, as follows:

14.3.1.1. First Generation Biofuel

First-generation biofuels derived from dietary sources such as corn, sugarcane, etc., the biomass of that feedstocks are directly related to first-generation biofuel. Whey, barley, potato wastes, soybean, and sugarbeets are some additional minor feedstocks that considered generating first-generation bioethanol. Brazil is the country that uses sugarcane as a universal feedstock for biofuel production (Lee and Lavoie, 2013).

Biodiesel, which is mainly produced by animal fat, oil crops such as palm, corn, rapeseed, soybean, and castor bean is a non-toxic, renewable and biodegradable fuel, and can be a potential alternative to fossil fuel. Still, this feedstock has low oil yield and requires a high demand for land, water, and fertilizer.

14.3.1.2. Second Generation Biofuel

Second-generation biofuels feedstocks are contrasting to the first generation biofuel, which are non-edible feedstocks such as lignocellulosic waste, municipal solid wastes, agricultural, and forest sediments (Lee and Lavoie, 2013).

Bioethanol may be another potential fuel to replace fossil-derived petrol (Prasad et al., 2007). It is produced through the fermentation process using a range of sugars, for example, sugarcane and sorghum. Bioethanol from lignocellulosic feedstock is also developed (Dwivedi et al., 2009).

Lignocellulosic feedstock includes woody sources, agricultural wastes (Huang et al., 2009), as well as dairy and cattle manures (Chen et al., 2004).

14.3.1.3. Third Generation Biofuel

Third-generation biofuels produced from the algal biomass, however, to a certain level, be correlated to consumption of CO_2 as its feedstock (Lee and Lavoie, 2013).

Algae-based biofuels mean those produced from heterotrophic (photosynthetic organisms) and autotrophic (which need carbon from sugar to grow) microalgal biomass. Microalgae are the oldest living micro-organisms on earth, are very tiny aquatic plants like structures that grow at an exponential rate (double their biomass in less than one day (Lam and Lee, 2012). They are the active energy components that can be used as biomass feedstock in fermentation and thermochemical (liquefaction, gasification, and pyrolysis) conversion pathways. (Davis et al, 2011). Ethanol, methanol, syngas, jet-fuel, bio-oil, and biodiesel are some energy sources derived from both microalgae and macroalgae. (Suganya et al., 2016).

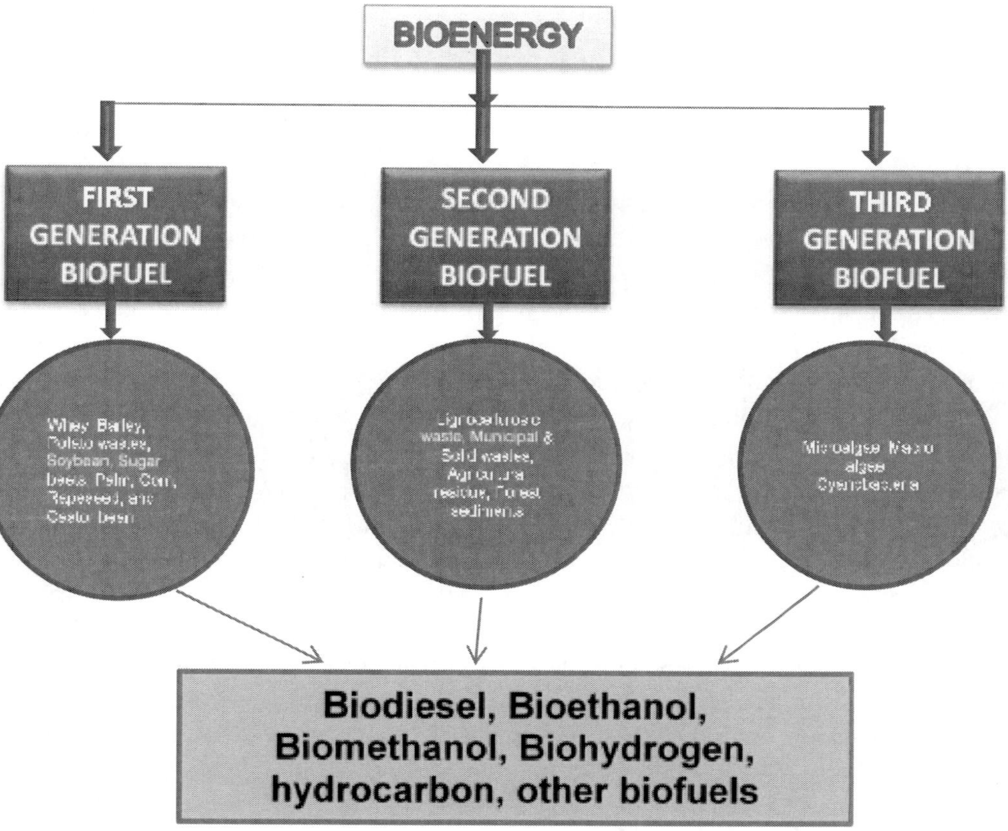

Figure 1. Schematic diagram for biofuel generation.

Scenedesmusobliqutus, Chlamydomonasrheinhardii, Dunaliellasalina, Chlorella pyrenoidosa, Scenedesmusquadricauda, Spirogyra sp., Scenedesmusdimorphus, Anabaena cylindrical, Spirulina maxima, Spirulinaplatensis, Chlorella vulgaris, Porphyridiumcruentum, Tetraselmis maculate, Euglena gracilis, Dunaliellabioculata, Prymnesiumparvum, and Synechoccus sp. are some oil-rich algae.

Some specific microalgal strains have a property to store CO_2 from photosynthesis as lipids, and it can be up to 30% of the dry matter content (IFP énergiesnouvelles). Microalgae have the potential to yields per unit area, then hundreds to thousands of times higher than most oil plants (Cheng &Timilsina, 2011). The process of producing biofuel comprises several steps, i.e., growth, harvest, extraction, and conversion into the final product (Gallagher, 2011).

14.3.3. Biogas Production from Algal Biomass

Biogas production uses all kinds of wastes, for example, municipal solid wastes, fruits, and vegetable waste leaves the wood, grasses, seaweeds, seagrasses (Pham et al., 2006 and Zamalloa et al., 2012). Microalgal biomass is considered a promising feedstock for renewable energy production. The natural lipids from dried biomass extracted by solvent like hexane then transesterification process performed to synthesis fatty acid methyl ester (FAME or

Biodiesel). In later stages, the algal cake formed that is composed of mostly protein, carbohydrates, and polar lipids that used as feedstock for batch biogas experiment, and there was a significant increase in methane production after lipid extraction than before extraction. It shows the importance of pretreatment for anaerobic biodegradation of microalgae to produce biogas.

Biogas is one of the well-thought-out renewable energy sources that is used in electricity, heat, transport fuel and can be stored for future usage (Milledge and Heaven 2017; Matzenbergeret al., 2015).

Biogas is a mixture of carbon dioxide (CO_2), methane (CH_4), and trace gases such as hydrogen sulfide (H_2S) or ammonia (NH_3) that are produced through decomposition by anaerobic bacteria and has a volume fraction of 54-75% CH_4.

The anaerobic fermentation is a biological process in which organic carbon converted from its oxidized form (CO_2) to its reduced form (CH4). This process contains various intermediate steps like hydrolysis (the cleavage of substrate polymers), acidogenesis (the acid-formation level), acetogenesis (acetic acid formation) and methanogenesis (methane formations) (Barbotet al., 2016: Meier et al., 2015; Zhu et al., 2014).

Biogas proposed as one of the cleanest energy-efficient and environmentally friendly. Promised technology for bioenergy production as beside It reduces greenhouse gases, emits fewer pollutants to the atmosphere, generates low carbon dioxide per unit energy compared to fossil fuels by using locally available substrates, the reliable remainder product from anaerobic degradation used as organic fertilizer (Herrmann et al., 2016; Montingelliet al., 2015; Dębowski et al., 2013).

Microalgae are photosynthetic units that use accessible inorganic nutrients such as water, CO_2, sunlight, and produce complex organic compounds such as carbohydrates, proteins, and lipids (Meier et al., 2017; Talukder et al., 2012). Microalgae are characterized by high growth rate in a concise time, as they can duplicate their biomass within 24 hours, and they can grow in different ecosystems including non-drinking water and uncultivated land that makes them an ideal feedstock for different biofuels production (Montingelli et al., 2015; Mussgnug et al., 2010; Vergara-Fernandez et al., 2008)

The microalgal cell wall composed of the complex structure of cellulose, hemicelluloses, and pectin that makes it hard to digest by bacteria. Therefore, the pre-treatment of biomass is necessary before anaerobic biodegradation (Mahdy et al., 2016; Koch et al., 2015; Passos et al., 2013).

Researchers have used various strategies for pre-treatment steps such as physical, chemical, and biological processes and found successful in improving the biodegradation of the microalgal biomass on anaerobic digestion (Neves et al., 2016). Finally, pulling out the lipids from microalgae showed the best pretreatment methods that could give rise to higher biogas production when compared with the whole microalgae biomass without extraction (Koutra et al., 2017).

14.3.4. Net Energy Ratio (NER) Analysis

The potential balance of microalgal biodiesel in energy varies widely depending on the cultivation system, energy savings realized by the co-products, and so on. The energy utilized may include pressure, electricity, heat, and other forms of energy required to run technologies. It found that the energy consumed in microalgal oil extraction was 0.37 MJ/kg biodiesel for wet extraction and cultivation under low-nutrient conditions. The energy consumed for transesterification was 0.0024 MJ/kg biodiesel, predicting HHV of 38 MJ/kg microalgal biodiesel (Lardon et al., 2009).

The net energy ratio (NER) measurement plays a vital role in the current study. The NER is associated with microalgal biodiesel production to validate the energy balance of the process. The NER is the total biodiesel produced over the energy required during operations (Singh and Gu 2010)

NER = Σ Energy produced / ΣEnergy required.

14.4. Improvement of Microalgal Biomass

14.4.1. Culture Techniques

For commercial cultivation of microalgae, the photoautotrophic mode widely accepted for extensive cultivation; it only needs lesser nutrition and some crucial abiotic circumstances which are species-specific. Photo-bioreactors invented from the old commercial cultivation method. Because of renewable biofuel demand, some advanced cultivation techniques achieved. Still, now the competition was going on for commercializing the biofuel at an affordable cost. The culture condition and the nutrition present in the growth media can manipulate some fundamental properties of products such as pigment, lipid, and some other fatty acids (Saha and Murray, 2018). The conditions for the cultivation of microalgae generally classified into four major types:

1. Photoautotrophic cultivation
2. Heterotrophic cultivation
3. Mixotrophic
4. Photoheterotrophic.

14.4.1.1. Photoautotrophic Cultivation

Photoautotrophic mode of cultivation is the method in which the light and carbon source used. Continuously supplement the additional inorganic carbon source (CO_2 in air or sodium bicarbonate as direct source) with the combination of sunlight or artificial lights (radiation ranges from 400nm-700nm), which are useful for photosynthesis (Saha et al., 2013, Wang et al., 2007). Each light range influences the enhancement or limitation of a particular product. Addition of 2% CO_2 to the culture medium enhanced biomass and lipid productivity (Chiu et al., 2008).

Scenedesmusobliquus SIT06 was an oleaginous microalga selected as potential feedstock for biodiesel production because of its lipid-rich nature and appropriate fatty acid composition for biodiesel production with high cetane number. Response surface methodology (RSM) used to optimize the media composition for photoautotrophic cultivation mode of microalgae. The highest biomass 1.99 ± 0.12 g L−1 with a high lipid content of 40.86 ± 0.32% obtained under the photoautotrophic condition in *Scenedesmusobliquus* SIT06 (Srinuanpan et al., 2018).

In photoautotrophic cultivation, in the 14 SK isolates, a different linear tendency was noticed between biomass and lipid productivity. The most dynamic strain for lipid production *Scenedesmus* sp.-AMDD was under cultivation in TAP media, which showed the rate of biomass productivity is 80, and fatty acid productivity 30.7 mg / L/ day. The important lipid producing strain in *Chlamydomonas debaryana*-AMLs1b in B3NV media revealed the quantity of biomass as 51.7mg and the production of fatty acid 5.9 mg/ L/ day−1 (Park et al., 2014).

14.4.1.2. Heterotrophic Cultivation

In the heterotrophic mode of cultivation, the energy generation not happened due to the lack of photosynthetic mechanisms (Minhas et al., 2016). The source of energy given through glucose, sucrose, lactose, fructose, etc. (organic carbon source), and the light source was absent here. Glucose is the first preference for many microalgae because of natural uptake and also energy-rich compound, which influence the growth improvement other than sucrose, lactose, fructose, etc. (Minhas et al., 2016). Biomass and lipid content was also improved in some algal species and decreased in some species due to different mode of nutrition (Kobayashi et al., 1992) (Xu et al., 2006). In this mode of food, the possibility of contamination was much higher than in other (Huang et al., 2010).

Biomass production of some algae improved by maintaining the cultures in heterotrophic conditions. Biomass improved up to 3.4 folds in *Chlorella protothecoides* (Shi et al., 2002), 4.8 folds in *C. vulgaris* (Choix et al., 2012a;), and 3.3 folds in *C. sorokiniana* (Zheng et al., 2012). In open pond, algal biomass production was 0.06-0.1g/L/Day (Pulz, 2001), and in closed photobioreactors, the biomass production was 0.36g/L/ Day. Still, in the heterotrophic cultivation mode, the range of biomass between 4 and 20 g/ L/ day was obtained (Li et al., 2008).

The lipid content of C.*protothecoides* is 15% in autotrophic culture mode, but in a heterotrophic way, the same algal strain produce 55.2% in heterotrophic mode (Xu et al., 2006). In the same heterotrophic mode, nitrogen starved cultures give more lipid and quicker growth than autotrophic culture (Leyva et al., 2014).

14.4.1.3. Mixotrophic Cultivation

Here both organic and inorganic carbon sources are used for algal growth; moreover, they can carry out photosynthesis with these compounds. *Spirulina* was grown well, and *H. pluvialis*also give maximum biomass and higher in astaxanthin production (Park et al., 2014; Wang et al., 2004). So the mixotrophic mode primarily influenced the biomass, carotenoid, and lipid content.

In secondary municipal wastewater (MCWW), 11 of the isolates screened for the quantity of biomass productivity between 21 and 33 mg/L/day with *Scenedesmus sp.*-AMDD being the most productive. In mixotrophic cultivation in MCWW supplemented with glycerol, three different strains *Micractinium sp.*-RB1b, *Chlamydomonas debaryana*-AMB1, and *Chlorella sorokiniana*-RBD8 showed increases in biomass productivity. In this cultivation mode with the other supplement, acetate high oleic acid content detected in 10 algae out of 14. There was no noticeable effect on the fatty acid profiles in cells cultivated mixotrophically in glycerol-supplemented MCWW. There was no noticeable effect on the fatty acid profiles in mixotrophically cultivated cells which are glycerol-supplemented MCWW. This data designates that the mixotrophic cultivation condition enhanced the productivity of both biomass and lipid (Park et al., 2012).

14.4.1.4. Photoheterotrophic Cultivation

In the photoheterotrophic mode of cultivation, microalgae necessitate light as a source of energy at the same time it using crude organic nutrients as the carbon supplementation (Chojnacka, 2004). The cultivation parameter of the mixotrophic and photoheterotrophic mode of cultivation is almost equal. Still, the tiny dissimilarity is associated with mode of nutrient supply in mixotrophic culture is organic compounds and in photoheterotrophic cultures want luminosity as the energy source. In the different cultivation conditions, the parameters such as light and composition of nutrients are drastically dependable for the biomass productivity and specific product formation (Chojnacka, 2004).

14.5. Biodiesel Production

Now a day inexhaustible, carbon-neutral, petroleum fuels are essential for ecological and financial sustainability. Biodiesel obtained from oil crops is a potential renewable and substitute to transport fuels. Unfortunately, biodiesel from oil crops, cuisine waste oil, and fat from animal sources cannot assure even a small amount of the existing demand for transport fuels. As established here, microalgae emerge to be the only source of renewable biodiesel that can be meeting the global market for fuels which are commonly used for transport. Like plants, microalgae utilize sunlight to produce oils, but they do very efficiently than crop plants (Chisti, 2007).

The oil productivity of many microalgae greatly exceeds the oil productivity of the best producing oil crops. Approaches for making microalgal biodiesel economically competitive with petrodiesel discussed. Some microalgal species contain about 60% of lipid or algal oil, the reason behind this mechanism was algae more efficiently utilize carbon dioxide, dissolved nutrients, sunlight and water for photosynthesis (Sheehan et al., 1998; Banerjee et al., 2002). These algal oils converted into biodiesel, and this is very important for rural peoples at a low cost.

Cyanofactories did direct conversion of carbon dioxide to high-value chemicals and potential biofuel production, they are genetically modified cyanobacteria (Zhou et al., 2016) (Zhou et al., 2016). The stored bioenergy precursors are now produced as excreted products because of the carbon flow by altering the targeted metabolic pathway of the cyanobacterial genome. United States alone required 0.53 billion m3 of biodiesel per annum for replacing all the transport fuel. The demand was not satisfied with other biofuel sources such as plant oil, animal fat, and waste cooking oil. Only 50% of the application was satisfied with soybean oil; still, the remaining 50% requirement was in crisis level (Chisti, 2007).

Microalgae only produce more oil, which is more or equal to other crop plants, but it requires only a small amount of cropping area compared to all the crop plants used for biodiesel production. Many microalgae are grown rapidly and contain more oil content; consequently, they only wholly replace fossil fuel other than all crop plants (Chisti, 2007).

Rich oil content microalgae are suitable for the production of biodiesel added to this; the doubling time of algae was much lesser. In some oleaginous algae, more than 50% of dry cell weight contains oil content noticed, so microalgae are an appropriate source for this (Spolaore et al., 2006). Microalgae produce different kinds of hydrocarbon, pigment, lipid, and complex oils depend upon the species changes (Banerjee et al., 2002). Oil-producing microalgae will not give privilege to other value-added products such as pigment, proteins, carbohydrates, feeder, PUFA, food, and other value-added products.

14.5.1. Enhancement of Lipid Productivity in Microalgae

Lipids usually accumulated in microalgal cells for the duration of the stress environment. Any inauspicious situation can guide to the synthesis of lipids and carbohydrates for their endurance. Nitrogen supply is required for the microalgal cells to engage in the cell division process dynamically. When the same cells exposed to nitrogen starved medium, they start synthesizing lipids (Illman et al.,2000), observed that Chlorella vulgaris, *C. minutissima, C.pyrenoidosa, and C. pyrenoidosa* can synthesize more lipids up to 40%, 57%, 23% and 63% of the dry cell weight of algal biomass in nitrogen starved medium.

Certain species are more specific for the accumulation of lipid, thus making them a potential feedstock for biodiesel production. Among lots of available microalgal species, *Isochrysisgalbana*had accumulated extra lipid when a comparative study was executed by (Li et al., 2008), which included 37 strains of 8 different genus of microalgae. It is also well-known for the lipid productivity in *Isochrysisgalbana* were elevated at some stage in the early stationary phase (Roopnarain et al., 2015).

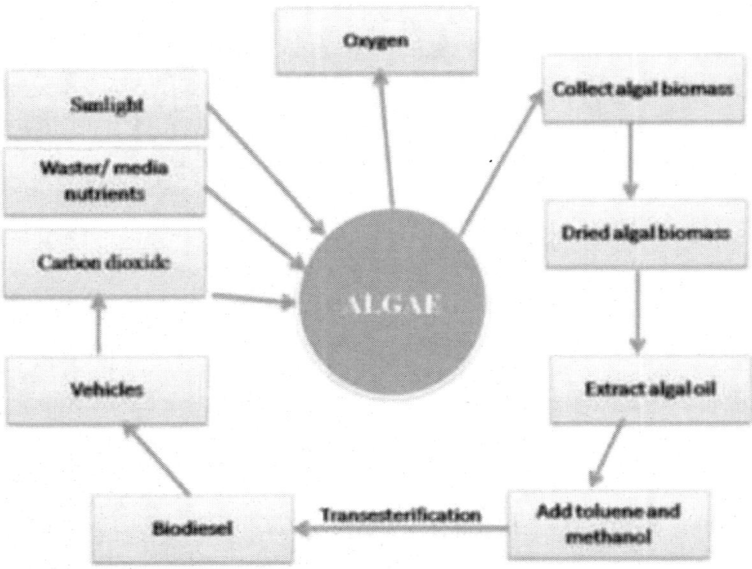

Figure 2. Algal bio-processing for biodiesel production.

Phosphorus is crucial for the cellular processes, and its undersupplied situation in media also influences the regular operation of cell development as phosphorus forms the backbone for the genetic material. The quantity of lipid synthesis improved with the decrease in phosphorus concentration in the case of *Isochrysisgalbana* (Roopnarain et al., 2015).

In nitrogen and phosphorus starved medium with the supplementation of glucose and vitamin B12 Chlamydomonasreinhardtii CC1010 contain 61% lipid content, it is 2.34 fold superior to the standard growth medium(Triis-Acetate-Phosphate). More than 80% of SFA and MUFA content was exposed by acid methyl ester (FAME) profiling of C. reinhardtii CC 1010 (Karpagam et al., 2015).

14.5.2. Lipid Extraction

14.5.2.1. Conventional Method

The algal biomass was collected at the stationary phase by centrifugation and dried it in a hot air oven. The dried algal biomass is preweighed, and ground well-using mortar and pestle extract the lipid by the combination of solvents in the ratio of 2:1 chloroform: methanol (Folchet al, 1957). The lipids are pulling out using these solvents by centrifuge the grounded crude extract at 8000rpm for 10minutes, the supernatant separated in separating funnel by 1% NaCl solution. The lower organic phase was dried and transesterified the lipid into biodiesel by using the amalgamation of alcohol (methanol) and catalyst. Finally, the gravimetrical method used to measure the lipid (Karpagam et al., 2015).

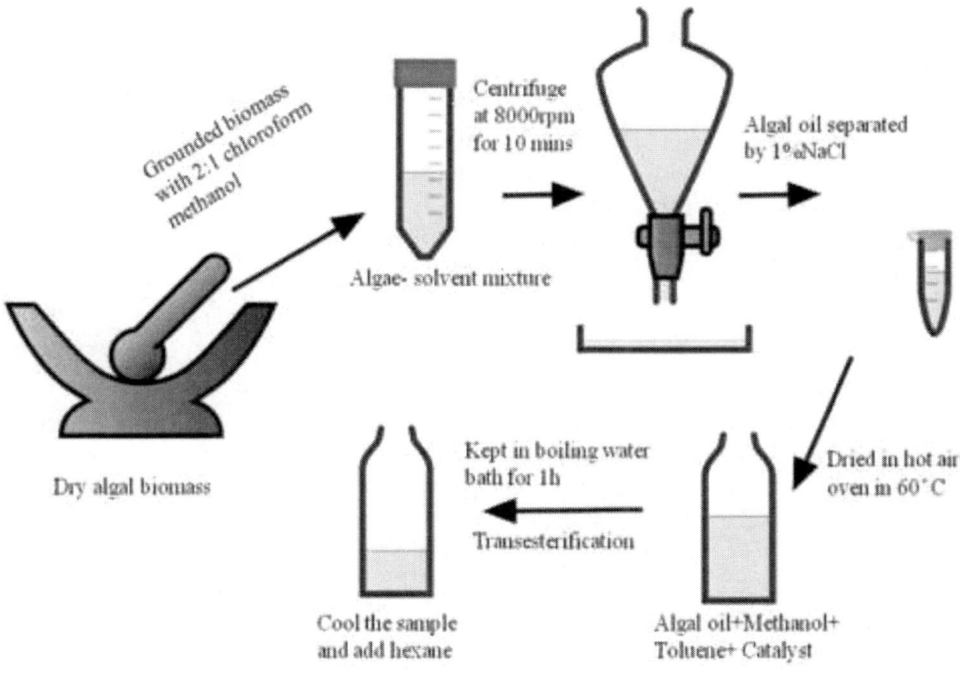

Figure 3: Schematic representation of transesterification for biodiesel production.

14.5.2.2. One Step Transesterification

The stationary phase cells are centrifuged, collected, and dried using a hot air oven. The final dry weight of algal biomass was weighed and used for one-step in situ transesterification. For this method, four essential ingredients added along with the dry cell, the components are acid catalyst, biologically derived nanoparticles, biochar, and commercially available nanoparticles taken for performing transesterification. For example, 40 μL of 37% con HCl and 30ml of methanol is essential for 180mg of dry cell; the mixture heated for 80°C for four h. 9% catalyst beside with 40μL of 37% con HCl used for the bio-derived and biochar mediated transesterification and heating the mixture for 80°C for the same time mentioned above. 30 mL hexane added after cooling and kept in shaking for FAME removal at 100 rpm for 12 h. Here the standard for FAME was Pentadecanoic acid methyl ester (Anto et al., 2019).

14.5.3. Hydrocarbon

Biohydrogen is a clean biofuel type since it used as a fuel cell with water as the only exhaust product and without any pollutant emissions. Previous research has shown that microalgae are therefore a versatile feedstock for the production of biofuels including biodiesel, bioethanol, biogas, biohydrogen, and many other fuel types like biobutanol, bio-oil, syngas, and jet fuel (Li et al., 2008; Koller et al., 2012) via thermochemical and biochemical methods.

Microalgae are well-thought-out fascinating applicants for biofuel production because of their faster reproduction rate. Compared to the animal manure, municipal waste, agriculture wastes, forest wastes, and energy crops algae as a source of Biofuel has paying attention due to many reasons such as faster growth rate and utilize atmospheric carbon dioxide as carbon source (Babich et al., 2011; Li et al., 2008). Intake of water and cropping area are also lesser than the terrestrial plant and algae consuming nitrogen and phosphorus because it significantly reduces the eutrophication. Most of the algae are thermophilic and can survive in extreme environmental conditions (Li et al., 2008).

Research on the production of bio-oil through fast pyrolysis and improvement to hydrocarbon from algal biomass (Suchithra et al., 2012).The major challenge for researchers is the utilization of biomass, which is improbability on the nonstop supply of feedstock for hydrocarbon production.

Hydrocarbon fuels produced by cleaving the C-C bonds in bio-oil by the activity of some specific catalyst are known as catalytic pyrolysis. Dehydration, decarbonylation, and decarboxylation are some cracking reactions involved in the production of aromatic hydrocarbons, carbon dioxide, and water. Catalytic pyrolysis is the less expensive method for improved quantity oil production compared to the hydrodeoxygenation process, in which catalytic pyrolysis does not require hydrogen gas (Williams and Nugranad, 2000).

14.6. Applications

Water bodies are gets polluted by a human being through wastewater from textile, industry, large scale laboratory waste, and hazardous heavy metal waste. Fascinatingly, microalgae provide better solutions for wastewater management; they afford tertiary bioremediation with the combination of many efficient product formations. Microalgae flourished in the wastewater by utilizing nitrogen and phosphorus as nutrition. By growing algae in polluted water allow cleaning the water with the combination of algal biomass productivity (Petr et al., 2018). The algal biomass from wastewater treatment was not appropriate for food-related applications; they can use for biofertilizers, bioplastics, and biofuel applications (Chen et al., 2011). Algae is the best potential non-edible biofuel resource, rich in lipid content 40-80 percentage of the dry cell contains the only lipid. On the basic of cropping area, microalgae generate more biodiesel 15-300 fold (Dry cell wt of algae contain 40-80% of lipid) than the conventional crop plants (Chisti, 2007). Change the composition of chemicals in the algal growth medium may influence the biomass and lipid content of the cells (Meher et al., 2006). Fast multiplication RATE in short time and elevated growth rate; e.g., doubling time within 24 h (Rittmann, 2008). A significant carbon source of algae is the carbon dioxide present in the atmosphere (Schenk et al., 2008).

CONCLUSION

Alternative energy resources are primarily needed to explore due to the depletion of nonrenewable energy sources and increasing environmental pollution globally. Increased production of biofuels will help conserve our natural resources and to save our environment. The microalgal based biodiesel is renewable biodiesel that can completely replace the liquids fuels derived from petroleum.

Algal technology has enormous potential for algae-based biofuels, and also for feed, food, renewable chemicals. We need to develop a suitable photobioreactor, and to design an innovative upstream and downstream processing method for enhancement of algal biofuel. Additionally, genetically engineered algal species can be a boom to produce a higher amount of biodiesel in a short time with less energy utilization.

REFERENCES

Anto, S., Karpagam, R., Renukadevi, P., Jawaharraj, K., Varalakshmi, P., 2019. Biomass enhancement and bioconversion of brown marine microalgal lipid using heterogeneous catalysts mediated transesterification from biowaste derived biochar and bionanoparticle. *Fuel 255* (2019) 115789.

Antoni, D., Zverlov, V.V., Schwarz, W.H., (2007). Biofuels from microbes. *Applied Microbiology and Biotechnology* 77, 23–35.

Avni, A., &Blazquez, M.A., 2011. Can plant biotechnology help in solving our food and energy shortage in the future? *Current Opinion in Biotechnology* 22, 220–223.

Babich, I.V., van der Hulst, M., Lefferts, L., Moulijn, J.A., O'Connor, P., Seshan, K., 2011. Catalytic pyrolysis of microalgae to high-quality liquid bio-fuels. *Biomass Bioenerg*. 35, 3199–3207.

Banerjee, A., Sharma, R., Chisti, Y., Banerjee, U.C., 2002. "*Botryococcusbraunii*: A Renewable Source of Hydrocarbons and Other Chemicals." *Critical Reviews in Biotechnology*. 22 (3): 245–279.

Barbot, Y.N., Al-Ghaili, H., Benz, R., 2016. A Review on the Valorization of Macroalgal Wastes for Biomethane Production, *Mar Drugs* 14(6): 120. doi: 10.3390/md14060120.

BP., 2012. *Statistical Review of World Energy June 2012*. Available from: http://www.bp.com/sectionbodycopy.do?categoryId=7500&contentId=7068481.

Chen, C.Y., Yeh, K.L., Aisyah, R., Lee, D.J., Chang, J.S., 2011. Cultivation, photobioreactor design and harvesting of microalgae for biodiesel production: A critical review. *Bioresour. Technol.* 102, 71–81.

Cheng, J. J., Timilsina, G. R., 2011. Status and barriers of advanced biofuel technologies: A review. *Renewable Energy, 36*(12), 3541-3549.

Chisti, Y., 2007. Biodiesel from microalgae. *Biotechnol. Adv.* 25, 294–306.

Chiu, S.Y., Kao, C.Y., Chen, C.H., Kuan, T.C., Ong, S.C., Lin, C.S., 2008. Reduction of CO_2 by a high-density culture of Chlorella sp. in a semicontinuousphotobioreactor. Bioresour. Technol. 99, 3389–3396. *Fermentation* 2018, 4, 46 17 of 17.

Chojnacka, K., Marquez-Rocha, F.J., 2004. Kinetic and stoichiometric relationships of the energy and carbon metabolism in the culture of microalgae. *Biotechnology*, 3, 21–34.

Christianssen, K. L., 2011. "*Cost structures and life cycle impacts of algal biomass and biofuel production*," Iowa State University, pp. 1-173.

Choix, F.J., de-Bashan, L.E., Bashan, Y., 2012a. Enhanced accumulation of starch and total carbohydrates in alginate-immobilized *Chlorella* spp. induced by *Azospirillumbrasilense*: II. Heterotrophic conditions. *Enzyme MicrobTechnol*51:300–309.

Davis, R.., Aden, A., Pienkos, P.T., 2011. Techno-economic analysis of autotrophic microalgae for fuel production. *Applied Energy, 88*(10), 3524-3531.

Dębowski, M., Zieliński, M., Grala, A., Dudek, M., 2013. Algae biomass as an alternative substrate in biogas production technologies — Review, *Renew Sust Energ Rev* 27: 596–604.

Demirbas, M.F., Balat, M. Balat, H., 2009. The potential contribution of biomass to the sustainable energy development. *Energy Conversion and Management* 50, 1746–1760.

Dwivedi, P., Alavalapati, J.R.R., Lal, P., 2009. Cellulosic ethanol production in the United States: conversion technologies, current production status, economics, and emerging developments. *Energy for Sustainable Development* 13, 174–182.

EC., 2008. 25/048/2008). *EU strategy for biofuels, from Directive 2009/28/EC of the European Parliament and of the Council of 23 April 2009*. http://europa.eu/legislation_summaries/energy/renewable_energy/l28175_en.htm.

EIA (The United States Energy Information Administration), 2013.

Eichlseder, H., Wimmer, A., 2003. Potential of IC-engines as minimum emission a propulsion system. *Atmospheric Environment* 37, 5227–5236.

European Expert Group on Future Transport Fuel, E. (2011). *Future Transport Fuels: European Commission.*

Fedorov, A.S., Kosourov, S., Ghirardi, M.L., Seibert, M., 2005. Continuous H2 photoproduction by Chlamydomonasreinhardtii using a novel two-stage, sulfate-limited chemostat system. *Appl Biochem Biotechnol* 2005, 121124:403–12.

Folch, J., Lees, M., Sloane Stanley, G.H., 1957. A simple method for the isolation and purification of total lipides from animal tissues. *J Biol Chem*. May, 226 (1):497-509.

Fosten, J., 2012. Rising household diesel consumption in the United States: A cause for concern? Evidence on asymmetric pricing. *Energy Economics* 34, 1514–1522.

Frac, M., Tys, S.J., Tys, J., 2010. "Microalgae for biofuels production and environmental applications": A review., *African Journal of Biotechnology*, 9 (54), pp. 9227-9236.

Gallagher, B.J., 2011. The economics of producing biodiesel from algae. *Renewable Energy, 36*(1), 158-162.

Halim, R., Danquah, M.K., Webley, P.A., 2012. Extraction of oil from microalgae for biodiesel production: A review, *Biotechnology Advances*, 30, pp. 709–732.

Herrmann, C., Kalita, N., Wall, D., Xia, A., Murphy, J.D., 2016. Optimized biogas production from microalgae through co-digestion with carbon-rich co-substrates, *Bioresource Technol* 214: 328-337.

Hill, J., Nelson, E., Tilman, D., Polasky, S., Tiffany, D., 2006. Environmental, economic, andenergetic costs and benefits ofbiodiesel and ethanol biofuels. *Proc. Natl. Acad. Sci. U.S.A.* 103, 11206-11210.

Huang, G.H., Chen, F., Wei, D., Zhang, X.W., Chen, G., 2010. Biodiesel production by microalgal biotechnology. *Appl. Energy* 2010, 87, 38–46.

IEA., 2009. Transport, Energy and CO_2: Moving toward Sustainability. Paris, France: IEA.IEA. (2012b). *World Energy Outlook*, 2012.IFPénergiesnouvelles.

IEA., 2012b. *World Energy Outlook*, 2012. IFPénergiesnouvelles.

Illman, A.M., Scragg, A.H., Shales S.W., 2000. Increase in Chlorella strains calorific values when grown in low nitrogen medium. *Enzyme and Microbial Technology* 27 631–635.

IPCC., 2013. *Climate Change 2013: The Physical Science Basis*. Cambridge, United Kingdom and New York, NY, USA: Cambridge University Press.

Johnson, M.B., Wen, Z., 2010. Development of an attached microalgal growth asystem for biofuel production. *Applied Microbiology & Biotechnology* 85, 525– 534.

Kapdan, I.K., Kargi, F., 2006. Bio-hydrogen production from waste materials. *Enzyme Microb Technol* 38:569–82.

Karpagam, R., Preeti, R., Ashokkumar, B., Varalakshmi, P., 2015. Enhancement of lipid production and fatty acid profiling in Chlamydomonasreinhardtii, CC1010 for biodiesel production. *Ecotoxicology and Environmental Safety*.

Kobayashi, M., Kakizono, T., Yamaguchi, K., Nishio, N., Nagai, S., 1992. Growth and astaxanthin formation of Haematococcuspluvialis in heterotrophic and mixotrophic condition. *J. Ferment. Bioeng*. 1992, 74, 17–20.

Koch, K., Helmreich, B., Drewes, J.E., 2015. Co-digestion of food waste in municipal wastewater treatment plants: Effect of different mixtures on methane yield and hydrolysis rate constant, *Appl Energ* 137: 250-255.

Koller, M., Salerno, A., Tuffner, P., et al., (2012). Characteristics and potential of microalgal cultivation strategies: a review. *Journal of Cleaner Production* 37, 377–388.

Koutra, E., Grammatikopoulos, G., Kornaros, M., 2017. Microalgae post-treatment of anaerobically digested agro-industrial wastes for nutrient removal and lipids production, *Bioresource Technol* 224: 473-480.

Lam, M.K., Lee, K.T., 2012. Microalgae biofuels: A critical review of issues, problems and the way forward. *Biotechnology Advances, 30*(3), 673-690. doi: 10.1016/j. biotechadv.2011.11.008.

Lardon, L., Helias, A., Sialve, B., Steyer, J.P., Bernard, O., 2009. Life-cycle assessment of biodiesel production from microalgae. *Environmental Science & Technology* 43, 6475–6481.

Lee, R.A., Lavoie, J.M., 2013. From first- to third-generation biofuels: Challenges of producing a commodity from a biomass of increasing complexity. *Animal Frontiers* April 2013, Vol. 3, No. 2.

Lewis, N.S., Nocera, D.G., 2006. Powering the planet: Chemical challenges in solar energy utilization. *Proc. Natl. Acad. Sci. U.S.A.* 103, 15729-15735.

Leyva, L.A., Bashan, Y., Mendoza, A., de-Bashan, L.E., 2014. Accumulation of fatty acids in *Chlorella vulgaris* under heterotrophic conditions in relation to activity of acetyl-CoA carboxylase, temperature, and co-immobilization with *Azospirillumbrasilense. Naturwissenschaften* 101:819–830.

Li, Y., Horsman, M., Wu, N., Lan, C.Q. Dubois-Calero, N., 2008. Biofuels from microalgae. *Biotechnology Progress* 24, 815–820.*ogy* 102, 1149–1158.

Mahdy, A., Mendez, L., Tomás-Pejó, E., Morales, M.D., Ballesteros, M., González-Fernández, C., 2016. Influence of enzymatic hydrolysis on the biochemical methane potential of Chlorella vulgaris and Scenedesmus sp., *J Chem Technol Biotechnol* 91: 1299–1305.

Matzenberger, J., Kranzl, L., Tromborg, E., Junginger, M., Daioglou, V., Goh, C.S., Keramidas, K., 2015. Future perspectives of international bioenergy trade, *Renew Sust Energ Rev* 43: 926–941.

Meher, L.C., Vidya Sagar, D., Naik, S.N., 2006. Technical aspects of biodiesel production by transesterification – a review. *Renew Sustain Energy Rev* 2006, 10:248–68.

Meier, L., Barros, P., Torres, A., Vilchez, C., Jeison, D., 2017. Photosynthetic biogas upgrading using microalgae: Effect of light/dark photoperiod, *Renew Energ* 106: 17-23.

Meier L., Perez R., Azocar L., Rivas M., Jeison D. (2015): Photosynthetic CO_2 uptake by microalgae: An attractive tool for biogas upgrading, *Biomass Bioenerg*73: 102–109.

Milledge, J.J., Heaven, S., 2017. Energy Balance of Biogas Production from Microalgae: Effect of Harvesting Method, Multiple Raceways, Scale of Plant and Combined Heat and Power Generation, *J Mar Sci Eng* 5: 9. doi: 10.3390/jmse5010009.

Minhas, A.K., Hodgson, P., Barrow, C.J., Adholeya, A., 2016. A review on the assessment of stress conditions forsimultaneous production of microalgal lipids and carotenoids. *Front. Microbiol.* 2016, 7, 546.

Montingelli, M. E., Tedesco, S., Olabi, A.G., 2015. Biogas production from algal biomass: A review, *Renew Sust Energ Rev* 43: 961–972.

Mussgnug, J.H., Klassen, V., Schlüter, A., Kruse, O., 2010. Microalgae as substrates for fermentative biogas production in a combined biorefinery concept, *J Biotechnol* 150: 51–56.

Neves, V.T.D.C., Sales, E. A., Perelo L. W., 2016. Influence of lipid extraction methods as pretreatment of microalgal biomass for biogas production, *Renew Sust Energ Rev* 59: 160-165.

Ong, H.C., Mahlia, T.M.I., Masjuki, H.H., 2011. A review on energy scenario and sustainable energy in Malaysia. *Renewable and Sustainable Energy Reviews* 15, 639–647.

Park, J.C., Choi, S.P., Hong, M.E., Sim, S.J., 2014. Enhanced astaxanthin production from microalga, Haematococcuspluvialis by two-stage perfusion culture with step wise light irradiation. *Bioprocess Biosyst. Eng.* 37, 2039–2047.

Park, K.C., Whitney, C., McNichol, J.C., Dickinson, K.E., MacQuarrie, S., Skrupski, B.P., Zou, J., Wilson, K.E., O'Leary, S.J.B., McGinn, P.J., 2012. *Mixotrophic and photoautotrophic cultivation of 14 microalgal isolates from Saskatechewan, Canada:*

Department of Energy's Aquatic Species Program-Biodiesel from Algae Office of Fuels Development. Program 1-4.

Passos F, Sole M, Garcia J, Ferrer I 2013. Biogas production from microalgae grown in waste water: effect of microwave pretreatment, *Appl. Energ* 108: 168-175.

Pham, T.H., Rabaey, K., Aelterman, P., Clauwaert, P., De Schamphelaire, L., Boon, N. & Verstraete, W. 2006. Microbial fuel cells in relation to conventional anaerobic digestion technology. *Eng. Life Sci.* 6, 285–292.

Pulz O., 2001. Photobioreactors: production systems for phototrophic microorganisms. *Appl. Microbiol. Biotechnol* 57:287–293.

Prasad, S., Singh, A., Jain, N. & Joshi, H.C. 2007. Ethanol production from sweet sorghum syrup for utilization as automotive fuel in India. *Energy and Fuels* 21, 2415–2420.

Prasanna, R., Nayak S. 2007. Influence of diverse rice soil ecologies on cyanobacterial diversity and abundance. *Wetl. Ecol. Manage.* 15:127–134.

Rittmann, BE., Krajmalnik-Brown, R., Halden, R.U., 2008. Pre-genomic, genomic and post-genomic study of microbial communities involved in bioenergy. *Nature Reviews Microbiology* 604-612.

Roopnarain, A., Sym, S. D., & Gray, V. M. 2015. Time of culture harvest affects lipid productivity of nitrogen-starved Isochrysisgalbana U4 (Isochrysidales, Haptophyta). *Aquaculture. 438, 12–16.*

Saha, S.K., McHugh, E., Hayes, J., Moane, S., Walsh, D., Murray, P. 2013. Effect of various stress regulatory factors on biomass and lipid production in microalga Haematococcuspluvialis. *Bioresour. Technol.*, 128, 118–124.

Saha, S.K., Murray, P. 2018. Exploitation of Microalgae Species for Nutraceutical Purposes: Cultivation Aspects. *Fermentation* 4, 46,

Schipper, L. & Fulton, L. 2013. Dazzled by diesel? The impact on carbon dioxide emissions of the shift to diesels in Europe through 2009. *Energy Policy* 54, 3–10.

Sheehan J., T. Dunahay, J. Benemann, P. Roessler. 1998. A look back at the U.S. Department of Energy's Aquatic Species Program – biodiesel from algae. National Renewable Energy Laboratory: Golden, Colorado. *NREL/TP-580-24190*, p. 1–328.

Shi X, Jiang Y, Chen F 2002. High-yield production of lutein by the green microalga *Chlorella protothecoides*in heterotrophic fed-batch culture. Biotechnol Prog. 18:723–727.

Singh, J. & Gu, S. 2010. Commercialization potential of microalgae for biofuels production. *Renewable and Sustainable Energy Reviews* 14, 2596–2610.

Spolaore P, Joannis-Cassan C, Duran E, Isambert A. 2006. Commercial applications of microalgae. *J Biosci Bioeng*, 101:87–96.

Srinuanpan, S., Cheirsilp, B., Prasertsan, P., Kato, Y., & Asano, Y. 2018. Strategies to increase the potential use of oleaginous microalgae as biodiesel feedstocks: Nutrient starvations and cost-effective harvesting process. Renewable Energy, *122, 507 516.*

Thangalazhy-Gopakumar, S., Adhikari, S., Chattanathan,S.A., Gupta, R.B. 2012. Catalytic pyrolysis of green algae for hydrocarbon production using H+ZSM-5 catalyst. *Journal of Bioresource Technology* 118 (2012) 150–157.

Suganya, T., Varman, M., Masjuki, H.H., Renganathan, S., 2016. Macroalgae and microalgae as a potential source for commercial applications along with biofuels production: A biorefinery approach. *Renew. Sustain. Energy Rev.* 55, 909–941.

Tabatabaei, M., Tohidfar, M., Jouzani, G. S., Safarnejad, M., and Pazouki, M., 2011. "Biodiesel production from genetically engineered microalgae: future of bioenergy in Iran," Renew. Sustain. Energy Rev. 15, 1918–1927.

Talukder MMR, Das P, Wu JC 2012. Microalgae (*Nannochloropsissalina*) biomass to lactic acid and lipid, *Biochem. Eng. J.* 68: 109-113.

UNFCCC. 2010. *Report of the Conference of the Parties on its fifteenth session, held in Copenhagen from the 7 to the 19* December 2009.

Vergara-Fernandez A, Vargas G, Alarco´n N, Velasco A 2008. Evaluation of marine algae as a source of biogas in a two-stage anaerobic reactor system, *Biomass and Bioenerg* 32, 338–344.

Wang, C.Y., Fu, C.C., Liu, Y.C. 2007. Effects of using light-emitting diodes on the cultivation of Spirulinaplatensis. *Biochem. Eng.* J., 37, 21–25.

Wang, S.B., Chen, F., Sommerfeld, M., Hu, Q. 2004. Proteomic analysis of molecular response to oxidative stress by the green alga Haematococcus pluvialis (Chlorophyceae). *Planta*, 220, 17–29.

Williams, P.T., Nugranad, N., 2000. Comparison of products from the pyrolysis and catalytic pyrolysis of rice husks. Energy 25, 493–513.

Xu, H., Miao, X.L., Wu, Q.Y. 2006. High-quality biodiesel production from a microalga Chlorella protothecoides by heterotrophic growth in fermenters. *J. Biotechnol.*, 126, 499–507.

Zamalloa, C., Vulsteke, E., Albrecht, J. & Verstraete, W. 2011. The techno-economic potential of renewable energy through the anaerobic digestion of microalgae. *Bioresour. Technol. 102, 1149-1158*

Zhang, N., Lior, N. & Jin, H. 2011. The energy situation and its sustainable development strategy in China. *Energy* 36, 3639–3649.

Zheng Y, Chi Z, Lucker B, Chen S 2012. Two-stage heterotrophic and phototrophic culture strategy for algal biomass and lipid production. *Bioresour. Technol.* 103:484–488.

Zhou J, Zhu T, Cai Z, Li Y 2016. From cyanochemicals to cyanofactories: a review and perspective. *Microb. Cell Fact.* 15: 1-9.

Zhu LD, Hiltunen E, Antila E, Zhong JJ, Yuan ZH, Wang ZM 2014. Microalgal biofuels: Flexible bioenergies for sustainable development, *Renew. Sustain. Energy Rev.* 30: 1035-1046.

Zhu, L., Hiltunen, E. & Takala, J. 2012. Microalgal biofuels beat the first and second-generation biofuels. *Appl. Mech. and Mater.* 197, 760–763.

In: Applied Algal Biotechnology
Editors: M. Arumugam, S. Kathiresan et al.
ISBN: 978-1-53617-524-0

Chapter 15

BIOFILM OF CYANOBACTERIA: ENVIRONMENTAL APPLICATIONS

Shailen Bhakat[1], Sikha Mandal[2] and Jnanendra Rath[1,*]
[1]Department of Botany, Visva-Bharati University, Santiniketan, India
[2]Department of Botany, Sree Chaitanya College, Habra, India

ABSTRACT

Biofilms are a cluster of microorganisms enclosed in an extracellular polymeric substance and have immense significance in environment, industry, agriculture, and health. Biofilms have the ability to persist and function even in extreme environment and can be explored for the various ecological role and can be explored for biotechnological applications. Biofilm forming microorganisms are diverse and form a very complex structure which has not been explored so far. Biofilm systems are especially suitable for the treatment of recalcitrant compounds because of their high microbial biomass and ability to immobilize compounds and various such environmental applications. Biofilms are emerging as a cost-effective means of neutralizing toxic constituents and sequestration of heavy metals from industrial wastewaters. An equally immerging interest is also growing on its application on soil conservation and application of nutrients to plants in poorly manure land. Glycans are important constituents of biofilm and recently growing interest in its biotechnological importance as antitumor antibiotics, cardiac glycosides, and gangliosides. A recent finding of eDNA having a great impact on biofilm formation and manipulation and have great potential for the development of new solutions to the environmental problems and biotechnological applications. In this review, we attempt to bring together information on these fascinating biofilms and associated microorganisms and their potential role in the various biotechnological application.

Keywords: biofilm, eDNA, biodegradation, glycotherapy, biotechnological application

[*] Corresponding Author's Email: jrath@visva-bharati.ac.in.

15.1. INTRODUCTION

A biofilm is an assemblage of surface-associated microbial cells that is enclosed in an extracellular polymeric substance matrix. Basically, microbial biofilm is composed of extracellular polysaccharides, lipid, protein, and extracellular DNA (eDNA) that form a three-dimensional polymer network attaching the biofilm to surfaces and holding cells in close proximity resulting in a synergistic community (Flemming and Wingender, 2010). Extra-polymeric substances (EPS) are responsible for cohesive and adhesive interaction whereas protein uses as carbon and energy source, and extracellular DNA plays a significant role for microbial biofilm formation. EPS may account for 50% to 90% of the total organic carbon and can be considered the primary matrix material of the biofilm. Bacterial EPSs are protected from desiccation, environmental stress, heavy metals and that are why many microorganisms produce biofilms to prevent themselves from these stresses (Sutherland, 2001). Biofilms may form on a wide variety of surfaces, including living tissues, indwelling medical devices, industrial or potable water piping system, or natural aquatic systems. Some of these polysaccharides are neutral or polyanionic, as is the case for the EPS of gram-negative bacteria. The presence of uronic acids (such as D-glucuronic, D-galacturonic, and mannuronic acids) or ketal-linked pyruvates confers the anionic property and biofilms have great environmental importance. This property is important because it allows the association of divalent cations such as calcium and magnesium, which have been shown to cross-link with the polymer strands and provide greater binding force in a developed biofilm. In some gram-positive bacteria, such as the staphylococci, the chemical composition of EPS may be quite different and maybe primarily cationic.

Accidental and deliberate crude oil spills have been a significant source of environmental pollution, as well as metal-based and metal finishing industries. Determination and characterization of the nature of combined pollution is a problem of current interest. Waterways, rivers and ultimately the seas are contaminated with organic waste containing high concentrations of heavy metals, which are of general concern as they are persistent and able to accumulate in the ecosystem creating serious environmental and health hazard. Conventional processes used for removal of heavy metals from industrial waste waters include chemical precipitation, oxireduction, filtration, electrochemical techniques and sophisticated separation processes using membranes. These processes are usually expensive and this characteristic stimulates the use of alternative biotechnologies, due to its reduced cost and lower aggressiveness to the environment. Phototropic biofilms are used to eliminate the nutrient from the wastewater because nitrogen and phosphorus are being used for their growth (Boelee et al., 2012). Extracellular polymeric substances play a very important role in biosorption of heavy metals. They contain ionizable functional groups such as carboxyl, phosphoric, amine, and hydroxyl groups, which enable EPS to sequester heavy metals. Besides the ecological role and environmental application recently there is a growing interest in biotechnological application like various application in psychotherapy, nanomedicine, and drug delivery devices, etc. In this review we have given a comprehensive account of different biofilm-producing cyanobacteria, factors influence the biofilm formation the structural diversity of biofilm and various environmental and biotechnological application of biofilm.

15.1.1. Biofilm Producing Cyanobacteria

Many bacterial species produce biofilms, are surface-attached, by densely packed cells of microbial communities and secreting polymers surround themselves. Bacteria can able to adhere to natural or artificial surfaces and form sessile communities. Both gram-positive and gram-negative species form biofilms. Many cyanobacteria also produce biofilm. Biofilm formation on clean surfaces usually starts with phototrophic organisms (algae, cyanobacteria) which use CO_2 from the atmosphere and sunlight as their carbon and energy source. Heterotrophic organisms (mostly bacteria and fungi) need some organic source for their growth, and this is provided by the metabolites of phototrophic organisms or by airborne deposition. Cyanobacteria produce exopolysaccharides which help bacteria and fungi to colonize on it and further growth of the biofilm occurs. *Aphanothece halophytica* is a halotolerant cyanobacterium that can grow in a wide range of salinity conditions as well as at alkaline pH and produce a large amount of biofilm that exhibits xanthan-like physical properties and is of industrial importance (Pereira et al., 2009). Besides many other cyanobacteria like *Phormidium autumnale*, *Anabaena augstumalis*, *Leptolyngbya foveolarum*, *Phormidium corium*, *Leptolyngbya foveolarum*, *Leptolyngbya tenuis*, *Cyanospira capsulate*, etc. also form biofilms. Many biofilms were dominated by a specific photosynthetic microorganism. However, some biofilms such as ones formed by *Dunaliella* (Chlorophyta) only appear in the most acidic localities and have a clear association with fungal communities. In addition to the vertical distribution of the same organism, different biofilms are distributed horizontally under the same environmental conditions.

15.1.2. Factors Influence the Formation of Biofilm

Biofilm formation is a complex developmental process involving attachment and immobilization on a surface, cell-to-cell interaction, microcolony formation, the formation of a confluent biofilm, and development of a three-dimensional biofilm structure (O'Toole et al., 2000). Bacteria in a biofilm behave differently from their free-floating (planktonic) counterparts. The regulation of bacterial gene expression in response to cell population density, called quorum sensing, is accomplished through the production of extracellular signal molecules called autoinducers (Miller and Bassler, 2001). Biofilm production is regulated by quorum sensing systems in several bacterial pathogens. Biofilms are notoriously difficult to eradicate and are a source of many chronic infections. According to the National Institutes of Health, biofilms are medically important, accounting for over 80% of microbial infections in the body (Lewis, 2001). A mature biofilm can tolerate antibiotics at concentrations of 10–1000 times more than are required to kill planktonic bacteria. Bacteria in biofilms are resistant to phagocytosis, making biofilms extremely difficult to eradicate from living hosts (Lewis, 2001).

Different factors were responsible for the biofilm formation and their structural differences. Several studies have demonstrated that age is a determining factor of biofilm. Biofilm structures with the more loosely structured mats representing a younger state in biofilm development (Horodyski et al., 1977). Maximum attachment depends upon high surface free energy, therefore, the biofilm also produces more on the surface with high surface free energies. Microorganisms have been found to attach more rapidly to hydrophobic and non-polar surfaces than hydrophilic surfaces (Mayer et al., 1999). The three types of forces involved in this process are electrostatic interactions, hydrogen bonds and London dispersion forces (Mayer et al., 1999). Physico-chemical characteristics of aquatic medium such as *p*H, nutrient levels, ionic strength, temperature, etc. also play an important role in the rate of microbial attachment to the surface and biofilm formation. It is also found that an increase in the concentration of several cations such as sodium, calcium, ferric ions affects the attachment by reducing the repulsive forces between the cell and the surface of attachment. Nutrient contents of the growth medium, such as glucose, availability of iron and CO_2, osmolarity, pH, and temperature, influence biofilm production among different cyanobacteria. Carbohydrate metabolism regulates biofilm production among various Gram-positive bacteria, including *Enterococcus faecalis* (Pillai et al., 2004). One study has shown that tryptic soy broth (TSB) medium with 1% glucose supplementation enhances biofilm production in *E. faecalis* compared to TSB without glucose (Baldassarri et al., 2001). The same study also observed greater biofilm production in media supplemented with 0.5% glucose compared to that with 0.2% glucose. The environmental factors have a very important role in biofilm growth. The variations of temperature and salinity influence cyanobacterial growth, metabolism and cell attachment on artificial surfaces. Cell to cell signaling plays an important role in cell attachment. Intracellular communication between cyanobacteria is generally carried out by metabolites that are able to diffuse away from one cell and enter into another cell and is one of the factors for biofilm formation. Changes in the osmotic strength also affect biofilm formation. A study showed that biofilm production was abolished by exposure to a medium to high osmolarity (2–3% sodium chloride) without affecting the growth of the bacteria (Kristich et al., 2004). In oligotrophic environments, mature biofilms may consist of little more than a sparse covering of cells with little structural complexity. Given that biofilms can take long periods to develop, care must be taken not to misinterpret retardation in the initial events of attachment and biofilm development as the total suppression of biofilm formation.

15.1.3. Structural Diversity of Biofilm

Biofilms are complex associations of cells, extracellular products, and detritus either trapped within the biofilm or released from cells which have lysed as the biofilm ages (Christensen, 1989). The major component in the biofilm matrix is water – up to 97%, microbial cells 2-5%, polysaccharides 1-2%, proteins < 1% and DNA and RNA < 1% (Zhang et al., 1998), and the characteristics of the solvent are determined by the solutes dissolved in it. The exact structure of any biofilm is probably a unique feature of the environment in which it develops. The exopolysaccharides (EPS) synthesized by microbial cells vary greatly in their composition and hence in their chemical and physical properties. Some are neutral macromolecules, but the majorities are polyanionic due to the presence of either uronic acids (D-glucuronic acid being the commonest, although D-galacturonic and D-mannuronic acids are also found) or ketal-linked pyruvate. Inorganic residues, such as phosphate or rarely sulphate, may also confer polyanionic status (Sutherland, 1990). A very few EPS may even be polycationic, (Mack et al., 1996). In some of these polymers, the backbone composed of sequences of 1,4-β- or 1,3-β- linkages may confer considerable rigidity. Other linkages in polysaccharides may yield more flexible structures. These can be exemplified by the 1,2-α- or 1,6-α-linkages found in many dextran's.

Most cyanobacterial biofilms are characterized by the presence of uronic acids, pentoses, a polypeptide moiety, or other monosaccharide components, such as organic (e.g., acetyl, pyruvyl, succinyl group) or inorganic (e.g., sulfate or phosphate group) substituents (De Philippis and Vincenzini, 1998). Therefore, cyanobacteria are regarded as a very abundant source of structurally diverse glycans, a feature rarely found in the polymers released by strains belonging to other microbial groups (Sutherland, 1994). Cyanobacterial released polysaccharide often shows the presence of one or two pentose sugar that is usually absent in another polysaccharide of prokaryotic origin (Sutherland, 1994). Most release polysaccharides (RPS) synthesized by cyanobacteria is quite complex, being composed of six or more monosaccharides. This is the striking difference from the polymers synthesized by other bacteria or by microalgae, in which the number of monomers is usually less than four. Usually, glucose is the dominant monosaccharide of the RPS although ironic acid, xylose, arabinose, fucose, rhamnose, and mannose are the dominant monosaccharides in some cyanobacterial RPSs. Galactose is the dominant sugar only in *Anabaena sphaerica*. Ribose, Osamine, methyl sugar, and unidentified residues are also present in several cyanobacterial RPSs. The polysaccharides produced by *N. commune*, *N. insulare* and *Cyanothece* sp. are composed of repeating units of six, four, and three monosaccharides, respectively (Volk et al., 2007). On the other hand, the structures proposed for the EPS produced by *Mastigocladus laminosus* and *Cyanospira capsulate* are far more complex, with repeating units of fifteen and eight monosaccharides, respectively (Gloaguen et al., 1999). For *Spirulina platensis*, no structure was proposed, but it was demonstrated that its RPS repeating unit contains at least 15 sugar residues (Mouhim et al., 1993). A large number of different monosaccharides in only one polymer make many structures and architectures possible (Atkins, 1986), thus increasing the chance of having a polymer with peculiar properties, not common to currently utilized products. The rheological properties of RPSs' aqueous solutions make them useful as thickening agents for water solutions, together with their ability to stabilize the flow properties of their own solutions under drastic changes of pH, temperature, and ionic strength (Sutherland, 1998).

The released polysaccharides in many cyanobacteria attached to proteins and characterized by the presence of pyruvate and acetate groups. Polypeptides enriched with glycine, alanine, valine, leucine, isoleucine, and phenylalanine have been reported in the RPS of *C. capsulate* (Marra et al., 1990). A most interesting feature is the presence of sulfate groups in the RPSs and CPSs of many strains of cyanobacteria. Differences in the sulfur content and in the monosaccharides composition reported for the sheath and RPS of some cyanobacteria strongly support the hypothesis of their origin from different biosynthetic pathways (Micheletti et al., 2008). The sulfate groups and the uronic acids contribute to the anionic nature of the RPS, conferring a negative charge and a "sticky" behavior to the overall macromolecule (Nichols et al., 2005). All these structural factors, i.e., the large numbers of constitutive monosaccharides, the variety of their linkages in the polymeric chains, the variability of functional groups, as well as the protein content give rise to very complex structures of cyanobacterial glycans and impotence in glycomics.

15.2. Environmental Application of Biofilm

Industrial processes and intensive agricultural practices often result in the release of various heavy metals into the terrestrial and aquatic environment. Heavy metals are stable and persistent environmental contaminants since they cannot be degraded. Therefore, their toxicity poses major environmental and health problems and requires a constant search for efficient, cost-effective technologies for detoxification of metal-contaminated sites. Biofilm systems are especially suitable for the treatment of recalcitrant compounds because of their high microbial biomass and ability to immobilize compounds. Bioremediation is also facilitated by enhanced gene transfer among biofilm organisms and by the increased bioavailability of pollutants for degradation as a result of bacterial chemotaxis. Biofilms have multiple environmental applications includes a role in soil conservation, nutrient availability to the plant community in biodegradation, wastewater treatment, etc.

15.3. Role of Biofilm in Soil Conservation

Water movement is a critical process in the shaping of arid landscapes, where, as a consequence of the water-retaining capacity of dry soils, nutrient-poor and nutrient-rich patches characterized by large differences in biological activity typically alternate in a mosaic pattern (Boeken et al., 1995). While in some cases biological crusting of soil is associated with a decrease in water infiltration rate (Bisdom et al., 1993) or to an increase in run-off yields (Kidron et al., 2003). In others studies, the opposite effect is reported (Belnap et al., 2001). In studies carried out in Australia, Biological Soil Crust (BSC)-covered soils displays reduce infiltration in comparison with bare soil in sandy-loam sites (Graetz and Tongway, 1986), while in degraded massive red earth soils; the crusts enhanced water penetration (Greene and Tongway, 1989). Extracellular Polymeric Substances (EPS) of microbial origin are commonly regarded as important for water movement in BSC covered soils (Belnap et al., 2001). It has been postulated that EPS reduce infiltration due to their capacity to clog soil pores (Kidron et al., 1999), especially in well-developed BSCs of sandy soils. Polysaccharides can positively affect soil porosity, increasing the number of micropores, which are known to promote water infiltration (Eldridge, 2003). Direct experimental assessments about the role played by this polysaccharide matrix are few and not definitive. Exopolysaccharides have been reported to possess hydrophobic properties and to clog the surface (Kidron et al., 1999). Mazor et al., (1996) found that small concentrations of EPS had a strong retarding effect on water movement through the sand. They demonstrated that the EPS not only arrest the movement of water through the sand, but it also slows down water absorption by the sand and reduces the amount of water that could penetrate the sand before the onset of water run-off. Chenu, (1993) reported that pure polysaccharides are able to hold more than 50 to 70 g of water per gram of dry matter while maintaining structural coherence. In a model experiment under water-saturated conditions, even small amounts of EPS significantly reduce the hydraulic conductivity of the porous medium (Vandevivere and Baveye, 1992).

15.4. Biofilm and Nutrient Availability to the Plant Community

Biofilms, attached to the plant roots of some crops, help in the cycling of nutrients as well as the biocontrol of pests and diseases, results in improved agricultural productivity (Seneviratne, 2003). It was observed that fungal-rhizobial biofilm (FRBs) fixed nitrogen (N_2) biologically, as revealed by nitrogenase activity and N accumulation, in contrast to when a rhizobial strain was grown as a monoculture (Jayasinghearachchi and Seneviratne, 2004a).

Application of biofilm inoculant of the Fungal-rhizobial biofilm (FRB) can significantly increase N_2 fixation in soybean (by ca. 30%) compared to a rhizobium-only (conventional) inoculant (Jayasinghearachchi and Seneviratne, 2004b). Another study shows the contribution of biofilm inocula in enhancing the release of organic acids and plant growth-promoting (PGP) substances that led to a ca. 25% increase in plant dry weight in the early growth of rice compared to conventional monoculture inocula. FRBs may act as 'pseudo-nodules', fix N_2 biologically on the roots of non-legumes. These inocula stimulate plant growth through a range of mechanisms that improve nutrient acquisition and inhibition of fungal plant pathogens (Toljander et al., 2006). When applied directly to the soil, FRBs helped to increase N and P availabilities by ca. 2-fold and 15-fold, respectively, and showed a high nitrogenase activity, even under a very high soil NO_3 - concentration, compared to the monocultures (Seneviratne and Jayasinghearachchi, 2005).

The biofilm inocula can be effectively used in biosolubilisation of rock phosphate. This was demonstrated by developing biofilms from *Penicillium* sp., *Pleurotus ostreatus* and *Xanthoparmelia mexicana*, a lichen fungus which able to increase the phosphate solubilization up to ca. 230% compared to the fungus-only cultures (Seneviratne and Indrasena, 2006). Furthermore, the biofilm inocula can also be used for the successful establishment of beneficial micro-organisms in plants for biocontrol of diseases. This was confirmed *in vitro* by a *Pleurotus ostreatus – Pseudomonas fluorescens* biofilm that increased endophytic colonization of tomato by *P. fluorescent*, a controlling agent, by over 1,000% compared to inoculation with *P. fluorescens* alone (Jayasinghearachchi and Seneviratne, 2006).

15.5. Role of Biofilm in Biodegradation

Oil spillage and oil pollution in the marine environment becomes a major threat to the ecosystem including the ocean life as well as to the human being through the transfer of toxic organic materials like polycyclic aromatic hydrocarbons (PAHs) into the food chain (Sei and Fathepure, 2009). Presence of polycyclic aromatic hydrocarbons (PAHs) in soil and water is a major problem as environmental contaminants and most of these PAHs are recalcitrant in nature. Physical and chemical methods like volatilization, photooxidation, chemical oxidation, and bioaccumulation are rarely successful in rapid removal and in cleaning up PAHs (Prince, 1997) and also these methods are not safe and cost-effective when compared to microbial bioremediation.

Microbes have a natural tendency to form multi-cellular aggregates being glued to form a biofilm (Wimpenny, 2000). The potential of microbial aggregates in the biofilm communities for bioremediation is always a safer and more adept method than planktonic microorganism as the biofilm matrix protects them during stress, and therefore, the organism gets a better chance of adaptation (Decho, 2000). Interestingly, Klein et al., 2010 reported that hexadecane assimilation by *Marinobacter hydro-carbonoclasticus* SP17 occurs through the formation of a biofilm at the alkane-water interface and how the cell behavior changes with the presence of utilizable or not- utilizable alkanes at the interface. The biofilm-based reactors furnish high microbial biomass accessible for better microbial activity than planktonic cells for other biological activities like biomineralization (Barkay and Schaefer, 2001). The efficiency of biofilm-associated cells in the degradation of naphthalene over planktonic had been elucidated for strain *Pseudomonas stutzeri* T102, and the survival of cells in petroleum-contaminated soil is well documented (Shimada et al., 2012). Chandran and Das 2011 demonstrated 97% degradation of diesel oil over a period of 10 days by yeast biofilm on gravel particle. Faster and intense depletion of linear and branched hydrocarbon was observed in biofilm microbial community of *Alcanivorax borkumensis* (Gertler et al., 2009). Microbial consortia on gravel particle were found as conducive tools for self-cleaning of oily gulf coast throughout all the sites and season (Mahmoud et al., 2010).

Bioremediation of hazardous pollutants by biofilms has gained more attention recently as they are resistant to changing environmental conditions and so resilient to variation in toxicity concentrations (Halan et al., 2011). Biofilm-mediated biodegradation is considered to be more advantageous because it is efficient, safe, cost-effective and environmentally friendly approach (Paul et al., 2005). In addition, there are some reports stating that the presence of secondary carbon source enhances (LaPat-Polasko et al., 1984) the degradation of xenobiotic compounds. Nisha et al. 2015 isolated two bacterial strains capable of degrading N, N-dimethylformamide (DMF) from the effluent and sludge samples of textile and tire industries which can be degraded 55% and 46% of DMF after 120 h of growth. Therefore, biofilm has a great potential in bioremediation.

15.6. Biofilm in Wastewater Treatment

Treating wastewater is vital. Less than 3% of the world's total water supply is available as freshwater, and saltwater is costly to desalinate. Due to industrialization, many freshwaters is now contaminated with heavy metals. Industrial processes and intensive agricultural practices often result in the release of various heavy metals such as copper, lead, and zinc and are the typical pollutants in freshwater and wastewater environment. Nowadays the most commonly used processes for metal removal are an addition of chemicals for metal precipitation or exchange of resins to bind them. Other less frequently used methods include activated carbon adsorption, electro-dialysis and reverse osmosis. It is generally accepted that technologies based on naturally occurring biological processes have a number of advantages over presently available Physico-chemical techniques for remediation of sites contaminated with toxic heavy metals. Moreover, metals adsorbed on cell wall surfaces of cyanobacterial biomass can be recovered and the sorbing material can be restored for reuse (Volesky, 2001).

Cyanobacteria are very abundant in the natural environment and are well adapted to a wide range of habitats e.g., fresh, and seawater, domestic and industrial effluents, salt marshes and constructed wetlands. They have a remarkable ability to take up and accumulate heavy metals from their surrounding environment (De Filippis and Pallaghy, 1994). Their ability to sequester various metal ions such as copper, cadmium, nickel, gold, and chromium is well documented (Chong et al., 2000; Tam et al., 2001). A study was performed by Teitzeland Parsek 2003 to examine the effects of these heavy metals on biofilm and planktonic *P. aeruginosa*. By binding metal ions to the EPS matrix, a biofilm is capable of eliminating heavy metals from the surrounding liquid (Teitzel and Parsek et al., 2003). Cyanobacterial biofilm can be used as biosorption because of low cost, easily growing on natural environment and high metal sorption capacity. Cyanobacterial EPS of *Gloeothece sp.* PCC 6909 and sheathless mutant Gloeothece sp. CCY 9612 both have specific Cu^{2+} binding sites and removes copper from aqueous solution (Micheletti et al., 2008). *Oscillatoria sp.* has potential role as flocculants to removes the heavy metals from the contaminated water. Cyanobacterium *Oscillatoria sp.* effectively removes Pb, Cd, Cu, Zn, Co, Cr, Fe, and Mn from the water column (Bender et al., 1995). Cyanobacteria *Nostoc PCC 7936* and *Cyanospira capsulate* produce a high amount of EPS, to removes Cu more than Zn, and Ni from the multi-metal system containing Cu, Zn, and Ni (de Philippis et al., 2007). According to Pradhan and Rai, (2001), Copper has high specificity active site for metal binding capacity in *Microcystis* sp., *Anabaena spiroides* polysaccharides have high affinity for Mn(II), then Cu(II), Pb(II), and Hg(II). Teitzel and Parsek, 2003; Huang, 2008 showed that bacterial biofilm Pb(II) is accumulated better than Cu(II) when both present in the bimetal system. In *Pseudomonas sp.* EJ01 and mutant *Pseudomonas sp-3055* produces EPS was able to tolerate Cd^{2+}upto 7 mM (Chien et al. 2013). Choudhary and Sar, (2009) reported that *Pseudomonas sp.* resistance to Ni^{2+}, Co^{2+}, Cu^{2+}, and Cd^{2+}.

Cyanobacterial cells can be viewed as natural ion-exchange material because they have many anionic groups on their cell surface (Kratochvil and Volesky, 1998) and this enables them to fix metal ions, mainly by means of an ion-exchange mechanism (Schiewer and Volesky, 1996). In this connection it has been suggested that exopolysaccharide producing cyanobacteria could be used for metal biosorption, since most of the polysaccharide envelopes that surround cyanobacterial cells are anionic (De Philippis et al., 2001), and recent studies have indeed reported that the capsulated biomass of two filamentous cyanobacteria *Cyanospira capsulate* and *Nostoc* PCC 7936, have good metal sorption capacity (De Philippis et al., 2003). The cyanobacterial cells surrounded by thick polysaccharide capsules or slimy possess a large number of binding sites for metal ions compared to non-capsulated strains. Indeed, most cyanobacterial polysaccharides have abundant uronic acid subunits (De Philippis and Vincenzini, 1998) which owing to their carboxyl groups and efficiently binds metal ions. Thus the use of the biomass of RPS-producing cyanobacteria for trapping metal ions seems quite promising.

15.7. Role of Biofilm in Biofouling and Its Environmental Consequences

Biological fouling ("biofouling") of engineered materials has been a significant problem for military and civilian oceangoing vessels. Materials deterioration, losses in heat-transfer efficiency, and mechanical blockages of fluid transport systems can result from biological fouling activities. These problems can also influence fuel consumption. In addition to the direct economic problems created by the activities of micro- and macro fouling organisms, the operational readiness of military vessels is influenced by the frequency of repairs and preventive maintenance activities that result from biological fouling. The biocide business in the United States is a multibillion-dollar business, and there are a number of fortune 500 companies with products and services designed to control biofouling in industrial systems. Bacterial biofilms are the root cause of biofouling in most industrial systems.

Biofilms clearly settle in different stages. The first step, occurring within minutes of immersion, corresponds to the adsorption of organic macro-molecules (exopolymers, proteins, humic acids) and/or inorganic molecules, salts, silts already present in the environment or produced by the micro-organisms (Compère et al., 2001). These adsorbed molecules form a conditioning layer. This non-continuous layer (Pradier et al., 2000) induces modifications of the surface properties (surface tension, surface energy (Fletcher et al., 1985) polarity, wettability which will permit a subsequent adhesion. The bacterial attachment to the substrate is first reversible but becomes irreversible due to the secretion of extracellular polymers (acid polysaccharides) developing polymeric bridging between the cell and the substrate (Féra, 1985). Two theories are generally considered to describe and predict micro-organisms attachment on solid surfaces: the DVLO theory, based on thermodynamics (Bos et al., 1999). The DVLO theory involves electrostatic interactions and Van Der Waals forces to explain the reversible adhesion without considerations of acid-basic interactions. The thermodynamic approach neglects the electrostatic interactions and considers the bacterial adhesion as equilibrium and is based upon the free energy of adhesion. The irreversible attachment of micro-organisms involves high energy interactions. These forces include dipole-dipole interactions, Debye interactions, ion-dipole interactions, hydrogen bonds, and polymeric bridging. Extracellular structures such as polysaccharides or proteins have multiple roles, permitting the irreversible attachment as well as the retention of nutrients, organic and inorganic substances in their immediate environment.

Marine cyanobacteria entrapped in their own exopolymers after attachment on stainless steel surfaces. Once film attachment has occurred and if the physicochemical conditions at the interface are adequate, bacteria grow on the surfaces as micro-colonies (Mills et al., 1981). Ideally, preventing biofilm formation would be a more logical option than treating it. However, there is presently no known technique that able to successfully prevent or control the formation of unwanted biofilms without causing adverse side effects. The main strategy to prevent biofilm formation is to clean and disinfect regularly before bacteria attach firmly to surfaces Biofilm detectors were already developed to monitor the surface colonization by bacteria and allow the control of biofilms in the early stages of development (Pereira et al., 2008). They developed a mechatronic surface sensor able to detect biofilms in the early stages of development. This sensor also able to detect the presence of cleaning products on a surface to identify when it was biologically and chemically cleaned and to measure the rate of cleaning (Pereira et al., 2009). Other preventive strategies attempted to identify materials that do not promote or can even suppress biofilm formation (Rogers et al., 1994). This study ranked different materials according to their biofilm growth propensity concluding that there is hardly any material that does not allow biofilm formation (Rogers et al., 1994). Moreover, biofilm formation may vary with the microbial species present and with the environmental conditions (Simo˜es et al., 2007). Frank and Chmielewski, (2001) tested the influence of surface finish on the Inhibition of biofilm formation by limitation of the carbon source is a virtually impossible procedure, as ultra-pure water systems have been found to support the formation of biofilms (Griebe and Flemming, 1998). Another approach is to supply the microorganisms with growth factors, so surface attachment is no more a benefit for them (Meyer, 2003).

Several attempts have been made to avoid biofilm formation by the incorporation of antimicrobial products into surface materials (Park et al., 2004), by coating surfaces with antimicrobials (Thouvenin et al., 2003) or by modifying the surfaces physicochemical properties (Rosmaninho et al., 2007). Gottenbos et al., (2001) demonstrated a reduction in infection rate using silicone rubber implants with covalently coupled quaternary ammonium coatings. Other authors reported biofilm formation inhibition by coating surfaces with a silver (Hashimoto, 2001). Due to the increased resistance of biofilms to conventional disinfection processes, novel means for their control are constantly sought through the control of environmental factors on the process line and the use of new control strategies.

15.8. Biotechnological Application

15.8.1. Biofilm for Glycotherapy

Glycans are saccharides that can be attached to a wide variety of biological molecules and is the important and major component of biofilm. Glycans play many important roles in biological processes, including involvement in the processes of cellular trafficking, cellular differentiation, cell-cell adhesion, hormone–cell recognition, and viral-host cell and bacterial-host cell attachment. Glycans and glycoconjugates are linked to a number of serious medical conditions such as malignancies, Creutzfeldt–Jakob disease, cystic fibrosis, and inflammatory diseases, cardiovascular diseases, microbial infections, and lysosomal storage diseases (Axford, 2001). Glycans also have important constituents of medicinal important agents, such as the antitumor antibiotics, cardiac glycosides, and gangliosides. The oligosaccharide moieties of the antitumor antibiotics ciclamycin and calicheamicin are important in the DNA recognition critical for the activity of the compound. Cardiac glycosides, such as digoxin are steroids conjugated to a trisaccharide. Gangliosides are lipid-oligosaccharide conjugates and have been used to treat spinal cord injury. Sulfated homopolysaccharides and heteropolysaccharides isolated from biofilms of a number of cyanobacteria have demonstrated potent activity against retroviruses, including human immunodeficiency virus (HIV) and herpes simplex virus (Schaeffer and Krylov, 2000). Scytovirin (SVN) isolated from aqueous extracts of the cultured cyanobacterium *Scytonema varium* binds to a specific tetrasaccharide substructure of the high-mannose oligosaccharide. SVN was active against HIV-1 laboratory strains and clinical isolates with EC_{50} values between 0.3 and 7 nM and 0.4 and 394 nM, respectively (Alexandre et al., 2010).

15.8.2. Glycans of Biofilm as Drug Delivery Device

Glycans of biofilm have great potential as binders, fillers, coatings and slow-release agents for orally administered drug preparations. Several common problems are encountered in attempts to successfully administer drugs by conventional oral or nasal routes. Bioadhesive drug delivery is a novel approach in which the drug is coupled to a carrier molecule that adheres to the biomembrane of the target cell. Glycans of biofilm are useful in this respect as they could be attached specifically to gut or nasal mucosa cells through the recognition of glycoconjugates present at the cell surface. This adhesion could allow the drug time to cross the plasma membrane successfully. Glycans of biofilms can be also used for directing drugs to the surface of bacteria responsible for human disease and this approach could be extended to allow more general carbohydrate-directed therapies (Duncan, 2003). A novel strategy, lectin directed enzyme-activated prodrug therapy (LEAPT), has recently been developed to allow directed delivery of drugs to specific organs. By altering the carbohydrate attached to the glycosidase enzyme, it is possible to target different cell types (Davis, 2002). Hydrogels are three-dimensional, hydrophilic, and polymeric networks capable of imbibing large amounts of water or biological fluids. Hydrogel-based systems have been exploited for protection and delivery of low-molecular-weight drugs and macro-molecular payloads (Agnihotri et al., 2004), such as peptide and protein drugs (insulin, melatonin, heparin, haemoglobin, parathyroid hormone, and calcitonin, etc. He et al., 1999), nucleic acids (DNA, siRNA; Borchard, 2001), and antigens (from pathogens responsible of influenza, pertussis, diphtheria, and tetanus, etc. Illum et al., 2001).

15.8.3. Glycans of Biofilm as Vaccines

Glycan antigens are diverse and range from large, elaborate capsular polysaccharides to small monosaccharide tumor antigens. For several types of bacterial infections, glycoconjugate vaccines can be based on fragments of capsular polysaccharides. The availability of methods to prepare specific oligosaccharide structures opened the possibility to explore the relationship between the oligosaccharide chain length and the potency of the glycoconjugate as a vaccine. In a study of a vaccine against *Shigella dysenteriae*type 1, a series of specific, oligosaccharides was prepared on the basis of the tetrasaccharide repeating unit of the O-specific polysaccharide. The oligosaccharides were coupled to human serum albumin and the conjugates were applied as vaccines (Pozsgay et al., 1999). Glycans are ubiquitous and the ability to understand and control their functions is going to be vital to pioneering future biological and therapeutic breakthroughs. The identification of potential carbohydrate epitope targets is typically the first step in the development of a carbohydrate vaccine. Emerging technologies include oxidative foot-printing of carbohydrate-binding surfaces, which complement epitope mapping techniques, such as saturation transfer difference NMR (DeMarco and Woods, 2008). Another promising approach for three-dimensional carbohydrate-protein complex determination combines partially oriented NMR spectroscopy with computational simulations (DeMarco and Woods, 2008). Carbohydrate antigens are important targets for the development of vaccines and pathogen detection strategies. Vaccines based on isolated glycan structures have helped to prevent deadly infectious diseases and to approach the nearly complete eradication of certain pathogens, therefore, glycans of biofilms have a great potential for future vaccine development.

15.9. Future Prospective

The discoveries of numerous types of biofilm have been documented from many bacterial, fungal and algal species. However, only a few have been industrial and medicinal relevance and have commercial potential. The limitation of the applications of some of these potential biofilm has been largely due to the cost of production relative to their commercial value. To address this problem cheaper substrates, improving product yield by optimizing fermentation condition or development of higher-yielding strains and or genetic and metabolic manipulation need to be carried out.

Recently many researchers focused on another component of biofilm matrix the extracellular DNA (eDNA). eDNA having a major role in biofilm formation. In the biofilm of *Pseudomonas aeruginosa*, *Streptococcus intermedius*, and *Streptococcus mutans* presence of eDNA was reported (Montanaro et al., 2011). Thomas et al., (2008) reported that *Enterococcus faecalis* produces extracellular DNA, which has a great relationship with serine protease and gelatinase. Autolysis is the main concept behind the excrete of extracellular DNA (eDNA). Autolysis occurs may be in two modes either by an altruistic suicide or a fratricide killing of the microbial population. Altruistic suicide that means microbial population committed suicide by the program cell death, which was found in *Staphylococcus aureus*. In fratricide (*Enterococcus faecalis*, *Streptococcus pneumonia*, *Bacillus subtilis)* microbial population differentiate into two forms attackers and targets. Attackers attack the target population and release the killing factors. The attackers are protected from self-destruction by releasing the specific immune protein. (Montanaro et al., 2011). Extracellular DNA (eDNA) and dead cells promoting the biofilm formation in pH 6.0. In these stress conditions, the biofilms are antibiotic tolerance. In *Streptococcus*, *epidermidis* biofilms increased the level of eDNA in the sub-minimal inhibitory complex (sub-MIC). Early-stage of biofilm formation occurs in log and stationary phase due to extracellular DNA but phase biofilms were promoting by exogenous addition of competence stimulating peptide (CSP) (Kostakioti et al., 2013). In *Pseudomonas aeruginosa* eDNA is released by lysis of the cell population. Lysis is controlled by quorum sensing system, which was based on *Pseudomonas quinolone* signaling and acyl-homoserine lactone (AHL). In this system, mutants are not released extracellular DNA during the stationary growth phase and restored upon supplementation with signal molecules. Moreover, eDNA secretion is inhibited by furanone (a specific inhibitor) of the QS system (Montanaro et al., 2011). These finding on eDNA will have a great impact on biofilm control and manipulation and in the future, it will lead to some great solution to the environmental problems.

Conclusion

Biofilm shows a great diversity in its structure and function and its production is not limited by taxa. There is a great diversity in biofilm-producing organisms and the factors affected by their formation are also quite diverse. Structurally they are very complex and vary greatly in their composition, physical and chemical properties. Some are neutral macromolecules, but many are polyanionic or ketal-linked pyruvate. Their functions could be summarized as intrinsic including morphological, structural, protective functions and have potential application is seen in human usage, medical, cosmetic, pharmaceutical, dairy product and other forms of industrial and environmental applications. Though production cost largely has been a limiting factor in the non-industrial applicability a growing interest is coming on their environmental applications. Biofilms are emerging as a cost-effective means of neutralizing toxic constituents and sequestration of heavy metals from industrial wastewaters. An equally immerging interest is also growing on its application on soil conservation and application of nutrients to plants in poorly manure land. Biofilms are the root cause of biofouling and have a great impact on the shipping industry and the marine environment. While the manipulation of fermentation conditions, genetic and metabolic engineering as well as the exploration of cheap fermentation substrates for their production are suggested tools for improving the chances of commercial-scale production and field application. New findings like the discovery of eDNA will have a great impact on biofilm control and its manipulation will lead to developing new solutions to the environmental problems in future.

REFERENCES

Agnihotri, S. A., Mallikarjuna, N. N. & Aminabhavi, T. M. (2004). Recent advances on chitosan-based micro- and nanoparticles in drug delivery. *J. Control. Release.*, *100*, 5-28.

Alexandre, K. B., Gray, E. S., Lambson, B. E., Moore, P. L., Choge, I. A., Mlisana, K., Karim, S. S., McMahon, J., O'Keefe, B., Chikwamba, R. & Morris, L. (2010). Mannose-rich glycosylation patterns on HIV-1 subtype C gp120 and sensitivity to the lectins, griffithsin, cyanovirin-*N*, and scytovirin. *Virology.*, *402*, 187–196.

Atkins, E. D. T. (1986). Biomolecular structures of naturally occurring carbohydrate polymers. *Int. J. Biol. Macromol.*, *8*, 323–329.

Axford, J. (2001). The impact of glycobiology on medicine. *Trends Immunol.*, *22*, 237–239.

Baldassarri, L., Cecchini, R., Bertuccini, L., Ammendolia, M. G., Iosi, F., Arciola, C. R., Montanaro, L., Di Rosa, R., Gherardi, G., Dicuonzo, G. & Orefici, G. (2001). *Enterococcus* spp. produces slime and survives in rat peritoneal macrophages. *Med. Microbiol. Immunol.*, *190*(3), 113-120.

Barkay, T. & Schaefer, J. (2001). Metal and radionuclide bioremediation: issues, considerations, and potentials. *Curr. Opin. Microbiol.*, *4*, 318–323.

Belnap, J., Prasse, P. & Harper, K. T. (2001). Influence of biological soil crusts on soil environments and vascular plants. In: Belnap, J., Lange, O. L. (Eds.), Biological Soil Crusts: Structure, Function, and Management. Springer-Verlag, Berlin, Germany (ISBN: 9783540437574).

Bender, J., Lee, R. F. & Phillips, P. (1995). A review of the uptake and transformation of metals and metalloids by microbial mats and their use in bioremediation. *J. Ind. Microbiol.*, *14*, 113–118.

Bisdom, E. B. A., Dekker, L. W. & Schoute, J. F. T. (1993). Water repellency of sieve fractions from sandy soils and relationships with organic material and soil structure. *Geoderma.*, *56*, 105–118.

Boeken, B., Shachak, M., Gutterman, Y. & Brand, S. (1995). Patchinessand disturbance: plant community responses to porcupinediggings in the central Negev. *Ecography.*, *18*(4), 410-422.

Boelee, N. C., Temmink, H., Janssen, M., Buisman, C. J. & Wijffels, R. H. (2012). Scenario analysis of nutrient removal from municipal wastewater by microalgal biofilms. *Water.*, *4*(2), 460-473.

Borchard, G. (2001). Chitosans for gene delivery. *Adv. Drug. Deliv. Rev.*, *52*, 145–150.

Bos, R., Van der Mei, H. C. & Busscher, H. J. (1999). Physico-chemistry of initial microbial adhesive interactions-its mechanisms and methods for study. *FEMS Microbiol. Rev.*, *23*, 179- 230.

Chandran, P. & Das, N. (2011). Degradation of diesel oil by immobilized Candida tropicalis and biofilm formed on gravels. *Biodegradation.*, *22*(6), 1181-1189.

Chenu, C. (1993). Clay or sand-polysaccharide associations as models for the interface between micro-organisms and soil: water-related properties and microstructure. *Geoderma.*, *56*, 143–156.

Chien, D. T., Bahri, S., Szardenings, A. K., Walsh, J. C., Mu, F., Su, M. Y., Shankle, W. R., Elizarov, A. & Kolb, H. C. (2013). Early clinical PET imaging results with the novel PHF-tau radioligand [F-18]-T807. J. *Alzheimer's Dis*, *34*(2), 457-468.

Chong, A. M. Y., Wong, Y. S. & Tam, N. F. Y. (2000). Performance of different microalgal species in removing nickel and zinc from industrial wastewater. *Chemosphere.*, *41*, 251–257.

Choudhary, S. & Sar, P. (2009). Characterization of a metal resistant Pseudomonas sp. isolated from uranium mine for its potential in heavy metal (Ni2+, Co2+, Cu2+, and Cd2+) sequestration. *Bioresour. Technol.*, *100*, 2482– 2492.

Christensen, B. E. (1989). The role of extracellular polysaccharides in biofilms. *J. Biotechnol.*, *10*, 181–202.

Compère, C., Bellon-Fontaine, M. N., Bertrand, P., Costa, D., Marcus, P., Poleunis, C., Pradier, C. M., Rondot, B. & Walls, M. G. (2001). Kinetics of conditioning layer formation on stainless steel immersed in seawater. *Biofouling.*, *17*, 129-145.

Davis, B. (2002). Preparation of enzyme-conjugated to carbohydrate moiety which binds to lectin directed prodrug delivery system. PCT Int. Appl. Patent No. WO02/080980.

De Filippis, L. F. & Pallaghy, C. K. (1994). Heavy metals: source and biological effects. In: Rai, L. C., Gaur, J. P., Soeder, C. J., eds. Algae and water pollution. Stuttgart, Germany: E. Schweizerbart'sche Verlagsbuchhandlung, 31–77. (ISBN 9783510470433).

De Philippis, R. & Vincenzini, M. (1998). Exocellular polysaccharides from cyanobacteria and their possible applications. *FEMS Microbiol. Rev.*, *22*, 151–175.

De Philippis, R., Sili, C., Paperi, R. & Vincenzini, M. (2001). Exopolysaccharide-producing cyanobacteria and their possible exploitation: a review. *J. Appl. Phycol.*, *13*, 293–299.

De Philippis, R., Paperi, R., Sili, C. & Vincenzini, M. (2003). Assessment of the metal removal capability of two capsulated cyanobacteria, *Cyanospiracapsulata* and *Nostoc* PCC7936. *J. Appl. Phycol.*, *15*, 155–161.

De Philippis, R., Paperi, R. & Sili, C. (2007). Heavy metal sorption by released polysaccharides and whole cultures of two exopolysaccharide-producing cyanobacteria. *Biodegradation.*, *18*, 181–187.

Decho, A. W. (2000). Microbial biofilms in intertidal systems: an overview. *Cont. Shelf Res.*, *20*, 1257– 1273.

DeMarco, M. L. & Woods, R. J. (2008). Structural glycobiology: a game of snakes and ladders. *Glycobiology.*, *18*, 426–440.

Duncan, R. (2003). The dawning era of polymer therapeutics. *Nat. Rev. Drug. Discov.*, *2*, 347–360.

Eldridge, D. J. (2003). Biological soil crusts and water relations in Australian deserts. In *Biological Soil Crusts: Structure, Function, and Management*, Belnap, J., Lange, O. L. (eds). Springer: Berlin. 315–326. (ISBN 9783540437574).

Flemming, H. C. & Wingender, J. (2010). The biofilm matrix. *Nat. Rev. Microbiol.*, *8*, 623-633.

Fletcher, M. & Pringle, J. H. (1985). The effect of surface free energy and medium surface tension on bacterial attachment to solid surfaces. *J. Colloid Interface Sci.*, *104*, 5-14.

Frank, J. & Chmielewski, R. (2001). Influence of surface finish on the cleanability on stainless steel. *J. Food Prot.*, *68*, 1178-1182.

Gertler, C., Gerdts, G., Timmis, K. N., Yakimov, M. M. & Golyshin, P. N. (2009). Populations of heavy fuel oil-degrading marine microbial community in presence of oil sorbent materials. *J. of Appl. Microbiol.*, *107*, 590–605.

Gloaguen, V., Morvan, H., Hoffmann, L., Plancke, Y., Wieruszeski, J. M., Lippens, G. & Strecker, G. (1999). Capsular polysaccharide produced by the thermophilic cyanobacterium *Mastigocladuslaminosus*. Structural study of a disaccharide obtained by lithium degradation. *Eur. J. Biochem.*, *266*, 762–770.

Gottenbos, B., van der Mei, H. C., Klatter, F., Nieuwenhuis, P. & Busscher, H. J. (2001). *In vitro* and *in vivo* antimicrobial activity of covalently coupled quaternary ammonium silane coatings on silicone rubber. *Biomaterials.*, *23*, 1417–1423.

Graetz, R. D. & Tongway, D. J. (1986). Influence of grazing management on vegetation, soil structure and nutrient distribution and the infiltration of applied rainfall in semi-arid chenopod shrubland. *Aust. J. Ecol.*, *11*, 347-60.

Greene, R. S. B. & Tongway, D. J. (1989). The significance of (surface) physical and chemical properties in determining soil surface condition of red earth in rangelands. *Aust. J. Soil Res.*, *27*, 213-225.

Griebe, T. & Flemming, H. C. (1998). Biocide-free antifouling strategy to protect RO membranes from biofouling. *Desalination.*, *118*, 153-156.

Halan, B., Schmid, A. & Buehler, K. (2011). Real-Time Solvent Tolerance Analysis of Pseudomonas sp Strain LB120ΔC Catalytic Biofilms. *Appl. Environ. Microbiol.*, *77*, 1563–1571.

Hashimoto, H. (2001). Evaluation of the anti-biofilm effect of a new antibacterial silver citrate/lecithin coating in an *in-vitro* experimental system using a modified Robbins device. *Kansenshogaku Zasshi*, *75*, 678–685.

He, P., Davis, S. S. & Illum, L. (1999). Chitosan microspheres prepared by spray drying. *Int. J. Pharm.*, *187*, 53–65

Horodyski, R., Bloeser, J. & VonderHaar, S. (1977). Laminated algal mats from a coastal lagoon Laguna Mormona, Baja California Mexico. *J. Sediment. Petrol.*, *47*, 680–696.

Huang, B., Wang, W., Bates, M. & Zhuang, X. (2008). Three-Dimensional Super-Resolution Imaging by Stochastic Optical Reconstruction Microscopy. *Science*, *319*, 810-813.

Illum, L., Jabbal-Gill, I., Hinchcliffe, M., Fisher, A. N. & Davis, S. S. (2001). Chitosan as a novel nasal delivery system for vaccines. *Adv. Drug Deliv. Rev.*, *51*, 81–96.

Jayasinghearachchi, H. S. & Seneviratne, G. (2004a). Can mushroom fix atmospheric nitrogen? *J. Biosci.*, *29*, 293-296.

Jayasinghearachchi, H. S. & Seneviratne, G. (2004b). A bradyrhizobial-Penicillium spp. biofilm with nitrogenase activity improves N2 fixing symbiosis of soybean. *Biol. Fertil. Soils.*, *40*, 432-434.

Jayasinghearachchi, H. S. & Seneviratne, G. (2006). Fungal solubilization of rock phosphate is enhanced by forming fungal-rhizobia biofilms. *Soil Biol. and Biochem.*, *38*, 405–408.

Kidron, G. J., Yaalon, D. H. & Vonshak, A. (1999). Two causes of runoff initiation on macrobiotic crusts: hydrophobicity and pore-clogging. *Soil Sci.*, *164*, 18-27.

Kidron, G. J., Yair, A., Vonshak, A. & Abeliovich, A. (2003). Microbiotic crust control of run-off generation on sand dunes in the Negev Desert. *Water Resour. Res.*, *39*, 1108–1112.

Klein, B., Bouriat, P., Goulas, P. & Grimaud, R. (2010). Behavior of *Marino bacter hydrocarbon oclasticus* SP17 cells during initiation of biofilm formation at the alkane-water interface. *Biotechnol. Bioeng.*, *105*, 461–468.

Kostakioti, M., Hadjifrangiskou, M. & Hultgren, S. J. (2013). Bacterial biofilms: development, dispersal, and therapeutic strategies in the dawn of the postantibiotic era. *Cold Spring Harb Perspect Med.*, *3*(4), a010306.

Krasowska, A. & Sigler, K. (2014). How microorganisms use hydrophobicity and what does this mean for human needs? *Front. Cell Infect. Microbiol.*, *4*, 112

Kratochvil, D. & Volesky, B. (1998). Advances in the biosorption of heavy metals. *Trends Biotechnol.*, *16*, 291–300.

Kristich, C. J., Li, Y. H., Cvitkovitch, D. G. & Dunny, G. M. (2004). Esp-independent biofilm formation by *Enterococcus faecalis*. *J. Bacteriol.*, *186*, 154–163.

Lapat-polasklo, T., Mccartyp, L. & Zehnde, J. B. (1984). Secondary substrate utilization of methylene chloride by an isolated strain of *Pseudomonas* sp. *Appl. Environ. Microbiol.*, *41*, 825-830.

Lewis, K. (2001). Riddle of biofilm resistance. *Antimicrob. Agents Chemother.*, *45*, 999–1007.

Mack, D., Fischer, W., Krokotsch, A., Leopold, K., Hartmann, R., Egge, H. & Laufs, R. (1996). The intercellular adhesin involved in biofilm accumulation of *Staphylococcus epidermidis* is a linear *β*-1,6-linked glucosaminoglycan: purification and structural analysis. *J. Bacteriol.*, *178*, 175–183.

Mahmoud, H., Al-Hasan, R., Khanafer, M. & Radwan, S. (2010). A microbiological study of the self-cleaning potential of oily Arabian Gulf coasts. *Environ. Sci. Pollut. Res.*, *17*, 383–391.

Marra, M., Palmeri, A., Ballio, A., Segre, A. & Slodki, M. E. (1990). Structural characterization of the exocellular polysaccharide from *Cyanospiracapsulata*. *Carbohyd. Res.*, *197*, 338–344.

Mayer, C., Moritz, R., Kirschner, C., Borchard, W., Maibaum, R., Wingender, J. & Flemming, H. C. (1999). The role of intermolecular interactions: studies on model systems for bacterial biofilms. *Int. J. Biol. Macromol.*, *26*, 3-16.

Mazor, G., Kidron, G. J., Vanshak, A. & Abeliovich, A. (1996). The role of cyanobacterial exopolysaccharides in structuring desert microbial crusts. *FEMS Microbiol. Ecol.*, *21*, 121-130.

Meyer, B. (2003). Approaches to prevention, removal, and killing of biofilms. *Int. Biodeterior. Biodegradation.*, *51*, 249–253.

Micheletti, E., Colica, G., Viti, C., Tamagnini, P. & De Philippis, R. (2008). Selectivity in the heavy metal removal by exopolysaccharide-producing cyanobacteria. *J. Appl. Microbiol.*, *105*, 88–94.

Miller, M. B. & Bassler, B. L. (2001). Quorum sensing in bacteria. *Annu. Rev. Microbiol.*, *55*, 165–199.

Mills, A. L. & Maubrey, R. (1981) Effect of mineral composition on bacterial attachment to submerged rock surfaces. *Microb.. Ecol.*, *7*(4), 315-322.

Mouhim, R. F., Cornet, J. F., Fontaine, T., Fournet, B. & Dubertret, G. (1993). Production, isolation and characterization of the exopolysaccharide of the cyanobacterium *Spirulina platensis*. *Biotechnol. Lett.*, *15*, 567–575.

Montanaro, L., Poggi, A., Visai, L., Ravaioli, S., Campoccia, D., Speziale, P. & Arciola, C. R. (2011). Extracellular DNA in biofilms. *Int. J. Artif. Organs.*, *34*, 824-831.

Nichols, C. M., Guezennec, J. & Bowman, J. P. (2005). Bacterial exopolysaccharides from extreme marine environments with special consideration of the southern ocean, sea ice, and deep-sea hydrothermal vents: a review. *Mar. Biotechnol.*, *7*(4), 253-271.

Nisha, K. N., Devi, V., Varalakshmi, P. & Ashokkumar, B. (2015). Biodegradation and utilization of dimethylformamide by biofilm-forming Paracoccus sp. strain MKU1 and MKU2. *Bioresour. Technol.*, *188*, 9-13.

O'Toole, G., Wplan, H. B. & Kolter, R. (2000). Biofilm formation as microbial development. *Annu. Rev. Microbiol.*, *54*, 49–79.

Park, A. I., Daeschel, M. A. & Zhao, Y. (2004). Functional properties of antimicrobial lysozyme–chitosan composite films. *J, Food Saf.*, *69*, 215– 221.

Paul, E. S., Harding, E. J. & Mendl, M. (2005). Measuring emotional processes in animals: the utility of a cognitive approach. *Neurosci. Biobehav. Rev.*, *29*, 469–491.

Pereira, S., Zille, A., Micheletti, E., Moradas-Ferreira, P., De Philippis, R. & Tamagnini, P. (2009). The complexity of cyanobacterial exopolysaccharides: composition, structures, inducing factors and putative genes involved in their biosynthesis and assembly. *FEMS Microbiol. Reviews.*, *33*(5), 917-941.

Pereira, A., Mendes, J. & Melo, L. F. (2008). Using nanovibrations to monitor biofouling. *Biotechnol. Bioeng.*, *15*, 1407–1415.

Petersen, F. C., Pecharki, D. & Scheie, A. A. (2004). Biofilm model of growth of Streptococcus intermedius favored by a competence stimulating signaling peptide. *J. Bacteriol.*, *186*, 6327- 6331.

Pillai, S. K., Sakoulas, G., Eliopoulos, G. M., Moellering, R. C., Jr. Murray, B. E. & Inouye, R. T. (2004). Effects of glucose on fsr-mediated biofilm formation in *Enterococcus faecalis*. *J. Infect. Dis.*, *190*, 967–970.

Pozsgay, V., Chu, C., Pannell, L., Wolfe, J., Robbins, J. B. & Schneerson, R. (1999). Protein conjugates of synthetic saccharides elicit higher levels of serum IgG lipopolysaccharide antibodies in mice than do those of the O-specific polysaccharide from *Shigella dysenteriae* type 1. *Proc. Natl. Acad. Sci. U. S. A.*, *96*, 5194–5197.

Pradhan, S. & Rai, L. C. (2001). Biotechnological potential of *Microcystis* sp. in Cu, Zn and Cd biosorption from single and multimetallic systems. *BioMetals.*, *14*, 67

Pradier, C. M., Bertrand, P., Bellon-Fontaine, M. N., Compère, C., Costa, D., Marcus, P., Poleunis, P., Rondot, B. & Walls, M. G. (2000). Adsorption of proteins on an AISI 316 stainless steel surface in natural seawater. *Surf. Interface. Anal.*, *30*, 45-49.

Prince, R. C. (1997). Bioremediation of marine oil spills. *Trends Biotechnol.*, *15*, 158–160.

Rogers, J., Dowsett, A. B., Dennis, P. J., Lee, J. V. & Keevil, C. W. (1994). Influence of temperature and plumbing material selection on biofilm formation and growth of Legionella pneumophila in a model potable water system containing complex microbial flora. *Appl. Environ. Microbiol*, *60*, 1585–1592.

Rosmaninho, R., Santos, O., Nylander, T., Paulsson, M., Mu¨ller-Steinhagen, H. & Melo, L. (2007). Modified stainless steel surfaces targeted to reduce fouling evaluation of fouling by milk components. *J. of Food Eng.*, *80*, 1176–1187.

Schaeffer, D. J. & Krylov, V. S. (2000). Anti-HIV activity of extracts and compounds from algae and cyanobacteria. *Ecotoxicol. Environ. Saf.*, *45*, 208–227.

Schiewer, S. & Volesky, B. (1996). Modeling multi-metal ion exchange in biosorption. *Environ. Sci. Technol.*, *30*(10), 2921-2927.

Sei, A. & Fathepure, B. Z. (2009). Biodegradation of BTEX at high salinity by an enrichment culture from hypersaline sediments of Rozel Point at the Great Salt Lake. *J. appl. Microbiol.*, *107*, 2001–2008.

Seneviratne, G. (2003). Development of eco-friendly, beneficial microbial biofilms. *Curr. Sci.*, *85*, 1395–1396.

Seneviratne, G. & Jayasinghearachchi, H. S. (2005). A rhizobial biofilm with nitrogenase activity alters nutrient availability in soil. *Soil Biol. Biochem.*, *37*, 1975–1978.

Seneviratne, G. & Indrasena, I. K. (2006). Nitrogen fixation in lichens is important for improved rock weathering. *J. of Biosci.*, *31*, 639–643.

Shimada, K., Itoh, Y., Washio, K. & Morikawa, M. (2012). Efficacy of forming biofilms by naphthalene degrading *Pseudomonas stutzeri* T102 toward bioremediation technology and its molecular mechanisms. *Chemosphere.*, *87*, 226–233.

Simo˜es, L. C., Simo˜es, M. & Vieira, M. J. (2007). Biofilm interactions between distinct bacterial genera isolated from drinking water. *Appl. Environ. Microbiol*, *73*, 6192–6200

Sutherland, I. W. (1990). Biotechnology of Exopolysaccharides. Cambridge: Cambridge University Press. (ISBN: 9780511525384).

Sutherland, I. W. (1994). Structure-function relationships in microbial exopolysaccharides. *Biotech. Adv.*, *12*, 393–448.

Sutherland, I. W. (1998). Novel and established applications of microbial polysaccharides. *Tibtech.*, *16*, 41–46.

Sutherland, I. W. (2001). Biofilm exopolysaccharides: a strong and sticky framework. *Microbiology.*, *147*, 3–9.

Tam, N. F. Y., Wong, J. P. K. & Wong, Y. S. (2001). Repeated use of two *Chlorella* species, *C. vulgaris* and WW1 for cyclic nickel biosorption. *Environ. Pollut.*, *114*, 85–92.

Teitzel, G. M. & Parsek, M. R. (2003). Heavy Metal Resistance of Biofilm and Planktonic Pseudomonas Aeruginosa. *Appl. Environ. Microbiol.*, *69*, 2313- 320.

Thomas, V. C., Thurlow, L. R., Boyle, D. & Hancock, L. E. (2008). Regulation of autolysis-dependent extracellular DNA released by *Enterococcus faecalis* extracellular proteases influences biofilm development. *J. Bacteriol.*, *190*, 5690–5698.

Thouvenin, M., Langlois, V., Briandet, R., Langlois, J. Y., Guerin, P. H., Peron, J. J., Haras, D. & Vallee-Rehel, K. (2003). Study of erodable paint properties involved in an antifouling activity. *Biofuels*, *19*(3), 177-186.

Toljander, J. F., Artursson, V., Paul, L. R., Jansson, J. K. & Finlay, R. D. (2006). Attachment of different soil bacteria to arbuscular mycorrhizal fungal extraradical hyphae is determined by hyphal vitality and fungal species. *FEMS Microbiol. Lett.*, *254*, 34–40.

Vandevivere, P. & Baveye, P. (1992). Effect of bacterial extracellular poly-merson the saturated hydraulic conductivity of sand columns. *Appl. Environ. Microbiol.*, *58*, 1690-1698.

Volesky, B. (1990). Removal and recovery of heavy metals by biosorption. In: *Biosorption of Heavy Metals*, ed. B Volesky, CRC Press, Boca Raton (ISBN 9780849349171).

Volk, R. B., Venzke, K. & Blaschek, W. (2007). Structural investigation of a polysaccharide released by the cyanobacterium *Nostocinsulare*. *J. Appl. Phycol.* *19*, 255–262.

Wimpenny, J., Manz, W. & Szewzyk, U. (2000). Heterogeneity in biofilms. *FEMS Microbiol. Rev.*, *24*, 661– 671.

Zhang, X. Q., Bishop, P. L. & Kupferle, M. J. (1998). Measurement of polysaccharides and proteins in biofilm extracellular polymers. *Water Sci. Technol.*, *37*, 345–348.

In: Applied Algal Biotechnology
Editors: M. Arumugam, S. Kathiresan et al.
ISBN: 978-1-53617-524-0

Chapter 16

ALGAE: CO_2 SEQUESTRATION AND BIOREFINERY

Jibu Thomas* and Pandian Sureshkumar
Algae Biomass Research Laboratory, Department of Biotechnology,
Karunya Institute of Technology and Sciences, Coimbatore, India

ABSTRACT

The significant environmental challenge is global warming triggered by increased levels of greenhouse gases in the environment. Multiple mitigation measures are required, including energy saving, efficient utilization of available energy, renewable energy use and sequestration for reducing global CO_2 emissions. In most instances, CO_2 is considered as waste to be disposed of. However, carbon sequestration can become a value-added proposition with advanced technology. There are various potential opportunities that render sequestration economically viable. Several efforts have been taken to solve these problems, but present methodologies are still concerned with the costs and sustainability. Cultivation of photosynthetic microorganisms like algae appears as a sustainable solution as (i) they can effectively utilise CO_2 and nutrients and (ii) the resulting biomass can be converted into value added bioproducts. This review discusses latest trends and technological advances in CO_2 emissions and microalgae sequestration. It also provides a clear view of the appropriate technology for future CO_2 sequestration.

Keywords: CO_2 sequestration, global warming, algae, value added products, biomass

16.1. INTRODUCTION

The growing atmospheric concentration of anthropogenic carbon (CO_2), which could have disastrous implications for the environment and climate, are the main cause of global warming (Chiu et al., 2009). Understanding the global C cycle and its disruption by anthropogenic activities were considered as a major factor for the development of sustainable methods for climate change mitigation. The rate of future increases in atmospheric CO_2

* Corresponding Author's Email: jibuthomas.t@gmail.com.

concentration depends on anthropogenic activity, biogeochemical interaction and climate mechanisms on global C cycle and interaction between the main C pools. Different abiotic (physical) techniques, including injection into geological formations/deep seas or the use in absorbent products, were assessed in order to decrease their atmospheric concentration (Kumar et al., 2010). Biological mitigation of atmospheric CO_2 was considered to be a sustainable one compared to the physical method (Kumar et al., 2011). However, the method of CO_2 sequestration by plant crops were considered as an insufficient mitigation strategy especially due to the low growth rates of terrestrial plant vegetation and its contribution to CO_2 capture was estimated to be just 3-6% of the fossil fuel emitted (Wang et al., 2008).

Alternatively, microalgae have gained renewed attention because of their high growth rate and higher photosynthesis performance than terrestrial crops (Chiu et al., 2009; Dragone et al., 2011). It is widely accepted that algae might be an interesting option to reduce CO_2 emissions. However, there are still a number of restrictions on current algal culture methods were seen. One of them is efficient CO_2 feeding, particularly at large scale cultivation. Current techniques of feeding CO_2 into algal cultivation were based on sparing pure CO_2 or on flue gas directly. The limiting factor in this system is the solubility of CO_2 in water, which requires significant energy for efficient gas to be transferred into fluid and leads to atmospheric losses.

16.2. CO_2 Capture and Storage Methods

There are currently many physico-chemical carbon sequestration and capture techniques, which are combined and classified as carbon capture and storage (CCS) methods. CCS operates through three main steps: capture, transport and storage of carbon dioxide. CO_2 capture is carried out from large point sources like power plants and cement manufacturing plants. CO_2 separation and capture are usually performed via chemical absorption, physical adsorption, cryogenic distillation from the other exhaust components (Pires et al., 2011, 2012). This high concentrated CO_2 is then compressed and transported via pipelines or ship to storage points (McCoy and Rubin, 2008). The next step is to store captured CO_2 in stockpiles, e.g., geological storage, oceano activity where CO_2, saline formations, aquifers or depleted oil and gas pits, are directly injected deep into the ocean. Apart from physical and chemical CCS, the biological path can be followed through natural sinks for CO_2 capture through (a) forestation (Cheah et al., 2016), (b) ocean fertilization (Williamson et al., 2012) and (c) microalgae cultivation (Yadav and Sen, 2017; Zhou et al., 2017).

16.3. CO_2 Mitigation Strategy

The sequestration of carbon can be described as the capture and safe storage of carbon that would otherwise be emitted to, or remain in the atmosphere (Herzog and Golomb, 2004). Yang et al. (2008) reported the complete emission of CO_2 into the atmosphere can be reduced by three ways via reducing use of energy intensity, carbon intensity and enhancing the sequestration of CO_2. Herzog and Golomb (2004) categorized CO_2 capture processes from power production in three general categories. The major concern on the separation of flue gas based on chemical absorption using monoethanolamine (MEA), diethanolamine (DEA) and methyldiethanolamine (MDEA). However, this method can only be achieved when the captured CO_2 is used for commercial commodity, so that the absorption method is productive due to the commercial CO_2 cost. After absorption processes, the heat input for ammonium regeneration requires decomposition of ammonium bicarbonate and ammonium carbonate. Maroto-Valer et al., (2005) estimated that up to 60% energy was saved compared with MEA, and also ammonium sulfate, ammonium nitrate and ammonium bicarbonate were obtained as by-products in that method.

16.4. Carbon Capture by Algae

The capacity to capture CO_2 in ponds in a nongaseous form as bicarbonate to fertilize algal growth is an advantage in aquatic carbon capture and biomass production. The predominant form of CO_2 in water is bicarbonate at moderate pH (>7) and temperatures (under 30°C). Algae usually have active bicarbonate pumps and can concentrate cell bicarbonate. Later these bicarbonates were spontaneously dehydrated and the resulting CO_2 captured in the form of algal biomass by means of the Calvin cycle (Herzog and Golomb, 2004). Algal ponds can be used to harvest a wide range of industrial sources of CO_2. However, their CO_2 levels and other contaminating molecules will vary from anthropogenic CO_2 sources, which will influence the design of CO_2 distribution in the cultivation ponds. Typically, fossil-fuel flue gas has elevated levels of CO_2 ranging from 10-20% but also contains biologically important quantities of nitrous and sulfur oxides (NO_x and SO_x). The injection of flue gasses in power stations into algal ponds was shown to increase the 3 fold of algal biomass, but at a high energy (Jeong et al., 2003). Depending on the algal physiology, pond chemistry and temperature, the effectiveness of CO_2 capture by algae can differ. Keffer and Kleinheinz (2002) reported that CO_2 emissions up to 80 to 99% can be achieved at optimal conditions with less gas residence times up to two seconds. Locating ponds close to CO_2 sources offers a range of cost-effectiveness and energy-saving benefits. Integrated power plant-algal ponds would reduce CO_2 transport expenses, generate restricted waste heat for winter warming ponds from the power plant and could provide the utilities with carbon credits. However, it could be difficult to locate ponds close CO_2, depending on the accessibility of soil and the site's climatic suitability. In addition, in the dark, it is impossible to capture CO_2 from biomass production.

16.5. Carbon Sequestration by Algae

CO_2 must not only be captured, but also sequestrated over long period of time (geological) in order to have the greatest effect on greenhouse gas accumulation. Yang et al., (2008) considered a strategic approach to mitigating CO_2 from point sources as the direct injection of CO_2 into geological formations. But it is hard to ensure long-term safe storage of CO_2 in geological formations. The escape into Earth's surface by large quantities of CO_2 can be disastrous (Benson and Cook 2005). The chemical conversion of CO_2 into stable fluids or solids is a possibly reduced risk approach for sequestering CO_2. The burial of carbonate salts as biomass is another approach for carbon sequestration. This is an especially appealing approach, as solar energy is used for biomass production. The proposition to fertilize open oceans to boost the development of phytoplankton was one of the more contentious manifestations of biomass carbon sequestration (Stepan et al., 2002). Algal growth and carbon capture can be greatly improved by fertilizing oceans with nutrients (e.g., fertilising iron). However, the environmental effects of oceans fertilization are largely unclear.

The controlled production of algal biomass in confined ponds or closed production systems optimized for biomass accumulation and CO_2 capture is an alternative strategic approach to open-ocean fertilization of phytoplankton. There are two possible algal carbon sequestration schemes were considered under this strategy: (1) permanent burial of total algal biomass in deep geologic formations, and (2) burial of extracted or processed carbon-rich fractions from algal biomass. The simplest technological strategy and energy effective way is direct burial of complete algal biomass, since no dewatering is required after processing. The downside is that considerable amounts of inorganic nutrients would also be buried along with the algal biomass.

The transformation of biomass into biochar is another strategy for geologically stable decreased carbon (Hielmann et al., 2010). Biochar is also a stable carbon (more than 90% carbon) which can last millions of years in soils (Amonette et al., 2007) and is a by-product of pyrolysis in the presence of catalyst at anaerobic conditions (Hielmann et al., 2010). Pyrolysis also generates gaseous products including hydrogen, methane, and CO_2, which may be combusted to drive the pyrolytic reaction. Inorganic components may be present in biochar may be added directly to soil as a supplement instead of burial, thereby reduced energy costs and increasing the range of possible applications.

16.6. Limitations of Current Carbon Sequestration Technologies

16.6.1. Chemical Absorption

Chemical absorption-desorption processes, in which the MEA solution is normally used to dissolve the CO_2 are the most prevalent way to separate CO_2 from industrial flue gas (Pires et al., 2011). After the solvent has reached to its maximum absorption capacity, it is heated to 100-150°C to release almost pure CO_2 for storage. Knudssen et al., (2009) reported that ~3.7 GJ/ton of CO_2 of intensive energy required during MEA solution regeneration. Provided the energy needed to absorb 1 ton of CO_2 by MEA solution is obtained from burning coal, approximately 352 kg of CO_2. Another problem in connection with the MEA absorption-desorption method is the fact that SO_2 (co-existing CO_2 in a typical coal-fired power plant) can be readily reacted to by MEA, leading in irreversible MEA degeneration with stable salt (Coleet al., 2011).

16.6.2. Solid Adsorbents

Another technology with low energy inputs and cost-effectiveness seems to be the industrial carbon capture by means of solid adsorbents (Pires et al., 2011). The main disadvantage of using strong CO_2 adsorbents is the need to pretreat flue gasses before they can be transformed into adsorbers. The elevated moisture content of the flue gas and the presence of pollutants (e.g., SO_x and NO_x) can cause adsorbents to intoxicate and thus require a common regeneration of adsorbents. For instance, owing to modifications in heat profile of the adsorbent bed and slow CO_2 desorption, the existence of humidity in flue gas has decreased CO_2 recovery by 78.5% to 60% (Li et al., 2008).

16.6.3. Membrane Technology

Membrane functions as a filter that separates required gas from a gas stream depending on membrane permeability and selectivity. There have been extensive reports of the application of polymer membranes to distinguish CO_2 from gases, and energy consumption is significantly lesser when compared with the amine absorption processes (Bounaceur et al., 2006). However, the operating temperature is always influenced by the performance of polymeric membranes and thus the cooling of warm flue gas is crucial if optimum process segregation is to be maintained (Pires et al., 2011). Other constraints in CO_2 separation membrane technology include high membrane costs of production, fouling impact and elevated membrane surface requirements for elevated industrial flue gas flow rates.

16.6.4. Cryogenic Fractionation

In many industries, cryogenic technology has well known for its high effectiveness in air separation. Cryogenic processes are based on cooling and condensing of multigaseous components at different temperatures. The primary drawback of this technology is the high energy requirement for adequate process refrigeration, resulting in significant efficiency decreases in the application of capture of flue gases from power plants. In order to avoid ice plug or unacceptable high rise in pressure drop during operation, moisture must be removed before the flue gas are channel to the refrigerator system (Tuinier et al., 2010).

16.7. Recent Technologies to Enhance CO_2 Solubility

16.7.1. Microbubble Technology

The most prevalent technique of supplying carbon source for microalgae cultivation used in the industrial sector is sparging CO_2 as a macrobubble (diameter 1-2 mm) (Zimmerman et al., 2011a). However, because of the slow mass transfer rate of CO_2 to water, constant air bubbling is necessary to ensure that the microalgae always have adequate energy sources for its accessibility. While supplying CO_2 could reduce energy and pumping costs, a considerable quantity of CO_2 is anticipated to return to the environment and only a small quantity dissolves into water which is utilised by microalgae for growth. The production of microbubbles instead of macrobubbles seems to be a feasible approach for improving the mass transfer rates of CO_2 into water to overcome these constraints. A range of distinctive features of microbubbles include a high surface-to-volume ratio, slower rising velocity in the liquid phase and faster dissolution rate of gas to liquid can be advantageous (Terasaka et al., 2011; Xu et al., 2011). Zimmerman et al., (2011b) first studied how microbubbles are applied to the microalgae cultivation system. The research used flue gas as a carbon-powered source for the cultivation of *Dunaliella salina* in an aerial photobioreactor aircraft scale (2200 L). The flue gas has been mediated by a fluidic oscillation system that produces microbubbles, which are continuously spread over a microporous diffuser and then generate uniform photobioreactor flow (Zimmerman et al., 2011a). Under this conditions, 23% of CO_2 with fixation rate of 2.4 g/L/day was found to be tolerable.

16.7.2. Membrane-Photobioreactor

Because microbubble technology is still in growth in the microalgae sector, the incorporation of the membrane into a photobioreactor type of airlift or column appear as a potential method for the creation of fine bubbles to improve mass transfer of CO_2 into water. Fan et al., (2008) found that the membrane includes micropores that produce smaller, but bigger bubbles than those of macrobubbles, with diameter (5.5-10.1 mm). Fine bubbles were more advantageous that won't damage the microalgae cells because doesn't collapse under water and doesn't produce a shear stress which usually the damages the cells. Furthermore, owing to high surface-to-volume ratio and slow increasing velocity (longer residential time), CO_2 dissemination in water by thin bubbles can be enhanced (Erikseen, 2008). Rahaman et al., (2011) had recommended certain criterion for an ideal membrane photobioreactor for microalgae cultivation system:

1. hydrophobic membrane with high resistance to acidic or alkaline conditions,
2. fine bubbles with uniform size,
3. even distribution of CO_2 to avoid O_2 accumulation and
4. minimum membrane fouling.

The cloudy dispersion of fine membrane bubbles, reducing light penetration into the photobioreactor is one of the constraints of membrane photobioreactor. This will influence microalgae growth indirectly (Fan et al., 2008). The membrane can be separated from and become an individual system to overcome this restriction. Through the implementation of this approach, mass transfer of CO_2 to water is enhanced by the development of fine membrane bubbles while retaining excellent light penetration.

Summary and Future Perspective

In conclusion, some laboratory and pilot scale studies suggested that capturing CO_2 by algae is a sustainable strategy for mitigating CO_2 emissions from anthropogenic sources. Currently there are few technologies that were implemented for continuous long-term biomass production as well as CO_2 sequestration and capture systems using algae. The long-term scenario for large-scale CO_2 capture and sequestration using algae will depend on the ways of regulation on promoting carbon- capture technologies and the economics of algal pond systems. Several studies reported that CO_2 capture technologies using algae have the sufficiently reduced greenhouse gas emissions and reports also addresses the challenges of algae on competent and economical combination of CO_2 biofixation, wastewater treatment and biofuel production.

References

Amonette, J., Lehmann, J., and Joseph, S. (2007). Terrestrial carbon sequestration with biochar: A preliminary assessment of its global potential. *American Geophysical Union; Fall Meeting 2007*, abstract #U42A-06.

Benson, S., and Cook, P. (2005). Underground geological storage. In: *Carbon Dioxide Capture* and Storage (Eds. Metz, B., Davidson, O., de Coninck, Loos, H. M., and Meyer, L.) 12. *Intergovernmental Panel on Climate Change*. pp 197-278.

Bounaceur, R., Lape, N., Roizard, D., Vallieres, C., and Favre, E. (2006). Membrane processes for post-combustion carbon dioxide capture: a parametric study. *Energy* 31, 2556-2570.

Brennan, L., and Owende, P. (2010). Biofuels from microalgae - a review of technologies for production, processing, and extractions of biofuels and co-products. *Renew. Sust. Energ. Rev.* 14, 557-577.

Cheah, W. Y., Ling, T. C., Juan, J. C., Lee, D. J., Chang, J. S., and Show, P. L. (2016). Biorefineries of carbon dioxide: from carbon capture and storage (CCS) to bioenergies production. *Bioresour. Technol.* 215, 346-356.

Chiu, S. Y., Kao, C. Y., Tsai, M. T., Ong, S. C., Chen, C. H., and Lin, C. S. (2009). Lipid accumulation and CO_2 utilization of *Nannochloropsis oculata* in response to CO_2 aeration. *Bioresour. Technol.* 100, 833-838.

Cole, I. S., Corrigan, P., Sim, S., and Birbilis, N. (2011). Corrosion of pipelines used for CO_2 transport in CCS: is it a real problem? *Int. J. Greenh. Gas Con.* 5, 749-756.

Dragone, G., Fernandes, B. D., Abreu, A. P., Vicente, A. A., and Teixeira, J. A. (2011). Nutrient limitation as a strategy for increasing starch accumulation in microalgae. *Appl. Energy* 88, 3331-3335.

Eriksen, N. T. (2008). The technology of microalgal culturing. *Biotechnol. Lett.* 30, 1525-1536.

Fan, L. H., Zhang, Y. T., Zhang, L., and Chen, H. L. (2008). Evaluation of a membrane-sparged helical tubular photobioreactor for carbon dioxide biofixation by *Chlorella vulgaris*. *J. Membrane Sci.* 325, 336-345.

Herzog, H., and Golomb, D. (2004). Carbon capture and storage from fossil fuel use. *Encycl. Energy*. 1e19.

Hielmann, S. M., Davis, H. T, Jader, L. R, Lefebvre, P. A., Sadowsky, M. J, Schendel, F. J., von Keitz, M. G., and Valentas, K. J. (2010). Hydrothermal carbonization of microalgae. *Biomass Bioenergy*. 34, 875-882.

Jeong, M. L., Gillis, J. M., and Hwang, J. Y. (2003). Carbon dioxide mitigation by microalgal photosynthesis. *B. Korean Chem. Soc.* 24, 1763-1766.

Keffer, J. E, and Kleinheinz, G. T. (2002). Use of *Chlorella vulgaris* for CO_2 mitigation in a photobioreactor. *J. Ind. Microbiol. Biotechnol.* 29, 275-280.

Knudsen, J. N., Jensen, J. N., Vilhelmsen, P. J., and Biede, O. (2009). Experience with CO_2 capture from coal flue gas in pilot-scale: testing of different amine solvents. *Energy Procedia* 1, 783-790.

Kumar, A., Ergas, S., Yuan, X., Sahu, A., Zhang, Q., Dewulf, J., Malcata, F. X., van and Langenhove, H. (2010). Enhanced CO_2 fixation and biofuel production via microalgae: recent developments and future directions. *Trends Biotechnol.* 28, 371-380.

Kumar, K., Dasgupta, C. N., Nayak, B., Lindblad, P., and Das, D. (2011). Development of suitable photobioreactors for CO_2 sequestration addressing global warming using green algae and cyanobacteria. *Bioresour. Technol.* 102, 4945-4953.

Li, G., Xiao, P., Webley, P., Zhang, J., Singh, R., and Marshall, M. (2008). Capture of CO_2 from high humidity flue gas by vacuum swing adsorption with zeolite 13X. *Adsorption* 14, 415-422.

Maroto-Valer, M. M., Fauth, D. J., Yeh, J. T., Resnik, K. P., Rygle, K., and Pennline, H. W. (2005). Semi-batch absorption and regeneration studies for CO_2 capture by aqueous ammonia. *Fuel Process. Technol.* 86, 1533e1546.

McCoy, S. T., and Rubin, E. S. (2008). An engineering-economic model of pipeline transport of CO_2 with application to carbon capture and storage. *Int. J. Greenh. Gas Cont.* 2, 219-229.

Pires, J. C. M., Alvim-Ferraz, M. C. M., Martins, F. G., and Simoes, M. (2012). Carbon dioxide capture from flue gases using microalgae: engineering aspects and biorefinery concept. *Renew. Sustain. Energy Rev.* 16, 3043-3053.

Pires, J., Martins, F., Alvim-Ferraz, M., and Simões, M. (2011). Recent developments on carbon capture and storage: an overview. *Chem. Eng. Res. Des.* 89, 1446-1460.

Rahaman, M. S.A., Cheng, L. H., Xu, X. H., Zhang, L., Chen, H. L. 2011. A review of carbon dioxide capture and utilization by membrane integrated microalgal cultivation processes. *Renew. Sustain. Energy Rev.* 15, 4002-4012.

Stepan, D. J, Shockey, R. E, Moe, T. A, and Dorn, R. (2002). Subtask 2.3 - Carbon dioxide sequestering using microalgal systems. *Energy and Environmental Research Center*, University of North Dakota. 1-27.

Terasaka, K., Hirabayashi, A., Nishino, T., Fujioka, S., and Kobayashi, D. (2011). Development of microbubble aerator for waste water treatment using aerobic activated sludge. *Chem. Eng. Sci.* 66, 3172-3179.

Tuinier, M. J., van Sint Annaland, M., Kramer, G. J., and Kuipers, J. A. M. (2010). Cryogenic CO_2 capture using dynamically operated packed beds. *Chem. Eng. Sci.* 65, 114-119.

Wang, B., Li, Y., Wu, N., and Lan, C. (2008). CO_2 bio-mitigation using microalgae. *Appl. Microbiol. Biotechnol.* 79, 707-718.

Williamson, P., Wallace, D. W., Law, C. S., Boyd, P. W., Collos, Y., Croot, P., et al. (2012). Ocean fertilization for Geoengineering: a review of effectiveness, environmental impacts and emerging governance. *Process Saf. Environ. Prot.* 1, 475-488.

Xu, Q., Nakajima, M., Liu, Z., and Shiina, T. (2011). Biosurfactants for microbubble preparation and application. *Int. J. Mol. Sci.* 12, 462-475.

Yadav, G., and Sen, R. (2017). Microalgal green refinery concept for biosequestration of carbon-dioxide vis-à-vis wastewater remediation and bioenergy production: recent technological advances in climate research. *J. CO_2 Util.* 17, 188-206.

Yang, H., Xu, Z., Fan, M., Gupta, R., Slimane, R. B., Bland, A. E., and Wright, I. (2008). Progress in carbon dioxide separation and capture: A review. *J. Environ. Sci.* 20, 14e27.

Zhou, W., Wang, J., Chen, P., Ji, C., Kang, Q., Lu, B., et al. (2017). Bio-mitigation of carbon dioxide using microalgal systems: advances and perspectives. *Renew. Sustain. Energy Rev.* 76, 1163-1175.

Zimmerman, W. B., Tesar, V., Bandulasena, H. C. H., (2011a). Towards energy efficient nanobubble generation with fluidic oscillation. *Curr. Opin. Colloid. In.* 16, 350-356.

Zimmerman, W. B., Zandi, M., Hemaka Bandulasena, H. C., Tesar, V., James Gilmour, D., and Ying, K. (2011b). Design of an airlift loop bioreactor and pilot scales studies with fluidic oscillator induced microbubbles for growth of a microalgae *Dunaliella salina*. *Appl. Energy* 88, 3357-3369.

In: Applied Algal Biotechnology
Editors: M. Arumugam, S. Kathiresan et al.

ISBN: 978-1-53617-524-0

Chapter 17

PHYCOREMEDIATION OF WASTEWATER

P. Dutta[1], S. Bhakta[2,*] and A. K. Bastia[1]

[1]P. G. Department of Botany, North Orissa University, Baripada, Mayurbhanj, Odisha, India

[2]Botanical Survey of India, Western Regional Centre, Pune, Maharashtra, India

ABSTRACT

Modernized society, vigorously increased industrialization, developed automobiles, chemically programmed agriculture and population growth are the major sources of pollution. Nowadays water pollution has become an important concern for the society which should be resolved as soon as possible. Due to rigorous anthropogenic activity, the accumulation of heavy metals, oils, pesticides, toxins and radioactive compounds in the aquatic systems cause serious problems on environment and organisms affecting negatively the stability of many aquatic ecosystems and can also cause difficulties for animals and human health. So it is important to remediate the polluted factor from the water bodies to make it livable for the living beings. Over the years, many traditional water treatment methods have been introduced which were expensive and ineffective for the removal of heavy metals. On the other hand microalgae culture offers an interesting step for wastewater treatments, because of their contribution to tertiary biotreatment and production of potentially valuable biomass that can be used for several purposes. Microalgae cultures offer an elegant solution to tertiary and quaternary treatments due to the ability of microalgae to use inorganic nitrogen and phosphorus for their growth. And also, for their capacity to remove heavy metals, as well as some toxic organic compounds, therefore, it does not lead to secondary pollution. In the current review, we will emphasize on the role of micro-algae in wastewater treatment by remediating various hazardous water pollutant.

Keywords: microalgae, pollution, heavy metals, wastewater, phycoremediation

---

* Corresponding Author's Email: skmrbhakta@gmail.com.

17.1. INTRODUCTION

Bioremediation is a pollution control technology that uses biological systems to catalyze the degradation or transformation of various toxic chemicals to less harmful forms by using the microbes to clean up contaminated soil and groundwater such as bacteria, algae, fungi that live naturally in the environment. It stimulates the growth of certain microbes that use contaminants as a source of food and energy. Some types of microbes eat and digest contaminants, usually changing them into small amounts of water and harmless gases like carbon dioxide and methane. For bioremediation to be effective, the right temperature, nutrients, and food also must be present. Proper conditions allow the right microbes to grow and multiply and eat more contaminants. If conditions are not right, microbes grow too slowly or die, and contaminants are not cleaned up that time conditions may be improved by adding "amendments." Amendments range from household items like molasses and vegetable oil to air and chemicals that produce oxygen. Amendments are often pumped underground through wells to treat soil and groundwater in situ (in place). There are many forms of bioremediation given as bioleaching, bio-venting, phytoremediation, phyco-remediation, land-farming, composting, rhizo-filtration, bio-absorption, bio-augmentation, mycoremediation and bio-reacting which may be natural or intrinsic bioremediation. Algae that carry out bioremediation are called as bioremediation and the process is called phyco-remediation:

- Microbial bioremediation uses microorganisms (bacteria) to break down contaminants by using them as a food source.
- Phytoremediation uses plants to bind, extract, and clean up pollutants such as pesticides, petroleum hydrocarbons, metals, and chlorinated solvents.
- Mycoremediation uses fungi's digestive enzymes to break down contaminants such as pesticides, hydrocarbons, and heavy metals.
- Phyco-remediation uses algae for removing pollutants from wastewater and improving water quality.

Phycoremediation is a biological treatment system to destroy or reduce the concentration of hazardous waste from a contaminated site by the use of algae. Nowadays various industries have been developed which lead to increased emission of pollutants into ecosystems. The effluents from residential and industrial discharge constitute a major source of water pollution because the industrial effluents are discharged into open drains which finally join the rivers (Kumari et al., 2006). Not only this but also in urban and semi-urban colonies, sewage disposal has become an ecological problem. It has been observed that the entry of metal pollutants into the food chain is easy if food crops are cultivated in heavy metal-contaminated soil which is really harmful to human beings by causing growth and developmental abnormalities, neuromuscular control defects, carcinogenesis, mental retardation, renal malfunction and wide range of other illnesses. Besides this, the elevated levels of metal ions are generally toxic and cause major damage to cell (Inouhe et al., 1996). In that case, phycoremediation is a cost-effective and efficient method of decontamination that has become increasingly popular nowadays to reduce environmental pollution.

The term "micro-algae" describes the different unicellular eukaryotic or aquatic thallophytic organisms belonging *Cyanophyceae/*Cyanobacteria/Oxygenic photosynthetic bacteria (Tomaselli, 2004), *Chlorophyceae, Euglenophyceae, Xanthophyceae, Bacillariophyceae* and *Rhodophyceae* having different photosynthetic assimilatory pigments. Due to modern molecular techniques, more than 40,000 species known to date and new species are still being discovered. Microalgae are found in both marine and freshwater environments having different structure and mechanism. Compared to land-based plants, microalgae are generally more efficient in converting solar energy to biomass due to presence of sufficient water, CO_2 and other nutrients in an aqueous environment. In addition, photosynthetically produced oxygen can relieve biological oxygen demand (BOD) in the wastewater. In the last two decades, researchers have extensively studied the accumulation of metals by microorganisms (algae, bacteria, fungi, and yeast). But it has been observed that algae are gaining more attention than other microorganisms because algae, particularly marine algae, are a rich source in the oceanic environment able to accumulate high metal content. Although adsorption on the cell surface is the dominant mechanism both surface adsorption and internal diffusion are involved in the uptake of metals by algae. Conventional physicochemical methods such as electrochemical treatment, ion exchange, precipitation, reverse osmosis, evaporation, and absorption for heavy metal removal from waste streams are not cost-effective and hence biological approach has been considered as alternative remediation for heavy metal contamination. Metal concentrations absorbed by algae were influenced by many environmental variables. It was strongly pH-dependent and the presence of co-cations generally reduced the uptake of the target cations by algae.

Algae can be cultivated in unproductive habitats and poor quality waters, which include municipal, agriculture and industrial wastewaters. Treatment of paper industry waste water with the indigenous algae lead significant result (Dutta et al., 2019). Cultivation of algae in these habitat helps to reduce the pollution load, carbon footprint of treatment and production cost of the algal biomass. Microalgae require a high amount of nutrients (N, P) for their growth, due to the high content of proteins in their biomass. Cultivation of microalgae with the use of the freshwater medium with chemical fertilizers increases the cost of production besides raising the questions regarding their environmental impact (Chisti, 2013). Studies have shown that harness of wastewater can be decreased either by dilution (Singh et al., 2012; Renuka et al., 2013) or preliminary treatment with UV radiation and so on (Cho, 2011). Therefore, wastewaters rich in nutrients (N, P) can be an environmentally friendly low-cost cultivation media for algal biomass production, generating high-value algal biomass useful for various industrial applications with an additional advantage of bioremediation (Chinnasamy et al., 2014).

17.2. Bioremediation by Microalgae

17.2.1. Bioremediation of Heavy Metals

In recent years the presence of heavy metals in the environment is a major concern due to their high toxicity to flora and fauna. As a result, the recovery of heavy metals from industrial waste streams is becoming progressively more important as civilization realizes the inevitability for recycling and conservation of essential metals. Nowadays due to rapid population growth, there is a vulnerable increase in industrial wastes, geochemical structure, and mining of metals which create a potential source of heavy metal pollution in the aquatic environment (Gumgum et al., 1994). Rapid urbanization, industrialization, fertilizer, and pesticide use has resulted in heavy metal (Cu, Zn, Cd, Cr and Ni etc) pollution of land and water resources (marine, ground, fresh) that has caused imbalance in aquatic ecosystems and the biota growing under such habitats which in turn, are being assimilated and transferred within food chains by the process of magnification (Pergent et al., 1999). As heavy metals are not eco-friendly in nature so it tends to accumulate in living organisms causing lots of problems. In general there are two types of toxic metals having the first group consists of metals (As, Cr, Co, Cu, Ni, Se, Va and Zn), essential as nutritional requirements at trace amount for many organisms but are toxic when present in greater amounts and the second group consists of metals (Pb, Hg, Cd, Ur, Ag and Be) having highly poisonous activity and are not known to have any nutritional value (Inthron et al., 2001).

Table 17.1. Microalgae used for bioremediation of heavy metals

Microalgae	Reference	Metal studied
Scenedesmus acutus, Chlorella vulgaris	Travieso et al., 1999	Cd, Zn and Cr
Scenedesmus sp., Chlorococcum sp., Chlorella vulgaris var. vulgaris and *Fischerella sp., Lyngbya spiralis, Tolypothirx tenuis, Stigonema sp., Phormidium molle, Lyngbya heironymusii, Gloeocapsa sp, Oscillatoria jasorvensis, Nostoc sp. Scenedesmus acutus, Scenedesmus acutus*	Inthorn et al., 2001	Cd, Pb, and Hg
Spirulina platensis	Gannikar, 2002	Cu, Hg and Pb
Padina sp.	Kaewsarn, 2002	Cu
Scenedesmus obliquus, Chlorella pyrenoidosa and *Closterium lunula*	Yan and Pan, 2002	Cu
Dunaliella and *Chlorella*	Muhaemin, 2004	Pb
Tetraselmis chuii	Yilmaz et al., 2005	Cu
Oscillatoria sp., Spirogyra sp.	Diwan, 2007	Ni(II)
Spirulina (Arthrospira) platensis	Arunakumara et al., 2008	Pb
Anabaena variabilis, Aulosira sp., Nostoc muscorum, Oscillatoria sp. and *Westiellopsis sp.*	Parameswari et al., 2010	Cr(VI), Ni (II)
Scenedesmus bijuga, Oscillatoria quadripunctulata	Ajayan et al., 2011	Cu, Co, Pb, Zn
Spirogyra sp., Nostoc commune	Mane et al., 2011	Se
Chlorella minutissima	Singh et al., 2011	Cr(VI)
Chlorella pyrenoidosa	Yao et al., 2012	Zn, Cu, As, Pb, Cd, Cr, Ni, Hg
Spirogyra hyalina	Nirmal Kumar and Cini, 2012	Cd, Hg, Pb, As and Co

To remove heavy metals from wastewaters various economic, effective and safe methods are needed which resulted in the search for alternative materials that may be useful in reducing the heavy metal levels in the environment. The ability of biological materials to remove and accumulate heavy metals from the aqueous environment is called as biosorption that received considerable attention in recent years because of using cheaper materials such as naturally abundant microalgae or biosorbents (Davis, 2000). The risk of accidental release of pollutants into the atmosphere causing health safety and environmental problems are avoided by physio-remediation (microalgae are employed for remediation). Algae have the power to take up poisonous heavy metals from the environment, resulting in higher concentrations than those in the surrounding water. The idea of the application of microalgae in bioaccumulation of heavy metal ions was proposed for the first time by Oswald and Gootas in 1957, but it has got attention only recently (Oswald, 1988; Doshi et al., 2007). Microalgae have preferentially developed the production of peptides capable to bind heavy metals and these molecules, as organometallic complexes, are further partitioned inside vacuoles to assist appropriate control of the cytoplasmic concentration of heavy metal ions, thus preventing or neutralizing their potential toxic effect (Cobbett and Goldsbrough, 2002). Though the absorption of metal ions by microalgae from the aqueous environment has spent much attention on wastewater treatment that may reduce the metal ion concentration.

17.2.2. Bioremediation of Oil

In the recent years, oil pollution due to oil spill has become a huge problem in the society due to the release of liquid petroleum hydrocarbon into the environment. It includes spills of refined petroleum products like gasoline, diesel, their by-products, the release of crude oil from tankers, offshore platforms, drilling rigs and wells. Not only this but also the heavier fuels used by large ships like bunker fuel or waste oil also cause pollution to the environment which is a big environmental concern. Oil pollution in the aqueous environment has a worse effect in the marine form than another aquatic source. The fact is from the middle of the 19th-century marine oil pollution has been receiving increasing attention with the amplification of marine tanker collisions, pollutant release from coastal refineries (Wake, 2005), tanker operations and oil use, continuous operative discharges from ships (Carpenter and Mac Gill, 2001). So to overcome this type of problem an easier and cheaper method should be monitored to clean up the spill. As a result, microalgae are taken as an experimented material to check whether the algal growth is favorable to purify the polluted water or not. Microalgae need nitrogen and phosphorus for their growth, so it can remove the excess amount of nitrogen and phosphorus from the wastewater including the farm runoff. Besides this, they have the ability to use carbon dioxide from coal-fired power plants, reduction of greenhouse gas by producing algal biomass that can be converted into biofuel. Not only this but also microalgae can degrade petroleum hydrocarbons found in crude and motor oils (Walker et al., 1975). Praepilas and Pakawadee (2011) investigated that the potential of microalgae (*Scenedesmus quadricauda* and *Chlorella* sp.) is better than the other microorganism for utilizing industrial wastewater as an inexpensive nutrient for their growth and oil accumulation. Numerous microalgae can build up lipids due to excess photosynthate and some species can accumulate lipids under heterotrophy or environment stress, such as nutrient deficiency or salt stress (Takagi et al., 2006). This suggested that the microalga are capable of degrading different oils to varying levels along with other microorganisms.

Table 17.2. Degradation of petroleum compounds and fuel components by different groups of microorganisms

Algae	Compound
Selanastrum capricornatum	Benzene, toluene, naphthalene, phenanthrene, pyrene
Cyanobacteria(Blue-green algae)	Benzene, toluene, naphthalene, phenanthrene, pyrene
Microcystis aeruginosa, Mixed cultures (Yeasts, molds, protozoa, bacteria, activated sludge)	Acrylonitrile
Activated sludge	Dibenzanthraceae
Sewage sludge	Fluoranthene
Acinetobacter calcoaceticus	Petroleum derivatives
Strains of *Pseudomonas putida*	Phenol cresols
Trichosporon pullulans	Paraffins
Aeromonium sp	Total petroleum hydrocarbons
Mycobacterium sp.	n-Undecane

17.2.3. Bioremediation of Pesticides

Contamination of aquatic systems due to pesticides is one of the serious problems that have been faced by the people, environment, and organism as a whole. A large variety of organochlorine compounds are exposed to an aquatic environment via domestic sewage, pesticide from agricultural lands and various industrial effluents which is a big concern to the society. Organochlorine pesticides are omnipresent environmental pollutants in the global ecosystem and due to their lipophilicity and persistence, they accumulate along the food chain causing a bad effect on the environment. (Loganathan and Kannan, 1991, Voldner and Li, 1995). Recent studies have indicated that extensive use of herbicides in agricultural land causes contamination to the aquatic system which is a risk to the environment (Moro et al., 2012). Once it was introduced to the environment these chemicals have the potential to create adverse effects on the ecosystem by causing various diseases (Moore et al., 2007). From various surveys it was found that in developing countries, due to spontaneous use of pesticides millions of people face death due to intoxication in a year (Duran-Nah and Colli-Quintal, 2000) and multiple methods including incineration and landfilling have been used to remove this class of pollutants, however, these physical methods are exclusive and ineffective. As a result to make this more effective and pollution-free researchers introduced aquatic halophytes as algae. Various studies have indicated that microalgal spices have different sensitivity to pesticides.

The response of pesticides depending upon the concentration used the duration of exposure and algal species tested. Recently it has observed that the microalgae *Scenedesmus quadricauda* was more effective in the removal of two fungicides (dimethomorph and pyrimethanil) and one herbicide (isoproturon) from their medium (Olette et al., 2010). Also, the green alga *Chlamydomonas reinhardtii* showed a great ability to accumulate and degrade the herbicide prometryne (Jin et al., 2012). Factors that affect biodegradation of pesticides are a microbial consortium, the optimum condition for their survival and activity of algae where the other one is the chemical structure of the pesticides. Some report showed that freshwater microalgae have the ability to convert bisphenol A into its simpler form which is not toxic (Nakajima et al., 2007) and two Chlorophyceae (*Pediastrum tetras, Ankistrodesmus fusiformis*), one diatom *(Amphora coffeaeformis)* are resistant to herbicides chlortoluron and mesotrione (Moro et al., 2012). In various experiments the interaction between pesticides and algae showing that not only microalgae are capable of bio-accumulating pesticides but also capable of bio-transforming of environmental pollutants.

17.2.4. Bioremediation of Toxins

Due to vigorous pollution, different types of pollutants are found to be seen in the environment which introduces various types of newly developed problems to the society. Compared to another pollutant, the toxin is a contradictory type of pollutant which causes environmental toxicity having a bad ecological effect on the organism. If the biomass of algae is available in more amounts then it can remediate the toxins from the sample product. It has seen that the availability of algal biomass is more than comparing to another present because it mainly found in the aquatic environment. So it was taken as a logical choice for aquatic toxicological studies, yet has been underutilized in this experiment.

17.2.5. Bioremediation of Radioactive Compounds

Due to the excessive increase in civilization, modern technologies have been developed to fulfill their demand and need in day-to-day life. Besides textile, paper, iron, steel, oil, pesticide and other industries, the nuclear industries are also developed in various cities which emits the excess amount of uranium into the environment. Uranium is one of the most seriously threatening heavy metals because of its high toxicity and some radioactivity that has a bad effect on the environment. Uranium causes pollution as a poisonous threat to the aqueous source mainly in surface and ground waters (Laul, 1992), so it is required to control the emission of heavy metal especially uranium into the environment. As a result, various predictable methods (e.g., precipitation and sludge separation, chemical oxidation or reduction, ion exchange, reverse osmosis, membrane separation, electrochemical treatment, and evaporation) were applied to remove radioactive materials from industrial effluents. But these methods are often ineffective and costly when applied to effluents either in dilute or very dilute forms (Aksu, 1998). Therefore the alternative methods are applied by using marine algae which are capable of absorbing the radionuclide such as radium, thorium, and uranium emitted from nuclear industry e.g., biosorption of uranium by brown algae (*Cystoseira indica*) biomass (Edgington et al., 1970) and experiment showed that this method is cheaper and easier than comparing to the other methods used before.

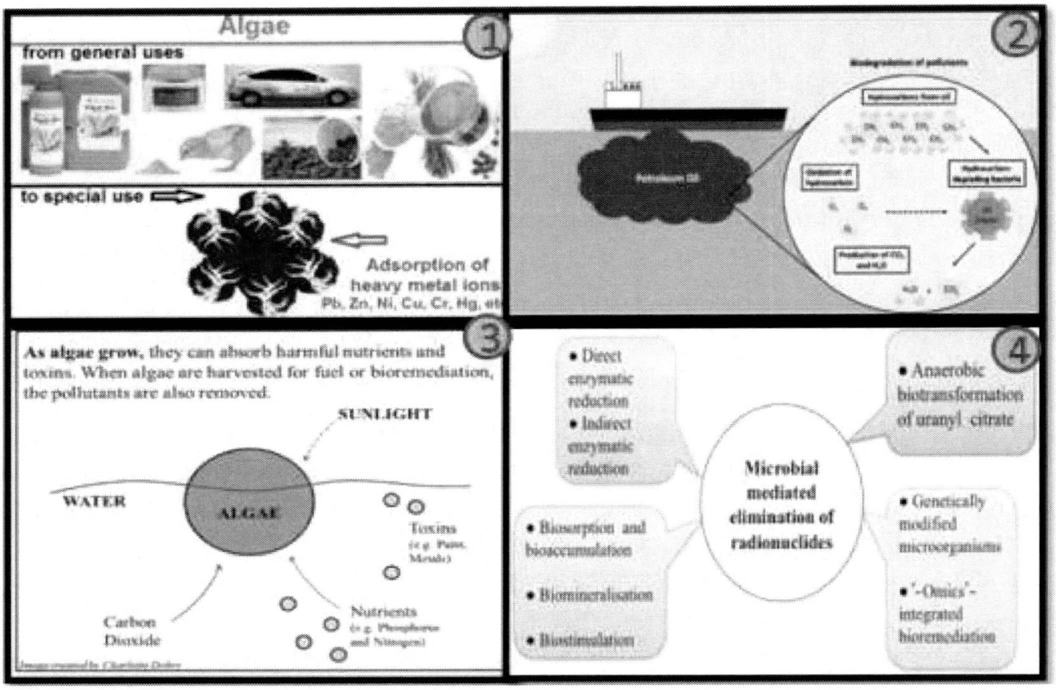

Figure 17.1. Bioremediation of heavy metals, oil, pesticides and radio active compounds by microalgae.

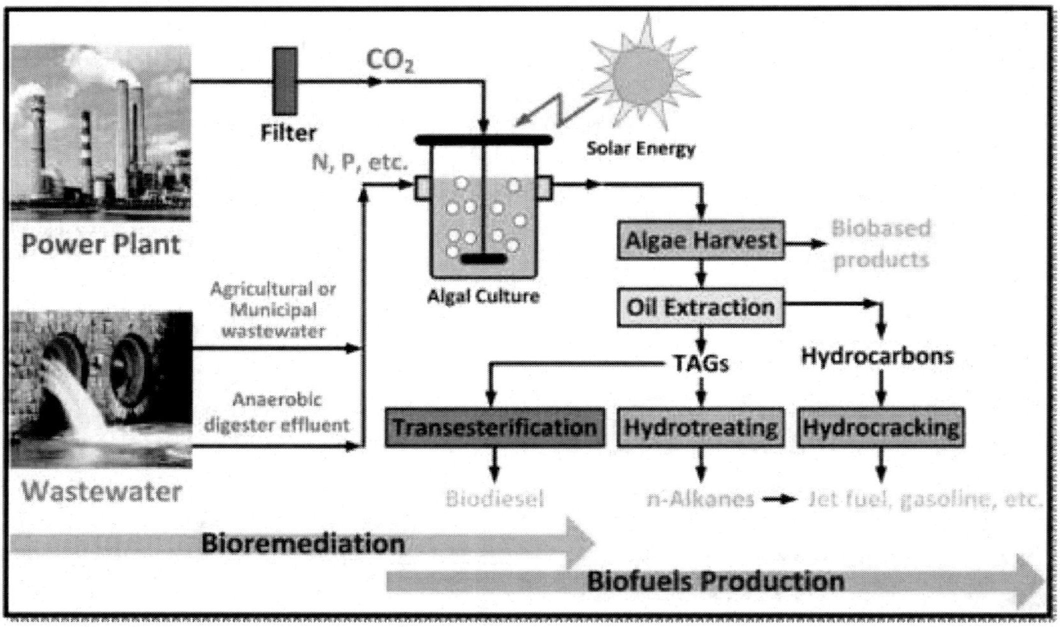

Figure 17.2. Bioremediation of wastewater by microalgae by using different methods with the production of biofuel (Shivakumar et al., 2012).

17.2.6. Methodology

Various methodologies are being proposed to access the treatment of waste water. Algal bioremediation has been well studied over the past 40 years by (Ryther, 1972; Kuyucak, 1988; Romero-Gonzalez, 2001 and Wang and Lan, 2011). Considerable research efforts have been devoted to the development of algal biosorbents to remediate pollutants, as algae are important bioremediation agents, and already being used in waste water treatment. The potential for algae in wastewater remediation is however much wider in scope than its current role (Volesky, 1990). Three types of ponds are employed in waste water treatment, anaerobic, facultative and aerobic pond. Anaerobic ponds are several meters deep (2.5-5 m), free of dissolved oxygen and have high BOD (400-3,000 kg/ha. d) removal rates. It is used as pre-treatment of high strength waste and light penetration is not important. The facultative ponds exhibit aerobic conditions on the surface due to photosynthetic oxygen production by algae and anaerobic conditions in the bottom layers and are shallow and completely oxygenated throughout. The depth of the pond is about 1-2 m and the aerobic layer acts as a good check against odor evolution from the pond. The aerobic ponds are shallow ponds with depth less than 0.5 m and BOD loading of 40-120kg/ha. d so as to maximize penetration of light throughout the liquid depth. Such ponds develop intense algal growth.

17.2.7. High-Rate Algal Ponds (HRAP)

Removal of the nutrient with the aid of algae compares very favorable method to other conventional technologies (Muthukumaran et al., 2005; De la Noue et al., 1992). In contrast to facultative ponds, HRAPs are designed to promote algae growth which is shallow (0.3-0.6 m) in order to allow maximum light penetration. They can operate at short hydraulic retention time (HRT) in the range of 4 to 10 days depending on climatic conditions reducing the required surface area. Continuous mixing is provided to keep the cells in suspension and to expose them periodically to light. The most common design that has proven successful at a large scale is the single loop paddlewheel mixed. More recently, a special flow pattern was introduced to improve the efficiency of this type of ponds. HRAPs are the most cost- effective reactors available for liquid waste management and for efficient capture of solar energy.

17.2.8. Cell Immobilization

In phyco-remediation, harvesting of micro algae is the costliest step and to avoid excessive harvesting cost, techniques have been developed by various research groups working in the field of cell immobilization. Green algae and cyanobacteria have been successfully immobilized for the removal of nutrients in phyco-remediation techniques. It is an important notice that chitosan and polyvinyl foams are low-cost polymers with a long term performance, but each one of these particular immobilization techniques would have come limitations at the industrial level. For example, hollow fibers are expensive as for techniques, loss of cell viability has been observed when using covalent immobilization and mass transfer limitations at a high degree have been encountered while using cell entrapment methods.

Thus it is recommended to choose matrices with physical and chemical resistance or long-term use and to avoid complex immobilization method.

17.3. Use of Strains with Special Attributes

Nutrient removal has been shown to be more efficient by using algae strains with special attributes. Such special attributes include tolerance to extreme temperatures, chemical composition with a predominance of high added value products, a quick sedimentation behavior, or capacity for growing mixotrophically. A *Phormidium* strain capable of removing nutrients more efficiently than a community of green algae below 10^0C was isolated from the polar environment. The authors suggested that this strain was appropriate for wastewater treatment in cold climates during spring and autumn. On the other hand, *Phormidium bohneri* was a good candidate for treating wastewater at high temperature (around 30^0C); additionally, such strain had a quick sedimentation behavior. *Spirulina (Arthrospira)* is one of the most favored microalgae for wastewater treatment. Like that the microalgae Scenedesmus quadricauda and Chlorella sp. are responsible for utilizing waste water as a cheap nutrient for their growth and oil accumulation (Praepilas and Pakawadee, 2011). It shows that different strains of algae have the different and peculiar character for remediation wastewater.

17.4. Micro Algae at Wastewater Pond Treatment in Cold Climate

Cold climate normally is not favorable for phyco-remediation due to the poor light availability and low temperature. But still, many workers in the past have successfully demonstrated the possibility of using certain microalgae with special attributes to treat wastewater. Gronlund (2004) working with wastewater treatment in Sweden used *Coelastrum*, *Chlamydomonas*, *Chlorella, Selenastrum, Scenedesmus* and *Micractinium pusillum* in HRAs with artificial illumination and covered with green house plastic. Comparative study with polar *cyanobacteria* and *chlorococcales* were carried out for tertiary wastewater treatment (Tang et al., 1997). *Phormidium sp* was considered more promising. The *cyanobacteria* are preferred due to their flocculation properties, active growth and nitrate and phosphate uptake at low temperature.

Conclusion

Growing microalgae in wastewater offer new insights for the microalgae industry as well as the wastewater treatment industry. The use of wastewaters for cultivating microalgae is necessary in order to reduce the cost of microalgae production. This is a requirement for microalgae to enter the energy market through biofuels. The wastewater treatment industry is facing challenges (such as micropollutants fate) that induce the development of other alternatives. Microalgae-related processes can be an interesting alternative to the conventional activated sludge process. Despite these two opportunities, many research and development challenges have still to be overcome in order to benefit from the full potential of the combination of microalgae production and wastewater treatment, namely in the development of robust, productive wastewater-adapted microalgal species, and in the improvement and innovation of cultivation and downstream processing systems which will allow for better growth, harvesting and conversion of the algal biomass. The cultivation of microalgae is known to be the most profitable business in the biotechnological industry. It is a wasteless, ecologically pure, energy and resource-saving process.

ACKNOWLEDGMENT

The authors thank full to the DST for providing INSPIRE fellowship, the Head, P. G. Department of Botany, North Orissa University, Odisha. Further, the authors are also thankful to the Director, Botanical Survey of India, Western Regional Centre, Pune, Maharashtra, India for providing laboratory facilities.

REFERENCES

Ajayan, K. V., Selvaraju, M. and Thirugnanamoorthy, K. (2011). Growth and heavy metals accumulation potential of microalgae grown in sewage wastewater and Petrochemical Effluents. *Pakistan J. Biol. Sci.* 14 (16), 805-811.

Aksu, Z. (1998). Biosorption of heavy metals by microalgae in batch and continuous systems. In: Wong, Y. S. and TAM, N. F. Y. eds. *Algae for wastewater treatment.* Germany. Springer. 164-182pp.

Arunkumar, KKIU, Zhang, X and Song, X. (2008). Bioaccumulation of PB2+ and its effects on Growth, Morphology and Pigment contents of *Spirulina (Arthrospora) platensis. J Ocean University China.* 7(4), 397-403.

Carpenter, A. and MacGill, S. 2001 Charging for port reception facilities in North Sea Ports: putting theory into practice. *Mar. Pollut. Bull.* 42, 257–266.

Chinnasamy, S., Sood, A., Renuka, N., Prasanna, R., Ratha, S. K., Bhaskar, S., Rengasamy, R. and Lewis, D. M. (2014). Ecological aspects of algae cultivation in wastewaters for recycling of nutrients and biofuel applications. *Biofuels.* 5, 141-158.

Chisti, Y. (2013). Constrains to commercialization of algal fuels. *J. Biotechnol.* 167, 201-214.

Cho, S., Luong, T. T., Lee, D., Oh, Y. and Lee, T. (2011). Reuse of effluent water from municipal wastewater treatment plant in microalgae cultivation for biofuel production. *Bioresour. Technol,* 102, 8639-8645.

Cobbett, C. and Goldsbrough, P. (2002). Phytochelatin and metallothioneins: Roles in heavy metal detoxification and homeostasis. *Ann. Rev. Plant Biol.* 53, 159–182.

Davis, T. (2000). Sargassum seaweed as biosorbent for heavy metals (Davis, T., Volesky, B., Vieira, R.). *Water Res.* 34, 4270-4278.

De La Noue J., Laliberté G. and Proulx D. (1992). Algae and Waste Water. *J. Appl. Phycol.* 4, 247-254.

Diwan, S. (2007). Removal of Ni(II) from Aqueous Solution by Biosorption using two green algal species *Oscillatoria sp.* and *Spirogyra sp.*, 5th WSEAS Int. Conf. *On environment, ecosystems, and development*, pp. 14-16.

Doshi, H., Ray, A. and Kothari, I. L. (2007). Bioremediation potential of live & dead Spirulina: Spectroscopic, kinetics and SEM studies. *Biotechnol. Bioeng.* 96 (6), 1051-1063.

Duran-Nah, J. and Colli-Quintal, J. (2000). Acute pesticide poisoning. *Mexican Pub. Heal.* 42, 53-55.

Dutta, P., Bhakta, S., Sahu, E., Bhuyan, P. and Bastia, A. K. (2019). Analysis of growth and biochemical contents of microalgae grown with wastewater effluent of Emami paper mill, Balasore. In: L. B. Sukla (eds), The Role of Microalgae in wasterwater treatment, *Springer Nature, Singapore Pte Ltd.*153-168pp. doi.org/10.1007/981-13-1586-2_12.

Edgington, D. N., Gorden, S. A. and Thommes, M. M. (1970). The concentration of radium, thorium, and uranium by tropical marine algae. *Limnol. Oceanogr.* 15(1), 945-955.

Garnikar, D. (2002). Accumulation of copper, mercury, and lead in *Spirulina* platensis studied in Zarrouk's medium. *J. KMITNB.* 12(4), 333-35.

Gronlund, E., Klang, A., Falk, S. and Hanaeus, J. (2004). Sustainability of wastewater treatment with microalgae in cold climate, evaluated with emergy and socio-ecological principles. *Ecol. Eng.* 22, 155-174.

Gumgum, B., Unlu, E. and Tez, Z. (1994). Heavy metal pollution in water, sediment, and fish from the Tiger River in Turkey. *Chemosphere.* 29 (1), 111-116.

Inouhe, M., Sumiyoshi, M., Tohoyanna, H. and Joho, M., (1996). Resistance to cadmium ions and formation of a cadmium binding complex in various wild type yeasts. *Plant Cell Physiol.* 37, 341-346.

Inthorn, D., Sidtitoon, N. and Silapanuntakul, S. (2001). Sorption of mercury, cadmium, and lead by microalgae. *Res. Art. Sci. Asia.* 28, 253-261.

Jin, Z. P., Luo, K., Zhang, S., Zheng, Q. and Yang, H. (2012). Bioaccumulation and catabolism of prometryne in green algae. *Chemosphere.* 87, 278-84.

Kaewsarn, P. (2002). Biosorption of copper(II) from aqueous solutions by pre-treated biomass of marine algae *Padina sp. Chemosphere.* 49, 471-476.

Kumari, S. B.., Kirubavathy, A. K. and Thirumalnesan, R. (2006). Stability and water quality criteria of open drainage municipal sewage water at Coimbatore used for irrigation. *J. Environ. Biol.* 27(4), 709-712.

Kuyucak, N. and Volesky, B. (1988). Biosorbents for recovery of metals from industrial solutions. *Biotechnol. Lett.* 10, 137–142.

Laul, J. C., (1992). Natural radionuclides in groundwater. *J. Radioanalyt. Nucl. Chem.* 156(1), 235-241.

Loganathan, B. G. and Kannan, K. (1991). Time perspectives of organochlorine contamination in the global environment. *Mar. Pollut. Bul,* 22, 582-584.

Mane, P. C., Bhosle, A. B. and Jangam, C. M. (2011). Bioadsorption of selenium by Pretreated Algal Biomass. *Adv. Appl. Sci. Res.* 2(2), 202-207.

Moore, M. T., Cooper, C. M., Smith, Jr. S., Cullum, R. F., Knight, S. S., Locke, M. A., Bennett, E. R. (2007). Diazinon mitigation in constructed wetlands: influence of vegetation. *Water Air Soil Poll.* 184, 313–321.

Moro, C. V., Bricheux, G., Portelli, C. and Bohatier, J. (2012). Comparative effects of the herbicides chlortoluron and mesotrione on freshwater microalgae. *Environ. Toxicol. Chem.* 31, 778-86.

Muhaemin, M. (2004). Toxicity and Bioaccumulation of lead in *Chlorella* and *Dunaliella. J. Coast. Develop.* 8 (1), 27-33.

Muthukumaran, M., Raghavan, B. G. and Subrahmanian, V. (2005). Bioremediation of industrial effluent using microalgae. *Indian Hydrobiol.* 7, 105 -122.

Nakajima, N., Teramoto, T., Kasai, F., Sano, T., Tamaoki, M., Aono, M., Kubo, A., Kamada, H., Azumi. Y. and Saji, H. (2007). Glycosylation of bisphenol A by freshwater microalgae. *Chemosphere*. 69, 934–41.

Nirmal Kumar, J. I., and Cini, O. (2012). Removal of heavy metals by biosorption using freshwater alga *Spirogyra hyalina*. *J. Environ. Biol.* 33, 27-31.

Olette, R., Couderchet, M., Biagianti, S. and Eullaffroy, P. (2010). Fungicides and herbicide removal in Scenedesmus cell suspensions. *Chemosphere*. 79, 117–123.

Oswald, W. J. (1988). Micro-algae and waste-water treatment. In: Borowitzka, M. A. and L. J. Borowitzka editors. Micro-Algal Biotechnology. Cambridge: *Cambridge University Press*. 305- 328 pp.

Parameswari, E., Lakshmanan, A. and Thilagavathi, T. (2010). Phycoremediation of heavy metals in polluted water bodies. *Elect. J. Environ. Agri. Food Chem.* 9(4): 808-814.

Pergent, C. and Pergent-Martini, C. (1999). Mercury levels and fluxes in *Podosonia oceanica* meadows. *Environ. Poll.* 106, 33–7.

Praepilas, D. and Pakawadee, K. (2011). Effects of Wastewater Strength and Salt Stress on Microalgal Biomass Production and Lipid Accumulation. *World Acad. Sci. Eng. Technol.* 60, 1163-1168.

Renuka, N., Sood, A., Ratha, S. K., Prasanna, R. and Ahluwalia, A. S. (2013). Evaluation of microalgal consortia for treatment of primary treated sewage effluent and biomass production. *J. Appl. Phycol.* 25, 1529-1537.

Romero-Gonzalez, M. E., Williams, C. J. and Gardiner, P. H. E. (2001). Study of the mechanisms of cadmium biosorption by dealginated seaweed waste. *Environ. Sci. Technol.* 35, 3025–3030.

Ryther, J. H., Tenore, K. R., Dunstan, W. M. and Huguenin, J. E. (1972). Controlled eutrophicationincreasing food production from the sea by recycling human wastes. *Bioscience*. 22, 144.

Singh, S. K., Bansal, A. and Jha M. K. (2012). An integrated approach to remove Cr (VI) using immobilized *Chlorella minutissima* grown in nutrient-rich sewage wastewater. *Bioresour. Technol.* 104, 257-265.

Sivakumar, G., Xu, J., Thompson, R. W., Yang, Y., Smith, P. R. and Weathers P. J. (2012). Integrated green algal technology for bioremediation and biofuel. *Bioresour. Technol.* 107, 1–9.

Takagi, M. and Karseno, Y. T. (2006). Effect of salt concentration on intracellular accumulation of lipids and triacylglyceride in marine microalgae *Dunaliella* cells. *J. Biosci. Bioeng.* 101, 223–226.

Tang, E. P. Y., Vincent, W. F., Proulx, D., Lessard, P. and Noue, J. (1997). Polar cyanobacteria versus green algae for tertiary waste-water treatment in cool climates. *J. Appl. Phycol.* 9(4), 371-381.

Tomaselli, L. (2004). The microalgal cell. In *Handbook of Microalgal Culture*, (ed. Richmond, A.). *Blackwell, Oxford,* 3-19pp.

Travieso, L., Canizares, R. O., and Borja. R. (1999). Heavy metal removal by microalgae. *Bull. Environ. Contamin. Toxicol.* 62,144-151.

Voldner, E. C. and Li, Y. (1995). Global usage of selected persistent organochlorines. *Sci. Total Environ.* 161, 201-210.

Volesky, B. (1990). Biosorption of heavy metals, *Boca Raton, Flo*: Press CRC, Florida. 3-6pp.

Wake, H. (2005). Oil refineries: a review of their ecological impacts on the aquatic environment. *Estuar. Coast. Shelf Sci*. 62, 131–140.

Walker, J. D., Colwell, R. R. and Petrakis, L. (1975). Degradation of petroleum by an alga, *Prototheca zop¢i. Appl. Microbiol.* 30, 79-81.

Wang B and Lan, C. Q. (2011). Biomass production and nitrogen and phosphorus removal by the green alga *Neochloris oleoabundans* in simulated wastewater and secondary municipal wastewater effluent. *Bioresour. Technol.* 102(10), 5639– 5644.

Yan, H. and Pan, G. (2002). Toxicity and bioaccumulation of copper in three green microalgal species. *Chemosphere,* 49, 471-476.

Yao, J., Li, W., Xia, F., Zheng, Y. Fang, C. and Shen, D. (2012). Heavy metals and PCDD/Fs in solid waste incinerator fly ash in Zhejiang province, China: Chemical and bioanalytical characterization. *Environ. Monitor. Assess.* 184 (6), 3711-3720.

Yilmaz, A. B., Isik, Oya and Sayin, S. (2005). Bioaccumulation and tosicity of different copper concentration in Tetraselmis chii. *E.U. J. Fish. Aqua. Sci.* 22 (3-4), 297-301.

In: Applied Algal Biotechnology
Editors: M. Arumugam, S. Kathiresan et al.

ISBN: 978-1-53617-524-0

Chapter 18

CYANOBACTERIA MEDIATED REMEDIATION OF ESTRONE: AN EMERGING POLLUTANT PRESENT IN AQUATIC BODIES

Neha Sami and Tasneem Fatma*
Cyanobacterial Biotechnology Laboratory, Department of Biosciences,
Jamia Millia Islamia, Delhi, India

ABSTRACT

Hormone activating agents form a hazardous group of pollutants called endocrine-disrupting chemicals (EDC). Natural estrogens, particularly the female estrogen, estrone (E1) has been considered as EDC. Humans and animals excreta, dumping of animal waste and domestic sewage effluents are the main sources of estrone in aquatic bodies. It bio-accumulates through the food chain and affects both aquatic and terrestrial organisms. It interferes with the signaling pathway of endogenous hormone and causes many deleterious effects including breast cancer, infertility, and animal hermaphroditism. Commonly used physical and chemical methods of remediation show residual toxicity besides being cost-ineffective. Thus, thorough toxicological and bioremediation studies of estrone are prerequisites for ensuring the good health of the organism as well as their complete degradation. Bioremediation through aquatic organisms like cyanobacteria may help in attaining the goals of the overall health of the environment and the organisms in an eco-friendly manner.

Keywords: cyanobacteria, endocrine-disrupting chemicals, steroid hormones, estrone, bioremediation

* Corresponding Author's Email: fatma_cbl@yahoo.com.

18.1. Introduction

Endocrine-disrupting compounds (EDC) have gained significant interest in recent decades as their exposure causes decreased fertility, altered sexual behavior, metabolic disorders, neurological disorders, cardiovascular diseases, cancer, disrupted fetal development and growth (Pelch et al., 2011; Vandenberg, 2014; Patisaul, 2017). The endocrine system of humans and animals are susceptible to EDC as it interferes with metabolism, growth, and reproduction. The embryonic and pubertal stages known as the "critical periods" are the most sensitive ones. Steroid hormones, especially the female estrogenic hormone from exogenous sources are one of the EDCs that act on various target sites and disturb many important physiological functions (Snyder et al., 2003; Meng et al., 2005; LaFleur and Schug, 2011).

18.2. Sources of Estrone

Estrone (E1) is one of the chemically stable natural fat-soluble steroid estrogens. The natural steroid estrogens excreted through faeces and urine (in free form or as inactive glucuronides and sulfate conjugates) are flushed down from home toilets, municipal, pharmaceutical and hospital effluents through which it enters wastewater treatment plants (WWTPs) and is then released into the aquatic environment (Belhaj et al., 2015; Ting and Praveena, 2017). The conjugated estrone gets readily bio-transformed to the free forms by enzymes like β-glucuronidase secreted by microorganisms (Koh et al., 2008). According to Andaluri et al., (2012) animal manure (poultry, horse and cow manure) is the largest source of estrone in the natural environment. The amount of estrone excreted through urine by humans vary according to age due to the use of oral contraceptives (Table 1). Municipal landfills are also sources of organic contaminants including steroid hormones that infiltrate into groundwater. The occurrence of female estrogens, such as estrone, estradiol, and progesterone in meat (fish, poultry, beef, and pork), milk and dairy products, vegetables and eggs, indicate that these compounds are present extensively in foods consumed by humans that make them prone to endocrine disruption (Hartmann et al., 1998). Among all the female estrogens, estrone shows greater estrogenic activity even at minute concentrations and has emerged as an important polluting agent (Spengler et al., 2001).

Table 1. The average amount of estrone excreted by humans

Age Group	Amount excreted (μg/day)	References
Adult male	3.50	Kostich et al., (2013)
Female child	0.60	Kostich et al., (2013)
Male child	0.63	Kostich et al., (2013)
Menstruating females	3.50	Hamid and Eskicioglu (2012)
Menopausal females	2.30	Hamid and Eskicioglu (2012)
Menopausal, no HRT	2.93	Kostich et al., (2013)
Menopausal, with HRT	31.50	Kostich et al., (2013)
Pregnant women	787	Kostich et al., (2013)
Average excretion per person	19.00	Laurenson et al., (2014)

18.3. Estrone as Hormone and as EDC

The endocrine system consists of many ductless glands that secrete hormones directly into the circulatory system to regulate different body functions. Estrone is produced in vivo from testosterone or androstenedione via estradiol, mainly in the ovaries, placenta, and in peripheral tissues (especially adipose tissue). It goes inside the cells of targeted tissues (e.g., pituitary, hypothalamus, and breasts) and interacts with the estrogen receptors. Estrone-bound estrogen receptors dimerize, translocate to the nucleus and bind to estrogen response elements (ERE) of genes. This induces alteration in the transcription rate of affected genes, by increasing or decreasing their expression. It increases the expression of thyroid-binding globulin (TBG), sex hormone-binding globulin (SHBG), and other serum proteins and suppresses the release of follicle-stimulating hormone (FSH) from the anterior pituitary under normal condition (https://pubchem.ncbi.nlm.nih.gov/substance/46505916).

Estrone is essential for normal animal physiology but it shows adverse effects when it accumulates in the environment and enters the food chain. Exposure to estrone effects male fish tremendously by feminizing them (change in the sex). During this process, increased concentrations of plasma vitellogenin occur that result in reduced testes size, lowering of the sperm count that affect the overall reproductive fitness and sustainability of fish populations (Rose et al., 2013; Arnold et al., 2014; Van Donk et al., 2016). In the case of cattle, estrone exposure changes the teat length and vulva color and induces infertility (Burton and Wells, 2002). Sometimes, it also impairs vision by varying intraocular eye pressure (IOP) (Shemesh and Shore, 2012). It induces premature menopause, affects reproductive development and also causes masculinization in young girls. Several studies have also reported that estrone reduces sperm counts that causes feminization of men and also disturb their reproductive system (Sumpter and Jobling, 2013). Not only these, but it also increases the incidence of cardiovascular diseases, breast, and prostate cancer if consumed above the safe threshold level (Nelles et al., 2011; Wocławek-Potocka et al., 2013; Moore et al., 2016).

18.4. Mechanism of Action of Estrone as EDC

Estrone (as EDC) binds and activates estrogen receptor (ER) i.e., it mimics the function of estrogen hormone required in the normal physiology of an organism. This disturbs the ER pathways viz., the genomic (nuclear-initiated) and the non-genomic (membrane-initiated) that regulate transcription of target genes. Estrone affects these pathways by directly interacting with ER or indirectly via other receptors like the thyroid receptor (TR) and aryl hydrocarbon receptor (AhR) (Shanle and Xu, 2010). In the genomic pathway, there occurs a competition between EDC and the endogenous estrone for ER binding sites. EDC directly binds to ER (either ERα or ERβ), behaves as an agonist and activates the receptor inappropriately. It then gets translocated to the nucleus and binds DNA at the estrogen response element (ERE) which consequently results in the abnormal expression of ERE-dependent genes. The high estrogenic activity of ER results in abnormal cell growth, disturbance in the metabolism and synthesis of other endogenous hormones that result in endocrine disruption as evident in many cancer cases (Aris et al., 2014; Maqbool et al., 2016).

In case of non-genomic pathway, estrone exerts its actions through a subpopulation of ER i.e., the plasma membrane ER (mER), Thyroid receptor (TR), Aryl hydrocarbon receptor (AhR), Peroxisome-proliferator-activated receptor (PPAR), novel G-protein coupled E2 receptors (GPER) that further activates various kinase signaling pathways (i.e., Ca^{2+}, cAMP, protein kinase cascades) which ultimately influences the downstream transcription factors. Alteration in these pathways is related to disorders like cardiovascular diseases, metabolic syndromes and neuronal degeneration (Diamanti-Kandarakis et al., 2009).

18.5. Degradation of Estrone

Different physical, chemical and biological methods have been used to mitigate estrone from the aquatic system (Table 2). The physical methods include the use of activated carbon and membrane filtration; whereas the chemical methods involve the use of advanced oxidation processes and ozonation (Liu et al., 2009). But, since they produce toxic end products and are relatively costly, research interests have shifted towards the biological methods that are environmentally friendly and depend on microbial metabolism for the degradation of estrone (Rosal et al., 2010). Moreover, bioremediation technology is less energy-intensive (Delrue et al., 2016). Coombe and colleagues (1966) first reported the estrogen biodegradation and isolated the bacterial strain involved in the degradation from the soil. Besides bacteria, fungus and algae are also able to degrade estrogenic compounds (Cajthaml et al., 2009; Yu et al., 2013; Liu et al., 2018).

Algae can prove to be a boon for the degradation of estrogenic compounds since they are photosynthetic and abundantly present in aquatic environments. They easily utilize micro-pollutants as the carbon source due to high surface area: volume ratio and rapid metabolism (Escapa et al., 2017a; Escapa et al., 2017b). Cultivation of algae in wastewater is considered an economical practice since the algal biomass left after the removal of micro-pollutant can be used as a source of other valuable products like biofuel, bioplastic, etc.

According to Wang et al., (2017), many green algae (*Chlamydomonas* sp., *Chlorella* sp., *Scenedesmus quadricauda*, *Selenastrum capricornutum,* and *Chlorella vulgaris*) are capable of degrading and removing estradiol (E2) and ethinyl estradiol (EE2). The biological removal of estrogen by them was found to be time-dependent and species-dependent. Among all the species, *Selenastrum capricornutum* showed the highest removal efficiency for both E2 (93.9%) and EE2 (75.3%). Lai et al., (2002) and Zhang et al., (2014) showed that *Chlorella vulgaris* and *Scenedesmus dimorphus* were able to remove female estrogenic hormones [E1, E2, estriol (E3), EE2 and 17α-estradiol]. Cyanobacterial ability to bio-remediate endocrine-disrupting compound still needs to be explored. *Anabaena cylindrical*, *Microcytis aeruginosa* Kutz. and *Spirulina* CPCC-695 have been reported to degrade estrone (Ge et al., 2009; Sami and Fatma, 2019). Adsorption process has been attributed to the bioremediation of estrone (Hom-Diaz et al., 2015; Liu et al., 2018). According to Ge et al., (2009), algae produce some secretions which produce hydroxyl radical that can enhance the degradation of estrogens.

Table 2. Different methods of degradation

Degradation Method	Examples
Physical Method	1. Activated carbon a. Powder Activated Carbon (PAC) b. Granular Activated Carbon (GAC) 2. Membranes a. Reverse Osmosis (RO) b. Nano-membrane c. Ultra-membrane d. Micro-membrane
Chemical Method	Advanced Oxidation Process (AOP) Ozonation
Biological Method	Bacterial degradation Fungal degradation Algal degradation Plant-based degradation
Enzymatic Method	Peroxidase Laccase Esterase Tyrosinase

Coombe et al., (1966) reported the estrone degradation pathway in the soil bacterium, *Nocardia* sp. E110. They proposed that the ring cleavage of estrone leads to its degradation. The ring cleavage occurs at the fourth carbon (C-4) of Ring A resulting in the formation of an intermediate product (4-hydroxy estrone) that undergoes meta-cleavage and forms carbon dioxide and water. Another pathway was proposed by Lee and Liu (2002) in sewage bacteria. They reported that during estrone degradation the sixteenth carbon (C-16) of Ring D gets hydroxylated, followed by its dehydration and ketonization resulting in the formation of an intermediate (X1) that has lactone ring which eventually gets completely degraded via tricarboxylic acid cycle (TCA cycle). Cyanobacteria are also prokaryotic therefore similar estrone degradation pathways might be involved. Based on the determination of product formed during degradation (as analyzed by mass spectroscopy) hydroxylation, glycosylation and ketonization are suggested to operate in *Scenedesmus capricornutum* (Della-Greca et al., 2008). The oxidoreductases enzyme-like laccase, peroxidase, and tyrosinase play an important role in the meta-degradation pathway of estrone to produce carbon dioxide and water. We have also found the role of these enzymes in cyanobacterial estrone degradation. Among all, laccase was found to have the highest activity (Data unpublished).

Other than estrogens, algae are also capable of degrading other EDCs like alkylphenols, phthalates, pharmaceuticals, etc. Zhou et al., (2013) reported that the freshwater green microalga *Scenedesmus obliquus* is capable of degrading octylphenol while other microalgae like *Phaeocystis globose*, *Nannochloropsis oculata*, *Dunaliella salina,* and *Platymonas subcordiformis* are capable to degrade nonylphenol (He et al., 2016; Wang et al., 2019). *Chlorella fusca* has also been reported to metabolize Bisphenol-A (BPA) (Hirooka et al., 2005; Nakajima et al., 2007).

The microorganisms can degrade estrone within their tolerance range only. At high concentrations, estrone becomes toxic to algae. As evident in case of green microalgae (*Pseudokirchneriella subcapitata* strain CPCC37 and *Chlamydomonas reinhardtii* strain

CC125) and cyanobacteria (*Microcystis aeruginosa* CPCC299 and CPCC632) where the EDCs like 4-octylphenol, 4-nonylphenol and β-estradiol affected the photosystem II energy fluxes which resulted in a reduction in photosynthesis and exhibited toxicity (Perron and Juneau, 2011).

Conclusion and Future Perspective

Bioremediation is considered as an efficient and environmentally safe technology for inexpensive decontamination of polluted water bodies. The present review highlights the female steroid hormone, estrone as an endocrine-disrupting compound when it enters artificially inside the body. The physical and chemical degradation techniques are not reliable due to their limitations. Cyanobacteria, being an aquatic organism, utilise estrone as a carbon source and degrades it into non-toxic end products. Thus, the potential of cyanobacteria as a bioremediation agent holds a great future in dealing with EDCs present in the aquatic bodies. Further study is required to optimize the degradation efficiency and elucidate the pathway involved in the degradation of estrone to identify the enzymes involved.

Acknowledgments

N. Sami sincerely thanks University Grants Commission (F1-17.1/2014-15/MANF-2014-15-MUS-JHA-47673/(SA-III/Website)) for Maulana Azad National Fellowship (MANF-SRF).

References

Andaluri, G., Suri, R. P., and Kumar, K. (2012). Occurrence of estrogen hormones in biosolids, animal manure, and mushroom compost. *Environ. Monit. Assess.* 184(2), 1197-205.

Aris, A. Z., Shamsuddin belhaj, A. S., and Praveena, S. M. (2014). Occurrence of 17α-ethynylestradiol (EE2) in the environment and effect on exposed biota: a review. *Environ. Int.* 69, 104-19.

Arnold, K. E., Brown, A. R., Ankley, G. T., and Sumpter, J. P. (2014). Medicating the environment: assessing risks of pharmaceuticals to wildlife and ecosystems. *Philos. Trans. Royal Soc.* B. 369, 20130569.

Belhaj, D., Baccar, R., Jaabiri, I., Bouzid, J., Kallel, M., Ayadi, H., and Zhou, J. L. (2015). Fate of selected estrogenic hormones in an urban sewage treatment plant in Tunisia (North Africa). *Sci. Total. Environ.* 505, 154-60.

Burton, J. L., and Wells, M. (2002). The effect of phytoestrogens on the female genital tract *J. Clin. Pathol.* 55(6), 401-407.

Cajthaml, T., Křesinová, Z., Svobodová, K., and Möder, M. (2009). Biodegradation of endocrine-disrupting compounds and suppression of estrogenic activity by ligninolytic fungi. *Chemosphere.* 75(6), 745-50.

Coombe, R. G., Tsong, Y. Y., Hamilton, P. B., and Sih, C. J. (1966). Mechanisms of steroid oxidation by microorganisms X. Oxidative cleavage of estrone. *J. Biol. Chem.* 241(7), 1587-95.

Della-Greca, M., Pinto, G., Pistillo, P., Pollio, A., Previtera, L., and Temussi, F. (2008). Biotransformation of ethinylestradiol by microalgae. *Chemosphere*. 70(11), 2047-2053.

Delrue, F., Álvarez-Díaz, P., Fon-Sing, S., Fleury, G., and Sassi, J. F. (2016). The environmental biorefinery: Using microalgae to remediate wastewater, a win-win paradigm. *Energies*. 9(3), 132.

Diamanti-Kandarakis, E., Bourguignon, J. P., Giudice, L. C., Hauser, R., Prins, G. S., Soto, A. M., Zoeller, R. T., and Gore, A. C. (2009). Endocrine-disrupting chemicals: an endocrine society scientific statement. *Endocr. Rev.* 30(4), 293-342.

Escapa, C., Coimbra, R. N., Paniagua, S., García, A. I., and Otero, M. (2017b). Comparison of the culture and harvesting of *Chlorella vulgaris* and *Tetradesmus obliquus* for the removal of pharmaceuticals from water. *J. App. Phycol.* 29(3), 1179-1193.

Escapa, C., Coimbra, R., Nuevo, C., Vega, S., Paniagua, S., García, A., Calvo, L., and Otero, M. (2017a). Valorization of microalgae biomass by its use for the removal of paracetamol from contaminated water. *Water*. 9(5), 312.

Ge, L., Deng, H., Wu, F., and Deng, N. (2009). Microalgae-promoted photodegradation of two endocrine disrupters in aqueous solutions. *J. Chem. Technol. Biot.* 84(3), 331-336.

Hartmann, S., Lacorn, M., and Steinhart, H. (1998). Natural occurrence of steroid hormones in food. *Food Chem.* 62(1), 7-20.

He, N., Sun, X., Zhong, Y., Sun, K., Liu, W., and Duan, S. (2016). Removal and biodegradation of nonylphenol by four freshwater microalgae. *Int. J. Environ. Res. Public Health.* 13(12), 1239.

Hirooka, T., Nagase, H., Uchida, K., Hiroshige, Y., Ehara, Y., Nishikawa, J. I., Nishihara, T., Miyamoto, K. and Hirata, Z. (2005). Biodegradation of bisphenol A and disappearance of its estrogenic activity by the green alga Chlorella fuscavar. *vacuolata. Environ. Toxicol. Chem.* 24(8), 1896-1901.

Hom-Diaz, A., Llorca, M., Rodríguez-Mozaz, S., Vicent, T., Barceló, D., and Blánquez, P., (2015). Microalgae cultivation in wastewater digestate: β-estradiol and 17α-ethynylestradiol degradation and transformation products identification. *J. Environ. Manage.* 155, 106-113.

https://pubchem.ncbi.nlm.nih.gov/substance/46505916.

Koh, Y. K., Chiu, T. Y., Boobis, A., Cartmell, E., Scrimshaw, M. D., and Lester, J. N. (2008). Treatment and removal strategies for estrogens from wastewater. *Environ. Tech.* 29(3), 245-267.

Kostich, M., Flick, R., and Martinson, J. (2013). Comparing predicted estrogen concentrations with measurements in US waters. *Environ. Poll.* 178, 271-277.

LaFleur, A. D., and Schug, K. A. (2011). A review of separation methods for the determination of estrogens and plastics-derived estrogen mimics from aqueous systems. *Anal. Chim. Acta*. 696(1-2), 6-26.

Lai, K. M., Scrimshaw, M. D., and Lester, J. N. (2002). Biotransformation and bioconcentration of steroid estrogens by *Chlorella vulgaris*. *Appl. Environ. Microbiol.* 68(2), 859-864.

Laurenson, J. P., Bloom, R. A., Page, S., and Sadrieh, N. (2014). Ethinyl estradiol and other human pharmaceutical estrogens in the aquatic environment: a review of recent risk assessment data. *AAPS J.* 16(2), 299-310.

Lee, H. B., and Liu, D. (2002). Degradation of 17 beta-estradiol and its metabolites by sewage bacteria. *Water Air Soil Pollut.* 134(1-4), 353-368.

Liu, W., Chen, Q., He, N., Sun, K., Sun, D., Wu, X., and Duan, S. (2018). Removal and Biodegradation of 17β-estradiol and diethylstilbestrol by the freshwater microalgae *Raphidocelis subcapitata*. *Int. J. Environ. Res. Public Health.* 15(3), 452.

Liu, Z. H., Kanjo, Y., and Mizutani, S. (2009). Removal mechanisms for endocrine disrupting compounds (EDCs) in wastewater treatment—physical means, biodegradation, and chemical advanced oxidation: a review. *Sci. Total. Environ.* 407(2), 731-748.

Maqbool, F., Mostafalou, S., Bahadar, H., and Abdollahi, M. (2016). Review of endocrine disorders associated with environmental toxicants and possible involved mechanisms. *Life Sci.* 145, 265-273.

Meng, Z., Chen, W., and Mulchandani, A. (2005). Removal of estrogenic pollutants from contaminated water using molecularly imprinted polymers. *Environ. Sci. Tech.* 39(22), 8958-8962.

Moore, S. C., Matthews, C. E., Ou-Shu, X., Yu, K., Gail, M. H., Xu, X., Ji, B. T., Chow, W. H., Cai, Q., Li, H., and Yang, G. (2016). Endogenous estrogens, estrogen metabolites, and breast cancer risk in postmenopausal Chinese women. *J. Natl. Cancer Inst.* 108(10), djw103.

Nakajima, N., Teramoto, T., Kasai, F., Sano, T., Tamaoki, M., Aono, M., Kubo, A., Kamada, H., Azumi, Y., and Saji, H. (2011). Glycosylation of bisphenol A by freshwater microalgae. *Chemosphere.* 69(6), 934-941.

Nelles, J. L., Hu, W. Y., and Prins, G. S. (2011). Estrogen action and prostate cancer. *Expert Revi. Endocrinol. Metab.* 6(3), 437-51.

Patisaul, H. B. (2017). Endocrine disruption by dietary phyto-oestrogens: impact on dimorphic sexual systems and behaviors. *Proc. Nutr. Soc.* 76(2), 130-144.

Pelch, K. E., Beeman, J. M., Niebruegge, B. A., Winkeler, S. R., and Nagel, S. C. (2011). Endocrine-disrupting chemicals (EDCs) in mammals. In: *Hormones and Reproduction of Vertebrates*, Academic Press, pp. 329-371.

Perron, M. C., and Juneau, P. (2011). Effect of endocrine disrupters on photosystem II energy fluxes of green algae and cyanobacteria. *Environ. Res.* 111(4), 520-529.

Rosal, R., Rodríguez, A., Perdigón-Melón, J. A., Petre, A., García-Calvo, E., Gómez, M. J., Agüera, A., Fernández-Alba, A. R. (2010). Occurrence of emerging pollutants in urban wastewater and their removal through biological treatment followed by ozonation. *Water Res.* 44(2), 578-88.

Rose, E., Paczolt, K. A., and Jones, A. G. (2013). The effects of synthetic estrogen exposure on premating and postmating episodes of selection in sex-role-reversed Gulf pipefish. *Evol. Appl.* 6(8), 1160-1170.

Sami, N., and Fatma, T. (2019). Studies on estrone biodegradation potential of cyanobacterial species. *Biocatal. Agric. Biotechnol.* 17, 576-582.

Shanle, E. K., and Xu, W. (2010). Endocrine-disrupting chemicals targeting estrogen receptor signaling: identification and mechanisms of action. *Chem. Res. Toxicol.* 24(1), 6-19.

Shemesh, M., and Shore, L. S. (2012). Effects of environmental estrogens on reproductive parameters in domestic animals. *Isr. J. Vet. Med.* 67(1), 6-10.

Snyder, S. A., Westerhoff, P., Yoon, Y., and Sedlak, D. L. (2003). Pharmaceuticals, personal care products, and endocrine disruptors in water: implications for the water industry. *Environ. Eng. Sci.* 20(5), 449-69.

Spengler, P., Körner, W., and Metzger, J. W. (2001). Substances with estrogenic activity in effluents of sewage treatment plants in southwestern Germany. 1. Chemical analysis. *Environ. Toxicol. Chem.* 20(10), 2133-2141.

Sumpter, J. P., and Jobling, S. (2013). The occurrence, causes, and consequences of estrogens in the aquatic environment. *Environ. Toxicol. Chem.* 32(2), 249-251.

Ting, Y. F., and Praveena, S. M. (2017). Sources, mechanisms, and the fate of steroid estrogens in wastewater treatment plants: a mini-review. *Environ. Monit. Assess.* 189(4), 178.

Vandenberg, L. N. (2014). Low-dose effects of hormones and endocrine disruptors. In: *Vitamins and Hormones,* Academic Press, 94: pp. 129-165.

Van-Donk, E., Peacor, S., Grosser, K., Domis, L. N., Lürling, M. (2016). Pharmaceuticals may disrupt natural chemical information flows and species interactions in aquatic systems: ideas and perspectives on a hidden global change. In: *Reviews of environmental contamination and toxicology Springer, Cham,* pp. 91-105.

Wang, L., Xiao, H., He, N., Sun, D., and Duan, S. (2019). Biosorption and Biodegradation of the environmental hormone nonylphenol by four marine microalgae. *Sci. Rep.* 9(1), 5277.

Wang, P., Wong, Y. S., and Tam, N. F. Y. (2017). Green microalgae in removal and biotransformation of estradiol and ethinylestradiol. *J. Appl. Phycol.* 29(1), 263-273.

Wocławek-Potocka, I., Mannelli, C., Boruszewska, D., Kowalczyk-Zieba, I., Waśniewski, T., and Skarżyński, D. J. (2013). Diverse effects of phytoestrogens on the reproductive performance: the cow as a model. *Int. J. Endocrinol.* 2013, 1-15.

Yu, C. P., Deeb, R. A., and Chu, K. H. (2013). Microbial degradation of steroidal estrogens. *Chemosphere*. 91(9), 1225-1235.

Zhang, Y., Habteselassie, M. Y., Resurreccion, E. P., Mantripragada, V., Peng, S., Bauer, S., and Colosi, L. M. (2014). Evaluating removal of steroid estrogens by a model alga as a possible sustainability benefit of hypothetical integrated algae cultivation and wastewater treatment systems. *ACS Sustain. Chem. Eng.* 2(11), 2544-2553.

Zhou, G. J., Peng, F. Q., Yang, B., and Ying, G. G. (2013). Cellular responses and bio removal of nonylphenol and octylphenol in the freshwater green microalga Scenedesmus obliquus. *Ecotoxicol. Environ. Safety.* 87, 10–16.

In: Applied Algal Biotechnology
Editors: M. Arumugam, S. Kathiresan et al.
ISBN: 978-1-53617-524-0

Chapter 19

MICROALGAL ZINC OXIDE NANOPARTICLES: CURRENT STATUS AND FUTURE PROSPECTS

***Nida Asif*[*], Md Nafe Aziz and Tasneem Fatma*[†]**
Department of Biosciences, Jamia Millia Islamia,
New Delhi, India

ABSTRACT

Biosynthesis of nanoparticles is an important area in the field of nanotechnology which has economic and eco-friendly benefits over chemical and physical methods. In particular, Zinc oxide based nanomaterials have emerged as a promising potential in biomedical applications due to abundance of zinc element and it's easy conversion in oxide form, excellent biocompatibility, and low cost. Zinc oxide nanparticles (ZnONPs) are one of the cheap materials utilized in cosmetic, nanofertilizers, bioimaging, biomedicine, targeted drug and gene delivery. Biogenic ZnONPs can be synthesized by using plants, bacteria, fungus and algae. In present review, biomedical applications of ZnONPs (antimicrobial, anti-inflammatory, anti-dandruff, anti-arthritic), along with their use in nanofertilizer industry, food preservation and packaging industry are highlighted with special reference to algal based synthesis.

Keywords: biosynthesis, zinc oxide nanoparticles, algae, cyanobacteria

19.1. INTRODUCTION

Nanotechnology is the branch of technology that deals with the nanoparticles (NPs) that are atomic or molecular that is characterized by size less than 100 nm (Kato, 2011). It provides new innovative solutions in different scientific disciplines and provides an opportunity to develop new tools for incorporation of NPs into existing functions and add new ones. The fabrication of nanoparticles is cheaper, simpler and the amount required for an

[*] Corresponding Author's Email: nidaasif13@gmail.com.
[†] Corresponding Author's Email: fatma_cbl@yahoo.com.

application is much lower. The properties of nanoparticles are dependent on their morphology and size. Morphology of zinc oxide nanoparticles (ZnONPs) depends on the process of its synthesis. They may be nanorods, nanoplates, nanospheres, nanoboxes (Wang and Lou., 2008; Kakiuchi et al., 2006; Mahmud et al., 2006), hexagonal, tripods, tetrapods (Mahmud et al., 2006; Shen et al., 2009; Ding and Wang, 2009), nanowires, nanotubes, nanorings (Wang, 2004-a & b), nanocages, and nanoflowers (Moezzi et al., 2011; Xie et al., 2009). ZnONPs have gained profound attention due to their wide range of applications in various scientific and biomedical fields (Wang, 2004- a&b; Akhtar et al., 2011; Siddiqi et al., 2018; Zhang et al., 2007).

19.2. Synthesis of Zinc Oxide Nanoparticles

Synthesis of ZnONPs has been carried out by physical, chemical and biological methods. Top down and Bottom up approaches has been adopted to synthesize nanoparticles. Top down, involves milling or attrition of large macroscopic particle, etching, laser ablation, grinding and sputtering (Koch, 2003). In top down approach, physical forces are involved in the attraction of nanoscale particles and formation of large, stable, well defined nanostructures (Koch, 2003). The bottom up approach involves chemical synthesis through colloidal dispersion method, sol gel process, laser pyrolysis, chemical reduction, spray pyrolysis, vapor condensation, amorphous crystallization, physical fragmentation and biological synthesis (Ahmed et al., 2016). Recently, scientists have started use of plant extracts and microorganisms for biogenic ZnONPs synthesis.

19.3. Biogenic Synthesis of Zinc Oxide Nanoparticle

During the last decade, development of biologically inspired technique evolved as an important branch in the field of nanotechnology and nanoscience (Ahmed et al., 2016). Conventional physical and chemical techniques require the involvement of perilous and hazardous chemicals for synthesis of NPs, besides these processes are quite expensive (Vijayan et al., 2016). In comparison to conventional methods biogenic synthesis of NPs are environment-friendly, cost effective, biocompatible and safe (Salam et al., 2014). So far, for biogenic ZnONPs synthesis many organism are used e.g., bacteria (Dhandapaniet al., 2014; Tripathi et al., 2014; Kundu et al., 2014; Jayaseelan et al., 2012), fungi (Chandraseekara and Pandurangan, 2016; Raliya and Tarafdar., 2014; Shamsuzzaman et al., 2013; Raliya and Tarafdar., 2013) and plants (Duraimurugan et al., 2019; Akhter et al., 2018; Colak and Karakose., 2017; Malaikozhundan et al., 2017).

Only few reports showed use of algae for synthesis of ZnONPs (Singh et al., 2014; Azizi et al., 2014; Mubarak et al., 2011). However algae have been used to synthesize silver nanoparticles (Aziz et al., 2014 & 2015; Husain et al., 2015) and gold nanoparticles (Luangpipat et al., 2011; Soleimani and Pirkoohi. 2017). It is suggested that potentially active compounds present in algal biomass (fucoxanthin, protein, lipids, polyphenols, fiber, vitamins, minerals, flavonoids, steroids and terpenoids) act as a reducing and stabilizing agents during nanoparticles synthesis (Raja et al., 2016). Macroalgae which has been used to synthesize ZnONPs are *Sargassum muticum* (Azizi et al., 2014), *Sargassum myriocystum* (Nagarajan and Kuppusamy., 2013), *Ulva lactuca* (Ishwarya et al., 2018), *Padina tetrastromatica* (Pandimurugan and Thambidurai., 2016 & 2017), *Brown algae* (Basha et al., 2016), *Gracilaria edulis* (Ramaramesh et al., 2014), while only few microalgae like *Anabaena strain L31* (Singh et al., 2014), *Chlamydomonas reinhardtii* (Rao and Gautam., 2016) are explored for ZnONPs synthesis (Table 19.1).

Table 19.1. Algae mediated synthesis of ZnONPs

Algal strain	Family	Shape/Morphology	Size (nm)	Application
Sargassum muticum	Sargassaceae	Hexagonal wurtzite	30-57	Highly biomedical and UV absorptive
Sargassum myriocystum	Sargassaceae	Spherical, radial, triangle,hexagonal,rod	20-39	Enhanced anti-bacterial activity against gram – negative bacteria
Ulva lactuca	Ulvaceae	Hexagonal, rods	10-50	Enhance photocatalytic degredation & antibiofilm activity.
Padina tetrastromatica	Dictyotaceae	Hexagonal wurtzite, rod, plate, star shaped	16-30	Enhanced photo-catalytic and high antibacterial activity.
Brown algae	Pheophyceae	Cubical or rod	20-40	Sensor - with high sensitivity
Gracilaria edulis	Gracilariaceae	Rod	66-95	Effective anti-cancer activity against PC3 cell lines.
Anabaena strain L31	*Cyanobacteriaceae*	Spherical	20-80	Non toxic sunscreen formulations
Chlamydomonas reinhardtii	Chlamy-domonadaceae	Nanorod, nanoflower, porous nanosheet	55-80	Enhanced photo-catalytic dye degredation.

Recently, we have screened 30 cyanobacterial strains for the synthesis of ZnONPs by using cell free extract and zinc precursor [zinc nitrate (0.01 M) or zinc acetate (0.05 M) or zinc sulphate (1M)]. Cell free extracts without metal precursor were taken separately to serve as positive and negative control. Optimization for ZnONPs synthesis was also done for pH and temperature (Data not shown).

Exact mechanisms explaining the biosynthesis of ZnO nanoparticles is still under research. However behind the biosynthesis of ZnO nanoparticles are the natural resources like phytochemicals (like saponins, polyphenols, terpenoids) which act as both reducing as well as stabilizing (capping) agents (Basnet et al., 2018). They reduce the metal (zinc) to the 0-valence state and then through calcinations, oxide may be added to the metal. Another very convincing mechanism is that zinc ions in the solution of the natural extract form complexation with the polyphenols (or other phytochemicals) in the form of Zn^{+2}. This is then followed by the formation of zinc hydroxide [$Zn(OH)_2$] through hydrolysis and finally after calcinations, the complex decomposes, thereby favouring the formation of ZnO nanoparticles.(Basnet et al., 2018). Many mechanisms are suggested for the stabilization/capping of the nanoparticles e.g., electrostatic stabilization, steric stabilization,

stabilization by hydration forces, and stabilization by van der Waals forces. The stabilization depends upon the function and application of sources used for synthesis of nanoparticles (Singh et al., 2014).

19.4. Applications of Zinc Oxide Nanoparticles

Biologically synthesized zinc oxide nanoparticles via green and eco-friendly route confer an array of advantages over synthetically synthesized. Their applications are:

19.4.1. Anti-cancer Activity

Among several biomedical applications, ZnONPs has been well explored to study anticancer property. ZnONPs illustrates their anticancer activity by inducing ROS generation, and apoptosis (Mishra et al., 2017). ZnONPs exhibit a different type of surface charge behavior because of neutral hydroxyl groups on their surface. Due to this the isoelectric point of these NPs is 9-10, which confers them to carry a strong positive surface charge under physiological conditions (Rasmussen et al., 2010; Degen and Kosec., 2000). By contrast cancer cells have higher concentration of anionic phospholipids on their surface due to which these cells have large negative surface charge (Abercrombie and Ambrose., 1962). Thus, the interaction of negatively charged cancer cells with positively charged ZnONPs by electrostatic interactions, phagocytosis is favoured that promotes the cellular uptake and cytotoxicity (Rasmussen et al., 2010). Above mentioned characteristics of ZnONPs make them a suitable candidate as synergistic cytotoxic agents with other chemotherapeutics like cisplatin (CPT), doxorubicin (DOX) and paclitaxel (PTX) (Sharma et al., 2016; Hackenberg et al., 2012). Researchers have showed selective toxicity of ZnONPs against cancer cells proliferation in C2C12 myoblastoma cancer cells and 3T3-L1 adipocytes by inducing apoptosis (Chandrasekaran and Pandurangan, 2016). They also demonstrated the cytotoxicity of NPs against malignant human T98G gliomas, KB epithermoids and non malignant normal HEK kidney cells. Results revealed that ZnONPs are most toxic to cancer cells while less toxic to normal human HEK cells (Wahab et al., 2013; Sasidharan et al., 2011).

19.4.2. Anti-diabetic Activity

Being micronutrient zinc plays an important role in insulin synthesis, storage and secretion, which makes ZnONPs a suitable candidate to explore its anti-diabetic activity. ZnONPs doses to rat showed improved glucose disposal, insulin levels and reversal of the diabetes induced pancreatic injury (Wahba et al., 2016; Nazarizadeh and Asrirezaie., 2016). ZnONPs used in combination with anti-diabetic drugs - red sandalwood and vildagliptin showed better results (Kitture et al., 2014).

19.4.3. Anti-bacterial Activity

Several studies have demonstrated that when bacterial cells come in contact with ZnONPs they absorb Zn ions, inhibiting the action of respiratory enzymes, thus leads to the ROS generation causing oxidative stress that is directly proportional to concentration of ZnONPs. This disrupts the bacterial cell membrane permeability, DNA and mitochondria leading to the cell lysis and ultimately death of bacterial cells (Dwivedi et al., 2014). According to Raghupati et al. (2011), anti-bacterial activity of the ZnONPs increase with gradual reduction in nanoparticles size and increase in its concentration. ZnONPs have been investigated to check their efficiency against conventional antibiotics ceftazidime & ciprofloxacin for *Acinetobacter baumannii* (Ghasemi and Jalal.2016). The results demonstrate that inhibitory concentration of ZnONPs, increase its antibacterial efficacy by increased fluidity, deformed cellular structure and ruptured cell membrane in the bacteria.

It is reported that ZnONPs were toxic to *Campylobacter jejuni* and *Escherichia coli* (both gram-negative bacteria causing diarrhea and dysentery) and multiple drug resistant *Staphylococcus aureus* (gram positive bacteria) causing range of illness, from minor skin infections, such as pimples, impetigo, boils, cellulitis, folliculitis, carbuncles, scalded skin syndrome, and abscesses, to life-threatening diseases. Such as pneumonia, meningitis, osteomyelitis, endocarditis, toxic shock syndrome, bacteremia, and sepsis (Manzoor et al., 2016; Ahmed et al., 2016). Kumar et al., (2017) synthesized the ZnONPs using *Passiflora caerulea* leaf extract that acted as an antibacterial agent against urinary tract infection pathogens *Klebsiella pneumonia*, *Bacillus subtilis*, *Escherichia coli*, *Serratia sp.* and *Streptococcus sp.*

ZnONPs possess an evident antimicrobial activity against many pathogenic organisms like *Escherichia coli, Pseudomonas aeruginosa, P. vulgaris, Campylobacter jejuni, Bacillus subtilis, Bacillus megaterium, Sarcina lutea, Klebsiella pneumonia, Candida albicans and Aspergillus niger* (Zhang et al., 2007; Siddiqi et al., 2018). ZnONPs also showed good bacteriostatic effect on *Streptococcus agalactiae* and *Staphylococcus aureus* which are two etiological agents of several infective diseases in humans (Huang et al., 2008).

Due to anti-bacterial activity, ZnONPs is applied to biodegradable polymeric metrics in the food preservation and packaging industry for improved preservation quality. It kills bacterial cells through formation of reactive oxygen species (Espitia et al., 2012).

19.4.4. Anti-fungal Activity

In addition to the anti-bacterial activity, ZnONPs also exhibit significant anti-fungal activity against many plant fungal pathogens like *Alternaria saloni* and *Sclerrotium rolfii* (Surendra et al., 2016), *Rhizopus stolonifera*, *Aspergillus flavus*, *Aspergillus nidulans* and *Trichoderma harzianum* (Gunalan et al., 2012). ZnONPs also exhibited concentration dependent toxicity on the viability of *Candida albicans,* an opportunistic human gut pathogen (Lipovsky et al., 2011). ZnONPs synthesized with aqueous plant extract *Lagenaria siceraria* inhibited the growth of dandruff causing fungi *Malassezia furfur* and *Malassezia pachydermatis* (Kalpana et al., 2017). Silver nanoparticles also showed anti-dandruff effect against dandruff causing fungi *Pityrosporum ovale* and *Pityrosporum folliculitis* (Pant et al., 2013).

19.4.5. Anti-inflammatory Activity

Biologically synthesized ZnONPs illustrates significant anti inflammatory activity in mouse model by exhibiting a visible decrease in *IL-1B* gene and enhancement in anti-inflammatory cytokine IL-10 in macrophages (Nagajyothi PC et al., 2015; Roy et al., 2014). ZnONPs penetrate deeper layers of allergic skin than bulk ZnO and suppress skin inflammation by producing the IgE antibodies (Ilves et al., 2014). Inflammatory arthritis is a synovial disease characterized by chronic inflammation of the joints and can result in disability owing to joint destruction. Kalpana et al., (2017) reported anti-arthritic potential of ZnONPs synthesized from *Lagenaria siceraria* pulp extract. Sheelarani et al., (2014) suggested that the ethanolic extract of *Myxopyrum serratulum* possesses anti-arthritic activity in mouse model.

19.4.6. Bio-imaging of Cancerous Cells

Due to the versatile surface chemistry, large surface area and phototoxic effect of ZnO NPs serve as multifaceted nano-platform for not only in optical bioimaging of cells but also in drug delivery. ZnONPs are highly toxic to cancer cells and leukemic T cells (Hanley et al, 2008; Wang et al., 2009; Wu et al., 2008). The inherent photoluminescence and ROS generation properties of ZnONPs make them applicable for bio-sensing applications and suitable candidate for molecularly targeted optical imaging of cancer cells. After the fictionalization ZnONPs remains water soluble, biocompatible and less toxic (Pan et al., 2011; Hong et al., 2011). Jiang et al., (2011) used ZnO nano-sheets for imaging the model leukemia K562 culture cells as these structures showed successful penetration in cells. Many studies illustrate the construction of ZnO-gated nano-platform with guided pH range for demand drug release to use them as contrast agents for trimodal imaging in in-vitro and in-vivo studies. This provides the detailed information in tumor diagnosis (Wang et al., 2015). Due to the extensive surface area nanoparticles serves as challenging prospects to overcome one of the main complications in the delivery system. Conjugate of ZnONPs with iron as $FeSO_4$ - ZnONPs provide the carcino-embryonic targeted antigens for dendritic cell (DC) immunotherapy (Cho et al., 2011).

19.5. Zinc Oxide Nanoparticles as Nano-Biofertilizer

Incorporation and application of nanoparticles in agricultural practices like nano-fertilizers and nano-pesticides has gained a lot of attention worldwide as an effective option for improvement of the agricultural yield. Uptake, translocation, accumulation and biotransformation of ZnONPs in plants depend upon the physical properties of NPs as well as the anatomy of the host plant. Nano-formulations can effectively be used as Zn fertilizer or as a foliar fertilizer for controlling worldwide Zn deficiency problem (Liu and Lal., 2015; Milani et al., 2012).

ZnONPs stimulate seed germination, plant growth and suppress plant diseases by its antimicrobial activity. The soil application of ZnONPs increase the Zn uptake e.g., in cucumber (Moghaddasi et al., 2017). Raliya and Tarafdar, (2013a; 2015b) reported that ZnONPs stimulated plant height, root length, biomass, chlorophyll and lycopene content in tomato. Foliar spray, soil mixing and seed priming methods are used for ZnONPs exposure to wheat plants (Rizwan et al., 2017).

Summary and Future Perspective

Day by day ZnONPs demand is increasing due to its various applications. Biocompatibility, easy accessibility, low cost and inherent ability to induce ROS generation by ZnONPs has opened its greater use in biomedical field for the cure of cancer and many deadly diseases. It is expected to appear as a smart weapon against multidrug resistant microorganisms. In addition ZnONPs are listed as GRAS (Generally recognized as safe) by the U.S. FDA. This further enhances its use as a drug carrier with increased therapeutic efficiency. The use of algal cell extract as well as bio-molecules, act as both reducing and stabilizing agents during ZnONPs synthesis. Considering above mentioned applications ZnONPs are expected to make more exciting contributions in diagnostic and therapeutic field.

Acknowledgments

The authors would like to thank UGC and DRDO New Delhi, for providing the financial assistance.

References

Abercrombie, M., and Ambrose, E. (1962). The surface properties of cancer cells: a review. *Cancer Research*. 22, 525-48.

Ahmed, S., Annu, Chaudhry, S.A., and Ikram, S. (2016). A review on biogenic synthesis of ZnO nanoparticles using plant extracts and microbes: A prospect towards green chemistry. *Journal of Photochemistry and Photobiology B: Biology*, doi:10.1016/j.jphotobiol.2016.12.011.

Akhtar, M.S., Ameen, S., Ansari, S.A., and Yang, O. B. (2011). Synthesis and characterization of zno nanorods and balls nanomaterials for dye sensitized solar cells, *Journal of Nanoengineering and Nanomanufacturing*. doi:10.1166/jnan.2011.1004.

Akhter, S.M.H., Mahmood, Z., Ahmad, S., and Mohammad, F. (2018). Plant-mediated green synthesis of zinc oxide nanoparticles using swertia chirayita leaf extract, characterization and its antibacterial efficacy against some common pathogenic bacteria. *Bionanoscience*. doi:10.1007/s12668-018-0549-9.

Aziz, N., Fatma, T., Varma, A., Prasad, R. (2014). Biogenic synthesis of silver nanoparticles using Scenedesmus abundans and evaluation of their antibacterial activity. *Journal of Nanoparticles*, 2014, 1-6. Article ID 689419. http://dx.doi.org/10.1155/2014/689419.

Azizi, S., Ahmad, M.B., Namvar, F., Mohamad, R. (2014). Green biosynthesis and characterization of zinc oxide nanoparticles using brown marine macroalga *Sargassum muticum* aqueous extract, *Mater. Lett.*116, 275–277, doi: 10.1016/j. matlet.2013.11.038.

Aziz, N., Faraz, M., Pandey, R., Sakir, M., Fatma, T., Varma, A., Barman, I., and Prasad, R. (2015). Facile algae-derived route to biogenic silver nanoparticles: Synthesis, antibacterial and photocatalytic properties. *Langmuir*, 31, 11605−11612. DOI: 10.1021/acs.langmuir.5b03081.

Basha, S.K., Lakshmi, K.V., Kumari, V.S. (2016). Sensing and bio-sensing research ammonia sensor and antibacterial activities of green zinc oxide nanoparticles, *Sensing and bio-sensing research,* 10, 34–40, http://dx.doi.org/10.1016/j.sbsr.2016. 08.007.

Basnet, P., Chanu, T.I., Samanta, D., Chatterjee, S. (2018). A review on bio-synthesized zinc oxide nanoparticles using plant extracts as reductants and stabilizing agents, *Journal of Photochem. & Photobiol., B: Biol.* 10.1016/j.jphotobiol.2018.04.036.

Chandrasekaran, M., Pandurangan, M. (2016). *In vitro* selective anti-proliferative effect of zinc oxide nanoparticles against co-cultured C2C12 Myoblastoma cancer and 3T3-L1 normal cells. *Biologi Trace Eleme. Res.* doi:10.1007/s12011-015-0562-6.

Cho, N.H., Cheong, T.C., and Min, J.H. (2011). A multifunctional core-shell nanoparticle for dendritic cell-based cancer immunotherapy. *Nature Nanotech.* doi:10.1038/ nnano.2011.149.

Çolak, H. and Karaköse, E. (2017). Green synthesis and characterization of nanostructured ZnO thin films using *Citrus aurantifolia* (lemon) peel extract by spin-coating method. *Journal of Alloys and Comp.* doi:10.1016/j.jallcom.2016.08.090.

Degen, A. and Kosec, M. (2000). Effect of pH and impurities on the surface charge of zinc oxide in aqueous solution. *Journal of the European Ceramic Society.* doi:10.1016/S0955-2219(99), 203-4.

Dhandapani, P., Siddarth, A.S., Kamalasekaran, S., Maruthamuthu, S., Rajagopal, G. (2014). Bio-approach: ureolytic bacteria mediated synthesis of ZnO nanocrystals on cotton fabric and evaluation of their antibacterial properties, *Carbohyd. Poly.* 103,448–455, doi: 10.1016/j.carbpol.2013.12.074.

Ding, Y., and Wang, Z. L. (2009). Structures of planar defects in ZnO nanobelts and nanowires. *Micron.* doi:10.1016/j.micron.2008.10.008.

Duraimurugan,J., Suresh, K.G., Maadeswaran, P., Shanavas, S., Anbarasan, P.M., Vasudevan, V. (2019). Structural, optical and photocatlytic properties of zinc oxide nanoparticles obtained by simple plant extract mediated synthesis, *Journal of Mate. Sci.: Mate. in Elect.*30, 1927–1935,doi.org/10.1007/s10854-018-0466-2.

Dwivedi, S., Wahab, R., Khan, F., Mishra, Y.K., Musarrat, J. and Al-Khedhairy, A.A. (2014). Reactive oxygen species mediated bacterial biofilm inhibition via zinc oxide nanoparticles and their statistical determination. *PLoS One.* doi:10.1371/ journal.pone.0111289.

Espitia, P.J.P., Soares, N.D.F.F., Coimbra, J.S.D.R., Andrade, D.N.J., Cruz, R.S., Medeiros, E.A.A. (2012). Zinc oxide nanoparticles: synthesis, antimicrobial activity and food packaging applications, *Food and Biopro. Technol.*5, 1447–1464.

Ghasemi, F. and Jalal, R. (2016). Antimicrobial action of zinc oxide nanoparticles in combination with ciprofloxacin and ceftazidime against multidrug-resistant *Acinetobacter baumannii. J. of Global Antimicrol Resis.* doi:10.1016/j.jgar.2016.04.007.

Gnanasangeetha, D. and Thambavani, S.D. (2013). One pot synthesis of zinc oxide nanoparticles via chemical and green method research, *J. of Mat. Sci.* 1(7), 1-8.

Gunalan, S., Sivaraj, R. and Rajendran, V. (2012). Green synthesized ZnO nanoparticles against bacterial and fungal pathogens. *Prog. in NatU. Sci. Mat. Int.* 2012. doi:10.1016/j.pnsc.2012.11.015.

Hackenberg, S., Scherzed, A. and Harnisch, W. (2012). Antitumor activity of photo-stimulated zinc oxide nanoparticles combined with paclitaxel or cisplatin in HNSCC cell lines. *J. of Photochem. and Photobiol. B: Biology*. doi:10.1016/j.jphotobiol.2012.05.014.

Hanley, C., Layne, J, and Punnoose, A. (2008). Preferential killing of cancer cells and activated human T cells using ZnO nanoparticles. *Nanotechnol.* doi:10.1088/0957-4484/19/29/295103.

Hong, H., Shi, J. and Yang, Y. (2011). Cancer-targeted optical imaging with fluorescent zinc oxide nanowires. *Nano Letter*. doi:10.1021/nl201782m.

Huang, Z., Zheng, X., and Yan, D. (2008). Toxicological effect of ZnO nanoparticles based on bacteria. *Langmuir*. doi:10.1021/la7035949.

Husain, S., Sardar, M., and Fatma, T. (2015). Screening of cyanobacterial extracts for synthesis of silver nanoparticles, *World J. of Microbiol. and Biotechnol.* DOI 10.1007/s11274-015-1869-3.

Ilves, M., Palomäki, J., and Vippola, M. (2014). Topically applied ZnO nanoparticles suppress allergen induced skin inflammation but induce vigorous IgE production in the atopic dermatitis mouse model. *Particle and Fibre Toxicol.*. doi:10.1186/s12989-014-0038-4.

Ishwarya, R., Vaseeharan, B., and Kalyani, S. (2018). "Facile green synthesis of zinc oxide nanoparticles using *Ulva lactuca* seaweed extract and evaluation of their photocatalytic, antibiofilm and insecticidal activity," *J. of Photochem. and Photobiol. B: Biol.*178, 249–258.

Jayaseelan, C., Rahuman, A.A., Kirthi, A.V., Marimuthu, S., Santhoshkumar, T., and Bagavan, A. (2012). Novel microbial route to synthesize ZnO nanoparticles using *Aeromonas hydrophila* and their activity against pathogenic bacteria and fungi, *Spectrochimica Acta Part A: Mole and Biomolcul. Spectros.* 90, 78–84, doi: 10.1016/j.saa. 2012.01.006.

Jiang, H., Wang, H., and Wang, X. (2011). Facile and mild preparation of fluorescent ZnO nanosheets and their bioimaging applications. *Appl. Surf. Sci.* doi:10.1016/ j.apsusc.2011.03.053.

Kakiuchi, K., Hosono, E., Kimura, T., Imai, H., and Fujihara, S. (2006). Fabrication of mesoporous ZnO nanosheets from precursor templates grown in aqueous solutions. *J. of Sol-Gel Sci. and Technol.* doi:10.1007/s10971-006-6321-6.

Kalpana, V.N., Payel, C., and Rajeswari, D.V. (2017). Lagenaria siceraria aided green synthesis of ZnO NPs: anti dandruff, Anti-microbial and Anti-arthritic activity. Res. *J. of Chem. and Environ.*21, 14–19.

Kato, H. (2011). *In vitro* assays: Tracking nanoparticles inside cells. *Nature Nanotechnol.* doi:10.1038/nnano.2011.25.

Kitture, R., Chordiya, K., and Gaware, S. (2014). ZnO nanoparticles-red sandalwood conjugate: a promising anti-diabetic agent. *J. of Nanosci. and Nanotechnol.* doi:10.1166/jnn.2015.10323.

Koch, C.C. (2003). Top-down synthesis of nanostructured materials: mechanical and thermal processing methods. *Rev. on Adv. Mat. Sci.*5 (2), 91-99.

Kumar, S.J., Kumar, S.V., and Rajeshkumar, S. (2017). Synthesis of zinc oxide nanoparticles using plant leaf extract against urinary tract infection pathogen. *J. of Ener. Resour. Technol.* doi:10.1016/j.reffit.2017.05.00.

Kundu. D., Hazra, C., Chatterjee, A., Chaudhari, A., and Mishra, S. (2014). Extracellular biosynthesis of zinc oxide nanoparticles using *Rhodococcus pyridinivorans* NT2: multifunctional textile finishing, biosafety evaluation and *in vitro* drug delivery in colon carcinoma, *J. of Photochem and Photobiol B: Biol.*140, 194–204, doi: 10.1016/j.jphotobiol.2014.08.001.

Lipovsky, A., Nitzan, Y., Gedanken, A., and Lubart, R. (2011). Antifungal activity of ZnO nanoparticles-the role of ROS mediated cell injury. *Nanotechnol.* doi:10.1088/0957-4484/22/10/105101.

Liu, R., and Lal, R. (2015). Potentials of engineered nanoparticles as fertilizers for increasing agronomic productions. *Sci. of the Tot. Environ.* doi:10.1016/j.scitotenv.2015.01.104.

Luangpipat, T.I., Beattie, R., Chisti, Y., and Haverkamp, R.G. (2011). Gold nanoparticles produced in a microalga. *J. of Nanopart. Res.*13, 6439–6445. https://doi.org/10.1007/s11051-011-0397-9.

Mahmud, S., Abdullah, M.J., Putrus, G.A., Chong, J., and Mohamad, A.. (2006). Nanostructure of ZnO fabricated via French process and its correlation to electrical properties of semiconducting varistors. *Synth. and Reacti. in Inorg. Met-Org and Nano-Met Chem.* doi:10.1080/15533170500524462.

Malaikozhundan, B., Vaseeharan, B., and Vijayakumar, S. (2017). Biological therapeutics of *Pongamia pinnata* coated zinc oxide nanoparticles against clinically important pathogenic bacteria, fungi and MCF-7 breast cancer cells. *Microb. Pathogen.* doi:10.1016/j.micpath.2017.01.029.

Manzoor, U., Siddique, S., Ahmed, R., Noreen, Z., Bokhari, H., and Ahmad, I. (2016). Antibacterial, structural and optical characterization of mechano-chemically prepared ZnO nanoparticles. *PLoS One*. doi:10.1371/journal.pone.0154704.

Milani, N., McLaughlin, M.J., and Stacey, S.P. (2012). Dissolution kinetics of macronutrient fertilizers coated with manufactured zinc oxide nanoparticles. *J. of Agricul. and F. Chem.* doi:10.1021/jf205191y.

Mishra, P.K., Mishra, H., Ekielski, A., Talegaonkar, S., and Vaidya, B. (2017). Zinc oxide nanoparticles: a promising nanomaterial for biomedical applications. *Dru. Disc. Tod.* doi:10.1016/j.drudis.2017.08.006.

Moezzi, A., Cortie, M., and McDonagh. (2011). A Aqueous pathways for the formation of zinc oxide nanoparticles. *Dalt. Transac.* doi:10.1039/c0dt01748e.

Moghaddasi, S., Fotovat, A., Khoshgoftarmanesh, A.H., Karimzadeh, F., Khazaei, H.R., and Khorassani, R. (2017). Bioavailability of coated and uncoated ZnO nanoparticles to cucumber in soil with or without organic matter. *Ecotoxicol. and Environ. Saf.* 144, 543.

Mubarak, Ali, D., Sasikala, M., Gunasekaran, M., and Thajuddin, N. (2011). Biosynthesis and characterization of silver nanoparticles using marine cyanobacterium, *Oscillatoria willei* NTDM01. *Dig. J. of Nanomat. and Biostru.*

Nagajyothi, P.C., Cha, S.J., Yang, I. J., Sreekanth, T.V.M., Kim, K.J., and Shin, H.M. (2015). Antioxidant and anti-inflammatory activities of zinc oxide nanoparticles synthesized using *Polygala tenuifolia* root extract. *J. of Photochem. and Photobiol. B: Biol.* doi:10.1016/j.jphotobiol.2015.02.008.

Nagarajan, S., and Kuppusamy, A.K. (2013). Extracellular synthesis of zinc oxide nanoparticle using seaweeds of gulf of Mannar, India, *J. Nanobiotechnol.* 11.39, doi: 10.1186/1477-3155-11-39.

Nazarizadeh, A., and Asri-Rezaie, S. (2016). Comparative Study of Antidiabetic Activity and Oxidative Stress Induced by Zinc Oxide Nanoparticles and Zinc Sulfate in Diabetic Rats. *AAPS Pharm. Sci. Tech.* doi:10.1208/s12249-015-0405-y.

Pan, Z.Y., Liang, J., Zheng, Z.Z., Wang, H.H., and Xiong, H.M. (2011). The application of ZnO luminescent nanoparticles in labeling mice. *Cont. Med. & Molecul. Imag.* doi:10.1002/cmmi.434.

Pandimurugan, R., and Thambidurai, S. (2016). Novel seaweed capped ZnO nanoparticles for effective dye photodegradation and antibacterial activity, *Adv. Powd. Technol.*27, 1062–1072, http://dx.doi.org/10.1016/j.apt.2016.03.

Pandimurugan, R., and Thambidurai., S. (2017). UV protection and antibacterial properties of seaweed capped ZnO nanoparticles coated cotton fabrics, *Int. J. of Biol. Macromol.* http://dx.doi.org/10.1016/j.ijbiomac.2017.07.097.

Pant, G., Nayak, N., and Prasuna, R.G. (2013). Enhancement of antidandruff activity of shampoo by biosynthesized silver nanoparticles from Solanum trilobatum plant leaf, *Appl. Nanosci.*3, 431-439.

Raliya, R., and Tarafdar, J.C. (2013). ZnO nanoparticle biosynthesis and its effect on phosphorous-mobilizing enzyme secretion and gum contents in cluster- bean (*Cyamopsis tetragonoloba L.*), *Agricul. Res.*2, 48–57, doi: 10.1007/ s40 0 03- 012- 0049- z.

Raghupathi, K.R., Koodali, R.T., and Manna, A.C. (2011), Size-dependent bacterial growth inhibition and mechanism of antibacterial activity of zinc oxide nanoparticles. *Langm.* doi:10.1021/la104825u.

Raja, R., Hemaiswarya, S., Ganesan, V., and Carvalho, I.S. (2016). Recent developments in therapeutic applications of Cyanobacteria. *Crit. Rev. in Microbiol.* doi:10.3109/1040841X.2014.957640.

Raliya, R., and Tarafdar, J.C. (2014). Biosynthesis and characterization of zinc, magnesium and titanium nanoparticles: an eco-friendly approach. *Int. Nano Lett.*4, 93.

Raliya, R., Nair, R., Chavalmane, S., Wang, W.N., and Biswas, P. (2015). Mechanistic evaluation of translocation and physiological impact of titanium dioxide and zinc oxide nanoparticles on the tomato *(Solanum lycopersicum L.)* plant. *Metallo.* doi:10.1039/c5mt00168d.

Ramaramesh, P.I., Govindaraj, P., Geetha, N., and Perumal, V. (2014). Microwave-mediated extracellular synthesis of metallic silver and zinc oxide nanoparticles using macro-algae (*Gracilaria edulis*) extracts and its anticancer activity against human PC3 cell lines, *Appl. Biochem. and Biotechnol.* 174, 2777–2790.

Rao, M.D., and Gautam, P. (2016). Synthesis and characterization of ZnO nanoflowers using *Chlamydomonas reinhardtii*: a green approach, *Environ. Prog. & Sust. Ener.*1–7, doi: 10.1002/ep.

Rasmussen, J.W., Martinez, E., Louka, P., and Wingett, D.G. (2010). Zinc oxide nanoparticles for selective destruction of tumor cells and potential for drug delivery applications. *Exp. Opi. on Dru. Deliv.* 20.

Rizwan, M., Ali, S., Qayyum, M.F., Ok, Y.S., Adrees, M., Ibrahim, M., Rehman, M.Z., Farid, M., and Abbas, F. (2017). J. of Haz. Mat. 322, 2.

Roy, R., Singh, S.K., Chauhan, L.K., Das, M., Tripathi, A., and Dwivedi, P.D. (2014). Zinc oxide nanoparticles induce apoptosis by enhancement of autophagy via PI3K/Akt/mTOR inhibition, *Toxicol. Lett.* 16, 227(1), 29-40. doi: 10.1016/j.toxlet.2014.02.024. Epub 2014 Mar 12.

Salam, A.H., Sivaraj, R., and Venckatesh, R. (2014). Green synthesis and characterization of zinc oxide nanoparticles from Ocimum basilicum L. var. purpurascens Benth.-Lamiaceae leaf extract. *Mat. Lett.* doi:10.1016/j.matlet.2014.05.033.

Sasidharan, A., Chandran, P., Menon, D., Raman, S., Nair, S., and Koyakutty, M. (2011). Rapid dissolution of ZnO nanocrystals in acidic cancer microenvironment leading to preferential apoptosis. *Nanos.*3(9), 3657-3669.

Shamsuzzaman, A., Mashrai, H., Khanam, R.N., and Aljawfi, R.N. (2013). Biological synthesis of ZnO nanoparticles using *C. albicans* and studying their catalytic performance in the synthesis of steroidal pyrazolines, *Arab. J. of Chem.* doi: 10. 1016/j.arabjc.2013.05.004.

Sharma, D., Sabela, M.I., Kanchi, S., Mdluli, P.S,, Singh, G., and Stenström, T.A. (2016). Biosynthesis of ZnO nanoparticles using Jacaranda mimosifolia flowers extract: synergistic antibacterial activity and molecular simulated facet specific adsorption studies, *J. of Photochem. and Photobiol. B: Biol.* 162, 199–207.

Sheelarani, T., Gopal, V., Seethalakshmi, S., and Chitra, K. (2014). *In vitro* anti-inflammatory and anti-arthritic activity of Selected Medicinal Plant, *Int. J. of Pharm. Sci. and Res.*28, 162-163.

Shen, L., Zhang, H., and Guo, S. (2009). Control on the morphologies of tetrapod ZnO nanocrystals. *Mater. Chem. Phys.* doi:10.1016/j.matchemphys.2008.10.004.

Siddiqi, K.S., ur Rahman, A., Tajuddin, and Husen, A. (2018). Properties of zinc oxide nanoparticles and their activity against microbes. *Nanos. Res. Lett.* doi:10.1186/s11671-018-2532-3.

Singh, G., Babele, P.K., Kumar, A., Srivastava, A., Sinha, R.P., and Tyagi, M.B. (2014). Synthesis of ZnO nanoparticles using the cell extract of the cyanobacterium, *Anabaena strain L31* and its conjugation with UV-B absorbing compound shinorine, *J. of Photochem and Photobiol. B: Biol.*138, 55–62.

Soleimani, M., and Pirkoohi, H.M. (2017). Biosynthesis of silver nanoparticles using *Chlorella vulgaris* and evaluation of the antibacterial efficacy against *Staphylococcus aureus*. *Avi. J. of medi. biotechnol.* 9(3), 120–125.

Stanier, R.Y., Deruelles, J., Rippka, R., Herdman, M., and Waterbury, J.B. (1979). Generic Assignments, Strain Histories and Properties of Pure Cultures of Cyanobacteria. *Microbiol.* doi:10.1099/00221287-111-1-1.

Surendra, T.V., Roopan, S.M., Al-Dhabi, N.A., Arasu, M.V., Sarkar, G., and Suthindhiran, K. (2016). Vegetable Peel Waste for the Production of ZnO Nanoparticles and its Toxicological Efficiency, Antifungal, Hemolytic, and Antibacterial Activities. *Nanos. Res. Lett.* doi:10.1186/s11671-016-1750-9.

Tripathi, R.M., Bhadwal, A.S., Gupta, R.K., Singh, P., Shrivastav, A., and Shrivastav, B.R. (2014). ZnO nanoflowers: novel biogenic synthesis and enhanced photocatalytic activity, *J. of Photochem. and Photobiol. B: Biol.*141, 288–295, doi: 10.1016/j. jphotobiol.2014.10.001 .

Vijayan, S.R., Santhiyagu, P., and Ramasamy, R. (2016). Seaweeds: A resource for marine bionanotechnology. *Enzy. Microb. Technol.* doi:10.1016/j.enzmictec.2016.06.009.

Wahba, N.S., Shaban, S.F., Kattaia, A.A.A., and Kandeel, S.A. (2016). Efficacy of zinc oxide nanoparticles in attenuating pancreatic damage in a rat model of streptozotocin-induced diabetes, *Ultrastru. Pathol.* doi:10.1080/01913123.2016.1246499.

Wahab, R., Kaushik, N.K., and Kaushik, N. (2013). ZnO nanoparticles induces cell death in malignant human T98G gliomas, KB and non-malignant HEK cells. *J. of Biomed. Nanotechnol.* doi:10.1166/jbn.2013.1652.

Wang, L.Z. (2004). Zinc oxide nanostructures: growth, properties and applications *J. Phys. Condens. Matter.*doi:10.1088/0953-8984/16/25/R01.

Lengke, M.F., Fleet, M.E., and Southam, G. (2007). Biosynthesis of silver nanoparticles by filamentous cyanobacteria from a silver(I) nitrate complex. *Langm.* doi:10.1021/la0613124.

Wang, Z. L. (2004). Nanostructures of zinc oxide. *Mater. Tod.* doi:10.1016/S1369-7021(04)00286-X.

Wang, T.X., and Lou, T.J. (2008). Solvothermal synthesis and photoluminescence properties of ZnO nanorods and nanorod assemblies from ZnO_2 nanoparticles. *Mater. Lett.* doi:10.1016/j.matlet.2007.11.083.

Wang, H., Wingett, D., and Engelhard, M.H. (2009.) Fluorescent dye encapsulated ZnO particles with cell-specific toxicity for potential use in biomedical applications. *J. of Mat. Sci. Mat. in Medic.* doi:10.1007/s10856-008-3541-z.

Wang, Y., Song, S., Liu, J., Liu, D., and Zhang, H. (2015). ZnO-functionalized upconverting nanotheranostic agent: Multi-modality imaging-guided chemotherapy with on-demand drug release triggered by pH. *Ange. Chemie. Int. Edi.* doi:10.1002/anie.201409519.

Wu, Y.L., Fu, S., and Tok, A.I.Y. (2008). A dual-colored bio-marker made of doped ZnO nanocrystals. *Nanotechnol.* doi:10.1088/0957-4484/19/34/345605.

Xie, J., Li, P., Li, Y., Wang, Y., and Wei,Y. (2009). Morphology control of ZnO particles via aqueous solution route at low temperature. *Mat. Chem. and Phys.* doi:10.1016/j.matchemphys.2008.11.007.

Zhang, L., Jiang, Y., Ding, Y., Povey, M., and York, D. (2007). Investigation into the antibacterial behaviour of suspensions of ZnO nanoparticles (ZnO nanofluids). *J. of Nanop. Res.* doi:10.1007/s11051-006-9150-1.

ABOUT THE EDITORS

Dr. Muthu Arumgam obtained his Doctoral degree in Microbiology and Cell Biology from the prestigious Indian Institute of Science (IISc), Bangalore, India. While doing PhD itself, he got an offer as Scientist at Defence Institute of Bio-Energy Research, DRDO, Govt. of India, and started his professional carrier in the field of renewable energy. During his stay at DRDO, he contributed to transforming the Algal biofuel R&D to commercial scale by developing a cost-effective growth medium for large-scale culturing of microalgae in open pond conditions and identification of potential algal strain for biofuel and enhancement of lipid productivity through cultural and biochemical engineering approach. Subsequently, he joined CSIR-NIIST in 2012 and continuing as Senior Scientist and Assistant Professor (AcSIR). His current work involves strategies to enhance the lipid, molecular causes for stress-mediated lipid accumulation and nutraceuticals from microalgae. He has published his research article in peer-reviewed journals of international repute and actively participated and delivered the presentation in national and international forums.

Dr. Arumugam, recipient of a fellowship for carrying out higher research in the area of life sciences from ICAR and CSIR. He is also a life member in the society of biological chemist, Indian Science Congress Association and Indian association of Biomedical Scientists. He is serving as onboard reviewer for many journals, including as an Associate Editor for Frontiers in Energy Research. Dr. Arumugam, received a young scientist grant from DST, Govt. of India, obtained major projects as principal investigator from DST, DBT, and DRDO including DST-Core Research Grant.

Dr. Shanmugam Kathiresan earned his BSc degree in Agriculture at Tamil Nadu Agricultural University, Coimbatore and Master Degree in Agriculture specialized in Plant Pathology at the University of Agricultural Sciences, GKVK Campus, Bangalore. He earned his PhD in Biotechnology at Central Food Technological Research Institute (CSIR-CFTRI), Mysore. During his PhD he has concentrated on carotenoid biosynthesis pathway in freshwater microalga – *Haematococcus pluvialis* and first described the *Agrobacterium*-mediated genetic transformation for the enhanced production of Astaxanthin, a ketocarotenoid. At present, he is serving as a faculty in the Department of Molecular Biology, School of Biological Sciences, Madurai Kamaraj University, Madurai, Tamil Nadu. Here, for about ten years, he is being dedicatedly involved in teaching for the PG students and PhD scholars for subjects like Plant Genomics, Gene Cloning, and Biotechnology, Recombinant DNA Technology, etc. His main area of research activities is metabolic pathway engineering in both plants and algae, especially in fatty acid biosynthesis. After joined in Madurai Kamaraj University, he has undergone several major research projects funded by UGC, DST, SERB, and CSIR. From his research findings, he isolated the specific genes related to high value polyunsaturated fatty acid from marine microalgae and also functionally characterized and expressed in prokaryotic and other heterologous host system and his research finding has been published in high impact peer-reviewed journals.

Dr. Nagaraj Subramani completed his Post Graduate degree (MSc) and Master of Philosophy (MPhil) in Botany from Madurai Kamaraj University, Tamil Nadu, India. Subsequently, he registered Doctor of Philosophy (PhD) in Botany (Specialization in Algology) under the mentorship of the legendary eminent Indian Phycologist Prof. R. Rengasamy in the year 2007. He had published 40 research articles in peer-reviewed journals with good impact factors and three books for his credits. He was awarded Doctor of Philosophy in 2011 besides cleared the State Eligibility Test for Lectureship (UGC-SET) in 2012. He also received the Indian prestigious award of Dr. D. S. Kothari Postdoctoral Fellow under University Grants Commission (UGC), New Delhi. Nagaraj Subramani was awarded INSA - Visiting Scientist Fellow, 2018 by Indian National Science Academy (INSA), New Delhi availed at Indian Institute of Science (IISc) Bangalore. At present, he is guiding four PhD, scholars they are working with algae and one more student has submitted a thesis. In addition, he has guided about 10 M.Phil. Scholars at CAS in Botany, University of Madras, India. Currently, he is working with three major research projects on algae funded by DST-SERB, UGC, and MoEF&CC, India.

Contributors

Prof. Tasneem Fatma, Department of Biosciences, Jamia Millia Islamia, Central University, New Delhi is actively engaged in Research activities pertaining to Environmental Biology and Biotechnology including Biomonitoring, Toxicology and Pollution Remediation.

Prof. Ruma Pal, Professor, Department of Botany, University of Calcutta, 35 Ballygunge Circular Road, Kolkata, West Bengal, India.

Dr. R. Sarada, Plant Cell Biotechnology Department, CSIR- Central Food Technological Research Institute, Mysuru, Karnataka, India. Academy of Scientific and Innovative Research (AcSIR), Ghaziabad, India.

Dr. Nagaraj Subramani, Assistant Professor, Centre for Advanced Studies in Botany, University of Madras, Guindy Campus, Chennai, Tamil Nadu, India.

Prof. A. K. Bastia, Professor, P.G. Department of Botany, North Orissa University, Baripada, Mayurbhanj, Odisha, India.

Dr. Jnanendra Rath, Professor, Department of Botany, Visva-Bharati University, Santiniketan, India.

Prof. G. M. Narasimha Rao, Professor, Department of Botany, Andhra University, Visakhapatnam, India.

Dr. Karuppasamy Kattusamy, Research Scientist, Institute of Soil, Water and Environmental Science, Agriculture Research Organization, Volcani Center, Rishon Lezion, Israel.

Dr. B. Vanavil, Associate Professor, Department of Biotechnology, Kalasalingam Academy of Research and Education, Krishnankoil, Tamilnadu, India.

Dr. Andreas Isdepsky, Algae Biomass and Energy System R&D Center (ABES), University of Tsukuba, Ibaraki, Japan.

Prof. Subburamu Karthikeyan, Professor and Head, Department of Agricultural Microbiology and Department of Renewable Energy Engineering, Agricultural Engineering College and Research Institute, Tamil Nadu Agricultural University, Coimbatore, Tamil Nadu, India.

Dr. Jibu Thomas, Associate Professor, Algae Biomass research laboratory, Department of Biotechnology, Karunya Institute of Technology and Sciences, Coimbatore, Tamil Nadu, India.

Dr. Muthu Arumugam, Senior Scientist and Assistant Professor (AcSIR) Microbial Processes and Technology Division, National Institute for Interdisciplinary Science and Technology (NIIST), Council of Scientific and Industrial Research (CSIR), Trivandrum, Kerala, India. Academy of Scientific and Innovative Research (AcSIR), India.

Dr. Shanmugam Kathiresan, Assistant Professor, Department of Molecular Biology, School of Biological Sciences, Madurai Kamaraj University, Madurai, Tamilnadu, India.

Dr. Gour Gopal Satpati, Assistant Professor, Department of Botany, Bangabasi Evening College, University of Calcutta, 19 Rajkumar Chakraborty Sarani, Kolkata, West Bengal, India.

Dr. Anila Narayanan, Department of Botany, St. Xavier's College, Aluva, Kerala.

Dr. Perumal Varalakshmi, Assistant Professor, Department of Molecular Microbiology, School of Biotechnology, Madurai Kamaraj University, Madurai, Tamil Nadu, India.

Dr. S. Bhakta, Botanical Survey of India, Western Regional Centre, Pune, Maharashtra, India.

INDEX

A

B

C

D

E

F

G

N

O

P

W

X

Y

Z